ENGINEERING GEOLOGY

Subinoy Gangopadhyay

Formerly, Senior Director
Geological Survey of India

OXFORD
UNIVERSITY PRESS

OXFORD
UNIVERSITY PRESS

Oxford University Press is a department of the University of Oxford.
It furthers the University's objective of excellence in research, scholarsh
and education by publishing worldwide. Oxford is a registered trademark
Oxford University Press in the UK and in certain other countries.

Published in India by
Oxford University Press
YMCA Library Building, 1 Jai Singh Road, New Delhi 110001, India

First published in 2013
Third impression 2015

ISBN-13: 978-0-19-808635-2
ISBN-10: 0-19-808635-0

Typeset in Times New Roman
by Cameo Corporate Services Limited, Chennai
Printed in India by Manipal Technologies Ltd., Karnataka 576104

*This book is dedicated
to the memory of my*
Mother
for her everlasting love.

Features of the Book

Each chapter begins with 'learning objectives' setting the theme for the chapter followed by a short introductory paragraph on the subject matter of discussion in the chapter.

A number of photographs of actual project sites as well as illustrative diagrams are interspersed in the text for better understanding of the theory.

Fig. 18.5 An arch bridge

18.11 CASE STUDIES ON BRIDGES INCLUDING A COLLAPSED BRIDGE

A bridge construction usually encounters several types of problems that adverse geology of the bridge site, especially in relation to the weak fou of the piers. Geotechnical investigation helps to identify the weakness of t specify correct treatment. The following case studies are presented to give of the problems faced at the bridge sites of varied geological set-ups and the

Case studies have been provided in relevant chapters to present real-life situations so that students are able to relate to the concepts discussed in the chapters.

Chapter 23 provides guidelines for preparing engineering geology reports. The field work required to collect basic data to be included in such reports is also discussed in this chapter.

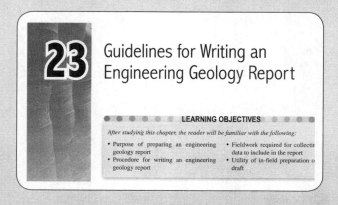

23 Guidelines for Writing an Engineering Geology Report

SUMMARY

is the application of
civil engineering practices
·sign, and construction.
ıg was known in India
·s when buildings and
ısing hard and durable
·d. Numerous buildings,
·, dams, and irrigation
·lia in early historic times.
· work started in India as

- The work of an engineering geologist at the initial
  stages of an engineering project includes geological
  mapping, subsurface evaluation by drillings, search
  for construction materials, and determination of
  rock/soil properties. At the construction stage, it
  involves the study of the conditions of foundations
  of engineering structures and suggestion of
  corrective measures for geological defects.
- Investigating natural hazards such as earthquakes
  and landslides and identifying measures for their

The summary at the end of each chapter revisits all the important points discussed in the chapter making for a fine guide for revision.

Multiple Choice Questions

Choose the correct answer from the choices given:

1. In the investigation of a site for construction of
   an engineering structure, an engineering geologist
   aims at:
   (a) stability of the structure
   (b) economy of its construction
   (c) both (a) and (b)

Review Questions

1. Discuss how the mutual relationship between the two
   disciplines—geology and engineering geology—
   resulted in engineering geology.

5. The structure that proves th
   the art of engineering even :
   (a) the sluice structure for wa
   (b) the embankments and
       century AD
   (c) both (a) and (b)
6. The rock type that was use
   Sanchi Stupa is:
   (a) pink granite

   weapons and domestic artic
   where Stone Age men lived
5. By whom and when was *Si*
   the guidelines mentioned iı
   the use of rock/soil for cons

In all chapters multiple choice questions and review questions are given to test students' understanding of concepts learnt from the chapters.

NUMERICAL EXERCISES

/ testing, while determining the
the weight of the density bottle was
. The soil sample after drying in the
ʒ was weighed as 35.38 g. The bottle
; then filled with distilled water and
20.21 g and the bottle with water to
d the specific gravity (*G*) of the soil

[Answer: *G* = 2.12]

maximum of 300 kg-cm to bring about failure of the
soil. Calculate the shear strength of the soil.
        [Answer: Shear strength = 0.379 kg/cm2]
3. In an analysis of 200 g of soil particles, the
   following result was recorded with respect to their
   size distribution: coarse gravel 80 g, fine gravel
   50 g, coarse sand 30 g, medium sand 20 g, fine sand
   15 g, and silt with clay 5 g. Draw a cumulative curve
   showing the distribution of particles.

Numerical problems along with answers are provided in those chapters where there is a requirement. In some cases hints are also given.

Two appendices on dams and tunnels in India have been provided at the end of the book. These appendices highlight the adverse geological conditions encountered during the construction of dams and the remedial measures undertaken.

APPENDIX
A
Geotechnical Problems of Dams and Their Solutions

APPENDIX
B
Geotechnical Problems of Tunnels and Their Solutions

Preface

It is said that one who learns from one's own experience is smart, but one who learns from other's experiences is smarter. This book is the outcome of the author's several years of experience of working on diverse projects in India and abroad. This book attempts to share his rich and varied experience with students—the future geologists and engineers—so that they are better able to assess and handle the geotechnical problems of different project sites effectively. The book is written in an Indian context using exhaustive examples from Indian case studies to elucidate the theoretical concepts discussed in the chapters and to help in solving practical problems. The investigative approach including techniques of site works discussed in this book are in line with those practised by the geologists of the Geological Survey of India (GSI) while collaborating with engineers in planning, designing, and constructing varied engineering projects across the country.

Initially, geological problems related to civil constructions were handled by geologists of GSI. But with time civil engineers started working hand in hand with geologists to find solutions to the numerous geotechnical problems encountered in different engineering projects. Thus, gradually engineering geology became an important subject of study especially for civil engineering professionals.

Like any new branch of study, engineering geology has also advanced at a fast pace. In fact, engineering problems and adverse conditions of project sites faced during construction of civil engineering projects were treated successfully by Indian engineers and geologists. The advances in the nature of geological works are also more pronounced compared to the early days. New methodologies such as satellite imagery and remote sensing techniques, and exploratory works using rotary drilling now provide accurate geological knowledge of both surface and subsurface strata. But our progress in using computers in geological works is yet to develop as in the western countries. In the West, IT based working in engineering geology is part of the curriculum. The future of engineering geology lies in the applications of geographic information system (GIS), remote sensing information provided by orbiting satellites, and IT based 3D imagery systems.

About the Book

This textbook aims to impart comprehensive knowledge of engineering geology to the undergraduate students of civil engineering and graduate and postgraduate students of geology, applied geology, and mining geology. It will also serve as a reference book for practising civil engineers and professional engineering geologists. The book deals with both the principles and practices of engineering geology. Because of the author's rich experience of working with various types of engineering projects such as dams, tunnels, bridges, and having been involved in disaster management activities in cases of natural calamities across the world such as earthquakes, landslides, and tsunamis, he has focussed on the practical aspects that will come in handy for geology students.

The subject matter of each chapter is elucidated with the help of maps, diagrams, and images including field pictures mostly taken by the author during his visits to different parts of the world. Factual data related to different aspects of discussions has been incorporated in the form of tables for ready reference. Solved examples wherever applicable have been included in the chapters. The important points covered in the chapters are recapitulated in the form of summary at the end of the chapters. Further, a section on exercises comprising review questions, multiple-choice questions, and numerical exercises is also provided at the end of each chapter.

Several books, websites, and literature on engineering geology including journals of the *Indian Society of Engineering Geology* (ISEG) and journals of the *International Association of Engineering Geology* (IAEG), and published records of GSI on geological works of the past till date were consulted while preparing the text of this book. The references of all these pertinent books/papers with author's name, year of publication, and other relevant details have been listed in the 'References' section at the end of the book.

Salient Features of the Book

- The book provides examples of various geological sites in the country to give an Indian context to the book. It also closely follows the works done by GSI, the premium institute in India involved in geological works in the country.
- The subject matter of each chapter has been illustrated by numerous figures and actual pictures of geological features taken from project sites to simplify the concepts discussed in the book.
- Chapter-end summary is provided in each chapter so that at a glance student can recollect the main aspects of the chapter.
- Geological reports are an important part of a professional geologists' work. A chapter is devoted to providing guidelines on writing good geological reports for designing project sites.
- The appendices at the end of the book include case histories of major Indian dams and tunnels. The discussion on the different types of hurdles which arose during their construction and the remedial measures adopted will be of special interest to practising civil engineers and professional geologists.
- Many parts of the world and several states of India—especially the eastern region and parts of central India—have problems related to karstic limestone, where special investigations such as speleological and tracer studies are necessary. This important aspect, which does not find mention in most other engineering geology books, is discussed as an exclusive chapter.

Content and Coverage

The book consists of 24 chapters and two appendices.

Chapter 1 introduces engineering geology with a discussion on the importance of engineering geology in civil engineering and the work-activities of engineering geologists. Chapters 2, 3, and 4 deal with rocks, minerals, and rock structures and weathering of rocks and its impact on constructions. Chapter 5 discusses the different types of soils and their classification. Chapter 6 deals with soil mechanics—laboratory analysis of soil, determination of soil permeability, density measurement, and consolidation and compactness tests.

Chapter 7 discusses the hydraulic parameters of a river followed by its activities and the utility of the river deposits. Chapter 8 highlights the geological works of oceans and coastal management. It also includes a case study on coastal erosion. Chapter 9 analyses the various aspects of subsurface water in relation to its engineering significance. The processes involved in the hydrologic cycle to maintain the earth's water equilibrium are also discussed.

Chapter 10 highlights the importance of rock mechanics in engineering geology and civil engineering works.

Chapter 11 discusses the different stages of investigation of engineering project sites. Different methods of geophysical explorations are also discussed. Chapter 12 discusses different types of rock materials including their characteristics and utilities. Petrological study of alkali-reactive minerals and other deleterious materials are also discussed. Chapter 13 discusses the different aspects of grouting in rocks and other materials. It also discusses the types of grouting, ingredients used in a grout mix, grouting patterns, and factors considered for finding the efficacy of grouting.

Chapter 14 explains the different aspects of dams, their utility, functions, investigation approaches during site selection, and construction stages as also post-construction stages. It also discusses different types of spillways and their working. Chapter 15 deals with reservoirs, their effect on the geo-hydrology of the neighbouring area, the sedimentation and degradation process of reservoir capacity, and the environmental effect due to the creation of a reservoir.

Chapter 16 classifies different types of tunnels, before explaining the geotechnical aspects of tunnel construction, problems in tunnelling and their solutions, and methods and machineries of tunnel construction. Chapter 17 deals with different types of powerhouses and their functions. Some case studies on power houses are also discussed. Investigation methods for nuclear and thermal powerhouses are also presented.

Chapter 18 on bridges explains the functions of superstructures and substructures of bridges. It describes the different types of bridges and the forces acting on them. The bridge components and supports systems are also elucidated. A case study on bridge foundation problems and their remedial measures are also presented. Chapter 19 deals with highways, canals, runways, power channels, and flumes. Chapter 20 discusses natural hazards such as earthquakes and tsunamis and explains the methods of forecasting them. The chapter also adds a note on the various remedial measures that can be adopted following a disaster.

Chapter 21 discusses landslide evaluation and mitigation. It also talks about geological, man-made, and natural causes of landslides and the preparation of landslide hazard zonation map and mitigation measures. Chapter 22 is on karstic terrain investigation. The origin of karst topography and processes responsible for formation of large cavities in surface and subsurface investigation methods including speleologist study of underground caverns are discussed in this chapter. The chapters also adds a note on the problems of creating reservoirs in karstic terrains.

Chapter 23 presents guidelines for writing geological reports. The field work required to collect basic data to be included in such reports is also discussed. Chapter 24 deals with physiography, stratigraphy, and ores and minerals of India. Without adequate knowledge of these aspects, the engineering geology study especially in the Indian perspective will be incomplete.

Appendices A and B include examples of some Indian dams and tunnels successfully built in various rock formations in different parts of India including the Himalayan terrain. These appendices highlight the adverse geological conditions encountered during the construction of these engineering structures and the remedial measures adopted. Chapter 14 on dams and Chapter 16 on tunnels may be read in conjunction with Appendices A and B respectively to impart knowledge on the adverse geological conditions of the construction sites of major dams and tunnels and their rectification measures.

Acknowledgement

I am indebted to many world-renowned engineering geologists and specialists in soil and rock mechanics for sharing their knowledge with me during my service life. Special thanks especially to Dr J.B. Auden, renowned engineering geologist and former Chief of Water Resource Development, UN, Dr N. Burton, expert in rock mechanics, and Dr L. Bjerrum, specialist in soil mechanics and former Head, Norwegian Geotechnical Institute, Oslo, with whom I had the rare privilege of working. I am indebted to his guidance and constant encouragement in my research on engineering geology and geotechnics.

I would like to thank my colleagues at GSI and engineers of various state and central government departments with whom I worked for a long period and got inspiration for writing this book. I am especially thankful to A.K. Chowdhury, N. Majumdar, M. Bandopadhyay, and S. Chakrabarty, retired directors and experienced engineering geologists of GSI, who have gone through the different chapters of the manuscript and provided valuable suggestions for improvement of the book. I am also indebted to D.N. Bhattacharya, C. Paul, and B.C. Poddar—my colleagues at GSI for providing me with several illustrative field photographs and fruitful discussions on specific subject matter related to this book.

This work would not have been completed without the dynamic effort of my daughter Shampa and son-in-law Asit Chakrabarty, both computer engineers in the US. They were a constant source of inspiration while writing this book and assisted me immensely in computer-processing of the entire manuscript including computer-drawing of illustrations and scanning of figures and photographs.

Subinoy Gangopadhyay

Contents

1

Introduction to Engineering Geology

● ● ● ● ● ● ● **LEARNING OBJECTIVES** ● ● ● ● ● ● ●

After studying this chapter, the reader will be familiar with the following:

- Application of engineering geology in civil engineering practice
- Knowledge that early Indians had on stone construction since prehistoric times
- History and development of engineering geology in India
- Work-activities of engineering geologists

1.1 INTRODUCTION

This chapter introduces the field of engineering geology and highlights its importance in civil engineering practice. It also brings to light the use of rocks by Indians for engineering works since prehistoric times by providing evidence from archaeological structures and writings in ancient books. The chapter traces the history and development of engineering geology and work-activities of geologists in the construction of dams, tunnels, and other engineering projects. The chapter further discusses the involvement of Indian geologists with engineers in various engineering projects of the country and their role in the remediation of natural hazards.

1.2 ENGINEERING GEOLOGY

Engineering geology is an applied branch of earth science. It involves the application of knowledge of geosciences to ensure safety, efficacy, and economy of engineering projects. It is primarily devoted to the study of rocks and soil and underground water. This is essential for proper location, planning, design, construction, operation, and maintenance of engineering structures. In addition, engineering geology is associated with the assessment and implementation of corrective measures for a wide variety of natural and man-made hazards.

1.2.1 Definition and Application

According to the statutes of International Association of Engineering Geology (IAEG) 1992, 'Engineering Geology is the science devoted to the investigation, study and solution of engineering and environmental problems which may arise as the result of the interaction between geology and the works and activities of man as well as to the prediction and of the development of measures for prevention or remediation of geological hazards'. Simply put, the science of geology applied to relevant aspects of engineering practice is *engineering geology*.

In general, engineering geology is concerned with the properties of materials such as soil and rocks used in engineering projects, including the quantitative assessment of their strength, permeability, and compactness. These properties help in the selection of sites for civil engineering structures. Engineering geology uses data, techniques, and principles of varied branches of geology to study the surface and subsurface materials for evaluating their origin, distribution, and effects on engineering constructions. 'Engineering geological investigations of Geological Survey of India (GSI) involve application of geomorphology, economic geology, geohydrology, stratigraphy, structural geology, petrology and mineralogy besides geophysics', observed Krishnaswamy (1972), the well-known engineering geologist and former Director General of GSI. In recent years, the field of engineering geology has expanded and encompasses rock mechanics, soil mechanics, satellite remote sensing, earthquake zoning, seismotectonics, and geohazard management.

1.2.2 Geology vs Civil Engineering

There are several branches of science that enables the study of the earth's features. Geology is the most important branch in the family of earth science. The different branches of geology deal with the history of earth's origin, the materials in its interior, and the physical processes that take place within it, especially plate tectonics. Knowledge of geology is necessary to study fossils in rock formations and to reconstruct the plant and animal world of remote past. However, geology for civil engineering work primarily involves activities related to rocks and soil occurring in the top crustal layer of the earth (to a depth of only a few hundred metres).

In a sense, the scope of a geologist and that of a civil engineer while working on a project are the same; both work towards construction of safe structures for the benefit of human beings. The mutual relationship between the two disciplines—geology and civil engineering—gave birth to the applied science of engineering geology that uses geological information combined with practice to assist civil engineers in solving problems that require such knowledge. Whereas geology deals with the history and structure of the earth, civil engineering focuses on the earth's surface.

In civil engineering practice of all major constructions, the geology of the sites and its bearing on engineering structures are properly evaluated and taken into consideration in the design for ensuring both economy and safety. 'Geology stands to Civil Engineering in the same relation as faith to work. The success or failure of an engineering undertaking depends largely upon the physical conditions within the province of geology and the *works* of the engineer should be based on the *faith* of the geologist', observed Professor Boyd Dawkins, FRS (Balasundaram, 1982).

1.3 IMPORTANCE OF ENGINEERING GEOLOGY IN CIVIL ENGINEERING

Knowledge of engineering geology is crucial to detect potential geological problems of a project and to identify their rational solutions. A country's progress depends on its technological development along with proper utilization of its natural resources. As a result, constructions of various engineering projects are envisaged having large financial investment. During construction activities, it is a common practice to seek advice regarding geological problems of the project site from personnel with adequate knowledge in geology. This helps to ensure safe construction of the projects.

Engineering geology contributes to the development activity of a country that necessitates construction of high dams, large reservoirs, long tunnels, railways, highways, and several other engineering structures. All these civil engineering constructions are closely associated with the geological environment, and as such engineering geological works are of utmost importance. The success and economy of engineering constructions depend upon the understanding of the degree and extent of geological problems and their solution.

There may be various geological attributes that create construction problems in a site for an engineering structure. Engineering geological work of a project site at various stages of construction aids engineers to evaluate detailed geological conditions of the site and assess the suitability of the site for design of the structure for construction. At the early stage of a project, it aims to find the suitability of the site for the proposed structure. During the planning stage, the work analyses and evaluates the site conditions by surface and subsurface investigations, including exploratory drill holes. At the construction stage, it provides constant help to civil engineering activities by identifying geological defects and suggesting measures for their rectification.

Thus, as Deere (1973) has rightly said, 'Engineering geologist is not responsible for the presence of adverse geology of a site but he is responsible for finding out that such adverse features do occur'. In fact, the engineering geological investigation is taken up to detect all the defects of a site and record them in a clear manner understandable to engineers for adopting suitable corrective measures and designing a stable structure.

Engineering geology strives to achieve stability, safety, and economy in constructing civil engineering structures. In the present set-up, engineering geology and civil engineering work as a team with mutual cooperation and the common goal of contributing their knowledge for the development work of the country.

1.4 ENGINEERING PRACTICE WITH ROCKS AND SOILS IN ANCIENT INDIA

Geology deals with rocks and soil; their proper utilization requires technical skill or knowledge of engineering. The following subsections provide evidence that Indians were adept in the art of engineering constructions using stone and soil even during the Stone Age and early historic time.

1.4.1 Prehistoric Time

Stone Age Indians were knowledgeable in selecting hard and durable stones. They developed techniques for making tools and weapons of various designs from these stones for their domestic use and hunting (Fig. 1.1). They used natural caves for shelter after removing the fallen rock chunks or excavating the rock ledges. Later, they learnt the techniques of making wall-like structures with the help of earth, boulder, and stone slabs.

The broad circular masonry structures constructed by the Stone Age man using stone slabs to protect against cold wind and wild animals can be seen even today at the Bhimbetka caves in Madhya Pradesh. Through the succeeding ages, the knowledge and skill passed to next generations who learnt the art of using cut and dressed stones and the techniques of fabrication for masonry construction.

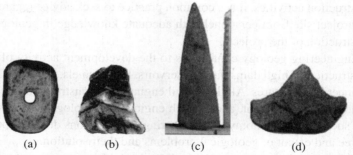

Fig. 1.1 Domestic tools and hunting weapons used by Stone Age Indians: (a) ring stone; (b) blade; (c) celt; and (d) hand axe

In fact, both rocks and earth (in the form of brick) were used for various construction purposes about 5000 years back, as is evident from the ancient cities of Sindhu–Saraswati (Harappa) civilization that extended over many parts of central, northern, and western India. Archaeological excavation has unearthed relics of brick houses, stone-lined drains, and irrigation channels from these ancient cities. It also brought to light a stone masonry dam at Dholavira, an ancient city in Gujarat, constructed in the third millennium BC for storage and supply of water to the people. The dam and storage structure (Fig.1.2) having a wide base with step-like features was made of hand-dressed stone blocks quarried from a local hillock and used local materials including clay as mortar (Gangopadhyay 2002a).

Fig. 1.2 Masonry dam and storage structure at Dholavira in Gujarat

1.4.2 Early Historic Time

Many engineering structures were built in India during the early historic times. Megasthenes (nearly 300 BC), the Greek ambassador to India wrote about the presence of sluice structures constructed for distribution of water to the people. Starting from the second century BC, Indians built hundreds of large rock-cut caves (underground chambers) (e.g., Ajanta, Ellora, and Elephanta) and numerous majestic monuments such as the *stupas*, temples, and monolithic pillars (e.g., Sarnath in Bihar and Sanchi in Madhya Pradesh). These structures stand testimony of the expertise of early Indians in the field of rock architecture and stone masonry works. Figure 1.3 shows the Sanchi stupa in Madhya Pradesh, which was constructed in the second century BC using hand-dressed Vindhyan sandstone blocks.

Embankments, bunds, and canals were constructed in the Cauvery valley as early as the first century AD. The Great Anicut across the Cauvery constructed by a Chola king for

Fig. 1.3 *Sanchi stupa* in Madhya Pradesh

irrigation and water supply is an example. The 12 m-high Mudduk Masur dam in South India built in AD 1500 is a notable example of earth dam construction. Archaeological studies have revealed that throughout the early historic time, royal palaces, temples, and forts of unique architectural design were constructed in India using dressed blocks of durable rocks and burnt-clay bricks.

1.5 HISTORY AND DEVELOPMENT OF ENGINEERING GEOLOGY IN INDIA

William Smith, who prepared a geological map in 1813 and also guided a canal construction in England, is known as the father of geology as well as of engineering geology. It was observed by Krynine and Judd (1957), that the collapse of St. Francis Dam in the United States in 1928, with loss of many lives, raised the awareness of geology in engineering practice in the US and other parts of the world. However, the concept of engineering geology was known in India since long ago as is evident from ancient books.

1.5.1 Concept of *Manasara Silpa Sastra*

There are several ancient books written in Sanskrit by Manasara, Mayamata, Visvakarma, Santakumara, and Srikumara that laid down rules on construction. Of these, *Silpa Sastra* on architecture and sculpture by Manasara (Acharya 1973) written in the sixth century AD contains guidelines on varied aspects of construction in the form of *sutras*. In this ancient book (*sutras* 198–211), rocks have been classified into three groups which is similar to the modern-day classification of rocks.

Silpa Sastra also contains guidelines for selection or rejection of rocks. For example, *sutras* 269–275 suggest that friable, porous, decomposed, stained, and easily erosive rocks are to be avoided for building purposes, whereas stones possessing sonic property equivalent to the sound of a jewel box or bell metal are to be considered as good quality stones for construction works. Compaction of soil and dewatering of foundation have been recommended in case of construction in soil. This work of Manasara is probably the earliest document in the world that deals with the concept of engineering geology with respect to properties of rocks and their use in engineering constructions (Gangopadhyay 2002b).

Indians have applied the techniques of engineering geology successfully in many of the civil engineering projects starting from mid-1800s. During this period, several long canals and dams were constructed for utilizing the water resources of the big rivers for irrigation and water supply, for example, the Ganga canal, Godavari canal, and Krishna canal were completed as far back as 1854–55. GSI has been associated with the engineering geological activity of the country since 1852 starting with Thomas Oldham's investigation of a railway alignment from Kolkata (then Calcutta) to Patna.

1.5.2 Activities of Geological Survey of India

Geological investigation for tunnel sites started in India as early as 1869, when T. Oldham conducted feasibility study of a traffic tunnel below river Sindh. In 1888, R.D. Oldham investigated land stability problems in the Nainital area of Kumaun Himalayas. His work on the Assam earthquake of 1897 was the first of its kind in the field of seismological study in the world (Oldham 1899). The Khadakwasla dam in Maharashtra (see Appendix A.6), a high masonry gravity structure, was a landmark construction in 1879 and helped irrigating the country for eight decades; it is considered to be the first of its kind in the dam-building history of the world.

Fig. 1.4 Dr J. B. Auden

The first hydroelectric project (micro hydel) of India was constructed at Darjeeling in West Bengal in 1897. Detailed engineering geological works in the construction of large dams in India date back to 1888–89 when GSI geologists carried out feasibility studies and rendered advice on the foundation conditions of the Mari Kanive dam site in Mysore and Bhavanisagar dam site in Tamil Nadu. As many as 105 engineering geological investigations were carried out in different types of projects during the first half of the twentieth century, covering many major dam projects such as the Bhakra, Thungabhadra, Yamuna, Mettur, and Pench. In addition, several railway alignments, bridge sites, and landslides were also investigated during this period.

While serving in GSI, Dr J.B. Auden (Fig. 1.4), the renowned engineering geologist, established the Engineering Geology Division of GSI in 1945 to assist in various civil engineering projects of the country. Since then, geological problems that arise in engineering projects are handled by the geologists of this specialized branch. Today, a team of over 100 trained engineering geologists of GSI are associated with all major civil engineering projects of the country. They are also involved in the evaluation and mitigation of natural/geological hazards such as earthquakes and landslides.

1.6 WORK-ACTIVITIES OF ENGINEERING GEOLOGISTS

Engineering geologists require specialized knowledge that is directly applicable in various engineering works involving rocks and soil. Their services are required in all large civil engineering constructions of the central and state governments. They are involved in selecting sites for engineering structures such as dams, reservoirs, power houses, bridges, and airports and in fixing alignments for construction of tunnels, highways, and railways, especially in hilly and hazardous geological terrains. Engineering geologists assess the conditions of the foundation for these structural sites based on the mechanical properties and stability of the rocks or unconsolidated materials that will bear these structures. Their main function is to find the potential geological hazards of these sites and suggest corrective measures.

As Balasundaram and Rao (1972) state in their study, in India, 'complex problems of site selection, evaluation of foundation condition and treatment, assessment of the availability of construction materials and study of reservoir competency necessitated constant attention and advice from geologist. For timely fulfilling of these tasks adequately, resident geologists were posted for the first time at the project site commencing from 1947'. In fact, in all major projects, resident engineering geologists work with the engineers from the initial stage of a project until its completion. They are to be associated with the day-to-day construction work of a site to detect the geological condition with the progress of excavation for the foundation and suggest remedy for immediate action. Foundation exploration is done under their guidance to reveal subsurface conditions and obtain specimens from depths.

The strength parameters of rocks and the soil of foundation including subsurface samples are determined by instrumental testing of specimens in laboratory and in special cases by in-situ tests. Such geotechnical data helps engineers in the planning and design of the structures and in the estimation of the overall cost of the project including possible treatment of geologically weak zones. Engineering geologists evaluate the terrain conditions of an area for development works including construction of building complexes and other utility structures. They study the dangers of landslide and subsidence and find measures to arrest potential slides. They also investigate areas affected by natural calamities such as earthquakes, floods, and tsunamis and advise the planners and administrators about the possible means of avoiding such dangers.

Engineering geologists investigate the groundwater or geohydrological condition of a terrain. The construction of a dam disturbs the groundwater regime. If a storage structure is constructed in areas with cavernous limestone or other highly permeable rocks, serious leakage may take place. Hence, engineering geologists study the possible changes in groundwater conditions due to the rise of groundwater level in storage areas and its effect on the project so that it may be considered in the planning and design stage. At times, they also advise on the probability of tapping groundwater for irrigation and drinking purposes. Engineering geologists should also determine whether overdrawing of groundwater will cause subsidence or sagging of land.

In recent times, a new dimension has been added to the role of an engineering geologist that involves the study of environmental problems associated with engineering constructions. Human actions due to engineering constructions or other development works bring changes in the geological environment of the landform. Development works change the land pattern of a country to suit the purpose of human society. For example, cutting a natural slope for road construction leads to immediate change in its land surface, sometimes creating conditions for landslides on the hill slopes. In fact, all construction activities create disturbances in natural equilibrium and cause artificial alteration of the surface.

Before undertaking a project engineers and engineering geologists need to understand the factors that control the geological processes so to minimize the destructive effects of human action. Large-scale destruction takes place due to mining and civil engineering constructions that change the landform and the natural environment. Controversy may arise when a certain engineering construction is aimed for the benefit of the public though it brings an imbalance or loss of equilibrium of natural environment/condition. In such an event, an engineering geologist may be the best judge as to how to create a harmony between the environment and the construction activity.

1.7 FORUM OF ENGINEERING GEOLOGISTS AND ENGINEERS

The Indian Society of Engineering Geology (ISEG), established in 1965, is devoted to developing a spirit of professional activities among engineering geologists and engineers of the country by way of interaction among them through several seminars and symposia. The Society also publishes a journal of its own. In the inaugural session of the Society in 1966, Dr D.N. Wadia expressed his confident hope that 'Indian Society of Engineering Geology will in the years to come play an important role in the national economy of India and it will provide a forum for Indian geologists and engineers to work together in framing a planned programme for building a new India'. Since then, ISEG has been publishing its journal that serves as a guide to understand the problems of civil engineering projects being constructed under varied geological conditions in different parts of the country. It also highlights the recent activities of the engineering geologists and engineers associated with various development works of the country.

ISEG also issues biannual newsletters aimed at promoting and encouraging the advancement of engineering geology. This is achieved by collecting, evaluating, and disseminating the data generated and results obtained in engineering, geological, and civil engineering activities and assessing the impact of such activities on the natural and living environment. ISEG, which has already completed 46 years, is also a national group affiliated to the International Association of Engineering Geology (IAEG).

1.8 RECENT ADVANCEMENTS IN ENGINEERING GEOLOGY

Recent years have seen advancement in the work approach and techniques in handling problems associated with engineering geology. Of late in India the trend in engineering geology work is on the application of digital techniques. The modern trend in engineering geology in advance countries such as the US, Europe, and Japan is dominated by digital data collection, utilization, and display. A famous US engineering geologist observes that 'engineering geologists will be challenged by the task of quantifying variability and uncertainties in observations and in interpretations. Interpreting and communicating the relevance and distribution of earth materials and identifying and quantifying hazardous natural processes takes on increasing importance for sustainable development'.

Today, sophisticated instruments are being developed for virtual geological mapping using three-dimensional laser scanning and terrestrial photogrammetry. Geographic information system (GIS) technology is being applied in the field even for small-scale projects and uses pen-based computers equipped with global positioning system (GPS), satellite receivers, and wireless communication for seamless data management. Traditional geological maps are being replaced by digital three-dimensional models attributed with geotechnical data.

Application of information technology (IT) to various aspects of engineering geology from data requirement, handling, and processing to numerical and GIS modelling and visualization is another modern trend well applied in countries such as the Netherlands, Sweden, and Denmark. With the advancement in the field of IT, these countries have developed digital methods to support effective and efficient execution of their work. Such methods are utilized in slope stability calculations and understanding the behaviour of soil and rock tunnelling and of earthworks.

Engineering geology is also used in creating database for large engineering geological project works, spatial data integration of two-dimensional modelling by GIS, and visualization techniques. In general, IT is being used in the following areas:

- Digital analysis of stream control transmission protocol (SCTP) data
- Digital description of borehole logs
- Digital field book for collection of engineering geology data
- Three-dimensional digital outcrop mapping

In Amsterdam, a tunnel project has been successfully constructed below the city by using IT techniques in the geological investigation. Moreover, in a tunnel and bridge project to link Denmark and Sweden, three-dimensional GIS technology was successfully used for excavation approach. However, India is yet to start applying digital techniques and IT in geotechnical works.

A very recent and encouraging trend in our country is that, in addition to Central and State governments, many companies providing engineering services have started appointing engineering geologists for their services related to engineering geology works. Some of these companies are involved in complicated geotechnical problem solving, which will necessitate the application of modern technology.

SUMMARY

- Engineering geology is the application of geological knowledge in civil engineering practices that include planning, design, and construction.
- The art of engineering was known in India from prehistoric times when buildings and even storage dams using hard and durable rocks were constructed. Numerous buildings, underground chambers, dams, and irrigation canals were built in India in early historic times.
- Engineering geological work started in India as early as mid-1800s when geologists of GSI were deputed to help the construction of railways, dams, and tunnels. From 1900 to 1947, engineering geological investigations were conducted in nearly 100 civil engineering projects of the country.
- A full-fledged Engineering Geology Division of GSI was established in 1945. The main objective of engineering geologists is to work with civil engineers and help them in evaluating the geological problems of the engineering project sites and suggest measures for solving them.

- The work of an engineering geologist at the initial stages of an engineering project includes geological mapping, subsurface evaluation by drillings, search for construction materials, and determination of rock/soil properties. At the construction stage, it involves the study of the conditions of foundations of engineering structures and suggestion of corrective measures for geological defects.
- Investigating natural hazards such as earthquakes and landslices and identifying measures for their remediation and conducting geohydrological studies to locate groundwater resources constitute the other activities of engineering geologists. They are also required to evaluate the stability of hill slopes for construction of highways and other communication systems.
- Submission of reports to the engineers with geological maps, borehole logs, laboratory and field test data, and so on is the most important part of the engineering geological activities to help in the planning, design, and safe construction of engineering structures.

EXERCISES

Multiple Choice Questions

Choose the correct answer from the choices given:

1. In the investigation of a site for construction of an engineering structure, an engineering geologist aims at:
   (a) stability of the structure
   (b) economy of its construction
   (c) both (a) and (b)
2. Primitive Indians knew to select hard and durable rocks as is evident from:
   (a) the stone weapons they used
   (b) the stone-made domestic articles
   (c) from both (a) and (b)
3. The ancient city of Dholavira was in:
   (a) Maharashtra
   (b) Gujarat
   (c) Andhra Pradesh
4. The relic found at Dholavira was constructed in:
   (a) the second millennium BC
   (b) the third millennium BC
   (c) the fifth millennium BC

5. The structure that proves that ancient Indians knew the art of engineering even in the distant past is:
   (a) the sluice structure for water supply dating 500 BC
   (b) the embankments and canals from the first century AD
   (c) both (a) and (b)
6. The rock type that was used in the construction of *Sanchi Stupa* is:
   (a) pink granite
   (b) red quartzite
   (c) buff sandstone
7. According to *Silpa Sastra* written in the sixth century AD:
   (a) rocks can be classified into three divisions
   (b) porous and decomposed stones are not suitable for construction purposes
   (c) both (a) and (b)
8. The first micro hydroelectric project of India was constructed in Darjeeling, West Bengal, as early as:
   (a) 1897
   (b) 1900
   (c) 1947

Review Questions

1. Discuss how the mutual relationship between the two disciplines—geology and engineering geology—resulted in engineering geology.
2. Enumerate the main work-activities of engineering geologists.
3. Highlight the importance of engineering geology in civil engineering works.
4. Cite evidence of the Stone Age man's knowledge of selecting hard and durable stones to make stone weapons and domestic articles. What was the habitat where Stone Age men lived?
5. By whom and when was *Silpa Sastra* written? List the guidelines mentioned in *Silpa Sastra* regarding the use of rock/soil for construction purposes.
6. Trace the history and development of engineering geology in India.
7. 'The art of engineering construction using rocks—the rock architecture—was known in India from early historic time'. Justify this statement citing examples.

Answers to Multiple Choice Questions

1. (c)   2. (c)   3. (b)   4. (b)   5. (c)   6. (c)   7. (c)   8. (a)

2 Rocks and Minerals

● ● ● ● ● ● ● **LEARNING OBJECTIVES** ● ● ● ● ● ● ● ●

After studying this chapter, the reader will be familiar with the following:

- Major types of rocks on the earth's crust and their genesis
- Mineral constituents of rocks and their properties including crystal forms
- Geological timescale and geological age of major rock formations of India
- Detailed descriptions of different rock types and their identifying characteristics
- Physiography and stratigraphy of India including the economic mineral resources

2.1 INTRODUCTION

The crust of the earth is composed of three major rock types. This chapter traces the genesis of these rocks providing examples of their occurrence in India. It provides the classification of the three major rock groups and a description of the rock types under each group. The chapter further provides an account of the various minerals including their physical properties and crystal forms. It illustrates the geological timescale of various rock formations of India along with equivalent absolute time in million years. The chapter also elucidates the identifying characteristics of all common rock types specified.

2.2 MAJOR ROCK TYPES AND THEIR ORIGIN

The earth's crust is composed of various types of rocks. These are of primary importance in engineering constructions because of their use as building materials and as foundation for engineering structures. The engineering properties of these rock types can be attributed to the different ways in which they originated. The rock types develop varied engineering properties because of their origin in different ways. Depending upon the nature of origin, the rocks of the earth are divided into three major groups—igneous rocks, sedimentary rocks, and metamorphic rocks.

2.2.1 Igneous Rocks and Intrusive Bodies

Igneous rocks are formed from the cooling and solidification of hot and fluid mass of rock called magma that exists in the interior part of the earth. Molten rock from rocks by detrital grains or by precipitation from solution in water help to easily distinguish them from the igneous and metamorphic rocks from textural grounds. This is because the grains composing them are non-interlocking fragments rather than interlocking crystals. The sediments formed by precipitation do have interlocking crystalline textures, but are made of minerals that are quite different from that of igneous rocks, notably calcite (in limestone) or evaporite minerals such as gypsum.

Sedimentary rocks provide information on the history of deposition and palaeo-environment. The knowledge of origin of plants, animals, and man is obtained from the study of fossils embedded in sediments. The most prominent primary structure of sedimentary rocks is the *bedding* or stratification of varying thickness as well as composition. Depending upon the depositional environment, several other primary structures are developed in sedimentary rocks such as *graded bedding*, *ripple marks*, *mud cracks*, and *cross laminations*. The secondary structures are provided by chemical actions forming *geodes*, *vugs*, *solution concretions*, and so on.

Fig. 2.2 Vindhyan sandstone hill, Madhya Pradesh

The initial dip of the sediments deposited on a nearly flat surface is usually horizontal. However, if the depositional surface is inclined, the sediments are deposited at an angle. The maximum initial dip of a bed formed by deposition of the sediments on an inclined surface is about 45°. Later deformation may create steeper dips. In engineering geological studies, the angle and disposition of dipping beds are important with respect to possible sliding. The sedimentary rocks of Vindhyan formation in India (Fig. 2.2) are famous for architectural works from the prehistoric time and are even now used as construction material in various engineering projects.

2.2.3 Metamorphic Rocks

Metamorphic rocks are formed by the metamorphism of pre-existing rocks under high temperature and pressure conditions or by chemically-active fluids bringing about both physical and chemical changes in the original rocks. A metamorphic rock may show remnants of the earlier rocks from which it has been formed under pressure–temperature conditions; such a rock is called *xenolith*. The pattern of change depends upon the grade of metamorphism controlled by the pressure–temperature conditions. The change is not as apparent in low-grade metamorphism as in high-grade metamorphism. In general, metamorphism takes place within a temperature range of 100°C to 900°C generated from the intrusion of plutonic rocks. The pressure comes from the huge thickness of overlying rocks or forceful intrusion of plutons and plate collisions. Hot water and other fluids associated with magma react with surrounding rocks to bring about chemical changes.

The two most important types of metamorphism are *contact* or *thermal* metamorphism and *regional* metamorphism. In contact metamorphism, country rocks react with intrusive igneous bodies causing changes in the surrounding rocks as a result of the heat generated from intrusion or injection of magmatic fluids. The rocks produced by contact metamorphism are restricted to areas close to plutonic (deep-seated) bodies. Regional metamorphism affects large regions such as the Precambrian shield and also cores of folded mountains. All common metamorphic rocks such as gneiss, schist, phyllite, and slate are produced by regional metamorphism.

There are several other types of metamorphism that produce rocks having restricted field occurrence. For example, pyrometamorphism is evident from the presence of xenoliths in some dykes of basalt. Cataclastic metamorphism is recognized in limited rock outcrops showing plastic flow or rupture without recrystallization or chemical change of original rocks. Retrogressive metamorphism brings about the conversion of high temperature minerals into an assemblage of stable minerals at low temperature. Turner and Verhoogen (1987) have mentioned about dislocation metamorphism that produces rocks such as mylonite and phyllonite found in areas of extreme deformation such as large fault zones. Such rocks produce adverse effects if present in the foundation of an engineering structure due to their genetic weakness caused by shearing effect, shown in Fig. 2.3; see also Figs 3.2 (a) and (b).

Fig. 2.3 Metamorphic rock (mylonite) showing pre-existing slip plane filled with quartz vein between two xenolith bodies (where a pen is kept)

During metamorphism, the effect of pressure causes growth of new crystals or reorientation of previously existing crystals to layers commonly known as *foliation*. Foliation causes splits in rocks and hence has an important bearing in engineering geology related to metamorphic rocks. Some wavy surfaces or cleavage may also develop. On the basis of foliation, metamorphic rocks are classified into the following types:

Slate   Fine grained; foliation is very closely packed.

Phyllite   Fine grained; wavy or crenulated surfaces having sheen caused by small flakes of mica or chlorite.

Schist   Coarse to medium coarse; undulating planes with abundant mica. Depending on the constituent minerals, this rock is designated as hornblende schist or garnet mica schist.

Gneiss   Coarse; widely spaced foliation than schist, mainly occurring with bands of dark and light minerals in thin layers.

Migmatite   Coarse; poor foliation showing signs of having begun to melt.

Metamorphic rocks of igneous origin do not develop foliation and the rocks are termed *hornfels*. New minerals do not always develop after metamorphism. Thus, quartzite formed from sandstone contains recrystallized quartz and marble contains interlocking calcites that were present in limestone.

Pre-existing rocks undergo important changes after metamorphism due to heat and pressure. This process of change is known as *metamorphic facies*. In thermal or contact metamorphism, when granite intrudes a rock with silty and clayey composition, it forms a rock which is hard and splintery called hornfels. Farther away from the edge of the granite body, a new mineral, sillimanite having the composition Al_2SiO_5, is developed by contact metamorphism with the sediments. Even farther, with changing pressure, temperature, and depth, another mineral,

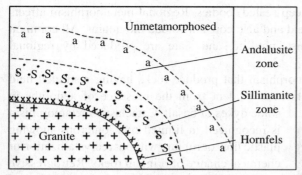

Fig. 2.4 Sketch showing metamorphic aureole of granite intruded in a rock of silty and clayey composition

andalusite, is formed having the composition Al_2SiO_3 but at a lower temperature (Fig. 2.4).

The mineral formation will change with change in the rock or sedimentary materials coming in contact during intrusion. A granitic rock intruding a limestone will form an iron-rich metamorphic mineral olivine, which is unknown in igneous rocks. Formation of sillimanite and andalusite requires a temperature above 600°C and 400°C, respectively. Another aluminium silicate mineral, kyanite (Al_2SiO_3), will be formed unmetamorphosed if the pressure is high enough (higher than 3 kilobars or 3000 atmospheres). Sillimanite, andalusite, and kyanite are typical minerals and their presence is indicative of their origin from metamorphic rocks including the approximate pressure–temperature conditions of origin. For example, the presence of kyanite indicates that metamorphism took place at a temperature range of 400°C to 800°C, pressure of more than 3 kilobars, and depth 10 km below the surface.

In general, thermal metamorphism is caused at the proximity of hot intrusion. In regional metamorphism, the pressure, and hence depth (as the two are related), and the temperature prevailing regionally are of importance. Low pressure–high temperature metamorphism occurs close to continental crust and lithosphere, but high pressure–low temperature metamorphism occurs near the plate boundary (the collision zone) with oceanic crust and upper lithosphere. The metamorphic minerals formed under different facies of regional metamorphism will depend mainly upon the original composition of the rock undergoing metamorphism. Table 2.1 gives a list of the different facies of regional metamorphism and the new minerals formed from the original rocks of silty and clayey composition (as in sedimentary rocks) and from basaltic igneous rocks:

Table 2.1 Facies of regional metamorphism

Facies name	Originally silty, clayey rock	Originally basaltic rock
Greenstone	Muscovite, chlorite, quartz, sodium rich plagioclase	Albite, epidote, chlorite
Amphibolite	Muscovite, biotite, garnet, quartz, plagioclase	Amphibole plagioclase
Granulite	Garnet, sillimanite, plagioclase quartz	Calcium-rich pyroxene, Calcium-rich plagioclase
Eclogite	Garnet, sodium-rich pyroxene, quartz	Sodium-rich pyroxene, garnet

2.3 CLASSIFICATION, DESCRIPTION, AND ENGINEERING USAGE OF IGNEOUS ROCKS

A detailed petrological classification that depends upon the chemical analyses and microscopic examination of rocks is beyond the scope of this chapter. 'No one classification of rocks can be regarded as ideal for all purposes. Schemes designed for use in petrological research are generally too sophisticated and rely on characteristics that are unsuitable for everyday practical use', observed Hatch, A.K. Wells and M.K. Wells (1984). It is, however, necessary to ensure that the nomenclature of a rock remains the same to an engineer as to a geologist. Hence, a simplified scheme of classification is presented in this chapter for each of the three main rock groups mentioning only the main rock types of each group (Tables 2.2–2.4). Igneous rocks are classified based on the mode of their formation with respect to depth (plutonic) and whether they are extrusive in character.

2.3.1 General Observation and a Simplified Classification

The main minerals in igneous rocks are quartz, feldspar, mica, amphibole, pyroxene, and olivine. Depending upon the mineral content and mode of occurrence, a simple classification of igneous rocks is presented following Blyth and Fereitas (1967) for practical use in engineering geology (Table 2.2). The rocks are arranged under four different headings, namely acid, intermediate, basic, and ultrabasic types. The acid type is composed of light-coloured rocks with dominance of quartz and subordinate feldspar. The rocks of basic type are dark coloured and contain mainly feldspar and ferromagnesian minerals. The intermediate type contains both light- and dark-coloured minerals and spare quartz. The two rocks under ultrabasic type are olivine-rich rocks. The details of these rocks (physical features, mineralogy, and so on) including their engineering use are given in the description.

Table 2.2 Classification of igneous rocks

Depth of origin	Acid type–light coloured (with quartz)	Intermediate type light-to-dark coloured (less or no quartz)	Basic type– dark coloured (no quartz)	Ultrabasic type– dark coloured (olivine rich)
Plutonic (deep seated)	Granite Granodiorite Pegmatite	Monzonite Diorite Syenite	Gabbro	Peridotite Picrite
Hypabyssal (moderate depth)	Granite-porphyry	Porphyrites (diorite porphyry)	Dolerite	
Volcanic (extrusive)	Rhyolite Dacite	Phonolite Trachyte Andesite	Basalt	

2.3.2 Colour and Texture

The mineral content in igneous rocks depends upon the nature of the magma from which they have originated. If the magma contains high silica, the rock will be light coloured with a dominance of quartz. On the contrary, rock formed from basic or low-acidic magma is dark coloured. If the rock has been formed in a deep-seated condition (plutonic), the slow cooling of the magma produces coarse-grained minerals with well-developed crystals. The sudden cooling of an extrusive rock results in fine-grained minerals and glass. In addition to mineral content, the size of the grain and the texture of the igneous rocks are characteristic features for their identification. Gabbro, dolerite, and basalt can be identified from their coarse, medium, and fine textures, respectively, though they are all dark-coloured rocks of nearly same composition.

The texture of an igneous rock refers to the pattern of grain arrangement. If the rock exhibits grains in the form of large crystals, it has *pheneric* texture. If the crystals are not well developed, the rock is fine grained, and if the ground mass is glassy in nature, it is said to have an *aphanitic* texture. The arrangement of feldspar in a basaltic rock is called *ophitic* texture. The texture is termed *porphyry* when large crystals called *phenocrysts* are embedded in a fine-grained matrix. The porphyry texture is generally found in rocks of hypabyssal origin. Slow cooling forms large grains, whereas sudden cooling as in the case of an extrusive rock results in fine grains and glass. Some structures such as vesicles, as seen in basalt, are characteristic features of rocks of volcanic origin.

2.3.3 Description and Engineering Usage

The following pages give a brief description of some of the important types of igneous rocks such as granite, pegmatite, dolerite, peridotite, and so on.

Granite

Granite is a medium-to-coarse grained, light-coloured rock with white or pink tint according to the colour of the feldspars (Fig. 2.5). When examined under the microscope, the rock is

found to contain essentially quartz, feldspar (orthoclase, plagioclase, and microcline), and ferromagnesian minerals such as hornblende and mica (Fig. 2.6). The accessory minerals include apatite, rutile, zircon, and magnetite. Granite with large crystals of potash feldspar embedded in comparatively fine-grained groundmass is known as *porphyritic* granite. Granitic rock in which feldspar is abundant is called *monzonite*. If plagioclase content is more than orthoclase, it is known as *granodiorite*. In engineering geological description, the common name 'granite' is used without making distinctions such as monzonite or granodiorite. The rock is hard and durable and possesses high strength; it is suitable for use as construction material. Fresh granite outcrop provides excellent foundation conditions for engineering structures. However, porphyritic granite with large feldspars when weathered and decomposed may cause weakness in the foundation. Granite occurs abundantly in peninsular India, eastern India, and central Himalayas.

Fig. 2.5   A polished block of pink granite

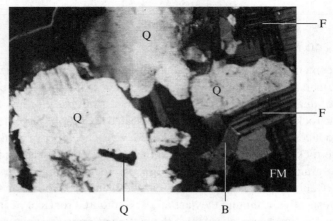

Fig. 2.6   Photomicrograph of granite (×70): Q, quartz; F, feldspar; FM, ferromagnesian mineral; B, biotite; and IO, iron oxide mineral

Pegmatite

Pegmatite is a very coarse-grained rock consisting mainly of feldspar, quartz, and mica. It occurs as a vein or dyke or as a large rock body intrusive into granite. Minerals found in large crystals in pegmatite result from the cooling of residual magmatic fluid. Large books of muscovite used in the industry are obtained from pegmatite. The pegmatite of Kodarma in Jharkhand is famous for the large yields of ruby mica (muscovite) that are used in the electrical industry. The decomposition of feldspar in pegmatite forms kaolin, which is very soft in nature and when present in a foundation is considered to be a weak feature. *Aplite* containing aggregates of sugary-grained quartz and feldspar are commonly associated with pegmatite.

Charnockite

Charnockite was first found in India and is named after Job Charnock of the British East Indian Company. This rock is similar to granite in physical and mechanical characteristics but differs in mineralogical composition. It is blue to bluish grey in colour and is a medium-to-coarse grained rock consisting mainly of quartz, feldspar (orthoclase and microcline), and pyroxene, and subordinately magnetite, apatite, zircon, and rutile. Charnockites may also be metamorphic in origin, in which case garnet, biotite, and ilmenite occur as accessory minerals. Charnockite occurs in the Nilgiris and Anamalai hills of South India and parts of Eastern Ghats in Tamil Nadu. It is well suited for foundation and construction purposes.

Syenite

Syenite is a medium-grained, light-coloured, and even-textured rock consisting essentially of feldspar, hornblende, and mica, or pyroxene. This rock is of deep-seated (plutonic) origin containing intermediate silica content and is high in alkali. A common variety of syenite is nepheline syenite, consisting of a large content of nepheline. Syenite appears similar to granite in hand specimen, but is not a common rock.

Diorite

Diorite is grey to greenish grey in colour. Diorite is a medium-to-coarse grained and even-textured rock. It consists essentially of plagioclase feldspar and ferromagnesian minerals such as biotite and hornblende or pyroxene. The percentage of ferromagnesian minerals present in the rock equals or exceeds the feldspar content and hence the dark colour. If the rock contains sufficient content of quartz, it is called quartz-diorite. Diorite is a commonly occurring rock and is used for foundation and construction purposes.

Porphyries

Igneous rocks (generally hypabyssal types) exhibiting large crystals (phenocrysts) embedded in fine-grained groundmass are called porphyries. Thus, there are granite porphyries, diorite porphyries, and so on.

Gabbro

Gabbro is a coarse-grained basic igneous rock consisting essentially of feldspar (plagioclase and oligoclase) and ferromagnesian minerals such as pyroxene (mainly augite with very little or no hypersthene). It is similar to diorite in appearance, but contains larger amounts of ferromagnesian minerals. If the hypersthene content in gabbro exceeds that of augite, it is called *norite*. If the mineral contents in gabbro are essentially plagioclase and labradorite with very little augite, the rock is *anorthosite*. When fresh, the rock chips can be used for construction of roads and for other engineering purposes. After decomposition, it develops weakness due to alteration of olivine to serpentine and chlorite, and such altered or weathered gabbro is not to be used for any engineering work.

Dolerite

Dolerite is a medium-grained basic igneous rock having the same composition as gabbro. It originates from moderate depth and occurs as dyke, being an intrusive body in country rocks. Weathering of the rocks with characteristic exfoliation gives rise to large spherical bodies (appearing like boulders) with thin concentric shells. Dolerite is a very hard rock and is used for engineering purposes, especially as road metal for its capacity to hold bitumen coating as a binder.

Peridotite

Peridotite is a medium- to coarse-grained, even-textured, and dark-coloured ultrabasic rock. Olivine is the chief constituent of this rock. Other minerals include augite, hornblende, biotite, and iron oxide. Felsic minerals are absent. Some varieties are composed entirely or predominantly of pyroxene, hornblende, or olivine. Depending upon the mineral content, the rocks are termed as *pyroxenite* (containing entirely pyroxene), *hornblendite* (containing entirely hornblende), and *dunite* (containing entirely olivine). These rocks contain several accessory minerals such as garnet, ilmenite, and chromite. The ultrabasic rocks may be rich in mineral content including ores; if present at any project site, the deposit requires thorough investigation to evaluate its economic importance prior to any construction in this rock.

Picrite

This ultrabasic rock is essentially made of olivine and augite or hornblende with little feldspar (<10 per cent). Ilmenite occurs as an accessory mineral in this rock. With the increase of feldspar and decrease in other constituents such as olivine, the rock grades into gabbro.

Basalt

Basalt is a fine-grained, dark-coloured basic igneous rock that originates from cooling of volcanic lava flow (Fig. 2.7). When examined under the microscope, it is found to contain mostly feldspars (plagioclase and oligoclase), pyroxene (augite), olivine, and iron oxide with

Fig. 2.7   A specimen of volcanic rock (basalt) from the Deccan trap

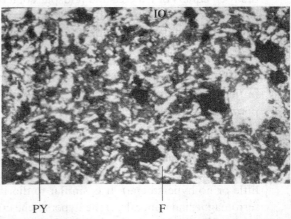

Fig. 2.8   Photomicrograph of basalt showing ophitic texture (×80): F, lathes of feldspar; PY, pyroxenes; and IO, iron oxide minerals)

ophitic texture (Fig. 2.8). Glass, which is present in some varieties, makes the rock vulnerable to chemical weathering. The Deccan basalt is also called *trap rock* or simply *trap*. There is a vesicular variety known as *amygdaloidal basalt* containing secondary minerals such as chalcedony, opal, agate, jasper, and zeolite inside the *amygdular* cavities. Basalt with these silica minerals in high content may cause alkali aggregate reaction and hence using it as a concrete aggregate should be avoided. The product of weathering of basalt is *laterite*, which is widely used in engineering constructions of road and buildings (see Section 19.1). Chemical alteration of the rocks produces certain minerals collectively known as 'green earth', which is a deleterious material and hence should not be used for construction purposes.

Rhyolite, dacite, trachyte, andesite, and phonolite

These volcanic rocks are commonly known as *felsites*. They are very dense and homogeneous and have flint-like or glassy appearance. The main constituent minerals of these different rock types are as follows: rhyolite—alkaline feldspars and quartz; dacite—lime-soda feldspar and quartz; trachyte—alkaline feldspar without or with very little quartz; andesite—soda-lime feldspar with little or no quartz; and phonolite—alkaline feldspar with nepheline. All these volcanic rocks are very fine grained or glassy and light to pale grey in colour except andesite, which is dark coloured. They are of common occurrence from volcanic eruption as lavas on the surface and also partial intrusion through cracks in the earth's crust.

Obsidian, pitchstone, and pumice

These are very dark-coloured glassy type of igneous rocks formed by rapid cooling of ejected lavas from volcanoes. The mineral grains cannot be distinguished in these glassy rocks. If the rock is vitreous in lustre and bright looking, it is called *obsidian*. When the lustre is dull and pitchy, it is *pitchstone*. The *pumice* has an earth texture and dull look. It is spongy and very porous. These rocks occur only in restricted areas, and if present in an engineering project site, their weathering characteristics are to be critically considered.

Tuff and volcanic breccia

The ashes ejected with fragments from volcanoes are deposited on the surface. Consolidation of these materials by natural processes form the rock called *tuff*. It is a light-coloured soft rock in which volcanic ash content is dominant. After mixing with large quantity of angular and uneven fragments from the surface, it forms *volcanic breccia* or *agglomerate*. The volcanic breccia commonly present in the Deccan trap is associated with tuff beds.

2.4 CLASSIFICATION, DESCRIPTION, AND ENGINEERING USAGE OF SEDIMENTARY ROCKS

The constituent particles of sedimentary rocks may contain silica-rich minerals or lime–magnesia-rich minerals. Depending upon their composition, the sedimentary rocks are grouped into two main divisions—clastic and non-clastic.

2.4.1 Simplified Classification

Sandstone and shale (clastic) and limestone (non-clastic) are the most common sedimentary rocks. It has been established that 99 per cent of the sedimentary rocks belong to these types. Depending upon the composition and size of the rock fragments or grains in clastic sediments and the content of minerals in non-clastic types, a simplified classification is derived (Table 2.3). Sandstones are further subdivided (Table 2.4) on compositional basis following Pettijohn (1957).

Table 2.3 Classification of sedimentary rocks

Group	Size of particle/ rock fragments (IS)	Type of constituent materials/rock fragments	Name of rock
Clastic	>300 mm	Boulder (rounded)	Boulder conglomerate
		Boulder (angular)	Breccia
	300–80 mm	Cobble	Cobblestone/breccia
	80–4.75 mm	Gravel/pebble	Pebblestone
	4.75–0.075 mm	Sand	Sandstone
	0.075–0.002 mm	Silt (non-plastic)	Siltstone
	<0.002 mm	Clay (plastic)	Claystone, shale
	>30 mm	Volcanic material	Agglomerate
	<4 mm	Volcanic ash	Tuff
	Unassorted large to small sizes	Rock fragments in clayey matrix	Tillite, till
Non-clastic	Crystalline	>50% calcite	Limestone
		>50% dolomite	Dolomite
	Organic matter	Plant remains	Peat, coal, lignite

Table 2.4 Subdivisions of sandstones

Cement matrix	Detrital matrix prominent; chemical cement absent	Detrital matrix absent or scanty; voids empty or filled by chemical cement		
Feldspar > rock fragments	Feldspathic greywacke	Arkose	Feldspathic sandstone	Orthoquartzite (Chert <5%)
Feldspar < rock fragments	Lithic greywacke	Subgreywacke	Proto-quartzite	Orthoquartzite (Chert >5%)
Quartz content	Generally <75%	<75%	75–95%	>95%

2.4.2 Texture of Sedimentary Rocks and Rounding of Particles

The texture, structure, and mineralogical and chemical composition of sedimentary rocks are related to their origin and subsequent modification. Texture is the pattern of packing of the particles in the rock. Packing enforces the strength of the rock and hence has an important bearing in engineering geology. The texture of the sedimentary rocks is related to the size of the grains. The various sizes of the sedimentary rocks and their corresponding terms (column 2, Table 2.3) are adopted following the Indian Standard. Rounding of the rock particles occur by one or more of the erosional agents. Current and wave action are responsible for smoothening the grain corners and rounding. The type of matrix or cementing material that binds the grains is an important criterion, especially in identifying sandstones (Table 2.4). Short descriptions of the various types of sedimentary rocks with special reference to sandstones of India are provided in the following subsections.

2.4.3 Description and Engineering Usage of Clastic Sedimentary Rocks

This section gives a description of sandstone and shale which are the major types of clastic sedimentary rocks and also of greywacke, orthoquartzite, arkose, siltstone, and conglomerate, which are subdivisions of sandstone.

Sandstones

Sandstones are the most dominant type of sedimentary rocks. They generally occur as layered or bedded strata and vary in grain size from gravel to fine-grained sandstone (Table 2.3). They contain quartz in varied proportions up to 95 per cent, with other minerals such as feldspar, iron oxide, mica, and chloride occurring in subordinate amounts. Sandstones vary widely with respect to their matrix, texture, porosity, and strength. If the matrix is siliceous, the rock will be hard, compact, and possesses high strength. Sandstone with argillaceous and calcareous matrix loses its strength if it remains saturated with water for a prolonged period.

Sandstone formations of India   There are several sandstone formations in India that are sources of construction material for various engineering projects. *Vindhyan sandstones*, which extend over a large part of central India, are found in various colours such as red, mauve, buff, grey, and mottle (Fig. 2.9). The rock consists of angular and subrounded quartz grains with tight packing and cemented in siliceous matrix (Fig. 2.10). Vindhyan sandstones possess high strength and low porosity and are used for foundation and construction purposes. *Siwalik sandstones* occur in the foothills of the Himalayas. They are porous and possess low strength. A variety of Siwalik sandstone called *sand-rock* is excessively porous and breaks down easily when saturated with water. *Gondwana sandstones* are found in many parts of eastern and central India. These are medium- to coarse-grained rocks, generally of low strength due to loose compaction of grains. *Tertiary sandstones* with low strength and high porosity are available in large parts of Assam. *Cretaceous sandstones* occur in the coastal belts of India. These sandstones have argillaceous and calcareous matrix that crumbles easily when saturated with water for a prolonged period.

Fig. 2.9   Large blocks and chips of Vindhyan sandstone

Fig. 2.10   Photomicrograph of Vindhyan sandstone showing angular and subangular quartz grains cemented in siliceous matrix (×70)

Greywacke

Greywacke is greenish grey or black in colour, resembling partially weathered basalt. It consists of angular-to-subangular grains of quartz and feldspar and fragments of phyllite, slate, and shale. The association of these rock fragments is a criterion of its identification. The grains are bound together in a fine-grained matrix. Greywacke may originate from sediments derived from plutonic and metamorphic rocks and sometimes from detrital matters of a sedimentary terrain.

Orthoquartzite

Orthoquartzite contains 90 per cent or more of quartz sand. It is a light-coloured rock and resembles quartzite. However, a close study may show that it is of sedimentary origin as it lacks the interlocking texture, characteristic of a metamorphic rock. Orthoquartzite is commonly called quartzose sandstone. The cementing material of the rock is mainly silica at places mixed with minor content of calcareous matter.

Arkose

Arkose is light grey to pale pink in colour and consists of coarse and angular grains of quartz and feldspar nearly in equal proportion. The grains are well sorted and derived from disintegration of granitic rock. The rock also contains mica, iron oxide, carbonate, and clay in subordinate amounts. If the feldspar content is more than 25 per cent, the rock is termed feldspathic sandstone.

Siltstone I

Siltstone is a light-coloured rock consisting mainly of sand particles with size varying between 2 microns and 75 microns. It also contains little amounts of coarser sand and clay. Siltstone has a gritty feel, which is an identifying character in a hand specimen. Siltstones are generally interbedded with shale and sandstone.

Shale

Shale is a soft sedimentary rock with thin layering or lamination along which it breaks easily. This easily breakable characteristic of shale is called *fissility*. Due to high compaction, some varieties of shale such as slate attain hardness. Shales vary widely in colour, from light grey to dark grey and also white, yellow, brown, and black. The rock contains particles of both silt and clay with clay-size particles (2 microns or less) being the maximum. If the clay percentage predominates and there is no layering, the rock is called *claystone* or mudstone. Claystone beds are frequent in the Siwalik formation. Shale and claystone contain clay minerals, which play a significant role in engineering geological investigations (Refer Sec. 5.8).

Conglomerate

This rock is formed from the consolidation of rounded boulders with siliceous, calcareous, or argillaceous type of cementing materials. If the matrix is siliceous, it is hard and of high strength, but when the rock contains other types of matrix such as clay, it becomes porous and breaks easily when saturated with water. Depending upon the dominance in size of these rounded rock fragments, the conglomerates are termed pebbly conglomerate or pebblestone, cobbley conglomerate or cobblestone, and boulder conglomerate (Table 2.3). When the majority of boulders in a sedimentary deposit are angular and uneven, the rock is termed *breccia*. When boulders in large numbers remain embedded in clay, the resulted deposit is called *tillite*. Conglomerate, breccias, and tillite are not suitable as foundation rocks as they crumble under pressure in saturated condition and permit leakage of water through them.

2.4.4 Description and Engineering Usage of Non-clastic Sedimentary Rocks

Limestone is the major type of non-clastic sedimentary rock. Calcite is the dominant constituent of limestone with dolomite and aragonite in subordinate amounts with occasional presence of few quartz grains. When dolomite (mineral) content is more, the rock is called *dolomite* (rock). The rock is found in different shades such as white, grey, yellow, brown, and black (Fig. 2.11). Fossil shells are frequent in many varieties of limestone; for example, *coral limestone* and *foraminiferal limestone* (Fig. 2.12). Limestone goes into solution and may form large caverns, which are of great concern when present in a reservoir project. If hard and free from cavities, the rock is suitable for foundation. Limestone occurs in Vindhyan, Cuddapah, and tertiary formations of India.

Fig. 2.11 Limestone beds showing solution effect along the bedding plane (see also Fig. 22.2)

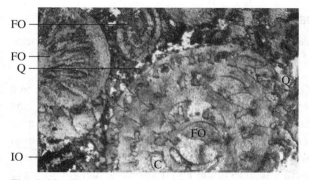

Fig. 2.12 Photomicrograph of foraminiferal limestone of tertiary age (×70): FO, shells of foraminifera; C, carbonate mineral; IO, iron oxide; and Q, quartz

Oolite, chalk, marl, and diatomaceous earth

Oolite is a variety of limestone consisting of spherical concretions. *Chalk* is another white variety comprising minute fossils of sea organisms. When the carbonate rock is mixed with a large percentage of clay, it is called *marl*. Diatoms are unique forms of algae that grow silica shells, which are preserved in water under sediments after death. The stratified product of diatoms is the *diatomaceous earth*. All these varieties of limestone and bio-fragmental deposits are generally soft and not suited for construction purposes.

2.5 CLASSIFICATION, DESCRIPTION, AND ENGINEERING USAGE OF METAMORPHIC ROCKS

Metamorphism of igneous and sedimentary rocks brings about changes in texture and causes recrystallization of the constituent minerals. Classification of metamorphic rocks is based mainly on this new fabric. Detailed description of the metamorphic rocks formed from the different pressure–temperature conditions are provided in the following subsections.

2.5.1 Simplified Classification

The parallel arrangements of the platy and elongated minerals develop foliation or banding structure, which is the most important characteristic feature of metamorphic rocks. Depending on the texture, structure, and mineral composition, a simple classification is outlined for metamorphic rocks as shown in Table 2.5.

Table 2.5 Classification of metamorphic rocks

Texture and structure	Characteristic features	Name of the rock
Banded	Fine- to coarse-grained; thin compositional bandings of platy and elongated minerals; breaks along bands into bulky pieces	Gneiss (e.g., granite gneiss and hornblende gneiss)
Foliated	Fine- to medium-grained; parallel layers of flaky or platy minerals; splits easily into flakes between layers	Schist (e.g., mica schist, and chlorite schist)
	Finely foliated showing sheen of flaky minerals like mica	Phyllite
	Very fine-grained; cleaves easily into thin sheets or flakes	Slate
Granular	Fine- to medium-grained particles of calcite or dolomite	Marble, dolomite
	Predominantly of fused quartz grains with interlocking texture	Quartzite

2.5.2 Fabrics

Metamorphism brings about changes in the pre-existing minerals, which form new shapes, generally flat or elongated, under pressure and temperature conditions. Rearrangement of the minerals in parallel orientation gives rise to the banded texture. If the minerals are platy or columnar, their parallel orientation by regional metamorphism results in the texture called *schistosity* (as in mica schists).When well-developed planes of schistosity alternate with less-developed schistose planes, the fabric produced is known as *foliation*, which is a very distinct feature of metamorphic rocks. The rock splits easily along this foliation plane.

Cleavage is a type of secondary foliation in metamorphic rocks consisting of closely spaced parallel planes of weaknesses of potential partings. Argillaceous rocks under moderate stress and increase in temperature form the *slaty cleavage*. Fabrics such as foliation, cleavage, and interlocking grains are features that distinguish metamorphic rocks from sedimentary rocks. The homogeneous texture of igneous rock changes when it is metamorphosed. Metamorphic rocks consisting of quartz and feldspar under regional metamorphism develop *granular* texture. Some minerals such as garnets have high strength of crystallization and develop large crystals, called porphyroblasts, for example, garnet porphyroblasts.

2.5.3 Description and Engineering Usage

The following is a description of the engineering usages of the major types of metamorphic rocks.

Gneiss

Gneiss is a medium- to coarse-grained and banded or foliated metamorphic rock. When observed under the microscope, it is found to comprise essentially quartz, feldspar, hornblende, and micas and accessory minerals such as garnet, tourmaline, apatite, zircon, sphene, and magnetite. Alternating bands of light and dark minerals are common in gneiss. The dark bands consist mostly of biotite and amphibole and the white bands of quartz and feldspar. The characteristic banding of this rock is due to the platy minerals such as micas and hornblende that are arranged in thin planes or bands along which the rock splits easily. The gneisses are designated as biotite gneiss, hornblende gneiss, or granite gneiss depending upon the mineralogical composition. In general, the rocks are derived from igneous rocks such as granites, but may also be from sedimentary rocks. The gneiss is a very hard rock and is suited for foundation of engineering structures. It is widely distributed in India and very old in age. Archaean gneisses occur in many parts of India, especially in the peninsular region and parts of the Himalayas.

Khondalite

Khondalite was first detected in India and has derived its name from 'Khond', a tribal area of Orissa. It is a light-coloured para-gneiss or para-schist consisting of quartz, sillimanite, garnet, and graphite. Due to schistosity, the rock can be split easily. Khondalite is subjected to alteration into *laterite* under natural processes. The rock occurs abundantly in the Eastern Ghats and contains manganese ores.

Granulite

Granulite is an even-grained rock comprising mainly quartz, feldspar, pyroxene, garnet, and small amount of mica. The rock is derived from high-grade metamorphism that imparts segregation banding and alignment of flat lenses of quartz or feldspars appearing like regular foliation to the rock. The rock with coherent grains attains high strength and is suited for use in engineering work.

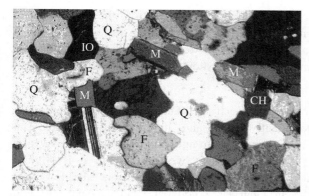

Fig. 2.13 Photomicrograph of schist (×80): Q, quartz; F, feldspar; M, micas; CH, chlorite; and IO, iron oxide mineral

Schist

Schist consists mainly of quartz, feldspar, muscovite, and biotite (Fig. 2.13). Various other minerals such as garnet, staurolite, kyanite, sillimanite, epidote, chlorite, and talc are also present in the rock. These rocks are named according to the presence of the flaky or platy minerals, for example, mica schist, talc schist, and chlorite schist. The rocks are of varied colours such as white, brown, yellow, and black. These are derived from the metamorphism of igneous and sedimentary rocks. In schist, the micas and other platy minerals lie with their cleavage planes parallel to each other to provide the fabric called schistosity. The rocks are generally of low strength and split easily along the planes of schistosity. However, it can be used for construction of houses, walls, and so on.

Phyllite

Phyllite is a grey-coloured rock characterized by glistering sheen derived from planes of partings or surfaces lined with fine-grained micas. Two types of phyllite are common—arenaceous (quartzitic) phyllite and argillaceous (shaley) phyllite. In the former, the grains are tightly packed by siliceous cement and hence the rock is very hard. In the latter, due to the presence of clayey matrix, the rock is soft though it has more sheen of mica. The rocks are derived from sandy and clayey shale by low-order metamorphism. Arenaceous phyllite can bear sufficient load and is safe for structural foundation.

Slate

Slate is a fine-grained rock with a characteristic cleavage called 'slaty cleavage', which is a new structure imposed by metamorphism. The rock can be split into big, smooth-surfaced sheets along the cleavage planes. The rock is generally grey, green, red, and black in colour and formed from intense metamorphism of clastic sedimentary rocks such as shale and claystone. Slate is a hard rock and is suited for foundation. However, if it occurs on hill slopes, it can cause damage as a result of breaking of large sheets of rock along the cleavage planes that slides down the slope. Large slabs of slates can be used for various purposes such as construction of walls and fencings.

Quartzite

Quartzite is white to light grey in colour. This metamorphic rock consists mainly of recrystal-lized quartz derived from metamorphism of sandstone. Quartz is the dominant content with iron oxide minerals present as impurity (Fig. 2.14). Some varieties contain more than 90 per cent

Fig. 2.14 Irregular blocks of quartzite

quartz with feldspar and micas in small amounts. With recrystallization, the boundaries of quartz become tight and interlocking assuming high strength. It occurs in various rock formations of India such as Dharwar, Aravalli, and Cuddapah. The rock is very hard and splits into sharp edges. It is suitable for foundation and is also the most durable construction material.

Marble

Marble is the metamorphic equivalent of limestone. It is found in various colours such as white, grey, and purple. It contains dominantly calcite and subordinately dolomite, iron oxide, graphite, and some quartz. Marble possesses good strength and durability and is resistant to meteoric weathering, but it is slightly susceptible to chemical erosion. Marble occurs in Dharwar, Cuddapah, and especially Raialo in Rajasthan. Indian marble specially mined from Rajasthan is well suited for use as building stones. Taj Mahal in Agra is made of chaste white Makrana marble from Rajasthan. Polished blocks of marble are used for decorative purposes (Fig. 2.15).

Fig. 2.15 Polished marble block

Mylonite

Mylonite is a fine-grained rock with flint-like appearance. The rock shows banding or streaking caused by extreme granulation and shearing of coarse-grained rocks. It is the product of dislocation metamorphism that disturbs the banding or other structure of original rocks but the groundmass may contain lenses of rocks in elongated condition, also see Fig. 3.2(b). The rock should not be used in construction sites due to its inherent weakness caused by shearing.

Phyllonite

Phyllonite resembles phyllite in its outward appearance. It shows silky films of newly recrystallized micas or chlorite along the planes of schistosity. Phyllonite originates from mechanical degradation or shearing by dislocation metamorphism of coarse-grained rocks such as granite, gneiss, and greywacke. The rock is not suitable for any engineering construction.

Metabasic rock

Metabasic rock is the metamorphic equivalent of a basic rock (e.g., basalt, dolerite, and gabbro). The fine-grained variety is the metabasalt, the medium-grained metadolerite, and the coarse-grained metagabbro. The metabasic rock is hard and possesses sufficient strength. It is also known as *epidiorite*. The mineral composition of the rock remains the same as in basic rock but the glass is altered forming minerals such as chlorophaeite and palagonite. When fresh and hard, it can be used for construction purposes.

Laterite

Laterite is extensively used in many parts of India for building houses, culverts, bridges, roads, and other engineering structures. It is a product of subaerial weathering in a monsoon climate having alternate dry and wet seasons. It occurs as a porous mass, which is red, yellow, or brown in colour, with a hard protective limonite crust on the exposed surfaces of the parent rocks. Laterite may form from varieties of rocks, the end product containing mainly the hydroxides of iron, alumina, and manganese. It is mostly found as an alteration product of the Deccan traps of peninsular India, khondalites of Orissa, and capping bauxite in the Eastern Ghats.

2.6 ORIGIN AND CHARACTERISTICS OF MINERALS

All rocks are composed of minerals, which are homogeneous substances with fixed chemical composition, crystal forms, and other distinctive characteristics such as colour, lustre, and hardness. The chemical elements and the crystal forms determine the properties of the minerals. The chemical elements depend on the composition of the rock in liquid state. Crystals are formed when a rock body passes from the liquid to the solid state. All minerals are solids with the exception of metallic mercury and water which is considered to be a mineral. The International Mineralogical Association (IMA) provides the following definition for a mineral: 'A mineral is an element or chemical compound that is normally crystalline and that has been formed as a result of geological processes.'

According to IMA, there are currently more than 4000 known minerals that occur in various types of rocks. These minerals are formed under varied environments and in different ranges of chemical and physical conditions such as pressure and temperature. The main group of mineral formations is found in igneous rocks, in which minerals occur as crystalline products formed from magmatic melt. In sedimentary rocks, the occurrence of minerals is the result of weathering, erosion, and sedimentation processes. In metamorphic rocks, new minerals are formed at the expense of the existing ones as a result of increasing temperature and pressure or both on some pre-existing rock types. Minerals are also formed by hydrothermal process in which they are chemically precipitated from solution. In the first three types, the minerals are interwoven with rocks as interlocking fabrics. Hydrothermal solutions tend to follow fracture zones with open spaces in which minerals are developed with good crystal forms.

Minerals may occur in rock formations in various forms such as metallic ores, industrial minerals, gemstones, and rare minerals whose origin, occurrence, and grade are studied in depth in economic geology. The rock formation of India has rich deposits of many metallic ores and industrial minerals, a description of which is given in Chapter 24.

2.6.1 Crystal Geometry of Minerals

The crystal form of a mineral is the outward expression of the internal molecular structure, which is dependent on the state of equilibrium due to the interatomic forces. Crystal geometry or crystal morphology deals with the symmetry, faces, and forms of a crystal. A crystal possesses the following three types of symmetry shown in Fig. 2.16.

Plane of symmetry   It divides a crystal into two halves such that one half is the mirror image of the other. A crystal may have one or many planes of symmetry or none.

Centre of symmetry   It is the central point about which every face and edge of a crystal is matched by one parallel to it on the opposite side of the crystal.

Axis of rotation symmetry   It is a line or an axis through a crystal about which the crystal can be rotated to bring it into an identical position a number of times in the course of one revolution.

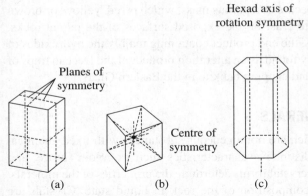

Fig. 2.16 Crystal symmetry: (a) plane of symmetry; (b) centre of symmetry; and (c) axis of rotation symmetry

For example, if during the course of full rotation of 360° about an axis of symmetry the crystal is brought into an identical position six times, the axis is a six fold. Such axes are termed *diad* (twofold), *triad* (threefold), *tetrad* (fourfold), and *hexad* (fivefold).

Crystal system

Crystal system is a category of crystals with reference to the position of their crystal faces and the relationship of the intercepts that the planes containing the faces make with three (or four) axes, intersecting at an origin (Fig. 2.17). In all, 32 combinations of symmetry elements are recognized under seven crystal systems depending upon the possible planes of symmetry and axes of symmetry. They are as follows:

  (i) Cubic (isometric): Three orthogonal tetrad axes of equal length, a_1, a_2, a_3
 (ii) Tetragonal: Three orthogonal axes, two horizontal diads of equal length and one vertical tetrad, a_1, a_2, c
(iii) Orthorhombic: Three orthogonal diad axes of unequal lengths, a, b, c
 (iv) Hexagonal: Four axes, three horizontal diads of equal length 120° apart and one vertical hexad at right angles, a_1, a_2, a_3, c
  (v) Trigonal: Four axes, three horizontal diads of equal length 120° apart and one vertical triad at right angles, a_1, a_2, a_3, c
 (vi) Monoclinic: Three unequal axes, one vertical, one horizontal diad, and a third making an oblique angle with the plane containing the other two, a, b, c
(vii) Triclinic: Three unequal axes, none at right angles, a, b, c

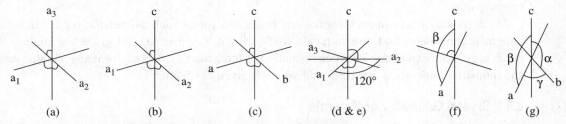

Fig. 2.17 Crystal systems showing intercepts of the crystal faces with three or four axes at the origin: (a) cubic; (b) tetragonal; (c) orthorhombic; (d & e) hexagonal and trigonal; (f) monoclinic; and (g) triclinic

Different crystal forms can be referred to the same set of crystal axes and hence belong to the same crystal system. For example, cube, rhombdodecahedron, and octahedron are different crystals of the same cubic system.

Each mineral crystallizes in one of the seven crystal systems. Figure 2.18 shows the outlines of these crystal forms with names of the common minerals, that is, isometric (fluorite), tetragonal (zircon), hexagonal (beryl), trigonal (tourmaline), orthorhombic (olivine), monoclinic (hornblende), and triclinic (albite feldspar). A mineral showing formation of crystals is termed *crystalline*. If no crystal of any geometric shape is visible in the constituent minerals of a

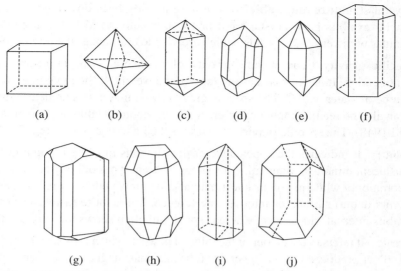

Fig. 2.18 Different systems of crystal forms with a representative mineral of each system: (a) & (b) isometric (fluorite); (c) & (d) tetragonal (zircon); (e) & (f) hexagonal (beryl); (g) trigonal (tourmaline); (h) orthorombic (olivine); (i) monoclinic (hornblende); (j) triclinic (albite)

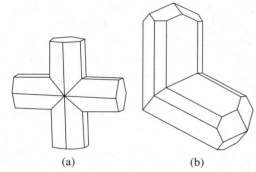

Fig. 2.19 Twinning of crystals: (a) staurolite; (b) rutile

rock in its thin section under the microscope, it is called *amorphous*. If only some small crystals are detected in the thin section of the rock under the microscope, it is termed *cryptocrystalline*.

Twinning of crystal

Two or more crystals of the same species can occur intergrown or in contact together in a particularly symmetric manner; such crystals are called twins. Twinning is found in different crystals of many minerals such as calcite, fluorite, gypsum, feldspar, staurolite, and rutile (Fig. 2.19).

2.6.2 Physical Characters of Minerals

All minerals are recognized by their physical characteristics, which are broadly divided into the following categories:

- Crystal structure
- Specific gravity
- Light-dependent characteristics such as colour, lustre, and transparency
- Sensory characteristics such as odour, feel, and taste
- Heat conductivity characteristics such as fusibility and malleability
- Electricity and magnetism
- Characters dependent on cohesion and elasticity such as cleavage, fracture, and hardness
- Others such as tenacity, fluorescence, and reactivity to acid

Crystal structure and habit   These have already been described in Section 2.6.1. When a mineral exhibits good crystal habit or form, it helps to identify the mineral. It may also be granular with very fine crystals, which are visible only under the microscope.

Specific gravity   The specific gravity of a mineral is a significant characteristic for its identification. It is expressed as the ratio of the weight of the mineral to the weight of an equal volume of water at 4°C. The specific gravity of a mineral can be determined in the laboratory. It can also be readily known by referring to the standard table of a mineralogy book (Dana and Ford 1948). The specific gravity of quartz is 2.65 and that of calcite 2.71.

Colour   It indicates the appearance of the minerals in reflected light or transmitted light for translucent minerals. Many minerals show different types of body colours. Quartz, for instance, is commonly white in colour but it may also be grey, yellow, or red. *Iridescence* is the play of colours due to surface or internal interference. Labradorite (a feldspar of plagioclase group) exhibits internal iridescence, whereas sphalerite often shows the surface effect.

Streak   It refers to the colour of the mineral powder, which is different from the body colour. It is the colour produced when a mineral is rubbed against unglazed porcelain. Magnetite is steel grey but its streak is cherry red. This property is generally observed in minerals with non-metallic lustre.

Lustre   It indicates the way a mineral's surface interacts with light. Based on lustre, minerals are broadly classified into metallic and non-metallic. The latter can be further classified into pearly, silky, silvery, dull, and so on. The soft minerals, calcite and serpentine exhibit pearly and silky lustre, respectively.

Odour, feel, and taste   Some minerals have typical *tastes* such as bitter, sour, alkaline, or saline. For example, common salt has a saline taste. Under certain circumstances, some minerals exhibit *odours* such as sulphurous, bituminous, or argillaceous. For example, friction or heat on pyrite generates a sulphurous odour. Serpentine and allied minerals when moistened give an argillaceous odour. A mineral may have a characteristic *feel* such as soapy, greasy, or harsh. Talc is well known for its soapy feel.

Fusibility and malleability   A mineral is called *fusible* if it melts on application of heat and *malleable* (e.g., gold) if it is flattened when it is mildly hammered.

Electric property   A mineral develops *electricity* by friction. Most of the gems are positively *electric* in the polished state. When a crystallized body develops electric charge under pressure, it is called *piezoelectricity*. Some crystals show positive and negative charges of electricity in different parts with change of temperature. For example, on heating, quartz shows positive electricity at the three alternative prismatic faces and negative electricity in the remaining three edges.

Magnetic property   Magnetite exhibits characteristics similar to natural magnet. The *magnetic polarity* is presumably developed in it by induction of the earth's magnet. Some minerals are attracted when brought close to a powerful magnet, for example, hematite, pyrrhotite, and native platinum.

Cleavage   It is the easily breakable plane along which the atoms are closely packed. When the mineral is subjected to a blow, it breaks along this plane leaving a smooth surface. The cleavage is termed basal, prismatic, pyramidal, and so on depending upon its parallelism with the crystal face. In topaz, cleavage is developed parallel to the basal face and so it is termed *basal* cleavage. Similarly, it is *prismatic* in amphibole and *pyramidal* in fluorite.

Fracture   A fracture is developed in a mineral by a sharp blow with a hard object. The resultant surface is characteristic of the mineral and may be of different types such as *conchoidal*, *jagged*, *fibrous*, or *irregular*. Thus, quartz develops conchoidal fracture that appears like a concave surface. The hardness of a mineral is measured on the Mohs scale, as stated in Section 2.6.3.

Brittle nature   A mineral is brittle when it forms fragments or powder (e.g., quartz, calcite). It is *flexible* if it bends without fracturing and returns to its normal shape after the removal of the force (e.g., some varieties of talc).

Elasticity   The elastic property of some minerals is derived by its capacity to resist any change in shape or volume when subjected to force. A muscovite is *elastic*, whereas phlogopite is *inelastic*.

Fluorescence and tenacity   Fluorescence is the response to ultraviolet light and tenacity is the response to mechanical-induced changes of shape or form. Acid reaction is seen is some minerals such as salcite, which produces effervescence with dilute hydrochloric acid.

2.6.3 Mohs Scale of Hardness

Of all the physical characters of a mineral, its *hardness* is of greatest significance in engineering geology. The strength and rigidity of a mineral is derived from its surface hardness, which is defined as the resistance to abrasion offered by it. Ten relatively common minerals are selected to represent various degrees of hardness in *Mohs scale of hardness*. Table 2.6 shows these ten minerals arranged in the order of increasing hardness.

Table 2.6   Mohs scale of hardness

Hardness (H)	Mineral	Character
1	Talc	Soapy to greasy feel
2	Gypsum	Can be scratched by finger nails
3	Calcite	Can be scratched by a copper coin
4	Fluorspar	Can be scratched by a penknife
5	Apatite	Can be scratched by window glass
6	Orthoclase	Can be scratched by a steel file
7	Quartz	Equivalent hard object porcelain
8	Topaz	Can be scratched by corundum but not by quartz
9	Corundum	Cannot be scratched by any mineral except diamond
10	Diamond	Cannot be scratched by corundum

To determine the hardness of a mineral, it is to be scratched by the minerals shown in the table to make a fine dent. For example, if a mineral cannot be scratched by fluorspar but can be scratched by apatite, its hardness is 4.5 in Mohs scale. In practice, the fingernail (H 2.6), a copper coin/plate (H 3.5), penknife (H 5), and window glass (H 5.5) are used to find the Mohs scale of hardness of a mineral.

2.7 CLASSIFICATION OF MINERALS BASED ON CHEMICAL COMPOSITION

Minerals may be classified based on their chemical composition. They are grouped and described under 17 chemical divisions such as native elements, sulphides, oxides, carbonates, silicates, phosphates, sulphates, borates, and nitrates as given by Dana and Ford (1948). Of these, silicates are the predominant minerals in rocks of the earth's crust. Carbonates, oxides, sulphides, and phosphates are the other commonly found minerals. The following list provides the classification of all the common minerals according to their abundance:

• Silicate class
• Carbonate class
• Sulphate, phosphate, chromate, tungstate, molybdate, and vanadate classes
• Halide class
• Oxide and hydroxide classes
• Sulphide class
• Native element class

The following paragraphs provide a description of these mineral classes.

Silicate class

The fundamental structural unit in all silicates is the (SiO_4) tetrahedron, in which a central silicon atom (Si) is surrounded by four oxygen atoms (O) in a tetrahedral arrangement. The bond between silicon and oxygen is very strong.

Silicates are the largest group of minerals found in rocks. Nearly 95 per cent of the rocks contain silicon and oxygen in addition to aluminium, magnesium, iron, and calcium. As shown in Table 2.7, the important silicate minerals are quartz, feldspars, pyroxenes, amphiboles, micas,

Table 2.7 Properties of silicate minerals

Name and composition	Colour	Crystal form	H	G	Cleavage	Lustre	Distinguishing characteristics
Nepheline $(Na, K)AlSiO_4$	Grey, brown	Hexagonal	6	2.6	Indistinct	Greasy, vitreous	Lustre and less hard than quartz
Leucite $K(AlSi_2)O_6$	Grey	Cubic	6	2.5	No cleavage	Dull, vitreous	Crystal form
Sodalite $Na_4(Si_3Al_3)O_{12}Cl$	Grey, green	Cubic	6	2.3	Indistinct	Vitreous	Colour change
Scapolite $(NaCa)_4(SiAl)_{12}$ $O_{24}(ClCO_3SO_4)$	White, grey	Tetragonal	5–6	2.5–2.6	Distinct	Vitreous	Crystal habit, cleavage
Stibnite $NaCa_4(Al_9Si_{27})$ $O_{72} \cdot 30H_2O$	Brown, grey	Monoclinic	3.5–4	2.2	Perfect	Vitreous	Sheaf-like aggregate
Lepidolite $K(LiAl)_3$ $(SiAl)_3O_{10}(FeOH)_2$	Lilac, pink	Monoclinic	3	2.8	Perfect	Pearly	
Riebeckite $Na_2(Fe^2Mg)_3(Fe^3)_2$ $Si_8O_{22}(OH,F)_2$	Blue, black	Monoclinic	6	3.4	Perfect	Vitreous	Colour, mode of occurrence
Cordierite $Mg_2Al_4Si_5O_{18}$	Blue, violet	Ortho-rhombic	7–7.5	2.6	Indistinct	Vitreous	Colour, gem variety

olivines, and garnets. The feldspar group alone (orthoclase, plagioclase [albite], oligoclase, labradorite, bytonite, and anorthite) constitutes about 60 per cent of the crust and quartz nearly 10 per cent. The silica minerals are predominantly found in igneous rocks and metamorphic rocks but are also found in sedimentary rocks.

Feldspar, pyroxene, amphibole, olivine, garnet, zircon, sillimanite, kyanite, staurolite, topaz, epidote, muscovite, biotite, and tourmaline talc are the silicates apart from the silica mineral quartz and opal (amorphous) varieties. The properties of many of these common silicate minerals are described in Table 2.15 under the heading rock-forming minerals. Table 2.7 describes the properties of the other minerals.

Clay minerals belong to a subclass of silicates called *phyllosilicates*, which are very fine grained and possess swelling properties. They can be identified by a study under electron microscope and also by X-ray diffraction method and differential thermal analyses. The properties of these clay minerals, which are of importance in engineering construction, are listed in Table 2.8.

Table 2.8 Properties of clay minerals under phyllosilicates

Name and composition	Colour	Crystal form	H	G	Cleavage	Lustre	Important characteristics
Kaolinite $Al_2Si_2O_5(OH)_4$	White, earthy	Triclinic	2	2.6	Perfect	Earthy	White, from decomposed feldspar
Montmorillonite $(NaCa)_{0.3}$ $(Al, Mg)_2Si_4O_{10}(OH)_2nH_2O$	Grey, green	Monoclinic	2–2.5	2–2.7	Perfect	Greasy	Expansive, has many uses as bentonite
Vermiculite $(Mg, Fe, Al)_3$ $(Al, Si)_4 O_{10}(OH)_2 4H_2O$	Brown yellow	Monoclinic	1.5	2.4	Perfect	Greasy	Highly expand, altered mica
Illite $(K, H_3O)Al_2(Si, Al)O_{10}$ $(H_2O, OH)_2$	Grey	Micro-crystalline	1.2	2.6	Perfect	Silky, dull	Main constituent of soil

Carbonate class

The carbonates (CO_3) class is isolated in structure and does not form chains, rings, or layers as in silicates. The important minerals of this group are calcite and aragonite (calcium carbonate,), dolomite (calcium/magnesium carbonate), and siderite (iron carbonate). Other carbonate minerals include magnesite (magnesium carbonate), rhodochrosite (manganese carbonate), smithsonite (zinc carbonate), witherite (barium carbonate), strontianite (strontium carbonate), and cerussite (lead carbonate).

Rocks containing carbonate minerals such as limestone and dolomite are found in the marine environment where shell and planktonic life are abundant. Carbonates are also found in evaporatic settings and in karst region where the dissolution and reprecipitation of carbonates form big cavities including stalactites and stalagmites (see Figs 22.2 and 22.4). Carbonate minerals also occur in metamorphic rocks as marble (calcium carbonate). The nitrates and borate minerals also contain carbonates, which are mostly formed by

Table 2.9 Properties of carbonate class* minerals

Name and composition	Colour	Crystal form	H	G	Cleavage	Lustre	Important characteristcis
Magnesite $MgCO_3$	White, grey, brown	Trigonal	4	3	Perfect	Vitreous	Cleavage, effervescence in HCl
Siderite $FeCO_3$	Grey, brown, steel white	Trigonal	4	4	Perfect	Vitreous	Low hardness effervescence in HCl
Aragonite $CaCO_3$	White, grey, blue	Ortho-rhombic	3.5–4	2.9	Distinct	Vitreous	Low hardness, effervescence in HCl
Strontianite $SrCO_3$	White, grey, yellow	Ortho-rhombic	3.5–4	3.8	Good	Vitreous	Density, effervescence in HCl
Witherite $BaCO_3$	White, grey	Ortho-rhombic	3.5–5	4.5	Distinct	Vitreous	
Cerussite $PbCO_3$	White, grey	Ortho-rhombic	3.5	6.6	Distinct	Ada-mantine	Density, colour, twin
Smithsonite $ZnCO_3$	White, brownish blue	Trigonal	4.4–5	4.4	Perfect	Vitreous	Effervescence in warm HCl

* For calcite and dolomite of the carbonate class, see Table 2.15.

evaporation from salt lakes. Table 2.9 provides a description of the common carbonate minerals.

Sulphate class

Sulphate is present in sulphate minerals as SO_4^{2-}. The (SO_4) tetrahedron is a complex anion or group in which four O atoms are placed around a central S atom at the corners of the tetrahedron. These tetrahedra are isolated in the structure and do not form groups, chains, or layers as in silicates. Sulphate minerals are commonly formed in evaporatic settings where slow evaporation of saline water allows the formation of both sulphates and halides at the water–sediment interface. They also occur in hydrothermal vein system as gangue minerals along with sulphide ore minerals. In addition, sulphates are obtained as secondary oxidation products of original sulphide minerals.

The common sulphate minerals are anhydrite (calcium sulphate), celestine (strontium sulphate), barium sulphate, and gypsum (hydrated calcium sulphate). The sulphate class also includes the phosphate, chromate, molybdate, vanadate, and tungstate minerals. Table 2.10 provides a description of the minerals of this class.

Halide class

The halide group of minerals include simple compounds such as the natural salts fluorite (calcium fluoride), halite (sodium chloride), and sylvite (potassium chlorite) in which the halogen element is bonded to an alkali metal. Halides such as sulphates are commonly found in evaporatic environment such as salt lakes and landlocked seas such as the Dead Sea and the

Table 2.10 Properties of sulphate class* minerals

Name and composition	Colour	Crystal form	H	G	Cleavage	Lustre	Important characteristics
Barite $BaSO_4$	Colourless, blue	Ortho-rhombic	3–3.5	4	Perfect	Vitreous	Colour, specific gravity
Celestine $SrSO_4$	Blue, green	Ortho-rhombic	3–3.5	4.5	Perfect		Crystal habit, colour
Anglesite $PbSO_4$	Green	Ortho-rhombic	2.5–3	6.3	Imperfect	Resinous	Colour, lustre
Crocoite $PbCrO_4$	Red, orange	Monoclinic	2.5–3	6	Distinct	Vitreous	Lustre, specific gravity
Wolframite $(FeMn)WO_4$	Black	Monoclinic	5	7.4	Perfect	Greasy	Red streak
Scheelite $CaWO_4$	Brownish green	Tetragonal	4.5–5	6.1	Distinct	Sub-metallic	Black streak
Apatite $Ca_3(PO_4)_2F$	Yellow, blue	Hexagonal	5	3.2	No	Greasy	Crystal habit, hardness
Wulfenite $FeMoO_4$	Yellow orange	Tetragonal	3	6.8	Perfect	Adamantine	Colour. lustre
Vanadimite $Pb_5(VO_4)_3Cl$	Red, orange	Hexagonal	3	6.9	No	Adamantine	Bright red or orange red

* Gypsum belonging to this class is described in Table 2.15.

Table 2.11 Properties of halide class minerals

Name and composition	Colour	Crystal form	H	G	Cleavage/ fracture	Lustre	Important characteristics
Halite $NaCl$	White, red	Cubic	2.5	2.5	Perfect	Vitreous	Salty taste
Sylvite KCl	Colourless, white	Cubic	2.5	2	Perfect	Vitreous	Same taste but bitter
Fluorite CaF_2	Blue, yellow	Cubic	4	3.2	Not clear	Greasy	Colour, hardness
Cryolite Na_3AlF_6	White, red	Mono-clinic	2.5	3.0	Perfect	Vitreous to greasy	Colourless or white
Carnallite (KNH_4) $MgCl_3 \cdot H_2O$	Colourless, white, red	Ortho-rhombic	2	1.6	No Conchoidal	Greasy	Colour, crystal habit

Great Salt Lake in the US. The halide group also includes the fluoride, chloride, bromide, and iodide minerals. Whereas the chlorides are formed mostly by evaporation from solutions, the fluorides are typically found in igneous rocks and their associated pegmatite and hydrothermal veins. Table 2.11 provides a description of the halide class minerals.

Oxide and hydroxide class

Many of the mineable ores belong to the oxide class containing metallic minerals that are compounds of oxygen and one or more metals. They are normally made up of chemical structures of closely packed large oxygen atoms with the small metal atom. They also carry the best record of changes in earth's magnetic field. They generally occur as precipitates close to the earth's surface and as oxidation products of other minerals in the near-surface weathering zone. They are also found as accessory minerals in igneous rocks and metamorphic rocks of the earth's crust and mantle.

Common oxide minerals include hematite, magnetite (iron oxide), chromite (iron chromium oxide), rutile (titanium dioxide), spinel (magnesium aluminium oxide), zincite (zinc oxide), pyrolusite (manganese oxide), uraninite (uranium oxide), cassiterite (tin oxide), and ice (hydrogen oxide). Owing to their great resistance to weathering and transport, they also occur in sediments where they can be concentrated in beds.

In the hydroxides, oxygen is completely or partly replaced by the OH group. Hydroxides are generally less hard and less dense than oxides. They typically occur in the upper weathering zone of ore deposits produced by the alteration of the primary minerals. Some of the common hydroxide minerals are gibbsite, $Al(OH)_3$; manganite, $MnO(OH)$; diaspore, $AlO(OH)$; and goethite, $FeO(OH)$. The properties of important minerals of this class are described in Table 2.12. Some of the common oxide minerals of this classes such as magnetite, hematite, ilmenite, and spinel are described in Section 2.8 under ore-forming minerals and precious stones.

Table 2.12 Properties of oxide and hydroxide class minerals

Name and composition	Colour	Crystal form	H	G	Cleavage	Lustre	Important characteristics
Cuprite Cu_2O	Black	Cubic	3.5–4	6.1	Distinct	Ada-mantine	Crystal habit, hardness, lustre
Chromite $FeCr_2O_4$	Black	Cubic	5.2	4.6	No	Metallic	Streak brown
Chrysoberyl $BeAl_2O_4$	Yellow brown	Ortho-rhombic	8.5	3.7	Distinct	Vitreous	Crystal habit, hardness, colour
Pyrolusite MnO_2	Steel grey	Tetragonal	6–6.5	5.1	Perfect	Metallic	Streak grey, hardness
Uraninite UO_2	Brown black	Cubic	5–6	6.5	No	Pitchy, silky	Specific gravity, colour, lustre
Goethite $FeO(OH)$	Dirty brown	Ortho-rhombic	5–5.5	3.3–4.4	Perfect	Dull	Yellow streak
Corundum Al_2O_3	Grey yellow	Trigonal	9	4.0	No	Vitreous	Hardness, parting, crystal habit
Cassiterite SnO_2	Brown yellow	Tetragonal	6–7	7	Indistinct	Ada-mantine	Hardness, specific gravity, lustre, crystal habit
Gibbsite $Al(OH)_3$	Dirty brown	Monoclinic	2.5–3.5	2.4	Perfect	Vitreous	Hardness, granular

Sulphide class

The sulphides include a large number of minerals many of which are economically important as metallic ores. A sulphide is basically an oxygen-free compound of sulphur and one or more metals. Most sulphides have metallic character with a strong colour and streak and metallic lustre. Most are also opaque. They have high densities, and in contrast to pure metal, they are mostly brittle.

The common sulphides are pyrite (iron sulphide), copper sulphide, chalcopyrite (copper iron sulphide), nickeline (nickel sulphide), sphalerite (zinc sulphide), galena (lead pyrite), and cinnabar (mercury sulphide). Table 2.13 provides the properties of minerals in the sulphide class. Many of these sulphide minerals such as chalcopyrite, galena, pyrite, and sphalerite are described in Table 2.16.

Table 2.13 Properties of sulphide class minerals

Name and composition	Colour	Crystal form	H	G	Cleavage/ fracture	Lustre	Important characteristics
Bornite Cu_5FeS_4	Reddish golden	Cubic	2.5–3	5.1	Indistinct Conchoidal	Metallic	Colour, hardness, fracture
Pyrrhotite $Fe_{1-8}S$	Bronze yellow	Hexagonal	4	4.6	Indistinct Brittle	Sub-metallic	Crystal habit, magnetic
Nickeline NiAs	Red, grey	Hexagonal	5.5	7.8	Distinct Conchoidal	Metallic	Colour, streak grey
Millerite NiS	Brass yellow	Trigonal	3–3.5	5.5	Perfect	Metallic	Colour, hardness
Covellite CuS	Indigo blue	Hexagonal	1.5–2	4.7	Perfect	Metallic	Colour, cleavage
Cinnabar HgS	Vermilion red	Hexagonal	2.5	8.1	Perfect	Ada-mantine	Colour, streak red
Realgar AsS	Red, orange	Monoclinic	1.5–2	3.6	Good	Resinous	Colour, lustre, hardness
Stibnite Sb_2S_3	Black	Ortho-rhombic	2	4.6	Perfect	Metallic	Streak lead colour
Marcasite FeS_2	Brass yellow	Ortho-rhombic	6–6.5	4.9	Distinct Conchoidal	Metallic	Crystal habit, hardness
Orpiment As_2S_3	Brownish yellow	Monoclinic	1.5–2	3.5	Perfect	Resinous	Colour, streak black
Argentine As_2S	Black	Cubic	2.25	7.3	Indistinct	Metallic	Silvery colour

Native element class

The chemical elements that occur in the earth's crust are few and none of them is in large quantities. Some of these minerals such as gold, silver, and diamond are well known because of their valuable properties. The element class includes native metals and intermetallic elements (gold, silver, copper, mercury, platinum, and iron), semimetals, and non-metals (arsenic, antimony, bismuth, graphite, sulphur, and tellurium). These metals are composed of spherical cations such as Au^+ in close packings, most of them in cubic closest-packed structure. This group also includes natural alloys such as electrum (a natural alloy of gold and silver), phosphides, nitrides, silicates, and carbides (which are found in nature in a few rare minerals). Table 2.14 gives a description of minerals of this class.

Table 2.14 Properties of minerals of native element class

Name and composition	Colour	Crystal form	H	G	Cleavage/ fracture	Lustre	Important characteristics
Gold Au	Yellow	Cubic	2.5–3	19.3	None/ Hackly	Metallic	Colour, malleability
Silver Ag	Silver white	Cubic	2.5–3	10.1– 11.1	None Hackly	Metallic	Colour, cleavage
Copper Cu	Copper red	Cubic	2.5–3	8–9	None Hackly	Metallic	Colour, hackly fracture
Platinum Pt	Steel grey	Cubic	4–4.5	21.5	None Hackly	Metallic	Colour, specific gravity
Arsenic As	Tin white	Trigonal	3.5	5.7	Good	Metallic	Hardness, odour garlic, streak grey
Antimony Sb	Tin white	Trigonal	3.5	6–7	Good	Metallic	Colour, hardness
Bismuth Bi	Silver white	Triclinic	3.5	9.7–9.8	Good	Metallic	Melt at 630° as globule
Sulphur S	Yellow	Ortho- rhombic	1.5–2.5	2.07	Indistinct	Resi-nous	Colour, hardness
Diamond C	Colour- less	Cubic	10	3.5	Conchoidal	Ada- mantine	Hardness, high refractive index
Graphite C	Black	Hexagonal	1–2	2.1–2.2		Metallic, dull	Black streak

2.8 ROCK-FORMING, ORE-FORMING, AND PRECIOUS MINERALS

2.8.1 Rock-forming Minerals

Silicates, oxides, and carbonates are the main rock-forming minerals. Minerals that are the dominant constituents of rocks are called *essential* rock-forming minerals. Quartz, feldspars, amphiboles, pyroxenes, micas, chlorite, nepheline, olivine, serpentine, talc, calcite, dolomite, gypsum, magnetite, and hematite are essential rock-forming minerals. In addition to these minerals, there are others that occur in rocks in minor proportions. These are known as *accessory* rock-forming minerals, which include garnet, tourmaline, epidote, zircon, apatite, rutile, ilmenite, magnetite, hematite, pyrite, staurolite, kyanite, and sillimanite. Though occurring as accessory minerals, many of them have an important bearing in identifying the rock. For example, the presence of any of the minerals such as garnet, tourmaline, staurolite, kyanite, and sillimanite is indicative of the rock being of metamorphic origin.

The rock-forming minerals can be identified from their physical characters such as colour, streak colour, lustre, crystal form, cleavage, hardness, and specific gravity. Though many of the common rocks can be identified by the naked eye or by use of a lens (×10), some rocks require to be studied in thin sections under the microscope. Microscopic study of a thin section of a rock helps its accurate identification from the determination of the constituent minerals, their relative proportions, and texture. The chemical composition, crystal forms, and other important characteristics of the rock-forming minerals are shown in Table 2.15.

Table 2.15 Composition and characteristics of rock-forming minerals (Dana and Ford 1948)

Name	Chemical composition	Colour	Crystal form	Lustre	Fracture	H	G
Quartz	SiO_2	White	Hexagonal	Glassy	Conchoidal	7	2.65
Feldspars (Albite, oligoclase, etc.)	$NaAlSi_3O_8$, $CaAl_2Si_3O_8$ to $KAlSi_3O_8$	White	Monoclinic, triclinic	Vitreous	Uneven	6–6.5	2.54–2.74
Amphiboles	Ca, Mg, Fe, etc. silicate	Black, green	Monoclinic	Vitreous, pearly	Subconchoidal	5–6	2.9–3.4
Pyroxenes	Ca, Mg, Fe, etc. silicate	Green, black	Monoclinic, orthorhombic	Vitreous	Uneven to conchoidal	5–6	3.3–3.6
Micas (muscovite, biotite)	$H_2KAl_3(SiO_4)_3$ $H_2K(Mg, Fe)_3$ $Al(SiO_4)_3$	White, black	Monoclinic	Pearly	Uneven	2.5–3	2.76–3
Chlorite	Hydrosilicates of Al, Fe, Mg	Green	Monoclinic	Pearly	Platy	2–2.5	2.7
Olivine	$(Mg, Fe)_2SiO_4$	Green	Ortho rhombic	Glassy	Conchoidal	6.5–7	3.27–3.37
Serpentine	$H_4Mg_3Si_2O_9$	Green	Monoclinic	Silky	Platy, fibrous	2–5	2.2–2.7
Talc	$H_2Mg_8(SiO_3)_4$	White	Monoclinic	Silvery	Soft, soapy feel	1	2.7–2.8
Calcite	$CaCO_3$	White	Hexagonal	Pearly	Conchoidal	3	2.71
Dolomite	$CaMg(CO_3)_2$	White	Hexagonal	Pearly, dull	Subconchoidal	3.5–4	2.8–2.9
Gypsum	$Ca\, SiO_4\, 2H_2O$	White	Monoclinic	Pearly	Conchoidal	1.5–2	2.32–2.33
Magnetite	Fe_3O_4	Black	Isometric	Metallic	Uneven	5.5–6.5	5.2
Hematite	Fe_2O_3	Black	Hexagonal	Metallic	Subconchoidal	5.5–6.5	4.9–5.3
Epidote	$Ca_2(Al, Fe)_3$ $(SiO_4)_3 (OH)$	Green	Prismatic	Glassy	Uneven	7	3.4–4.5
Fluorite	CaF_4	Colourless	Octahedron	Glassy	Conchoidal	4	3.0–3.3
Garnet	$(Ca, Fe, Mg)_3$ $Al_2(SiO_4)_3$	Red, brown	Dodecahedron	Glassy	Conchoidal	6–7.5	3.5–4.3
Sillimanite	$Al_2\, SiO_5$	Grey, brown	Ortho-rhombic	Satiny	Splintery	6–7	3.2–3.3

(Contd)

Table 2.15 (*Contd*)

Name	Chemical composition	Colour	Crystal form	Lustre	Fracture	H	G
Kyanite	Al_2SiO_5	Bluishgreen	Triclinic	Glassy	Splintery	5 prism 7 across	3.6–3.7
Andalusite	Al_2SiO_5	Grey, pink	Orthorhombic	Glassy	Conchoidal	7.5	3.1–3.2
Zircon	$ZrSiO_4$	Brown, grey	Prismatic	Adamantine	Conchoidal	6.5–7.5	4.0–4.5

After disintegration of a rock, many of its constituent minerals including accessory minerals of high hardness (H > 6) may remain in sediments. For example, quartz, kyanite, sillimanite, tourmaline, garnet, cordiarite, staurolite, zircon, alusite, beryl, spinel, topaz, chrysoberyl, and corundum possess hardness between seven and nine in Mohs scale of hardness. Such minerals of high hardness, if present in the river water used for hydroelectric power generation, may abrade and damage the turbine of a hydroelectric power plant. The suspended silts of such river water need petrological studies to determine the hardness of silt particles and its effect on turbine blades. This aspect has been specially addressed in Section 12.7.

2.8.2 Common Ore-forming Minerals

There are several ore-forming minerals. Table 2.16 describes the most important ones that are available in Indian rock formations (see Section 24.9) along with their characteristic features.

Table 2.16 Composition and characteristics of ore-forming minerals

Name	Composition	Hardness	Specific gravity	Distinguishing features
Hematite	Fe_2O_3 (iron oxide)	6	4.9–5.3	Black-to-dark red colour, may occur as kidney shaped, no cleavage
Magnetite	Fe_3O_4 (iron oxide)	5.5–6.6	5.2	Greyish black, commonly occurs in igneous rocks as opaque minerals
Chalcopyrite	$CuFeS_2$ (copper-iron sulphide)	3.5	4.1–4.3	Yellow, less brassy than pyrite, tendency to tarnish
Cassiterite	SnO_2 (tin oxide)	6	6.8–7.1	Black, tetragonal crystal, often found in stream placers or pegmatite
Ilmenite	$FeTiO_2$ (iron titanium oxide)	5–5.5	4.1–4.8	Black, tabular or massive crystals
Molybdenite	MoS_2 (molybdenum sulphide)	1	4.7–4.8	Bluish grey, platy or massive crystals
Pyrite	FeS_2 (iron sulphide)	6	5.0	Brassy yellow colour, often occurs as cubes, no cleavage (also called fool's gold)
Sphalerite	ZnS (zinc sulphide)	4	3.9–4.1	Brown, black, or red, tetrahedral crystals
Galena	PbS (lead sulphide)	2.5		Lead grey in colour, three good cleavage directions, occurs as cubes
Cobaltite	$CoAsS$ (Cobalt arsenic sulphide)	5.5	6.0–6.3	Tin white colour, cubic crystal, sparse cleavage

2.8.3 Precious and Semiprecious Minerals

Minerals used as precious and semiprecious stones are associated with crystalline schists and gneisses. Pegmatite also contains large crystals of gem minerals. Transparency, hardness, and most importantly colour are the significant factors in determining the use of minerals for precious and semiprecious stones. The mineral beryl produces emerald. Diamond, occurring in kimberlite, is composed of carbon but its hardness, transparency, lustre, and rarity makes it valuable. Amethyst is a violet-coloured variety of the abundantly occurring silica mineral quartz whose colour makes it a semiprecious stone. The common precious and semiprecious stones occurring in the rock formations of India are described in Section 24.9.4. Table 2.17 provides a description of the minerals.

Table 2.17 Composition and characteristics of precious and semiprecious stones

Name	*Composition*	*Hardness*	*Specific gravity*	*Distinguishing characteristics*
Diamond	C (pure carbon)	10	3.52	Cubic crystal, adamantine lustre, forms under high pressure, hardest and most-valued mineral
Emerald	$Be_2Al_2Si_6O_{18}$ (beryl)	8	2.6–2.8	Green-coloured beryl caused by minute traces of corundum
Ruby	Al_2O_3 (mineral corundum)	9	3.9–4.1	Red-coloured variety of corundum with chromium in traces for the colour, conchoidal fracture
Sapphire	Al_2O_3 (mineral corundum)	9	3.9–4.1	Blue-coloured variety of corundum; traces of titanium and iron provide the colour
Topaz	$Al_2SiO_4(FOH)_2$	8	3.5–3.6	Orthorhombic crystal, colourless to blue, red, or pink, good cleavage
Amethyst	SiO_2 (mineral quartz)	7	2.65	Violet-coloured variety of quartz, coloured due to minute amount of iron
Garnet	$(Ca, Fe, Mg)_3Al_2(SiO_4)_3$	6–7.5	3.5–4.3	Magnesium-rich red and iron-rich violet varieties are used as precious stones
Opal	$SiO_2\ nH_2O$	5–6	1.9–2.2	Hydrated silica, amorphous and non-crystalline, often fluorescent
Zircon	$ZrSiO_4$	6.5–7.5	4.0–4.7	Tetragonal crystal, adamantine, gem variety fluorescent
Spinel	$MgAl_2O_4$	7.5–8	3.5–4.1	Cubic, high hardness, multihued—orange, green, red, violet, blue

2.9 IDENTIFICATION OF MINERALS UNDER MICROSCOPE

The previous sections provided an elaborate description of the various properties of minerals. In fact, all common minerals under different mineral classes have been described by means of tables (Tables 2.7 to 2.17) providing their composition, crystal forms, colour, streak, cleavage, lustre, hardness, specific gravity, and other characteristics. Study of such minerals by naked eye and laboratory tests are possible if the minerals are sufficiently coarse. However, if they occur in rocks or in very fine grains below the range of observation by hand lens of $10\times$ magnification, then microscopic observation is necessary. Detailed discussion of the optical method of mineral identification is beyond the scope of this book. There are many books on optical mineralogy for study of non-opaque minerals such as Phillips W. R et al. (1979). This section provides a short description of the optical method.

A petrological microscope is necessary for the study of optical properties of minerals. In general, grains of less than 0.25 mm size are observed under microscope. They can be observed by immersion in a liquid on a glass slide with a cover slip. In fact, to study the minerals in a rock, the standard method is to prepare a thin slide of 0.03 mm thickness with a cover slip and then study it under a microscope having different magnifying lenses. The petrological microscope, called polarizing microscope, has a polarizer at the bottom and an analyzer at the top. An isotropic substance (having same physical properties in different directions) is found to be dark and remains so even after rotating the microscopic stage. However, an anistropic mineral shows interference on rotation by 90° but distinctive interference colour on the rotation of stage. Crystals belonging to other than cubic system are optically anisotropic, and light is transmitted at differing velocities along different directions within each crystal.

When a ray of light enters an anisotropic crystal, it is split into two rays, which travel in different directions and are polarized in perpendicular planes. Such crystals are said to exhibit double refraction or *birefringences*. The interference colour will depend upon the orientation. To obtain an interference figure, the nicols (polarizers) are required to be crossed. Two types of figures appear when the mineral is observed under crossed nicols. For a uniaxial substance the figure is a black cross, and for a biaxial substance it is a biaxial bar. The angle between the optic axes is known from the study of the curvature of the biaxial bar.

Each mineral has a fixed refractive index. This can be determined by the optical microscope and it helps to identify the mineral. One simple way is to place grains of the mineral under a microscope and observe under low-power lens with transmitted light. The mineral grains are placed on a microscopic slide immersed in a liquid of known refractive index. The liquid needs to be changed until the grains are no more visible. This happens when the refractive index of the mineral equals that of the liquid. From the refractive index of the liquid, the name of the mineral can be identified by referring to a standard chart of a mineralogy book.

Under the microscope, the distinctive characters that are to be observed for identification of the minerals include colour, cleavage, extinction angle, and birefringence of the grain. Birefringence is the difference between the least and the greatest refractive indices of an isotropic mineral. The higher the order of colour, the higher is the birefringence, which is often used in optical studies. However, it is controlled by the thickness of the slide in grain mount as the interference colour is dependent on the thickness of the grains through which light is transmitted.

X-ray diffraction method is another method of mineral identification wherein a very minute amount of material can be used to correctly detect the mineral. Small grains of gem and rare minerals and noble metals are identified by this method. Electron probe microanalyser is nowadays used for mineralogical analysis. In its simplest form, scanning electron microscope is used for the purpose. In addition, spectroscopy is also used for the identification of minerals.

2.10 GEOLOGICAL TIMESCALE AND UNCONFORMITIES BETWEEN ROCK FORMATIONS

Geology has recorded the age of rock formations worldwide and arranged them into columns of its own timescale. This geological timescale is, in fact, the break-up of history of the earth into hierarchical sets of divisions or units that contain eon, era, period, epoch, and age. Each of these different spans of time in geological timescale indicates many tens of or hundreds of million years.

2.10.1 Relative and Absolute Timescales

Timescale includes both *relative* and *absolute* timescales. Relative timescale is based on the relative age relationship expressed as vertical or stratigraphic positions and named on global basis. The geological timescale subdivisions are essential for determining the relative age of the strata. Absolute time is the numerical age (in million years) obtained from radiometric dating of rocks. The geological age is usually presented by a vertical table to be read from the bottom upwards, the oldest being in the bottom and the youngest at the top (Table 2.18).

Table 2.18 Geological timescale* and corresponding major rock formations of India

Era/period (life form of the time)	Period/epoch	Age MY**	Major rock formations of India
Quaternary period (origin of man)	Holocene	0.01	Newer alluvium
	Pleistocene	1.8	Older alluvium, Karewas of Kashmir
Tertiary period (extinction of dinosaur; arrival of mammals, flora, and modern fish)	Pliocene	5	Siwalik System, Assam sedimentaries
	Miocene	23	
	Oligocene	34	Murree and Barail series
	Eocene	55	Deccan trap
	Palaeocene	65	
Mesozoic era (mainly dinosaurs, first birds)	Cretaceous	141	South coast, Assam, and Himalayan cretaceous
	Jurassic	208	Gondwana system and Himalayan sedimentaries
	Triassic	251	
Palaeozoic era (invertebrates, amphibians, fish, and early flora)	Permian	299	Kashmir sedimentaries, Muth quartzites, and Simla slates
	Carboniferous	359	
	Devonian	416	
	Silurian	444	
	Ordovician	488	
	Cambrian	542	
Precambrian (sponges, diatoms, algae, bacteria)	Proterozoic	2500	Vindhyan, Cuddapah, Aravalli, Dharwar
	–	–	Gneissic complex, charnockite series,
	Archaean complex	3800	Bengal gneiss, and Bundelkhand gneiss

* After International Commission on Stratigraphy.

** MY (absolute age) starting time in million years.

In India rock formations of various ages can be found that extend over the entire geologic timescale starting from Archaean to Holocene. The older rocks are generally stronger and the oldest Archaean rocks (gneisses and granites), which have wide occurrence in India, are very good for foundation purposes and also for use in other engineering constructions. The major rock formations of India that find frequent reference in this book have been included in Table 2.18 showing their age in both relative and absolute geological times. Detailed descriptions of these rock formations are provided in Chapter 24.

Fig. 2.20 Unconformity (angular unconformity)

2.10.2 Unconformity of Rock Formations and Its Delimitation

The rock formations of an area are the specific rock units delimited from major geological evidences including tectonic and paleontological events. All these events make a mark on the rock formations. The earth's crust has been constantly changing during the geological time due to tectonic movements or structural disturbances such as upheaval, tilting, folding, and faulting. Igneous activity and metamorphism are also responsible for the changing patterns in rock formations. An abrupt change in the depositional regime of sedimentary rocks is marked by *unconformity* (Fig. 2.20), which is the main basis for differentiating two rock formations. It may be of three types, namely *disconformity*, *non-conformity*, and *angular unconformity*.

Between two rock formations, if the older and younger beds have structurally parallel disposition but the overlying bed has been formed after a long geological time compared to the underlying bed, it is termed as disconfirmity. In other words, disconformity is an unconformity between parallel layers of sedimentary rocks representing a period of erosion or non-deposition (Fig. 2.21). The term *non-conformity* is used when a deep-seated igneous rock body is overlain by sedimentary rocks or lava flow (Fig. 2.22). In another type of unconformity known as angular unconformity, horizontally layered sedimentary beds are deposited overlain by tilted and eroded layers of sedimentary beds exhibiting discordant relationship (Figs 2.23 and 2.20).

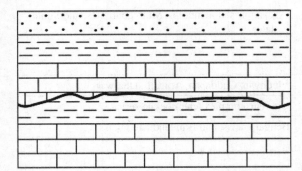

Fig. 2.21 Disconformity between parallel layers by erosion

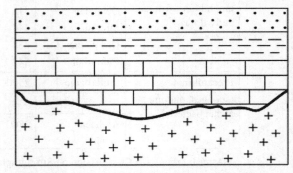

Fig. 2.22 Non-conformity between sedimentary bed and igneous rocks

In a sequence of conformable sedimentary beds, if the younger bed appears to be occurring beyond the limits of the older underlying bed, it is called *overlap* (Fig. 2.24). Such feature is found in the fringe of continental shelves caused during marine transgression. Palaeontology (study of fossils) provides major criteria for geologic divisions of rock formations on the basis of specific life forms during the geologic past. The life forms that started or became extinct during different geological times have been shown in column 1 of Table 2.18.

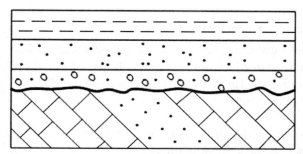

Fig. 2.23 Angular unconformity (overlying horizontal beds on eroded surface)

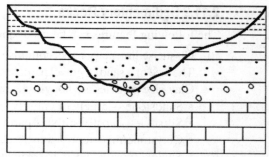

Fig. 2.24 Overlap (younger conformable beds occurring beyond limits of older underlying beds)

2.11 FIELD IDENTIFICATION OF COMMON ROCKS WITH SIMPLE ACCESSORIES

The megascopic characters of rocks help to a great extent in identifying common rocks. A hand lens ($\times 10$) is necessary to examine the mineral features and textures of rocks. A grain-size chart is to be used to measure (by comparison) the size of clastic particles of sedimentary rocks. This chart can be prepared by fixing particles of measured size on a small board by means of glue. Dilute hydrochloric acid is used to test limestone that effervesces with the acid. The following subsections provide the key features to be observed in hand specimens of igneous rocks, sedimentary rocks, and metamorphic rocks for their identification, following Wahlstrom (1974).

2.11.1 Igneous Rocks

The following features can be used to identify igneous rocks:

1. Light-coloured, coarse-grained rocks with phenaric texture (>50% minerals are seen by naked eyes)
   - Granite—contains both feldspar and quartz abundantly
   - Syenite—consists of abundant feldspar but no quartz

2. Light- or dark-coloured rocks with afenetic texture (<50% minerals are visible to naked eyes)
   - Rhyolite and trachite—light coloured, glassy
   - Basalt—dark coloured, entirely fine grained

3. Coarse-to medium-grained, dark-coloured rocks containing pyroxenes and abundant feldspars
   - Gabbro—coarse grained with tabular feldspar at places showing twin striations
   - Anorthosite—a variety of gabbro containing almost entirely calcic plagioclase
   - Dolerite—medium grained

4. Rocks with large crystals of quartz and feldspar
   - Pegmatite—size of crystals 3 cm or more
   - Aplite—sugary-grained quartz–feldspar aggregate found associated with pegmatite

5. Glassy rock containing few or no phenocrysts
   - Obsidian—vitreous lustre
   - Pitchstone—dull or waxy lustre

6. Rocks consisting of mostly olivine and/or pyroxene
   - Peridotite—abundant olivine
   - Pyroxenite—abundant pyroxene
   - Dunite—largely olivine and some pyroxene

2.11.2 Sedimentary Rocks

The following features can be used to identify sedimentary rocks:

1. Clastic sedimentary rocks comprise mainly quartz and feldspar with clay and/or silica matrix. The size of a sand or gravel particle can be measured by comparison with the particle size chart. Clay-size particles are not visible even with a hand lens, but the silt grains are just visible.
   - Sandstone—largely or entirely sand-size particles
   - Siltstone—mostly silt-size particles
   - Claystone/mudstone—mostly clay-size particles without exhibiting bedding
   - Shale—mostly clay-size particles showing bedding or fissility
   - Arkose—mostly angular sand-size grains (quartz and feldspars) with scanty or no clay matrix
   - Gravelly sandstone—gravel- and sand-size particles
   - Graywacke—angular sand-size grains of quartz, feldspar, and rock fragments packed in clay matrix
   - Quartzose sandstone—abundant rounded/subrounded quartz grains cemented by silica
   - Conglomerate—rounded boulders in fine matrix
   - Breccia—mostly angular boulders in a fine-grained matrix

2. Non-clastic sedimentary rocks consisting dominantly of anhedral to euhedral crystals of carbonate minerals and non-clastic fabric
   - Limestone—mainly calcite (effervesces with dilute HCl)
   - Dolomite—mostly dolomite (does not effervesce readily with cold and dilute HCl)
   - Calcarenite—entirely of sand-size calcite particles

3. Pyroclastic rocks formed by deposition of rock fragments and ash derived from volcanic eruption
   - Volcanic breccia and volcanic agglomerate—uneven rock fragments
   - Tuff—mostly volcanic ash (particle size <4 mm)

4. Further divisions in nomenclature of sedimentary rocks such as *sandy shale*, *calcareous sandstone*, *clayey sandstone*, and *limestone breccia* can be made depending upon the constituents and the cementing materials.

2.11.3 Metamorphic Rocks

The following features can be used to identify metamorphic rocks:

1. Foliated rocks containing abundant quartz and/or feldspar and other minerals to different extent and showing foliation or linear arrangement of minerals.
   - Gneiss—foliated rocks due to platy or parallel arrangements of minerals including micas (depending on the content of platy minerals, gneiss is subdivided as *granite gneiss*, *mica gneiss*, *hornblende gneiss*, etc.)
   - Schist—micaceous or platy minerals in subparallel arrangements (schistosity) (depending upon the major constituent mineral, schists may be *muscovite schist*, *biotite schist*, *hornblende schist*, etc.)

2. Rocks containing mostly fine-grained platy and micaceous minerals with tendency to split easily.
   - Phyllite—shows characteristic sheen, minerals cannot be distinguished by visual examination
   - Slate—individual grains not easily visible, rock splits easily in thin plates along the cleavage

3. Easily identifiable minerals that may be platy, elongate, or equidimensional. Foliation is scarcely developed or not developed at all.
   - Quartzite—mostly quartz grains rigidly packed in siliceous matrix
   - Marble—hard rock consisting mainly of calcite (recrystallized)

SUMMARY

- The earth's crust is composed of various types of rocks that are grouped under three categories, namely igneous, sedimentary, and metamorphic rocks.
- Igneous rocks were formed from cooling of magma and first occupied the crust of the earth. Clastic sedimentary rocks were derived from transportation, deposition, and later cementation of the clastic particles of disintegrated rocks. Soluble materials of pre-existing rocks taken by running water and deposited in the lake and sea floors after compaction produced the non-clastic sedimentary rocks. Change in igneous and sedimentary rocks under certain pressure–temperature conditions gave rise to the metamorphic rocks.
- All rocks are composed of minerals that can be identified from their chemical composition, specific gravity, and physical characteristics such as colour, crystal forms, lustre, fracture, and hardness. The main rock-forming minerals include quartz, feldspar, amphibole, pyroxene, micas, chlorite, nepheline, serpentine, talc, dolomite, gypsum, magnetite, and hematite.
- Depending upon the geological history, the rocks of the earth are divided into several hierarchical sets, namely eon, era, period, epoch, and age. Each of these time spans in geological timescale includes hundreds of million years.
- India possesses rock formations starting from the oldest Archaean to the youngest Holocene time dated to be 3800 million years to 0.01 million years, respectively, in absolute timescale. India has three distinct physiographic features—the peninsular shield, the extra-peninsular region, and the Indo-Gangetic plains—each of which has distinctive mountains, river systems, and rock formations. The older rocks such as Archaean gneisses and granites possess high strength and are very good for use as building stones and engineering constructions.
- Igneous rocks are classified based on colour (light or dark), presence or absence of quartz, and depth of origin. Grain size and mineral content (whether clastic or non-clastic) are considered in the classification of sedimentary rocks. Metamorphic rocks are subdivided on the nature of banding and foliation produced by flaky and platy minerals and also based on the granular characters and content of recrystallized calcite/dolomite and fused quartz.
- Among the various rock types classified under igneous rock group, granite, charnockite, diorite, gabbro, basalt, dolerite, rhyolite, and pegmatite are very common. Sandstone, siltstone, shale, and limestone are the main types of rocks classified under sedimentary rocks. The most common rock types of metamorphic rock group are gneiss, schist, quartzite, phyllite, slate, and marble.
- In hand specimen, the salient features that help in the identification of the igneous rocks include colour, texture, and mineral content. Examination of size, shape, and nature of clastic particles or carbonate mineral helps in the identification of sedimentary rocks. Banding, foliation characteristics, and mineral content including their arrangements help in recognizing the types of metamorphic rocks.

EXERCISES

Multiple Choice Questions

Choose the correct answer from the choices given:

1. An example of igneous rock is:
   (a) marble   (b) aplite
   (c) phyllite   (d) oolite

2. An example of metamorphic rock is:
   (a) gabbro   (b) arkose
   (c) quartzite   (d) diorite

3. An example of clastic sedimentary rock is:
   (a) limestone   (b) greywacke
   (c) pegmatite   (d) schist

4. An example of non-classic sedimentary rock is:
   (a) slate   (b) chalk
   (c) shale   (d) breccia
5. The mineral with a hardness of 9 in Mohs scale is:
   (a) quartz   (b) olivine
   (c) corundum   (d) topaz
6. The accessory minerals are:
   (a) olivine, pyroxene, micas, and dolomite
   (b) quartz, oligoclase, and hypersthene
   (c) plagioclase, calcite, and olivine
   (d) kyanite, sillimanite, garnet, and tourmaline
7. Crystallization of minerals can be grouped into:
   (a) ten systems   (b) six systems
   (c) five systems   (d) eight systems
8. Precambrian rocks include:
   (a) bundelkhand gneisses
   (b) gondwana system
   (c) siwalik system
   (d) vindhyan system
9. The minerals belonging to the silicate class are:
   (a) olivine and serpentine
   (b) diamond and sulphur
   (c) calcite and aragonite
   (d) both (b) and (c)
10. The halide class minerals are:
    (a) orthoclase and pyroxene
    (b) fluorite and sylvite
    (c) barite and apatite
    (d) pyrite and quartz
11. The oxide and hydroxide minerals are:
    (a) plagioclase and amphibole
    (b) graphite and diamond
    (c) cinnabar and realgar
    (d) goethite and uraninite
12. The ore-forming minerals are:
    (a) olivine and calcite
    (b) cobaltite and ilmenite
    (c) hematite and magnetite
    (d) both (b) and (c)
13. Some minerals of precious stone variety are:
    (a) quartz and feldspar   (b) cobaltite and fluorite
    (c) opal and amethyst   (d) topaz and spinel

Review Questions

1. Name the three major rock types of the earth's crust. Trace the origin of each of these rock types.
2. Distinguish between concordant and discordant bodies of igneous rocks. What are dykes and sills and how are they formed? Illustrate the following with block diagrams: batholith, laccolith, and lopolith.
3. Explain how sedimentary rocks provide information on palaeo-environment. What are the primary sedimentary structures?
4. What is metamorphism? Enumerate the processes responsible for different grades of metamorphism.
5. Explain the two important types of metamorphism and their effect. Name the other types of metamorphism and state how rocks formed by them are recognized.
6. Describe the mode of formation of a crystal. Classify the crystal forms into different systems and state the symmetry elements of each crystal system. Explain the terms crystalline, amorphous, and cryptocrystalline.
7. Enumerate the six main divisions of minerals based on their physical characteristics. Give a brief description of each division with examples.
8. What are the main rock-forming minerals? Describe their colour, crystal system, lustre, and hardness.
9. What are the identifying characteristics of unconformity between two rock formations? Explain non-conformity, disconformity, and overlap.
10. Describe briefly the following rock types: granite, pegmatite, dolerite, sandstone, limestone, conglomerate, marble, gneiss, schist, and phyllite.

Answers to Multiple Choice Questions

1. (b)	2. (b)	3. (b)	4. (b)	5. (c)	6. (d)	7. (b)	8. (a)	9. (a)
10. (b)	11. (d)	12. (d)	13. (c) & (d)					

3 Rock Structures and Their Engineering Significance

● ● ● ● ● ● **LEARNING OBJECTIVES** ● ● ● ● ● ●

After studying this chapter, the reader will be familiar with the following:

- Elastic and plastic deformation of rocks under stress and strain
- Principal types of folds and their characteristic features
- Different types of faults and clues for their identification
- Genesis and parameters of joints in various types of rocks
- Problems from rock structures in engineering constructions
- Primary structures of sedimentary rocks and their origin

3.1 INTRODUCTION

Under applied stress, rocks are deformed and give rise to different types of structures that have an important bearing on engineering constructions. This chapter mainly deals with the principal rock structures formed under deformation mechanism including folds, faults, and joints. It describes with illustrations all the major types of folds occurring in the earth's crust. The chapter also illustrates the characteristic features of different types of faults, their effects on underground structures such as tunnels, and the methods of identifying the faults. It traces the origin of different types of joints and analyses their parameters in relation to design of support system for preventing wall collapses and overbreaks of underground structures. The chapter further highlights the influence of folds, faults, and joints, especially their harmful effects, in the construction of engineering structures.

3.2 MECHANISM OF DEFORMATION OF ROCKS

A rock body is deformed when it is subjected to external forces. The factors responsible for deformation of rocks include pressure, temperature, rock composition, presence or absence of fluids, and type of stress and its rate of application. The most important among these factors are the type of stress, rate of stress, and temperature.

3.2.1 Effect of Stress and Strain—Brittle and Ductile Rocks

Stress is the force applied on the surface of a body and *strain* is the resultant effect that causes a change in the shape, size, or volume of that body. In other words, strain is the measure of material deformation such as the amount of compression when something is squeezed or elongation when it is stretched. A change in the volume of a body is termed as *dilation*, whereas an alteration of shape is known as *distortion*. Depending upon the resistance offered by a rock body, stresses are classified into three types—*compressive, shear,* and *tensile*. Compressive stress tends to compress the rock body resulting in a decrease in volume and shear stress tends to shear (i.e., break apart) one part from the other. Tensile stress tends to develop cracks in the body of the rock samples. Stress is measured as the total force per unit area and expressed as kg/cm^2 or $tonne/m^2$ in the metric system.

A material is said to be *brittle* if it deforms under stress by development of fractures. A rock body behaves typically brittle at low temperature and pressure. A *ductile* material deforms without breaking or cracking, except when it deforms too much and too quickly. A rock body behaves typically ductile at high temperature and pressure. A typical *stress–strain curve* of a brittle material will be linear and that of a ductile material non-linear as shown in Fig. 3.1(a).

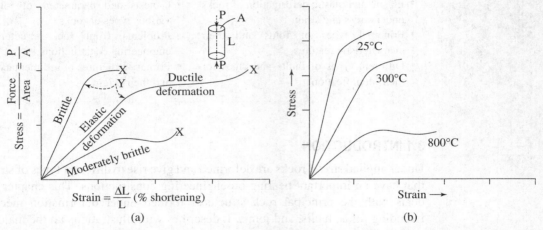

Fig. 3.1 Stress and strain: (a) curves showing stress–strain relation and elastic and ductile deformation of rocks (X—rupture, Y—yield point); (b) curves showing stress–strain relation of ranite at different temperatures (after Attewell and Farmer 1976)

3.2.2 Elastic and Plastic Deformation

Deformation of a rock is related to its elastic properties. A rock subjected to stress gets deformed, but on removal of the stress, it may return to its original form partially or wholly. This property of a rock is known as its *elasticity*. Under elastic deformation, a rock kept under load will restore to its original shape after it is relieved of the load. If it cannot restore to its original shape even after the removal of stress, the state of deformation of the rock is known as *plastic deformation*. Depending on the extent of applied stress, the shape of a rock under plastic deformation may be partially altered; however, with excessive stress or depending on the rock type, there may be permanent deformation of rock.

Under plastic deformation, a body may develop rupture when its cohesion is completely destroyed. When a rupture occurs, the discrete portions of the body are separated from the fractured surface. The stress difference required to develop the fractures is known as the *ultimate strength* of the material. Some substances such as clay may suffer from strain, but without destroying their cohesion. However, under compression, the water content of the clay is squeezed out and when it dries, it loses its plasticity. Thus, the deformation of a rock depends upon its physical properties and the stress difference under specific pressure–temperature conditions (Section 10.5).

Cataclasis is the deformation of rocks by the mechanical process of shearing and granulation. Such cataclastic rocks are said to have undergone dislocation metamorphism and range from coarsely broken breccia to intensely deformed mylonite. Cataclasite is a fault rock that consists of angular clasts in a fine-grained matrix. Cataclastic flow is the main deformation mechanism accommodating large strains above the brittle–ductile transition zone. It can be regarded as a ductile mechanism, although it takes place within the elastico-frictional regime of deformation. The deformation is accommodated by the sliding and rolling of fragments within the cataclastic rocks (Fig. 3.2).

(a)　　　　　　　　　(b)

Fig. 3.2 Cataclastic rocks: (a) intense fragmentation; and (b) intense granulation (mylonite)

3.3 FOLDS AND CAUSES OF THEIR FORMATION

The strata forming the earth's crust, when subjected to both horizontal and vertical forces, are bent or buckled. The structure thus developed is called *flexure* or *fold*. In most cases, folding involves the operation of forces tangential to the earth's surface. They are generally formed by horizontal forces acting at the two ends of a single bed or multiple beds of stratified rocks. However, when a horizontal bed is subjected to force in one end like a beam loaded at the end, it will bend with a simple flexure at a very low angle forming a monocline. The cross section of folds may not be ideally convex or concave, but at places they will be contorted or twisted with thickening and thinning in an irregular way. Figure 3.3 shows a fold (symmetric anticline) with innumerable joints (shear and tension joints) aligned parallel and transverse to fold axis formed during folding or bending of strata.

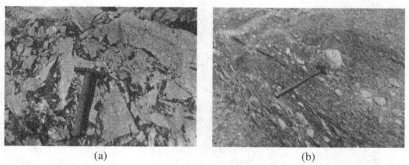

Axial joints　Fold axis　Transverse joints

Fig. 3.3 A fold with shear and tension joints (A–A fold axis and S–S minor slip)

There are many causes for the formation of folds, which appear in all rocks and in varied scales. Folds are formed when one or a stalk of originally flat and planar surfaces, such as sedimentary strata are bent or curved as a result of permanent deformation. Folds in rocks vary in size from microscopic crinkles to mountain-size folds. They occur singly as isolated folds or as extensive ranges of different sizes on a variety of scales. They may be formed under various conditions of stress, hydrostatic pressure, pore pressure, and temperature as evident by their presence in soft sediments, the full spectrum of metamorphic rocks, and even as primary flow structures in some igneous rocks. Figure 3.4 shows the initial stage of fold formation by bending of sedimentary strata in the Aravalli formation in Rajasthan.

Fig. 3.4 Initial stage of fold formation in Aravalli formation in Rajasthan

Fig. 3.5 Buckled fold in competent layers with incompetent matrix

Folds are commonly formed by shortening of existing layers, but may also be formed as a result of displacement on a non-planar fault or at the tip of a propagating fault by differential compaction or due to the effects of a high-level igneous intrusion. When a sequence of layered rocks is shortened parallel to its layering, deformation may be accommodated by a number of ways such as homogeneous shortening, reverse faulting, or folding. The response depends on the thickness of the mechanical layering and the contrast in properties between the layers. If the layering tends to make a folding, the nature of folds will also be guided by these properties. Isolated thick competent layers in a less-competent matrix control the folding and typically generate classic rounded buckle folds accommodated by deformation in the matrix (Fig. 3.5). In the case of regular alterations of layers of contrasting properties, such as sandstone–shale sequence, kink-bands, box-folds, and chevron folds (Fig. 3.6) are produced.

Fig. 3.6 Chevron fold in thin sandy and clayey layers

3.3.1 Anatomy of Folds

The different parts of a fold are termed differently. The highest point of an anticline is termed *crest*, see Fig. 3.7(a). When an anticline type of fold shows nearly equal dips in all sides with respect to its crest, it is known as a *dome*. The inclined parts of the strata where the anticline and syncline merge are called the *limbs* of the fold. The *axial plane* is the imaginary divisional plane separating the fold into two nearly equal parts. The *axis* or *axial line* is defined as the intersection of the axial plane of a fold with the ground surface. The *plunge* (also called *pitch*) of a fold is the angle made by the axial line with the horizontal in the axial plane, see Fig. 3.7(b).

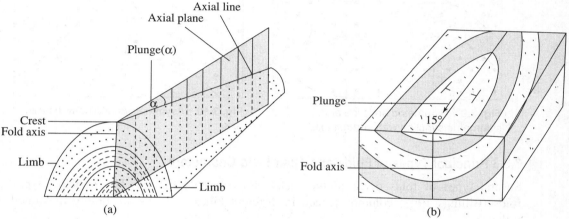

Fig. 3.7   Different parts of a fold: (a) an anticline; (b) a syncline

3.3.2 Anticline and Syncline—Symmetry and Other Features

An anticline is the upward convex flexure of a bed, whereas a syncline is the downward convex flexure. In an anticline, younger beds will be in the convex side and older beds in the core, whereas in a syncline, younger beds will be in the concave side and older beds in the core. Due to tension at the top of an anticline, joints or cracks are developed, which are termed as tension joints (Fig. 3.3). Similarly, the bottom part of a syncline is also likely to develop fractures due to tensional force. Figure 3.8 shows the limbs and fold axes for anticline and syncline of an asymmetrical anticline and syncline. In the figure, the pencil tip is on the crest of the anticline and its other end points to the syncline.

If the two limbs of an anticline and a syncline have equal slopes, they are termed *symmetric anticline* and *symmetric syncline*, respectively (Fig. 3.9). In a symmetric fold the axial plane is

Fig. 3.8   Limbs and fold axes of an asymmetrical anticline and syncline (pencil tip points to crest of anticline and its other end to syncline)

vertical, whereas in an asymmetric fold it is inclined (Fig. 3.10). The term *antiform* is used for any fold that is convex upward for which the relationship between the fold and the various strata is unknown. An anticline of regional scale consisting of a series of smaller anticlines and synclines is called *anticlinorium.* A vast syncline of regional scale with its strata further folded into subordinate synclines and anticlines is termed as *synclinorium.*

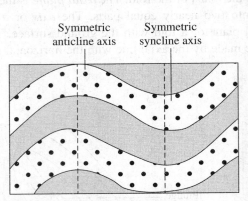

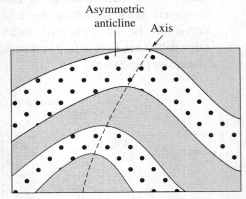

Fig. 3.9 Symmetric anticline and symmetric syncline with vertical axis

Fig. 3.10 Asymmetric anticline with inclined axis

3.3.3 Principal Types of Folds and Their Field Characters

Several types of folds occur in the rocks of the earth's surface. The principal types of folds (Billings 1997) commonly seen in different kinds of rocks are briefly illustrated as follows:

Plunging fold A fold may not continue for an indefinite distance, but dies away after a certain length. If a fold is traced along its axial line direction, it may be found that the amplitude of an anticline decreases until it merges with the unfolded beds, when it is called *plunging anticline,* see Fig. 3.7(a). In case of a *plunging syncline* the trough becomes shallower along its axis in one direction, as shown in Fig. 3.7(b).

Isoclinal fold In this type, both the limbs of a fold dip in the same direction and are equally inclined, as shown in Fig. 3.11(a). If the axial plane is vertical, the fold is termed *vertical isoclinal fold*; if it is inclined, it is called *overturned isoclinal fold.*

Recumbent fold Here, the axial plane of a fold is horizontal or makes a very low angle with the horizontal, as shown in Fig. 3.11(b).

Chevron fold This type of fold, also called the *zig-zag fold*, has straight or planar limbs and angular hinges. The bedding planes of the fold are parallel to the limbs, as shown in Figs 3.11(c) and 3.6.

Monocline It is a simple type of flexure formed by bending of a horizontal or very low-dipping bed with anticlinal bend at the top and synclinal bed at the bottom, as shown in Fig. 3.11(d).

Drag fold It is formed when a competent bed such as sandstone slides past an incompetent bed such as shale. The drag folds are related to fault movement. The relative direction of movement of the fault can be traced from the trend and inclination of the axial planes of the drag fold as shown in Fig. 3.11(e).

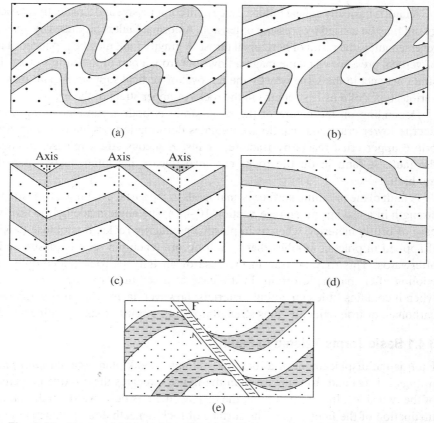

Fig. 3.11 Types of folds: (a) isoclinal fold; (b) recumbent fold; (c) chevron fold; (d) monocline; and (e) drag fold with a fault (F–F) along the downward direction

Fig. 3.12 Ptygmatic fold in quartzo-felspathic veins

Ptygmatic fold This type of fold is formed in weak beds yielding easily to deformation and assuming any shape impressed upon them by the surrounding rigid rocks. Ptygmatic fold may result from viscous flow of incompetent rocks under small stress difference and hence is also known as *flow fold*. Quartzo-feldspathic veins and pegmatites are commonly convoluted and buckled as ptygmatic folds during high-grade metamorphism see Fig. 3.12.

3.4 CAUSES AND MECHANISM OF FAULTS

A rock mass cannot separate into two parts and glide or flow past one other because of its rigidity and friction. This results in a build up of stress, which leads to the accumulation of potential energy, up to a level exceeding the strain threshold. Once the threshold energy is gained, the necessary relative motion is set in along a weak plane in the rock to cause fault movement. The relative motion of rocks on either side of a fault surface controls the origin and mechanism of the small faults as well as the tectonic fault related to plate movement.

A fault is basically a shear failure. Since both compressive and tensile stresses act simultaneously, a fault may be caused by pressure or tension. A normal fault is produced by vertical pressure and a reverse fault is the result of horizontal thrust. A gravity fault may be produced at a place by release of vertical pressure when the magma flows out from a part of the earth's crust. Tension in deeper strata may act in the same way by the removal of vertical support. A fault may occur because of various types of tensile strengths in the rocks of upper strata of the earth.

Depending on the nature of the rock, strain is both accumulative and instantaneous. The ductile lower crust and mantle accumulates deformation gradually via shearing, whereas the brittle upper crust reacts by fracture, or instantaneous stress release, to cause motion along the fault. A fault in ductile rocks can also release instantaneously when the strain rate is too high.

A fault plane or failure surface is analogous to the surface obtained by testing for compression strength in a test specimen (see Section 10.4.4). The angle made by the test is acute ($<45°$) in case of brittle material, whereas in plastic and ductile rocks, it is obtuse ($>45°$). This analogy if applied in practice implies that a fault occurs when the rock is plastic and becomes brittle afterwards. This may be true in the case of an igneous rock undergoing its final stages of cooling after magmatic origin. In the case of a sedimentary rock too, this may be the case when it contains little cementing materials; however, later, when it is combined with siliceous, carbonate or iron oxide, it gains strength to act as a brittle rock (Krynine and Judd 1957).

3.4.1 Basic Terms Related to Faults

Fault is the displacement of a rock mass along a weak plane (called *fault plane*) marked F–F in Figs 3.13(a) and (b). The arrow in the figure indicates the relative movement of the blocks of the two sides. In a normal fault, the displacement is downwards in the same direction as the inclination of the fault plane. The surface of rock beneath a fault plane is the *footwall* and that

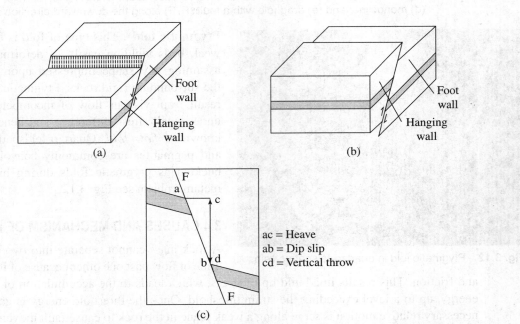

Fig. 3.13 Block diagrams: (a) a normal fault; (b) reverse fault showing hanging wall and footwall; and (c) slip, throw, and heave associated with a fault

above the fault plane is the *hanging wall*. In a reverse fault, the movement along the inclined fault plane is up-dip. The term *slip* denotes the relative displacement between two blocks; the vertical component of a slip is known as *throw* and its horizontal component as *heave*.

In reality, a fault plane is rarely a clean-cut plane or surface, but it is mostly a zone of complex deformation containing shattered rocks. This zone, also called the *fault zone*, varies in width from a few centimetres to several hundreds of metres. A *shear zone*, also called the *crush zone*, can also be very narrow to very large (few centimetres to even few kilometres) in width and it consists of many parallel fractures and crushed rocks formed by shearing of rock mass.

Shearing may also create zones with rock fragments of lenticular shape resembling elongated pebbles. Figure 3.14 illustrates sheared sandstone showing a lenticular body and cavity formation with erosion of crushed material. Intensive crushing of rock in the fault zone produces clayey material called *gouge*. There are two types of movements along a fault, namely *rotational* and *translational*. In rotational movement, there is rotation of one displaced block with respect to another. In translational movement, the two blocks remain parallel with respect to one another even after displacement.

Fig. 3.14 Sheared sandstone showing lenticular body and cavity formation

3.4.2 Illustrative Description of Different Types of Faults

The following is a description of the different types of faults depending on the nature of displacement:

Normal fault

Normal fault is the simplest type of fault in which the hanging wall block moves downwards, relative to the footwall block, see Fig. 3.15(a). Tectonic movement of the earth's crust may give rise to normal faults. The tensional stress is responsible for the displacement of crustal blocks in a normal fault.

Reverse fault or thrust

In reverse fault or thrust, the movement of the crystal block is such that the hanging wall moves upwards relative to the footwall, see Fig. 3.15(c). This fault brings about salient changes in the rock mass. If the fault plane slopes at an angle of more than 45°, it is *up-thrust*. If it is less than 45°, it is *over-thrust*. In the over-thrust, the hanging wall actually moves over the footwall. It is the reverse in the case of *under-thrust* when the footwall is pushed under the hanging wall. A very low angle (nearly horizontal) thrust is called *nappe*.

Strike-slip fault

In strike-slip fault, shown in Fig. 3.15(b), the movement is essentially horizontal under the action of shearing stresses. This type of fault is associated with folding and tearing. Figure 3.15(c) shows a strike-slip reverse fault where the movement is both in the strike

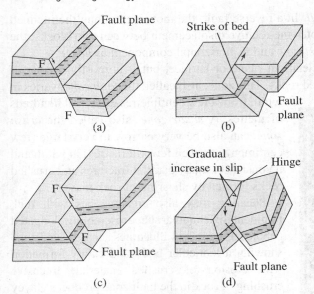

Fig. 3.15 Different types of faults: (a) normal (gravity) fault; (b) strike-slip fault; (c) strike-slip reverse fault; and (d) hinge fault (pivotal fault)

direction also horizontal which makes the right-hand block to move upwards.

Pivotal fault

In a pivotal fault, two blocks are joined at a certain part as a pivot. It appears like a normal fault on one side of the pivot, whereas on the other side it is a reverse fault. The interface between the two blocks is the pivotal part. The term *hinge fault* denotes the type of fault in which the relative displacement of the two blocks increases away from one end of the fault, which acts as the hinge of the two blocks, see Fig. 3.15(d).

3.4.3 Effect of Faulting in Brittle and Ductile Rocks

A fault brings about disruption of strata. The nature of disruption depends not only on displacement but also on the strike and dip of the fault plane, the strike and dip of the affected blocks, and the surface configuration. In the upper parts of the earth's crust (top 10 km), faulting results from brittle fracture and displacement of rock blocks. Shearing, crushing, brecciation, gouge, and *cataclastic rocks* (fragmented pre-existing rocks) are the characteristic features of *brittle deformation*. The process of brittle deformation is also evident from the presence of a glassy and cryptocrystalline material called *pseudotachylite*, which looks like a volcanic rock. This is produced as a result of melting by heat generated from frictional slide along the fault plane of brittle rocks such as gabbro, amphibolite, or gneiss.

At deeper parts of the crust (below 10 km), faulting is caused by *ductile deformation* as evident from the presence of *mylonite*. Large relict grains called *porphyroclasts* are found as asymmetrical tails in mylonites. A partially mylonitized rock called *protomylonite* may also be produced in the shear zone by ductile deformation. Linear structures such as rotated boudins and drag folds originated from deformation in ductile shear zones provide evidence of the relative displacement along the fault plane between the plutons and country rocks.

3.4.4 Clues for Field Identification of Faults

Some clues for field identification of faults are listed as follows:

- A valley running straight for a long distance is indicative of a fault. The straight feature of the valley may be clearly visible in aerial photographs.
- Juxtaposition of younger beds with older strata is an evidence of the presence of a fault.
- Offsets of sedimentary beds suggest the presence of a fault between the offset beds.
- Offsets of quartz stringers or pegmatite veins in igneous rocks indicate the presence of a fault.

The presence of a fault can be understood from features such as feather joints (tension joints), which also help to identify whether it is a normal or reverse fault. The acute angle between the fault plane and the feather joints indicates the direction of actual movement of the blocks.

The polished surfaces of faults are termed *slickensides* (Fig. 3.16), which may also form on joint surface. They contain small parallel groves or striations that resulted from friction.

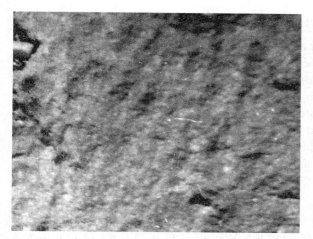

Fig. 3.16 Slickenside on a fault plane indicating the direction of fault from left to right

Slickensides are indicative of faults. The fault surface is also smooth and striated, having a step-like feature. Moving the fingers along the slickenside surface produces a smooth feeling along the slip direction and a harsh feeling in the reverse direction.

3.5 CAUSES OF JOINTING AND GENETIC TYPES OF JOINTS IN ROCKS

Since rock structures result from stresses, it can be said that a rock will eventually fracture if it is under stress. Joints are fractures with regularity and continuity with respect to each other or to the other elements of the rock masses. Typically, there is little or no lateral movement across joints. They normally have a regular shape related to either the mechanical properties of the individual rock or to the thickness of the layers involved. Joints occur as sets, with each set having sub-parallel arrangement of joints to each other.

Joints form in solid and hard rock that is stretched such that its brittle strength is exceeded (the point at which it breaks). When this happens, the rock fractures in a plane parallel to the maximum principal stress and perpendicular to the minimum principal stress (the direction at which the rock is being stretched). This leads to the development of a sub-parallel joint set. Continued deformation may lead to development of one or more additional joint sets. The presence of the first set strongly hampers the stress orientation in the rock layer, often causing subsequent sets to form at a high angle to the first set. The joint sets are commonly observed to have relatively constant spacing that is roughly proportional to the thickness of the layer. Depending upon their genesis, joints are broadly classified into the following types:

Tectonic joints These are formed during deformation episodes of rock formation when the differential stress is high enough to induce tensile failure in rock body, irrespective of the tectonic regime (Fig. 3.3).

Unloading joints These are formed when erosion removes the overlying rocks, thereby reducing the compressive load and allowing the rock to expand laterally (see Section 3.5.2).

Exfoliation joints These are in fact special cases of unloading joints formed parallel to the present land surface in rocks of high compressive strength.

Cooling joints These are formed by the cooling of the hot rock masses, particularly lava, and are commonly expressed as vertical columnar joints (see Section 3.5.2).

3.5.1 Orientation, Spacing, Roughness, and Other Features of Joints

Joints are planes of discontinuities in the rock mass along which very little or no displacement has taken place. They are to be distinguished from fractures, which are defined as planar surfaces along which the rock has no cohesion. Fractures may be irregular in their trend but joints have fixed orientation and attitudes such as strike and dip. Joints extend over a considerable part of a rock body generally showing more than one pattern or a set of joints.

The important features of joints are as follows:

- A group of parallel joints is called a *set* and several joint sets intersect to form a *joint system*.
- If the angle of intersection between the two sets is 90°, they are said to be *conjugate sets*. The joints are generally open and the openings at the surface are from a few millimetres to several centimetres. The openings of joints decrease and finally pinch out at the depths.
- Joints may be filled with different materials. In general, the fillings are clayey or crushed materials or the opening may be sealed by secondary silicification with quartz veins.
- Joints are frequently found to follow bedding and foliation planes or cleavage and accordingly called *bedding joints*, *foliation joints*, and *cleavage joints*.

Orientation

Attitudes such as direction and amount of dip and strike of joints provide the orientations of joints. It is expressed as E–W 40°➔S, meaning the strike of joint is east–west and it dips at an angle of 40° directed south (Fig. 3.17). The orientation of joints is measured instrumentally from the rock outcrops in the field and plotted on a geological map by symbols. In fixing the alignments of engineering structures such as tunnels, underground chambers, and dams, the orientation of joints in the rocks of these sites requires careful study.

Fig. 3.17 Joints in quartzite depicting a bedding joint or dip joint with strike E–W dip 40° South

Spacing

Spacing is the perpendicular distance between two adjacent joints. The spacing between two joints may vary from a few centimetres to more than a metre, depending on the rock type. Under the same stress, competent rocks will have more joints and smaller spacing than less-competent rocks. There is a relationship between joint spacing and bed thickness. Joint sets are commonly observed to have relatively constant spacing roughly proportional to the thickness of the bed.

Roughness

Roughness is the waviness or miniature stepping of the joint surface. A joint surface may possess a series of small steps showing that the surface is rather rough unless there is a slip along the surface. Any displacement along the joint surface during or after its origin makes it smooth and develops slickenside (Fig. 3.16). The presence of minute steps or waviness, called striations, provides evidence that a displacement or slip has taken place along the joint surface. Quantitative evaluation of this parameter is done in rock quality classification (see Section 10.9 and Table 10.6).

Frequency

The number of joints per metre is estimated as the frequency of the joints. The measurement is done from rock outcrops in the field and also from cores of drill holes. The measured frequency of joints from drill cores when correlated with water percolation test data provides information on the seepage or leakage problem of dam projects.

Aperture

Aperture is the perpendicular distance between the joint walls. In other words, it is the width of the joint opening. Apertures of many joints are wide open to allow large flow of water through them but some are tight and may cause seepage of water. Some fine joints are watertight but may not be airtight. Before selecting a site for underground rock chambers for gas storage, it is essential to thoroughly study the apertures with respect to their tightness to water and air.

Number of sets T

The number of joint sets is counted from the intersecting joint system. Vertically intersected two conjugate joint sets with horizontal or near joints create regular cubic blocks, the dimensions of which depend upon the spacing of the joints. Such blocks when formed by open joints are vulnerable to rock slides. In Fig. 3.18, there are three distinct sets of joints—vertical, horizontal, and inclined. In Fig. 3.19, there are two sets, that is, one inclined and one steeply inclined.

Fig. 3.18 Joint system containing one vertical set, two inclined sets, and one horizontal set, joints tight and joints open (nearly 10 cm), two open cross fractures

Fig. 3.19 Joint system containing two close-spaced tight vertical joints

Filling and coating

The fill materials in a joint vary widely. Many joints that originated with considerable openings are tightly filled with veins or dyke minerals. Secondary silica or hydrothermal quartz completely seals the joint openings. Filling by loose materials such as clay, silt, and micas is found in joints close to surface rocks, but with increasing depths there is a decrease in loose fillings.

3.5.2 Types of Joints in Different Rock Types and Their Origin

The origin of joints is associated with the formation of rock types and later tectonic movements. The joints associated with tectonic movement are very prolific in all the rock groups, namely igneous, sedimentary, and metamorphic rocks. The failure of a brittle body under compression takes place by shearing along two directions of least strain, which are nearly at right angles or at acute angles to each other. These are the *shear joints*. In ductile substances, the angle between the shear planes is obtuse. In addition to shearing forces, tensional forces (i.e., pull) also create joints, which are known as *tension joints*. An anticline develops innumerable tension joints arranged in its upper surface (see Fig. 3.3).

Joints in sedimentary rocks

Sedimentary rocks may undergo varying degrees of dislocation consisting of fissuring transverse to beds. The most common types of joints in sedimentary rocks are the *bedding joints* formed parallel to the bedding plane. In a dipping stratum, joints may be developed along the strike as well as dip directions of the bed and are known as *strike* and *dip joints*, respectively. In Fig. 3.17, the bedding joint and the dip joint are the same. Joints may be formed intersecting the bedding planes or layered surfaces at a very steep angle. In general, two sets of such steep-dipping to near-vertical joints occur in bedded or foliated rocks trending at right angles to each other, as may be seen in Fig. 3.17. The two vertical sets of joints in conjunction with the dipping bedding joints give rise to *block joints* (Fig. 3.20).

Fig. 3.20 Block joints in sandstone formed by sub-horizontal and vertical joints

Joints in igneous rocks

Igneous rocks form several types of joints when magma during its intrusion passes from the liquid to solid state. Hill (1953) has mentioned three types of joints, namely *cross joints*, *longitudinal joints*, and *flat-lying joints* that are commonly present in intrusive igneous rock (Fig. 3.21). Cross joints are aligned perpendicular to linear structures similar to the primary flow lines corresponding to the long axes of crystals. These joints are developed due to tension, and when present in rocks free from primary flow, they are called tension joints. The joint spaces are commonly filled with hydrothermal quartz or

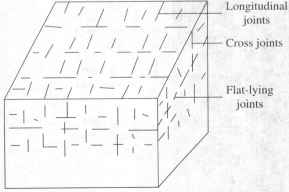

Fig. 3.21 Three types of joints commonly seen in intrusive igneous rock

veins of pegmatite and aplite. In contrast to cross joints, longitudinal joints are formed trending parallel to the flow lines following regional structural trend. These are best developed near the roof of large intrusive bodies such as dykes of basalt, pegmatite, and aplite. Flat-lying joints are formed mostly during early flow of magma occurring oblique to the flow lines. Dykes and veins generally occupy the flat-lying joints of primary flow origin. In addition to these three joints, *diagonal joints* are formed transverse (nearly 45°) to the flow lineation of the plutons. The open spaces of these joints are filled by hydrothermal minerals such as chlorite, epidote, and fluorite or occupied by veins and dykes of aplite and pegmatite.

Fig. 3.22 Columnar joints of Deccan

Fig. 3.23 Sheet joints in quartzite open to different basalt showing hexagonal pattern extents (Wikipedia 2012)

Columnar joints In homogeneous and extrusive types of igneous rocks, the force of contraction during cooling of lava gives rise to columnar joints (Fig. 3.22). This type of joints generally show hexagonal pattern but may be of rectangular, pentagonal, or other polygonal shape in cross section. The columnar joints are abundantly present in volcanic rocks such as basalt in the Deccan traps of India. Some dykes formed by solidification of near-surface magma also exhibit columnar joints. In a lava flow, these joints are developed perpendicular to the cooling surfaces. Both compression and tension act together to give rise to columnar joints.

Extension joints or sheets These joints result from the unloading of rocks by erosion of landmass. With the removal of super incumbent load, the rock expands vertically giving rise to these joints, which have a rather close spacing created by segmentation of the rock body with sheet-like appearance. The process is known as *sheeting*. The spacing of these joints varies between 10 cm and two metres closer to surface, but becomes wider with increasing depths. Sheet joints are mainly prevalent in granites but may also occur in quartzite and sandstone (Fig. 3.23).

Radial joints and ring fractures Joints arranged in a radial pattern are formed in lava flows around a volcanic crater. Due to withdrawal of magma from its chambers underlying the crater, the rock surface of crust close to the volcano's vent gets fragmented creating radial joints and also fractures of cylindrical pattern called *ring fractures*.

3.5.3 Diagrammatic Representation of Joints

It is common practice to present the joints measured in the site of an engineering structure by a diagram. A simple way to show the intensity of joints and their attitudes is by a *histogram*. Histograms indicate the percentage of the joints dipping at different angles. Nearly 200 joints are measured with respect to their dip direction and amount. The dips are grouped into 10°

or 15° intervals. Thus, there may be six divisions, 0–15°, 15–30°, 30–45°, 45–60°, 60–75°, and 75–90°. The percentages of joints measured in the field that fall within these six divisions are then plotted in a histogram to show the direction and intensity (percentage) of joints. Figure 3.24 is a histogram of 200 joints in hard rocks showing their dips in relation to concentration (by percentage) in different directions.

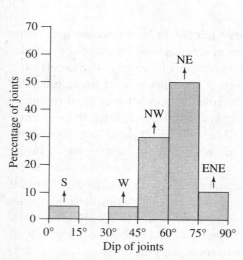

Fig. 3.24   Histogram of joints in rocks

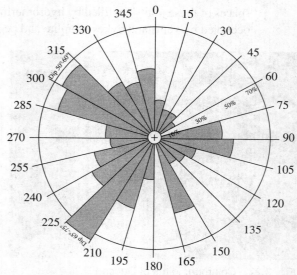

Fig. 3.25   Rose diagram of joints in rocks

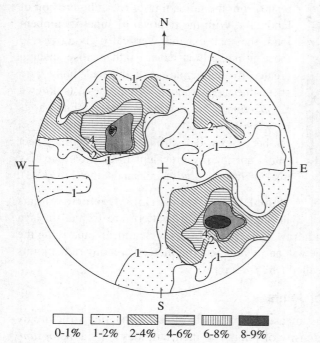

0-1%   1-2%   2-4%   4-6%   6-8%   8-9%

Fig. 3.26   Plot of joint poles on an equal-area net

The joints measured in the field can also be illustrated in a *rose diagram*, which shows the intensity (as percentage) of joints measured in the field grouped under 15° divisions and the directions covering the entire 360°. The amount of dip of the most-intensive joints is written by the side. Figure 3.25 is a rose diagram of over 200 joints in rocks showing maximum concentration (by percentage) in the southwest and northwest parts with steep dips. It is, however, preferable to plot the joint attitudes (joint poles) on an equal-area net and present the data by contour diagram. From this diagram, the parameters of divisional planes (joints) in rock with maximum concentration can be easily detected and used in engineering geological analyses of site conditions for founding engineering structures. Figure 3.26 depicts the plotting of joint poles (on the lower hemisphere of an equal-area net) of 200 joints showing the orientation of joints dissecting the rocks of a tunnel and their disposition with respect to the tunnel alignment.

3.6 PRIMARY STRUCTURES OF SEDIMENTARY ROCKS AND THEIR SIGNIFICANCE

The structures of sedimentary rocks are visible mostly in the outcrops of the rocks and scarcely in hand specimens. These features formed by sediments of both inorganic and organic matters are characteristic of varied depositional environment as stated by Pettijohn (1957). Figure 3.27 illustrates the various sedimentary structures. This is followed by brief descriptions of the different sedimentary structures analysing the geological merits and demerits of their presence in a project site.

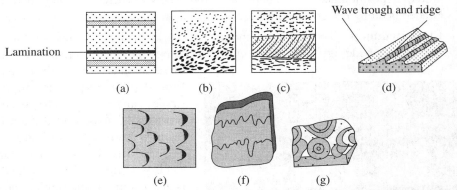

Fig. 3.27 Primary sedimentary structures: (a) lamination; (b) graded bedding; (c) cross bedding; (d) ripple marks of stream flow; (e) flute cast in shale; (f) stylolite in limestone; and (g) stromatolite in limestone

Bedding　The most prominent structural feature in sedimentary rocks is the bedding defined by texture and composition of the mineral grains (Fig. 3.28). A bed is formed by uniform deposition of sediments under nearly constant physical conditions. Thickness, composition, fabric, and orientation of the beds are important parameters of study in a sedimentary bed if present in an engineering site.

Mud cracks　These are shrinkage cracks found in fine-grain sedimentary beds consisting of siltstone and shale. The pattern of the cracks may appear nearly rectangular, pentagonal, or hexagonal in geometric shape, but are mostly irregular in their surface features (Fig. 3.29). When clay or silt is exposed to sun in an intertidal zone, these physical cracks are created by desiccation during low-tide period. With the incoming tide, the mud cracks remain buried and preserved. Erosion of the upper layers in this stratified intertidal deposit brings the mud cracks to the surface. As a result of drying by the sun, they tend to curl concave upwards and may indicate the top and bottom surfaces.

Fig. 3.28　Bedding in sandstone with low angle dip

Fig. 3.29　Mud cracks in clayey siltstone

Lamination   A minor 'change in the depositional condition during the formation of bedding may cause thin layering of silty or clayey matter, which is a distinctive feature between two thick beds. Such a layered feature is called *lamination*, see Fig. 3.27(a). These minor layers referred to as *lamina* vary in thickness from a few millimetres to one or two centimetres.

Graded bedding   It is formed when particles are deposited in a gradual variation in the vertical direction from coarser at the bottom to finer towards the upper parts, see Fig. 3.27(b). This kind of deposition takes place under slow movement of current causing good sorting of grains. Graded bedding is a good indicator of top and bottom beds in isoclinals or overturned folds.

Cross beddings   These are commonly present in medium- to coarse-grained sedimentary beds. Cross beddings (also called current beddings) are characteristic features of deposition of sandy aparticles in rivers, deltas, and dunes formed from downstream migration of current ripples and waves. Extension fractures are prevalent in cross-bedded strata. These structural features are indicators of stream flow direction in river deposits and wind direction in dune sands, see Figs 3.27(c) and 3.30. Cross bedding is also useful in determining the top and bottom of overturned beds in recumbent folds.

Fig. 3.30   Cross bedding (or current bedding)

Ripple marks   These features are found in sedimentary beds consisting of fine-grain sands and silts. The two common types of ripple marks are asymmetrical and symmetrical in profiles. The former type is formed by stream current and the latter by oscillation of waves. When a current flows over a bed under slow velocity, the sand moves forming ripples on the surface of sands. These features are also caused by wind action in dune sands. The ripple marks are characterized by alternative small troughs and ridges oriented at right angles to the direction of water or wind movement, see Fig. 3.27(d). As such, the features when preserved in rocks can be used to determine the direction of paleocurrent.

Sole marks, scour casts, flute casts, and groove casts   Sole marks are trough-like surfaces with stand-up borders formed by the action of current on clay beds by unequal loading or by the action of organisms. These are commonly found in claystone or shale underlain by sandstone and created by turbidity current. When a clay bed is scoured by erosion having a pot-like surface, the surface may be later filled by sandy material to form scour cast structures. Flute casts are the narrow ends of scour casts, see Fig. 3.27(e). Groove casts are formed by filling of grooves by sand in shale. All these structural features are used in deciding the top and bottom beds in intensely folded strata.

Stylolite   This structure appears as a suture or 'tracing of a style' and commonly occurs in sandstone, limestone, and marble, see Fig. 3.27(f). The sutures range from a fraction of a millimetre to a maximum of a centimetre in thickness. The relief of the stylolite surface has teeth-like projections produced by chemical solution under certain pressure conditions.

Stromatolite   This is a sedimentary structure of organic origin. Such a structure occurs only in limestone, see Fig. 3.27(g). It is formed from the assemblage of blue-green algae by means of precipitation but not as a product of fossils of other organisms.

3.7 POTENTIAL PROBLEMS FROM ROCK STRUCTURES IN ENGINEERING CONSTRUCTIONS

Study of the behaviour of the geological structures originated from deformation of rocks (such as fold, fault, shear zones, joints, and fractures) is very important in engineering geology. All these structures affect the strength and rigidity of rocks. The presence of fault, shear zone, and extensive joints in rocks weakens the foundation and poses problems in surface and subsurface engineering constructions. Before undertaking a project, it is therefore necessary to identify these geological structures and assess the problems that may be created due to their presence in various construction sites such as dams, tunnels, powerhouses, and underground chambers. The potential problems from rock structures are then to be reported to the engineers to facilitate the estimation of cost of treatment in the planning and design. Wherever possible, a site with highly adverse geological structures requiring very high cost of treatment should be avoided and a better site should be chosen. The problems that may arise due to the different types of geological structures in engineering constructions and in structures such as dams and tunnels are discussed here.

3.7.1 Problems of Folds

Folds are obvious features when visible in surface exposures of rock. However, in many sites, the folded features are obliterated due to land erosion. The presence of folds and their details can be easily detected from aerial photo study. Repetition of beds in ground surface is a criterion of identification of a fold structure. Plotting of rock attitudes such as strike and dip of beds in a geological map and preparation of a geological section helps in the determination of the types of folds. Closure of a fold is also known when attitudes are plotted in a geological map extending over a considerable area.

An anticlinal fold is commonly segmented by intensive joints and in some places associated with a fault. As such, the presence of a fold in a dam site may create foundation problems including leakage and requires thorough treatment adding to project cost. Infiltration of water along the joints further deteriorates the rock condition by weathering and erosion. When such a harmful feature occurs in a slope cut along a road, it may create problems of instability and rock slide. The synclinal fold also plays an important role in storing and migration of water through the planes of discontinuities. When present in the foundation or reservoir periphery, synclinal folds may create problems of seepage or extensive leakage. When drilling or excavation punctures water-bearing beds of a syncline with artesian condition, it leads to the problem of sudden rush of trapped water. This problem is of significant importance in a tunnel construction because of flooding of the tunnel cavity with water involves costly treatment. Figure 3.31 shows a dam abutment with folded (anticline), faulted, and jointed rocks. Erosion of the soft materials along the fault has resulted in a wide fissure.

The rocks of the Himalayan terrain are known for complex folding associated with faults and thrusts. Information on the complex folding and

Fig. 3.31 A dam abutment with folded (anticline), faulted, and jointed rocks

thrusting of the rocks of the Himalayas can be obtained by consulting the maps and publications of Geological Survey of India of the concerned area. Geological mapping of the area may then be taken up to reveal the details of the structures and their effect on the construction sites.

3.7.2 Harmful Effects of Faults

Among the different geological structures, checking the presence of a fault at a construction site is crucial because of its harmful effects. In many sites, the existence of a fault is not so apparent on the surface due to the cover of overburden materials, but after excavation of the site, the faulted zone is visible. A fault consists of shattered rocks that create weakness to the foundation, threatening differential settlement of the structure, seepage of water, and development of uplift pressure. A site traversed by a fault showing presence of breccia, sheared rock, and clayey zone called *gouge* involves expensive treatment of the foundation. Figure 3.32 shows a fault zone in the foundation of Umiam dam in Meghalaya with crushed rocks, clay gouge, and fault breccia. The intensity of the problem increases with the increase in thickness of the shattered rock and gouge zone and their depthwise extension. As such, wherever possible, a site with the presence of a major fault with thick zone of crushed rocks should be avoided. After a fault is detected from field investigation, it is necessary to find if it is active or not. An active fault will have originated in a geologically recent time (Holocene or Pleistocene). A Pleistocene fault may be identified from its effects on alluvial terraces. An active fault is liable to initiate further earthquake movement along its fault plane within the life of a project. Hence, a site with an active fault should not be considered for construction of heavy structures.

Fig. 3.32 A fault zone in the foundation of Umiam dam, Meghalaya

When a fault zone is postulated traversing the structural site of a project, it is required to take up pitting, trenching, or tunnelling along the probable fault zone to confirm its presence and study the effect. Whenever required, geophysical method is applied to prove its existence and find its condition at the subsurface. When a fault is detected in a project area from surface observation, parameters such as the width of shattering, the nature of crushed materials present (gouge, fault breccia, etc.), and its attitudes (such as trend and dip) with respect to important directions of the proposed engineering structures should be evaluated. The magnitude of the problem and economics of the treatment of the site affected by the fault are then considered to accept or reject the site.

3.7.3 Weakness of Rocks Caused by Joints, Fractures, and Other Features

Joints and fractures are universally present in all varieties of rocks. These are clearly visible in exposed rock after the excavation of a site for the foundation of engineering structures. Drill cores also indicate their presence in rocks at the depths. In engineering geological investigations, the nature, intensity, and permeability characteristics of the joints and fractures are studied from the outcrops as well as from drill cores at the planning stage of a project. The results of geological studies on joints along with field permeability test data are required by the engineers for designing appropriate measures to strengthen and consolidate the jointed rocks.

Joints are the primary means of movement of surface and subsurface water through all types of rocks. They provide the bulk porosity and permeability of rock mass. Weathering and alteration of joints create larger avenues for entry of surface water to the deeper parts of the rock body. *Clayey infillings* in joints enhance rock slide by lubrication effect of the clayey or soapy materials. Presence of intensive joints in foundation reduces the strength of the rock and poses problems of uplift pressure and leakage through dam foundation and reservoir.

A tunnel pierced parallel to trends of foliation planes or joints is liable to create extensive roof falls and overbreaks. As such, it should be aligned at right angle to the trends of joints, fractures, and foliations of rocks. Roads running parallel to steep-dipping bedding and foliation joints bring about rock slides and obstruction of vehicle movement (Fig. 3.33). A dam may fail by slip along the direction of low-dipping bedding joints of layered rocks. Infillings in joints require study with respect to the nature and tightness of the filled materials. Silica solution and veins of many minerals may completely seal the joints that enhance the strength of the rock and prevent any leakage through them. However, loose infilling of joints by clay, silts, and calcareous materials create weakness to the foundation. Clay minerals such as montmorillonite may cause heaving when soaked with water. Iron oxide coating of joints is also harmful and as such requires cleaning when present in the rocks of structural foundation.

Fig. 3.33 Extensive joints (along strike and dip) in hill slope rocks along a highway causing rock fall and consequent road blockage

A massive sandstone bed generally provides a good foundation condition if it has horizontal disposition. However if they contain block joints may cause rock slide or rock fall. Porous and permeable sedimentary beds such as poorly compacted sandstone and conglomerate influence the flow of water through the void spaces and development of uplift pressure. Sedimentary rocks with alternate shale and sandstone beds dipping at low angles may pose the problem of sliding.

Soluble rocks such as limestone and marl may initiate leakage of reservoir water through interconnected solution cavities formed by solution action along bedding planes, joints, and faults. Bedding planes containing clay seams and soft organic matter, when present in foundation, create the problem of settlement. Clayey siltstone having low compressive strength may initiate plastic deformation and shear failure. Clay minerals (e.g., montmorillonite) present in some clay or shale beds may result in swelling under saturated conditions.

Sedimentary rocks containing primary structures such as mud cracks, sole marks, and flute casts are vulnerable to easy weathering and erosion. Well-developed foliation of metamorphic rocks is a weak feature, as along the planes of foliation the rocks are more susceptible to decay. Presence of large quantities of mica in some varieties of schists renders the rock unsuitable for foundation purposes. Volcanic rocks containing soft breccia, amygdaloidal flow, or red bole if present in a dam foundation may cause seepage and foundation settlement.

- Deformation of rocks gives rise to various types of structures such as folds, faults, shear zones, joints, and fractures.
- An anticline and a syncline are the simple forms of a fold. Depending upon the deformation mechanism, the folds may be of different types such as plunging fold, isoclinal fold, recumbent fold, zig-zag fold, monocline, drag fold, and ptygmatic fold.
- Faults result from the displacement of a rock mass along some weak planes. Depending upon the nature or pattern of movement of the rock blocks along the slip planes, faults are designated as normal fault, reverse fault, strike-slip fault, and pivotal fault.
- Joints are segmentation of rocks mostly associated with tectonic movements and named as tension joints, shear joints, bedding joints, and foliation joints, depending upon their mode of origin or pattern of disposition. The origin of joints in igneous rocks such as cross joints, longitudinal joints, and flat-lying joints is related to the flow of magma at the primary stage of igneous activity.
- Geological structures such as folds, faults, joints, and fractures reduce the strength of a rock mass, and their presence in the foundation of an engineering structure involves expensive treatment.
- In folded rocks, the axial planes of anticlines are extensively jointed. Movement of water through these joints creates problems of uplift and leakage in a dam foundation. A syncline with artisan condition possesses trapped water, and when pierced by a tunnel, it affects tunnelling work due to constant seepage of water.
- Faults, shear zones, and clay gouge are the most undesirable geological features that endanger the stability of engineering structures and also create leakage problems. An active fault occurring close to an engineering site is vulnerable to earthquake, and as such the site should be avoided from construction of any heavy structure.
- Joints in hill slopes with dips towards slope direction are liable to cause landslides and slope failure.
- Presence of joints especially with clay infillings brings about weakness to the rocks with respect to their strength and rigidity. In dam foundations, joints and fissures create instability, uplift, and leakage problems. When present in tunnels, they cause overbreak and roof collapse, requiring extensive supports.
- Several other geological structures such as beds of sedimentary rocks with unfavourable dips and containing clay layers and organic matters, well-developed foliation planes of metamorphic rocks, and breccia and red bole of volcanic rocks when present in the foundation of an engineering structure may endanger the stability of the foundation.
- The solution cavities of limestone in a dam site or reservoir area may pose the problem of large-scale leakage of impounded water.
- At the planning stage of a project, it is necessary to investigate for the probable presence of all harmful geological features in the project site and consider them in the design for treatment with cost estimation.

EXERCISES

Multiple Choice Questions

Choose the correct answer from the choices given:

1. Folds are formed by:
   (a) plate tectonism (b) deformation of rocks
   (c) overloading
2. In a fault, one fault block moves up-dip of the fault plane relative to other blocks. The type of fault is:
   (a) normal fault (b) reverse fault
   (c) thrust fault
3. Columnar joints are formed by cooling of rocks and found in:

   (a) intrusive granite
   (b) basal of the Deccan traps
   (c) dolerite dyke
4. A fault can be identified from the presence of:
   (a) breccia (b) slickenside
   (c) both (a) and (b)
5. The top rocks of an anticline are found to have joints inclined to axial plane formed by:
   (a) shearing (b) tension
   (c) both (a) and (b)
6. In an engineering project, the presence of syncline creates problems of:
   (a) seepage and leakage in a dam foundation

(b) constant seepage and sudden rush of water and flooding inside

(c) both (a) and (b)

7. The joints in a dam foundation create the problem of:
(a) uplift and stability   (b) leakage
(c) both (a) and (b)

Review Questions

1. Explain graphically the mechanism of deformation and stress–strain relation of brittle and ductile rocks.
2. Give an account of all the major types of folds. Describe with illustration the differences between a recumbent fold and an isoclinal fold. Explain how drag folds help to trace relative directions of fault movement.
3. Explain the following with diagrams:
(a) Anticline and syncline
(b) Fold axis
(c) Axial plain of a fold
(d) Plunge of an anticline
4. Define a fault. Draw a sketch to show the hanging wall, footwall, and heave and throw of a fault.
5. Name the different types of faults and state their distinguishing features. What is the effect of a fault on ductile and brittle rocks? How do you identify faults in the field?

6. Discuss the types of joints found in three major rock types. What is the distinction between a joint and a fracture? Enumerate the important parameters of joints. Explain the following structural features: strike and dip joints, columnar joint, block joint, tension joint, and sheet joint.
7. How do you identify folded strata? What are the prime problems that may be created by folded rocks when present in a gravity dam, tunnel, and hill slope?
8. How do you detect a major fault in a site covered by overburden materials? What types of materials are likely to be present in a fault zone? Explain with examples how a fault affects the construction of a heavy engineering structure.
9. State why excessive joints in rocks are considered harmful features for engineering constructions. What are the infilling materials of joints and in what ways are they harmful?
10. Name an expansive clay mineral. Explain how its presence in soil is harmful when the soil is saturated with water. Enumerate the types of problems a soluble rock such as limestone can initiate. What types of volcanic rocks if present in a foundation may cause leakage?

NUMERICAL EXERCISES

1. In a dam site, the abutment rocks are affected by a fault. The throw and heave of the fault are measured and found to be 10 m and 5 m, respectively. Calculate the neat slip of the fault.

[Answer: 11.18 m]

2. During a field study, 200 joints are measured from rocks of a project site and then grouped in seven divisions with respect to their dips and dip directions as follows:

No. of joints	Dip of joints	Direction of joints
40	0–15°	N
30	15–30°	NE
10	30–45°	SW
25	45–60°	S
35	60–75°	NW
60	75–90°	SE

Represent this data of joints in a histogram.
Hint: Calculate the percentage of joints in each division and then proceed to plot on a graph paper following the illustration in Fig. 3.24.

Answers to Multiple Choice Questions

1. (b)   2. (b)   3. (b)   4. (c)   5. (b)   6. (c)   7. (c)

Weathering of Rocks and its Impact on Engineering Constructions

● ● ● ● ● ● ● **LEARNING OBJECTIVES** ● ● ● ● ● ● ● ●

After studying this chapter, the reader will be familiar with the following:

- Different processes associated with rock weathering
- Agents that exert stress to break the rocks in mechanical weathering
- Chemical weathering and the agents that help to bring about changes in rocks
- Living organisms that contribute to biological weathering of rocks
- Impact of weathering on engineering geology

4.1 INTRODUCTION

Weathering is the breaking down of rocks. No rock is immune to weathering. It includes two processes—physical or mechanical weathering and chemical weathering. Any process that exerts stress on a rock and eventually causes it to break into smaller fragments is mechanical weathering. Thermal stress is the main reason for mechanical weathering. Plant roots that develop along cracks and joints in rocks expand them, eventually resulting in their crumbling. Chemical weathering involves the direct effect of atmospheric chemicals or biologically produced chemicals in the breakdown of rocks. Chemical weathering by biological processes may cause complete breakdown of rocks including their minerals and also result in the production of new minerals. Hydrolysis, frost action, and salt crystal growth are associated with the chemical weathering of rocks. This chapter explains all the processes associated with weathering of rocks along with illustrations, including field photographs.

4.2 TYPES OF WEATHERING

Many pathways or processes are involved in weathering; however, they can be classified mainly into two groups, namely physical or mechanical weathering and chemical weathering. In fact, the structure and composition of a rock are altered by the physical and chemical factors present in the atmosphere such as water, air, organic acids from vegetable decay, and even minute organisms.

Mechanical weathering includes processes that involve the breakdown of rocks into fragments or their disintegration into smaller pieces without altering their mineral composition. Thus, mechanical weathering destroys a rock but leaves its chemical composition unchanged. Rocks on the earth's surface are not only subjected to chemical attacks, but also subjected to the abrasive and grinding actions of the atmospheric forces, including wind and water. Temperature variation may also cause drying and cooling and ultimate fracturing in exposed rocks. When water that seeps into cracks freezes to form ice, it widens the cracks, resulting in the crumbling of rocks.

Chemical weathering is the alteration of rocks into new minerals. Geological processes that bring about chemical weathering or result in the alteration of rocks may originate deep within the earth. Burning or melting of rocks by molten lava or washing of rocks by acidic water are examples. Water and minerals react and exchange atoms with the rock materials in chemical weathering.

The process of weathering may also be considered as the disintegration of rocks and their decomposition. Here, disintegration refers to the weathering process that involves physical agents such as temperature changes—freezing and thawing, and decomposition refers to the changes in rocks caused by chemical agents. Both the processes contribute to weathering, but one process may dominate the other. The type of rock and the climate of the place determine the kind of weathering that predominates. However, mechanical and chemical weathering can go hand in hand and cause destruction of rocks. Figures 4.1 and 4.2 depict the combined destructive action of both mechanical and chemical weathering.

The horizontally bedded sandstone of the Bagh Caves in central India is affected by both mechanical and chemical weathering and is so prominently seen on the rock surface, as shown in Fig. 4.1. Here, temperature-caused fracturing of surface rocks attributes to the mechanical weathering, whereas the deterioration of rocks due to rainwater (which is acidic to some extent) results in chemical weathering. Even the sculptures hewn out of rocks inside the Bagh Caves have not escaped the destructive effects of weathering. The sculptures hewn out of sandstones that have calcareous matrix are extensively deteriorated mainly by chemical weathering, as shown in Fig. 4.2.

Fig. 4.1 Extensive deterioration of the cretaceous sandstones of Bagh Caves by chemical and mechanical weathering

Fig. 4.2 Deterioration by chemical weathering of Bagh Cave sculptures

4.3 MECHANICAL WEATHERING

Mechanical weathering exerts stress on rocks that eventually causes it to break into smaller fragments. In fact, physical strain caused by the environment or living things is a constant contributor to mechanical weathering of rocks. It is active mainly in very cold and dry climates and the agents include sharp temperature changes, frost, drought, crystallization, and growth of plants along rock fractures.

Fig. 4.3 Mechanical weathering by freezing of water resulting in fracturing

Fig. 4.4 Selective weathering in a deformed (folded) rock mass along the trough parts

One of the most important forms of mechanical weathering is frost weathering, which is the process of water freezing in rocks, especially in cold climates. This is illustrated in Fig. 4.3. Cold weather causes the seeped-in water in rock joints to freeze and expand, thereby exerting pressure. This widens the joint openings and eventually causes the crumbling of the rock mass. Plant and tree roots that penetrate into the cracks can also widen them. Earthquakes and volcanoes, to a lesser extent, also contribute to the weathering process.

Deformation of rocks also creates avenues of weathering along selective planes of weakness, as shown in Fig. 4.4. Cracks or joints caused by earthquakes and those widened by plant roots are easily subjected to both mechanical and chemical weathering. In case of molten rocks or lava, the rocks undergo physical phase changes due to thermal stress.

Some of the important agents that are responsible for mechanical weathering are described in the following sub-sections.

4.3.1 Thermal Stress

Thermal stress results from the expansion or contraction of rocks, caused by temperature changes. It comprises two main types, namely thermal shock and thermal fatigue. In deserts, there is a significant difference between the day and night temperatures. The stress exerted on the outer layers of the rocks because of such temperature difference can cause those layers to peel off in thin sheets. Moreover, the cold and moisture of the night also enhance thermal expansion, though the day is the principal driver in this change process. There is a continuous alternation of compression and tensile stresses because of the temperature changes. If the pores of a rock contain water, then when the water freezes, it expands and the rock fails in tension further weakening the joints and cracks. Forest fires also cause weathering of rocks and boulders to a significant extent because the sudden intense heat can rapidly expand a boulder.

4.3.2 Spheroidal Weathering and Block Disintegration

Spheroidal weathering is the flaking of highly heated, exposed rock as it expands more than the cooler rock underneath it. This process produces rounded rock mass structures (Fig. 4.5) and sometimes exfoliation domes (Fig. 4.6). It is less common in sedimentary rocks than in igneous rocks. This process is predominant in granitic rocks where the process of disintegration happens

Fig. 4.5 Spheroidal weathering of charnockite showing rounded rock mass structures

Fig. 4.6 Image depicting exfoliation domes

Fig. 4.7 Block disintegration causing a large mass of exposed rock (charnockite) to split into several blocks

through a layer-by-layer removal to evolve towards rounded forms (see Fig. 4.5)

The process of *block disintegration* results from sharp temperature changes causing expansion and contraction of rocks, especially in very dry climatic conditions. This results in the enlargement of the joints in the rock, which ultimately splits into smaller blocks (Fig. 4.7). Sometimes, some peculiar features may also develop as a result of such mechanical weathering.

4.3.3 Frost Action

The process wherein snow or ice inside cracks cause their expansion and the ultimate fragmentation of the rock is known as frost weathering, frost wedging, or ice wedging (Fig. 4.8). This process causes freeze–thaw weathering and is found mainly in the cold mountainous regions such as the Himalayan terrains in India, where the temperatures rise above and fall below the freezing point. Water expands as it freezes and widens the crevices in rocks until they shatter into pieces. Freezing also causes granular disintegration of porous rocks such as chalk.

The end product of frost action may be cone-shaped deposits of slope materials called scree or talus, seen at the foothills (Fig. 4.9). A *talus cone* is developed by the accumulation of broken rock pieces in various shapes and sizes at the base of a mountain cliff or steep hill slopes. Landforms associated with these deposited materials may be very thick and are known as *talus piles*. They have a concave upward form while the maximum slope angle corresponds to the angle of repose of the debris. Such piles of debris may also be formed by thermal stress or the unloading processes of mechanical weathering and the hydration process of chemical weathering, thereby transforming the hill slope rocks.

4.3.4 Pressure Release

In the *pressure release* (also known as 'unloading') phenomenon, the overlying rock by erosion or other processes causes the underlying rocks to expand and develop fractures parallel to the surface. It also promotes sheeting or peeling of rock from the inner mass into a series of concentric shells. Retreat of an overlying glacier, which

Fig. 4.8 Widening of rock joints and cracks, which may crumble the entire rock mass (a slip plane is seen along a joint plane in the front)

Fig. 4.9 Deposits in the shape of talus cones at foothills

Fig. 4.10 Disintegration of rock mass by pressure release (unloading) with the retreat of glacier in parts of Alaska

is a mechanical weathering process of rocks, can also lead to disintegration of rock mass due to the pressure release phenomenon (Fig. 4.10).

4.3.5 Slacking and Haloclasty

Slacking is the process that causes the crumbling of rocks when exposed to air or moisture. It is more apparent in clay-rich sedimentary rocks as they dry out during drought. Crystallization of salts, which is also known as *haloclasty*, causes the disintegration of rocks when water (acidic solution) seeps into cracks and joints in rocks and evaporates, leaving the salt crystals behind. When heated up, the salt crystals expand and exert pressure on the confining rock. This process splits the rock and honeycomb structures develop on its surface. The salts that are most effective in disintegrating a rock are sodium sulphate, magnesium sulphate, and calcium chlorite. These three salts may expand up to three times their volume or even more.

4.3.6 Hydraulic Action

In coastal areas, when water from powerful waves rushes rapidly into the cracks on the rock face, hydraulic action takes place. This causes the trapping of a layer of air at the bottom of the cracks, which compresses them and weakens the rock. When the waves retreat, the trapped air is suddenly released with an explosive force. This causes widening of the cracks or crumbling of the rock, thereby hastening the process of weathering (Fig. 4.11).

Fig. 4.11 Widening of cracks in rocks, which crumble to pieces, because of hydraulic action caused by battering of waves

4.3.7 Tree Root Action

Tree roots can widen the joints and fractures in rocks as they grow up, causing weakness and ultimately the crumbling of the rock mass. This is a frequently observed process of physical weathering. The disintegration process is activated by the exposure of the rock, especially by the removal of soil cover. Vegetation in the absence of soil cover may appear to be a minor factor in rock fracturing or rock disintegration, but tree roots, where present, may actually widen the rock joints to a great extent and ultimately cause fragmentation of the jointed rock mass (Fig. 4.12).

4.4 CHEMICAL WEATHERING

Chemical weathering changes the composition of rocks, often transforming them when water interacts with their minerals forming various chemical reactions. Water plays a very important role in chemical weathering. First, it combines with carbon dioxide in the soil to form a weak acid called carbonic acid. Microbe respiration generates abundant soil carbon dioxide, and rainwater percolating through the soil provides the required water. Carbonic acid slowly dissolves the minerals in rocks, especially the carbonate minerals that make up limestone and marble.

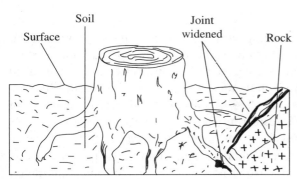

Fig. 4.12 Tree roots penetrate the joints in the bedrock causing their widening and fracturing

Second, water can be adsorbed onto the mineral lattice, which is known as the process of *hydration*. The conversion of anhydride into gypsum is an example. Finally, water can break up minerals through *hydrolysis*, which can take place both at the surface and at the shallow depths, wherever water is circulating through the rock. Large crystals of igneous and metamorphic silicate minerals decay into tiny flakes of clay minerals. These minerals are all hydrated, having an OH component derived from the water, and often water itself, H_2O. Thus, the most common group of minerals, the silicates, is decomposed by hydrolysis. The reactive hydrogen ions liberated from water attack the crystal lattices of the minerals. In the process, new and secondary minerals may develop from the original minerals of the rock.

In humid climates, rocks are affected aggressively by chemical weathering. The chief destructive agents are rain, water, and certain mineral contents. These dissolve some kind of rocks or rot the natural cements that bind the particles in the rock together. However, certain minerals such as quartz are more resistant than augite, biotite, hornblende, or orthoclase. The important ways of rock decay by chemical weathering are described in the following sub-sections.

4.4.1 Oxidation

Oxidation happens when atmospheric oxygen combines with the compound (minerals) in some rocks. The most commonly observed oxidation process is that of Fe^{2+} (iron) and its combination with oxygen and water to form Fe^{3+} hydroxides and oxides such as hematite, limonite, and goethite. This gives the affected rock a reddish-brown colouration on the surface, which crumbles easily and weakens the rock. This process is better known as *rusting*. Many other metallic ores and minerals oxidize and hydrate to produce coloured deposits, such as chalcopyrite as $CuFeS_2$ oxidizing to copper hydroxide and iron oxides.

4.4.2 Carbonation

Carbonation is a solution weathering process that is caused by the minerals in the rocks reacting with water containing a considerable amount of carbon dioxide. It may produce harmful effects in a limestone terrain. Limestone is readily dissolved by rainwater, which is acidified by carbon dioxide from the atmosphere or soil. The reactions are as follows:

$$CO_2 + H_2O = H_2CO_3$$

Carbon dioxide   Water   Carbonic acid

$$H_2CO_3 + CaCO_3 = Ca(HCO_3)_2$$

Carbonic acid   Limestone   Calcium bicarbonate

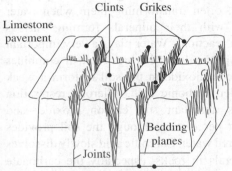

Fig. 4.13 Weathering of limestone platform with deep fissures (grikes) and ridges (clints) that are formed by carbonation

The resulting weak carbonate widens the rectangular joints in limestone and produces deep and wide fissures, which are called *grikes*. The surfaces or ridges are called *clints*. Thus, limestone pavements may form such features of grikes and clints as a result of carbonation, a process of chemical weathering (Fig. 4.13). The terms grikes and clints originated from the dialect of Northern England, where such features were first found in limestone pavements (Fig. 4.14). Carbonation is responsible for creating sink holes, swallow holes, and even large cavities in limestone. The formation of karst landscape of limestone terrain is also an instance of chemical weathering (see Sec. 22.1).

4.4.3 Hydration

Hydration involves chemical addition of water to the minerals of certain rocks. When some minerals take up water and expand, they result in the breakage of shells from the rocks. Water thus added is be distinguished from the type of water that promotes disintegration process.

Tor is generally formed in granite as a result of millions of years of weathering. The name 'tor' originates from Dartmoor in the UK where granite blocks of varying sizes, after weathering, form tors covering a large area. It is stated that the rocks overlying the granite plutons got eroded with the consequent release of pressure (unloading), which caused the formation of sub-horizontal and vertical

Fig. 4.14 Grikes and clints created by chemical weathering in a limestone pavement in Northern England (Source: Wikipedia, 2011)

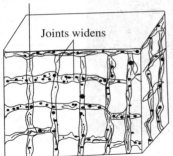

Granite splits
into blocks

Joints widens

Fig. 4.15 Weathering along joints in rocks (vertical and sub-horizontal) cause their widening and splitting of rocks resulting in formation of tors

joints in granite and tended to give them the shape of blocks. Then, chemical weathering by the process of hydration caused extreme breakage of the hard granite along these joints to create tors, which are cubic or sub-rounded blocks (Fig. 4.15). Such features are also found in the rocks of various countries. Figure 4.16 represents a tor in the rocks of Kenya. Tors are also prevalent in the granitic rocks of peninsular India, especially in Andhra Pradesh.

4.4.4 Hydrolysis

Hydrolysis involves the chemical addition of water to the minerals of rocks, which results in a chemical reaction. Water thus added is structured and is to be distinguished from water that directly affects disintegration. The process can cause feldspar in granite to decompose and produce kaolin, a white powdery substance, which when mixed with rainwater forms clay.

$$2KAlSi_3O_8 + H_2CO_3 + 9H_2O \longleftrightarrow Al_2Si_2O_5(OH_4) + 4H_4SiO_4 + 2K + 2CO_3$$

Orthoclase Carbonic acid Water Kaolinite Silicic acid Potassium Carbonate ion in solution

Hydrolysis contributes to the breakdown of micas, amphibole, and many metamorphic minerals. This process can take place on the surface as well as in sub-surface conditions where water is circulating through the rock. In some cases, a granite body at a depth may turn into a deposit of kaolinite of economic importance and can be quarried as a source of china clay. However, in some granite foundations of engineering structures, deep-seated conversion of feldspar into kaolin creates construction problems.

When pyroxene and olivine react with water, magnesium and silicon are carried away in solution but the remaining iron reacts with oxygen to form the solid residue hematite (Fe_2O_3), which is a mineral form of iron oxide. The hydrolysis process of chemical weathering is much common in rocks. Carbonic acid is consumed by silicate weathering, resulting in more alkaline solutions because of the bicarbonate. This is an important reaction in controlling the amount of carbon dioxide in the atmosphere and can affect the climate.

Fig. 4.16 Tor of Kenya formed by the process of hydration (Source: Wikipedia, 2011)

The only common mineral that is immune to chemical attack is quartz (SiO_2). Although it is soluble in acidic groundwater, this is applicable only for quartz released by chemical breakdown. However, quartz itself is insoluble under surface conditions.

4.5 BIOLOGICAL WEATHERING

In addition to mechanical and chemical weathering, rocks are also altered and decomposed by *biological weathering* (Fig. 4.17). The main agent in biological weathering is the organic acids released by organisms such as bacteria, lichens, mosses, and decaying plants of many types.

Fig. 4.17 Rocks attacked by biological weathering

The acids attack the rock-forming minerals. The mineral composition can also be initiated and accelerated by soil organisms.

The most common forms of biological weathering are the release of chelating compounds (i.e., organic acids) and of acidifying molecules (i.e., protons, organic acids) by plants so as to break down aluminium- and iron-containing compounds in the soils beneath them. The decaying remains of dead plants in soil may form organic acids, which when dissolved in water cause chemical weathering.

Some kind of fungi associated with the root systems can release inorganic nutrients from minerals such as biotite and transfer these nutrients to the trees, thus contributing to tree nutrition. It has also been evidenced that bacterial communities can impact mineral stability, leading to the release of inorganic nutrients.

4.6 EFFECT OF WEATHERING DUE TO CLIMATIC CONDITIONS AND GEOLOGIC TIME

The rate of weathering depends on many factors. First, weathering will be faster if there is more water in the system. This is why chemical weathering tends to be faster in the hot and humid tropics. Second, the higher the temperature, the faster is the weathering. Third, the more the mineral surface area is exposed in the rock by joints, the faster will be the weathering. The increased number of cracks in the rocks will allow the agents of water and oxygen to interact more intensely with the minerals. Finally, the mineral composition of the rock is also a factor of weathering. Felsic igneous rocks tend to weather much faster than the igneous rocks with more quartz grains.

What would happen to an outcrop of granite that was exposed to the agents of weathering in a humid and warm environment over millions of years? The agents would slowly deteriorate the rock and liberate the quartz grains, which are very resistant to chemical weathering and would be deposited as sand grains in the stream beds, beaches, and dunes. The feldspars would be decomposed as kaolin and will ultimately form clay and salt mixed with other decomposed minerals. The amphibole and biotite minerals would form iron oxides and clays. Most of the salt would be taken away by groundwater or washed away by surface water into the seas to render salinity to the sea water. Clays, over a long geologic time, will form shale, which is the most abundant of the sedimentary rocks.

Chemical weathering produces salts and clay and forms canyons and gorges. Mountains and islands are the result of massive mechanical weathering caused by huge shifts in continental plates that smash together to make mountains or break off and create islands. In a way, mechanical weathering produces mountains, cliffs, plateaus, and islands.

4.7 IMPACT OF WEATHERING ON ENGINEERING CONSTRUCTIONS

Weathering has great significance in engineering geological works. Weathered rock is a weak feature for use as foundation for engineering structures. Such features of rock disintegration and decomposition may remain hidden under the soil cover or at a depth, which could be detected when the excavation exposes the surface. This may involve a change of design or even the rejection of the project though it has started, resulting in increased project cost; hence, it requires the important consideration of the project management.

4.7.1 Engineering Geological Significance

As seen in the formation of blocks and joints, very hard rocks are also separated into small blocks due to weathering. When a heavy structure such as a concrete dam is constructed over such rocks, the blocks may move slowly over a certain period of time, and ultimately, the structure may fail. Hence, a cautious approach is needed while selecting the site for heavy structures because even apparently hard rocks can be affected by weathering and form blocks and joints.

At the investigation stage of a project, it is necessary to check whether the foundation of the engineering site contains any weathered rock. If a weathered rock portion is present at a shallow depth, it is to be thoroughly scaled out before laying the foundation for structures such as dams, powerhouses, and huge buildings. Even in sub-surface projects such as tunnels or underground powerhouses, or in a dam or bridge site covered by thick riverine deposit, it is necessary to confirm the depth of fresh rock under the weathered rock or river debris by geophysical survey or exploratory drilling. Accordingly, the structural foundation should be designed in fresh and sound rock.

It should be noted that a few scattered exposures of sound rocks do not prove the presence of sound rock conditions underneath, especially in a granitic terrain. A thorough examination, especially by seismic refraction method and drilling, is necessary to prove the presence of sound rock covering the foundation area of the engineering structure. In Orissa, during the excavation of a dam site, a large thick zone of kaolinite clay was found to be present underneath a shallow cover of river deposits. Several exposures of sound granite were present at the surface close to the foundation of the dam site. The depth and extent of the weathering of granite that had been altered to kaolinite clay found during the excavation was so extensive that the dam could not be constructed at the proposed site. The site had to be shifted.

Moreover, care should also be taken in the selection and use of construction materials for engineering structures. No portion of these materials should contain disintegrated or decomposed rocks. Use of even partly weathered rock either as a block or as a concrete aggregate should be avoided for engineering constructions, as such rocks may also rot because of long exposure to atmosphere and result in the failure of the structure.

SUMMARY

- Weathering is the process of alteration of rocks. By combining with a number of atmospheric agents of the earth, including water and air, rocks are altered mainly by mechanical and chemical weathering.
- Mechanical weathering results in the disintegration of rocks through physical processes such as temperature changes, pressure release, hydraulic action, and frost action. Formation of spheroidal rock bodies, block disintegration, and penetration of tree roots in rock joints causing their expansion and ultimate crumbling of the rock are examples of mechanical weathering.
- Chemical weathering caused by water, air, and other agents is responsible for the decomposition of rocks through processes such as oxidation, carbonation, and hydrolysis. Biological processes can also cause decay and decomposition of rocks by the organic acids produced from the decay of plants and animals. Formation of kaolin clay from

the decomposition of granitic rocks is an example of chemical weathering.
- Disintegration is more prominent in warm climate, whereas warm and humid climate can cause more decomposition of rocks. The combination of both disintegration and decomposition is more prolific in temperate climate.

- Weathering of rock has a great impact on engineering constructions. Disintegrated and decomposed rocks are not suitable for engineering construction purposes. Even partly weathered rocks should not be used for foundation of engineering structures or as construction materials.

EXERCISES

Multiple Choice Questions

Choose the correct answer from the choices given:

1. The process of mechanical weathering includes:
   (a) thermal stress   (b) spheroidal weathering
   (c) both (a) and (b)   (d) none
2. The process of chemical weathering includes:
   (a) hydrolysis   (b) hydraulic action
   (c) both (a) and (b)   (d) none
3. Tor type of weathering forms in granite by the process of:
   (a) release of pressure   (b) hydration
   (c) both (a) and (b)   (d) none
4. Grikes and clinks are formed in limestone pavement by:
   (a) oxidation   (b) carbonation
   (c) both (a) and (b)   (d) none
5. Tree roots that penetrate through the joints in rocks can cause:
   (a) widening of joints
   (b) fragmentation of the rocks
   (c) both (a) and (b)
   (d) none
6. Talus cones at the foot of a hill can be formed by:
   (a) hydraulic action   (b) bacterial action
   (c) both (a) and (b)   (d) none
7. Decaying remains of dead plants can cause:
   (a) mechanical weathering
   (b) biological weathering

(c) chemical weathering
(d) both (a) and (c)
8. Granite affected by hydrolysis process of weathering by groundwater when detected after excavation of foundation for construction of a heavy structure results in:
   (a) change in design
   (b) huge increase in construction cost
   (c) rejection of the site
   (d) any one of (a), (b), or (c)

Review Questions

1. Name two of the earth's atmospheric agents responsible for weathering.
2. Write four processes responsible for mechanical weathering.
3. Write four processes responsible for chemical weathering.
4. What is kaolin? What kind of weathering is responsible for its formation?
5. How may the weathering of granitic rock in the foundation of a concrete dam create construction problem?
6. What is the main agent of biological weathering? Write how rocks are decomposed by biological weathering.

Answers to Multiple Choice Questions

1. (c)   2. (a)   3. (c)   4. (b)   5. (c)   6. (d)   7. (c)   8. (d)

5

Soil Formation, Engineering Classification, and Description of Indian Soils

● ● ● ● ● ● ● **LEARNING OBJECTIVES** ● ● ● ● ● ● ●

After studying this chapter, the reader will be familiar with the following:

- Role of geological processes in changing landforms
- Geological works of ice and wind
- Formation of different types of soils and their engineering uses
- Components, structures, and descriptive analysis of soils
- Classification of soils as per Indian specification
- Indian soils including their origin and places of occurrence
- Clay minerals and their effects on the soils

5.1 INTRODUCTION

This chapter deals with the different geological processes and their activities that give rise to various types of soils on which many engineering structures such as buildings, highways, runways, and even dams and bridges are constructed. It explains with illustrations the destruction and different stages of weathering of rocks that eventually produce soil. The chapter also describes the geological processes such as erosion of rock and transportation and deposition of eroded particles that give rise to the soils. It further discusses the depositional environment giving rise to various types of soils including their engineering uses. It provides a detailed account of Indian soils with their origin and places of occurrence and elucidates the different types of clay minerals and their effects on the soils.

5.2 GEOLOGICAL PROCESSES RESPONSIBLE FOR SOIL FORMATION

Geological processes are responsible for the formation of soils of varied types, which are of importance in engineering constructions, such as a dam built across a river valley, a tunnel excavated through an escarpment, or a road aligned along a hill slope composed of soil and overburden materials. In fact, processes such as erosion, transportation, and deposition of soil have been actively associated with changing lands of the earth's surface since a long geological time. Water, wind, and ice are the main agents associated with the processes that produce soils and cause their transportation. An understanding of these processes that control the land patterns is necessary to examine the merits and demerits of a site while carrying out an engineering geological investigation.

5.2.1 Formation of Soil from Weathering and Decomposition of Rocks

Weathering of rocks of the uppermost crust of the earth is a complex process of disintegration and decomposition of rock body and has been elaborated in Chapter 4. The *disintegration* or breakdown of rocks results from alternate heating and cooling caused by periodical temperature changes and also from the actions of wind and ice. If the water in the fissures and pore spaces of rocks freezes during cold conditions, the rock body tends to expand and develops cracks, which hasten the disintegration process. Sometimes, tree roots also penetrate the fine fissures of rocks and develop big cracks causing further deterioration and disintegration. Landforms in warm and dry climate are prone to weathering mostly by disintegration of rocks. *Decomposition* of rocks results from chemical processes such as oxidation, hydration, carbonation, and leaching of organic acids. The decay of vegetation produces organic acids, which cause further decomposition of rocks, with the ultimate product being soil. This aspect of mechanical and chemical weathering has been described in detail in Chapter 4.

Decomposition of rocks produces soil with clay minerals. Expansive soil consisting of swelling type of clay minerals is harmful if present in the foundation of a structure. Weathering of rocks due to decomposition is more prolific in lands experiencing a warm and humid climate. The tropical climate of India with high rainfall (monsoon rains) is responsible for the excessive disintegration as well as decomposition of rocks that has produced rich deposits of soil conducive for agricultural growth of the country.

The works of several authors (Fookes et al. 1971; Knill et al. 1970) have established that the weathering of rocks that produce soil is the combined effect of physical disintegration, chemical decomposition, and solution of rocks. These zones of disintegration and decomposition are dependent on the topographic features, rock types, and structure of rocks in the concerned area. Rocks in slopes with uneven surfaces are prone to substantial weathering. When the rocks are subjected to structural disturbances due to folding, faulting, jointing, or fissuring, the process of weathering becomes intense giving rise to thick zones of soils (Fig. 5.1).

Weathering of rocks such as mudstone and tuff causes only iron stains along joint surfaces, but limestone is more decomposed and goes into solution more easily along joints and other structural planes. The solution action results in the following two types of conditions:

(i) When a soluble rock such as limestone is argillaceous in nature, the mechanical action and solution effect will both be operative in the weathering process resulting in the removal of more materials.

Fig. 5.1 Weathering and disintegration of rocks into soil

(ii) However, if limestone contains very little insoluble material, the solution of the rock will produce voids due to the removal of rock material by solution from the planes of discontinuities.

5.2.2 Residual and Transported Soils

Soil covering only the top part of the bedrock from which it has been derived is known as *residual soil*. In other words, a deposit of residual soil remains in the same position capping the parent rock since its origin. A bedrock undergoes different stages or grades of weathering to form residual soil (Fig. 5.2). These gradations of weathering, applicable for soil formation from all types of bedrocks, are based on the conditions of the structure, texture, and mineral contents from initial to advance stages of weathering. Table 5.1 provides a classification of the weathering grades.

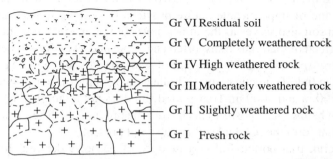

Fig. 5.2 Grades of rock weathering to form residual soil

Table 5.1 Classification of weathering grades (IAEG, 1981)

Grade I—Unweathered or fresh rock	No visible sign of weathering. Rock is fresh and crystal bright. Few discontinuities with slight weathering.
Grade II—Slightly weathered rock	Penetrative weathering developed on open discontinuity surfaces but only slight weathering of materials. Discontinuities are discoloured and discolouration can enter into rock up to a few millimetres from discontinuity surface.
Grade III—Moderately weathered rock	Slight discolouration extends through greater part of rock mass. The rock material is friable (except in the case of poorly cemented sedimentary rocks). Discontinuities are stained or filled with altered material.
Grade IV—Highly weathered rock	Weathering extends throughout the rock mass and the material is partly friable. The rock has no lustre. All materials except quartz are discoloured. The rock can be excavated with a geologists' pick.
Grade V—Completely or extremely weathered rock	The rock is totally discoloured and decomposed and is in a friable condition. Only fragments of the rock structure are preserved. The external appearance is that of a soil.
Grade VI—Residual soil	Soil material with complete disintegration of texture, structure, and mineralogy of parent rock.

Fig. 5.3 Extensive weathering and erosion of hill slope forming soil

The residual soil deposits are of limited thickness varying roughly between 15 m and 60 m and prevalent in hill slopes (Fig. 5.3) and also in parts of plateau regions. The black soil mantle of the Deccan basalts and the lateritic soil occurring in different parts of India are examples of residual soils. Residual soil formed of hard rocks such as granite and charnockite may be better for foundation than that derived from sedimentary rocks such as shale, sandstone, and limestone. In a limestone terrain, thick deposits of residual soil derived from limestone may possess caverns and sinkholes hidden below the soil. Hence, if a storage structure or reservoir is constructed in such a terrain, there may be leakage of reservoir water. Residual soil with good plasticity and of impermeable nature provides favourable foundation conditions for an earth dam. Such soil also serves as the impervious core of an embankment or rock-fill dam.

Soil formed of materials transported and deposited away from its original place of formation is called *transported* soil. The deposit of transported soil may be very thick. The alluvial soil covering the Indo-Gangetic plains of India is an example of transported soil varying in thickness between 100 m and 5000 m. Transported soil does not retain the ingredients of the parent rock even down below the deposits of such soil. The products of different stages of weathering of the bedrock that form the residual soil are also not present.

In general, transported soil is a mixture of particles derived from rocks of two or more regions and also of reworked sediments. This may be deposited in a homogeneous manner with well-sorted grains or in a heterogeneous manner with assorted particles depending upon the manner of their transportation and condition of deposition. Transported soils frequently contain lenses or layers of organic matters harmful for construction of engineering structures. However, due to the widespread occurrence of such soils, engineering structures such as roads, rails, airports, and other buildings are constructed on this soil after adopting suitable safety measures in the design of the structures.

5.2.3 Erosion, Transportation, and Deposition

Geological processes such as erosion, transportation, and deposition bring about changes in the landform by eroding the rock types that give rise to soil deposits. Structural habits, especially discontinuities in rocks, accentuate the process of erosion. Water flowing through soft rock causes rapid erosion. A fault zone having soft materials is subjected to easy erosion providing path for a river channel. In sedimentary rock formations with soft rocks or alternated with beds of hard and soft rocks, the softer beds may be easily eroded creating deep valleys or gorges. Igneous rocks such as granite and basalt are hard and less eroded compared to sedimentary rocks. In metamorphic rocks, soft bands such as mica schists are easily eroded by running water.

Rocks affected by joints and faults are subject to rapid erosion and transportation of the material on the downstream side. Sometimes, a large rock body gets eroded and forms soil but the hard core remains standing in the midst of the soil appearing like a sculpture (Fig. 5.4). Erosion is associated with weathering that brings about a superficial change of rock mass.

Fig. 5.4 Naturally formed sculpture, resembling a snowman wearing a cap

Chemical weathering of rock also results in rapid erosion, eventually leading to the formation of soil. Spheroidal weathering, common in igneous rocks, is a type of chemical weathering that creates rounding of rock mass and formation of boulders, thus hastening the process of soil formation.

Some rocks such as limestone and dolomite, in addition to erosion, are subjected to solution effect under acidic water. Rainwater passing through the soil and vegetation becomes slightly acidic and this causes the solution of the limestone. Sandstone with calcareous matter also gets eroded easily. The process of solution causes formation of big cavities in the surface and subsurface regions of limestone. Hence, the presence of limestone in a dam foundation or reservoir may create problems of leakage and subsidence. In the case of argillaceous limestone, solution and erosion processes act together resulting in rapid removal and deposition of the material in the form of transported soil.

Many of the river valleys in the mountainous terrains with folded rocks are formed in the anticline part of the fold due to easy erosion of this part containing extensive number of joints and fractures. In glaciated mountains, the ice sheet comes down eroding the bedrock at the bottom. Sometimes, sudden fall of a huge mass of ice (avalanche) takes place in the lower parts of the mountains with accumulation of debris from melting ice. This creates large deposits of glacial materials (drifts) in the high-altitude glaciated terrains. Such glaciated materials are prevalent in the high mountain slopes of the Himalayas.

Floods, earthquakes, and volcanic eruptions also bring about large-scale changes in land surface and help in the formation of soil deposits. Floods carry huge loads of silt that gets deposited in the rivers and surrounding terrains reducing the life of a reservoir if located there. Earthquakes cause catastrophic changes in landform features and complete destruction of man-made structures. Flow of lava due to volcanic outbursts destroys existing land surface and creates new landforms. Man also acts as an agent in changing the land space. Excessive steepening of a hill slope by cutting for a road may result in rock slide or debris slide. Deforestation for land development expedites erosion and brings about changes in the morphology within a short time along with a change in the atmospheric conditions. Detritus or loose soil is easily washed away and deposited elsewhere and the land becomes a desert-like terrain. This process of changing landform patterns and formation of detritus or soil deposits began in the geological past and still continues.

5.3 GLACIAL AND FLUVIOGLACIAL ACTIONS

Glaciers are thick ice sheets that move like a stream slowly along hill slopes. The ice sheets erode and abrade the uneven hill slopes to produce channels for their movement. When the ice melts, the eroded materials are deposited as glacial drifts in different structural forms.

5.3.1 Features of Glacial Activity Including Old Remnants

The most prominent features formed from the deposition of a glacier by melting of the ice sheet during its downward movement are *moraines*, *drumlins*, and *eskers* (Fig. 5.5). If the deposition of the materials is perpendicular to the direction of movement of ice sheets, it forms *terminal moraine*. When the accumulation of the heterogeneous materials takes place in front of the ice sheets, it is termed *ground moraine*. These are formed when the drifts are accumulated on depressions on the receding path of ice sheets. The deposition of drifts along the two edges of the glacial channel forms *terminal moraines*. A *drumlin* is formed when accumulation of sediments deposited by the glaciers takes an elliptical shape in its ground configuration. *Eskers* and *kames* are developed when streams flow in the glaciers forming irregular layers of particles brought by the stream and the ice sheets. The drifts creating eskers are deposited in a highly zig-zag manner. All these structures of glacial origin generally vary in thickness from 15 m to 60 m and stretch irregularly from a few hundreds to thousands of metres.

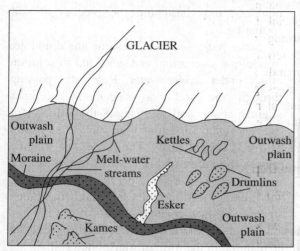

Fig. 5.5 Glacial drifts showing outwash, kettles, drumlins, eskars, and terminal moraine

In the geological past, India experienced several glacial invasions that left their existence in the form of heterogeneous mixtures of unconsolidated sediments. The Talchir tillites formed of glacial till is an evidence of a glacial invasion during the Gondwana age. The striations and polished surface characteristics of glacial movements are present in the rocks immediately below the Talchir beds. Pleistocene India experienced at least four glaciations as evident from the piled-up glacial deposits in the foothill regions of the north-west Himalayas. Several large glaciers are at present active in the high-altitude regions of the Himalayas. They carry piles of sediments and deposit the drifts in the downhill areas with the melting of ice sheets (Fig. 5.6).

Fig. 5.6 Deposits of glacial meltdown at the foothills of the Himalayan plane

5.3.2 Problems of Engineering Construction in Glacial Deposits

Construction of any engineering structure in deposits of glacial origin will encounter problems of permeability and stability. Due to heterogeneity and lack of compactness of the materials, founding a dam on this soil will lead to differential settlement. A reservoir spread on this soil may experience leakage due to the permeable nature of the material. Construction of a highway or road traversing hill slopes formed of glacial drifts may have to deal with the problem of slope failure for which extensive drainage arrangement is necessary in addition to compactness. For a major construction in the high-altitude terrain of the Himalayas close to a glaciated area, the feasibility stage investigation should include a study on the stability and permeability of the glacial drifts. In addition, glaciological investigation is necessary in uphill terrain to collect information pertaining to snow line, the rate of movement of ice sheets, the volume of load depositing in the downstream parts by melting ice sheets, the chance of avalanche striking the proposed structure, and other activities of the glaciers.

5.3.3 Fluvioglacial and Glacial Lacustrine Deposits

In some areas, both ice and water may take part in forming huge deposits. For example, the outwash materials of glaciers accumulated at the toe of the hills are transported by river water and deposited in rivers or lakes to form *fluvioglacial deposits*. These deposits exhibit the characteristics of alluvial deposits of simple river origin, but the heterogeneity of the sediments of glacial deposits is absent in them. Figure 5.7 shows the fluvioglacial deposits at the Himalayan foothills where melting of glacial ice forms huge debris comprising boulders and pebbles that are deposited on the sides as the water flows as streams.

Fig. 5.7 Fluvioglacial deposits at the Himalayan foothills

The deposition of glacial sediments in lakes due to glacial melt water results in *glacial lacustrine* deposits. These are formed by filling the depressions or blocking the course of a stream by ice sheets. The sediments of the glacial lacustrine deposits are rock powder, clay, and silt-size particles. The settlement of the particles takes place very slowly in the still water of the lakes, resulting in very thin and horizontally layered deposits called *varves*. It takes nearly a year to form one layer of silt and one layer of clay, each less than a centimetre in thickness.

Both fluvioglacial and glacial lacustrine deposits are permeable in nature. Laying roads on these deposits will have the same problem as that of the deposits of glacial origin. The size, texture, and properties such as porosity and permeability vary in the different layers of varves. The overall material has a low-bearing capacity having clay in its ingredients, which may create settlement problems if a heavy structure is constructed in such deposits.

5.4 WIND ACTION AND WINDBORNE DEPOSITS

Wind can carry small particles such as sand and smaller particles. The power of wind to erode depends on particle size, wind strength, and whether the particles are able to be picked up by the wind. Wind is an important force causing erosion. It is more prevalent in arid regions than humid regions. In arid regions, sand dunes are common. These are wind deposits that come in different shapes, depending on winds and sand availability.

5.4.1 Sand Dunes

Wind-blown loose sands from plains create *dunes*, which are found in different structural forms (Fig. 5.8). Sand-laden wind erodes the rock mass of hills and may form a flat terrain full of loose sand-size clastic particles of quartz, feldspar, and mica. If the area is windy, the sand-size particles will be swept away by the wind and deposited elsewhere to form dunes. The dunes may form geometric patterns of structures such as crescent- or ridge-shaped *Barchans* or irregular sand deposits. Dune deposits have a gentle slope (roughly 5–10°) in the wind direction but are rather steep (slope angle about 30°) in the leeward side. They attain a height of only a few metres to approximately 100 m in their vertical extent of deposition. Dunes are characterized by their mobility from one place to another caused by the sweeping of the sand due to high winds. Sand dunes with characteristic features of deposition are found in deserts and along seashores. The sands blown from the Arabian Sea coast were deposited in parts of Rajasthan giving rise to a desert-like landform with sand dunes. Stone Age artefacts were found in stratified dune deposits underlying the recent sands providing evidence of the starting of dune-forming processes tens of thousands of years ago.

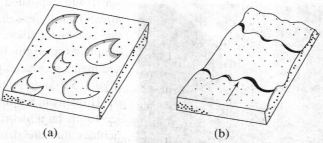

(a)   (b)

Fig. 5.8 Structure of a dune: (a) Barchan dune; and (b) Barchanoid ridge (→ indicates wind direction)

Due to the loose and highly permeable nature of sand, stability problems are encountered in the construction of roads and highways. Any engineering construction in terrains with sand dunes requires good maintenance. Maintaining the roads is difficult because of the migratory nature of sand dunes. Some species of conifers and long grass are planted in the area as a measure to create binding of the sands. Sand arrestors are also provided to guard the roads. No structure should be constructed in areas where sands are very mobile in nature. Canals constructed through the areas of sand dunes need to be lined due to the porous nature of the deposits.

5.4.2 Loess—Formation and Engineering Problems

Loess is an aeolian (windborne) deposit or soil that is porous and permeable in nature and is constituted of poorly graded fine sand, silt, and clay. It may contain a high content of silt or it may be a clay-rich soil. Loess shows the phenomenon of hydro-consolidation, that is, the soil mass will consolidate when it comes in contact with water. The permeability of horizontal layers in loess is more than that in vertical direction. One type of loess formed by wind-blown materials, remains in place without undergoing any decomposition. However, in another type, the soil undergoes chemical decomposition though the place of origin is not shifted. Loess can withstand very steep slope cuts, at places even upto vertical angles.

The main problem encountered while constructing a structure on loess is subsidence under wet conditions. A building constructed on loess may settle and develop cracks. Hence, construction of heavy structures on loess is avoided because of the threat of settlement. Pile driving is generally undertaken for constructing building structures on this soil. Grouting of loess is done to stabilize the soil before undertaking any construction. If the soil is present in a slope, it may undergo sliding when it comes in contact with water. Berms (step cutting) are provided as protective measures to loess slopes. Construction of a dam on loess needs compaction of the soil before founding the structure. To use this soil in the construction of an earth dam, the soil is to be rolled as per design requirement.

5.5 FORMATION OF TERRACES, TALUS, AND ORGANIC DEPOSITS

The grain-size (textural) distributions of surficial material deposits such as glacial till, end moraines, fines (silt and clay), sand, gravel, floodplain alluvium, swamp deposits, salt marsh deposits, and artificial fills are the main reasons of how the formation of terraces, talus, and organic deposits will take place. There is a relationship between the depositional origins and the distribution and character of the surficial material deposits. These deposits range from a few feet to several hundred feet in thickness, overlie the bedrock surface and underlie the organic soil layer of the site in consideration.

5.5.1 Glacial Terrace

Terrace may be developed in glacial areas when it is called *glacial terrace*. While flowing through a glacial terrain, a river may deposit its bed load in any place along its course due to the decrease in its rate of flow. This causes formation of glacial terrace by the narrowing of the stream course. A glacial terrace is very wide but is short in longitudinal direction. These terraces are made of loosely packed, poorly graded heterogeneous materials and any construction such as roads or highways through these terraces may experience stability problems.

5.5.2 River Terrace

River terraces are formed by deposition of riverborne materials in bench- or step-like slopes or benches made by the river itself (Fig. 5.9). A river in the mature stage flowing through plains may cut a new channel and the old channel is left out as a terrace composed of the river sediments. Even in the young stage, a stream flowing through a soft bedrock of a hilly terrain may gradually deepen its channel floor resulting in benching or terracing of the upper slopes. From the hill slope, loose materials may drop on these terraces and accumulate as talus. Thus, the flow of a river in both the young and mature stages creates river terraces. A terrace may also be formed by landslide in a sloping landform. A river terrace is rather long but is of narrow width. River terraces consist of sand- and gravel-size particles that are used as good construction materials. Those developed on the gentle slopes can be used as sites for constructing roads connecting towns or villages located along the river valleys.

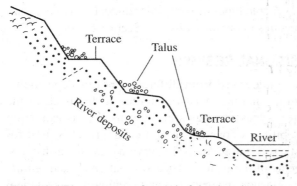

Fig. 5.9 River terraces formed of riverine deposits with talus on step-like slopes

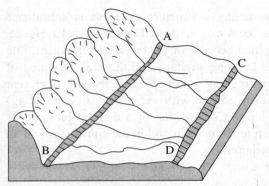

Fig. 5.10 Marine terrace showing former web cut (A–B) and earlier sea level (C–D)

5.5.3 Marine Terrace

Marine terraces are formed along continental coasts. Powerful waves are created in marine water due to postglacial rise in sea level, when glacial drift and alluvial material are deposited along the coastal belts. The deposition creates elevated shorelines. The marine cliffs and abraded platform along the coasts are raised abruptly resulting in the formation of marine terraces (Fig. 5.10). Erosional processes in a later time may be active along the terraces and bring their partial or complete obliteration.

5.5.4 Talus

Talus is formed of loose debris (called colluvial deposits) accumulated at the base of a cliff or the bottom of a slope. Here, gravity acts upon uphill materials and brings them down and deposits them at the toe of the ridge. Because of high porosity and the heterogeneous character of the deposits, engineering construction is generally avoided on this soil. Screening of the talus, however, provides gravel- and sand-size particles for use as construction materials.

5.5.5 Organic Soils Including Pits

Organic soils are formed in marshy lands, swamps, and also in shallow lakes. Various types of water plants, vegetables, and grasses are grown in marshy lands and swamps. In lake water, the organic materials including vegetable matter are carried and deposited by streams. All these plant and vegetable matter after decay form organic soils, which are black in colour and cohesive in nature under dry conditions. However, when wet, the soil loses its cohesiveness. In some swamps, peats are formed by humification of the plant remnants including seaweeds. The peats mixed with inorganic materials form peat soil, which is an organic soil showing no trace of the original ingredients of plants.

Organic soil is highly porous and possesses very low shear strength. It is also prone to slide and settlement and hence construction of a road or any heavy engineering structure should be avoided in this soil. The construction of even a light structure requires proper drainage arrangement. When organic soil is present in parts of a road or rail alignment, it is always better to remove the portion of the organic soil and backfill with suitable types of soils. Pile foundation is necessary in constructing buildings or other heavy structures in organic soils (see Section 6.16).

5.6 SOILS DERIVED FROM DIFFERENT DEPOSITIONAL REGIMES

Engineering structures are constructed on soils derived from different types of depositional environments such as river, glacial, lacustrine, marine, and windborne, as discussed in the preceding sections. Soil deposits of different types of origins are also used as construction materials depending upon their local availability. A coarse soil with clean gravel is used for concrete aggregates, and a fine-grained plastic soil is used for the impervious core of earth dams. Water, wind, and ice are the three main agents of the natural processes that *erode* and *abrade* weathered rock surfaces or soil deposits and transport the materials and eventually deposit them in water or on land. In some areas, gravity also acts as an agent to deposit materials to form soil.

The soils thus formed are designated differently based on their manner of transportation and place of deposition as follows:

Alluvial soil　Water is the main agent involved in transporting and depositing the sediments. The transported soil of riverine origin is called *alluvial deposit* or *alluvium*. The soil formed from the transported materials in a lake floor or under the sea is called *lacustrine soil* and *marine soil*, respectively. Soil deposits of alluvial, or marine, and lacustrine origin have characteristic depositional features such as layering of different-size particles, graded bedding, and ripple marks.

Glacial soil　The huge ice sheets of glaciers during their slow movements through the hilly terrains erode and abrade the rock mass of the hills. The eroded materials consisting of a heterogeneous mixture of rock powder, clay, silt, sand, gravels, and boulder carried by the glaciers are called *glacial drifts* or simply *drift*s. The material beneath the glacier when deposited by the ice sheet itself is called *till* and comprises mostly fine silt to clay-size particles. Both drift and till do not show any layered structure.

Aeolian soil　The deposition of wind-blown particles forms Aeolian soil deposits. Loess and sand dunes are examples of Aeolian deposits. Loess comprises mostly clay- and silt-size particles without any layering. Dunes are the accumulation of loose sands in different structural forms that may migrate from one place to another by the action of wind.

Colluvial soil　In hilly terrains, the force of gravity takes part in accumulating rock fragments and eroded materials of the uphill region at the foothills creating soil deposits of heterogeneous mixture called talus. This type of transported soil is known as colluvial soil.

Soil from volcanic ash　Another type of transported soil is the volcanic ash where volcanic eruption is the main agent of transportation. Several ash beds of geological time are found in the Deccan traps formed by volcanic activity. Ash deposits of recent origin exist in Hawaii where volcanoes are active and erupt even today (see Section 20.5).

Organic soil　This is derived from chemically decomposed vegetable matter mixed with sediments transported by streams and deposited in shallow lakes or ponds and swamps. This is a transported type of organic soil. Peat or peat soil, formed in marshy land and swamps created by the decay of plants, is a residual type of organic soil. This type of soil is avoided in engineering works as it may create problems of subsidence and heaving when saturated with water.

5.7 SOIL CLASSIFICATION FOR ENGINEERING CONSTRUCTIONS

Soil classification is aimed at describing the various types of soil in a systematic way and at dividing soils of distinct physical properties into groups or units. There are several soil classifications, but engineers are more concerned with the one best applied for the purpose of engineering constructions. Since Casagrande (1943) evolved the Unified Soils Classification System for army engineers in the US, it has been widely accepted for varied engineering purposes such as construction of dams, highways, and other civil engineering structures on soil. Indian Standard (IS) classification of soils is based on this unified soil classification after some modifications. This classification adopted for civil engineering works has taken into consideration the size of soil particles, texture of the constituent grains, organic content, and properties such as plasticity, grading, and compressibility. Soils are broadly grouped into three divisions, namely coarse-grained, fine-grained, and organic soils. In this classification, the symbol G stands for gravel, S for sand, M for silt, C for clay, and O for organic.

5.7.1 Coarse-grained Soils

In the coarse-grained soil group, more than 50 per cent of the material by weight will be retained above the 75-micron sieve. The basic components of soils of this division are boulder, cobbles, gravel, and sand. The size range of these components is given in Section 5.7.3. In the IS classification, the nature of grading and mixing of the soils under the divisions gravels and sands are designated by the symbols W for well-graded and clean, C for well-graded with good clay binding, P for poorly graded but fairly clean, and M for poorly graded mixed with finer materials. These symbols are used with group names GW, GP, GM, and GC in the grading of gravels. The group symbols SW, SP, SM, and SC describe the grading nature of sands in the IS classification.

5.7.2 Fine-grained Soils

In fine-grained soils, more than 50 per cent of the particles by weight are silt and clay that pass through the 75-micron sieve. The size of silt particles is 75 microns to 2 microns and that of clay is less than 2 microns. The fine-grained soils are grouped under three subdivisions, namely mixed silt and fine sands of inorganic nature, clay of inorganic type, and organic matter with silty and clayey particles.

   Compressibility of soil is also used in the classification of fine-grained soils. The index for compressibility is determined from the results of liquid limit tests as follows:

- Low plasticity—liquid limit of less than 35 represents silt and clay of low compressibility expressed by the symbol L.
- Intermediate or medium plasticity—liquid limit between 35 and 50 indicates silt and clay of medium compressibility designated by the symbol I.
- High plasticity—liquid limit more than 50 suggests high compressibility of silt and clay type soil denoted by the symbol H.

   In the IS classification, the compressibility of fine-grained inorganic soils under different subdivisions has been designated by the combination of symbols such as ML, MI, CL, and CI.

5.7.3 Organic Soils

Organic soils contain organic matter including peat (symbol PT) and decomposed vegetable products in large quantities. Soils comprising materials such as shells and concretions in large quantities are also included in this group. These soils have high compressibility. The symbols OL, OI, and OH are used for organic soils of low, medium, and high plasticity, respectively. The complete IS classification of the different types of soils indicated by the different group symbols is given in Table 5.2.

5.8 IDENTIFICATION OF DIFFERENT SOIL TYPES

The reason why it is so useful to identify different soil types is because each soil type has unique hydrological properties. Some soil horizons (surface) will have high hydraulic conductivities and thus have greater and more rapid fluctuations in soil moisture. Some will have greater bulk densities with lower effective porosities and thus have lower saturation values. Some will have clay films that will retain water at field capacity longer than other soil surfaces. Knowledge of these different soil surfaces will allow the engineering geologist to construct a more complete picture of the movement of water in the soil. In general, with increasing depth, the clay content increases, the organic matter decreases, and the base saturation increases. Soil surfaces can be identified by colour, texture, structure, pH, and the visible appearance of clay films.

Table 5.2 Engineering classification of soils

Main divisions	Divisions	Subdivisions	Group symbols	Characteristics of soils
Coarse-grained soils (more than 50% particles larger than 75 microns in size)	Gravels more than 50% grains more than 4.75 mm in size	Clean gravels	GW	Well-graded gravels, gravel–sand mixture, little or no fines
			GP	Poorly graded gravels or gravel–sand mixture, little or no fines
		Gravels and fines	GM	Silty gravels, poorly graded gravel–sand–silt mixture
			GC	Clayey gravels, poorly graded gravel–sand–silt mixture
	Sands more than 50% grains less than 4.75 mm in size	Clean sands	SW	Well-graded sands, gravelly sands, little or no fines
			SP	Poorly graded sands or gravelly sands, little or no fines
		Sands with fines	SM	Silty sands, poorly graded sand–silt mixtures
			SC	Clayey sands, poorly graded sand–clay mixtures
Fine-grained soils (more than 50% particles less than 75 microns in size)	Silts and clays, low compressibility, liquid limit less than 35		ML	Inorganic silts and very fine sands, silty or clayey fine sands, or clayey silts with none-to-low plasticity
			CL	Inorganic clays, gravelly clays, sandy clays, silty clays, lean clays of low plasticity
			OL	Organic silts of low plasticity
	Silts and clays, medium compressibility, liquid limit 35–50		MI	Inorganic silts, silty or clayey fine sands, or clayey silts of medium plasticity
			CI	Inorganic clays, gravelly clays, sandy clays, silty clays, lean clays of medium plasticity
			OI	Organic silts and organic silty clays of medium plasticity
	Silts and clays, high compressibility, liquid limit more than 50		MH	Inorganic silts of high compressibility, micaceous or diatomaceous fine sandy or silty soils, clastic silts
			CH	Inorganic clays of high plasticity, fat clays
			OH	Organic clays of medium to high plasticity
			PT	Peat, humus, and other highly organic soils with very high compressibility

5.8.1 Visual Observation to Identify Soil Types

Coarse-grained soil is identified in a hand specimen with respect to its grain size, shape, grading, and mineral contents. Spread over a flat surface, the percentage of gravel, sand, and fines in a soil can be roughly estimated from visual observation. Two charts, one made of glued particles of gravel with 50 per cent particles of greater than 4.75 mm size and another for sand with particle size 0.075–4.75 mm, help in the identification of gravel and sand by comparison. A hand lens (×10) will facilitate viewing clearly the mineral content and the characteristic shape of the grains such as angularity and roundness. The grading nature can be well understood with practice from observing the different grain sizes in a sample.

Poorly graded soil particles (Fig. 5.11) are predominantly of one size (GP and SP types of soil), whereas well-graded soil (Fig. 5.12) has a wide range in their particle size (GW and SW types of soils). For accurate measurement of the percentage of gravel and sand in the soil, a representative sample is taken in a graduated cylinder after mixing with water and allowed to settle when gravel, sand, silt, and clay particles are deposited one above another in sequence of coarseness to fineness. The percentage by volume is measured from deposited gravel- to clay-size particles in the cylinder. A substantial amount of clay-size particles mixed with gravel is indicative of GC and SC types.

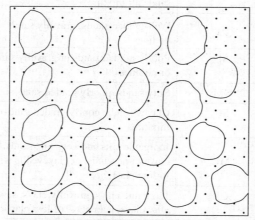

Fig. 5.11 Poorly graded soil with mostly single-size grains

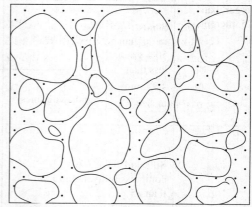

Fig. 5.12 Well-graded soil with varied-size grains

Identification of fine-grained soil is done by estimating its *plasticity*, *dilatancy*, and *toughness*. Plasticity can be determined by rolling a small quantity of the soil on the palm to make threads of approximately 3 mm diameter. In case the threads remain intact, the soil is of high plasticity, but if cracks develop, it suggests a soil of low plasticity. Dilatancy is estimated by holding about 1 cm³ soil sample on the palm of one hand and shaking it with the help of the other hand. If moisture beads are formed quickly on the surface, the soil type is inorganic fine sands, clayey fine sands, or clayey silts, which are non-plastic in nature, such as ML and MI. However, when formation of moisture beads is slow, the soil type is inorganic, organic silty, or clayey with medium-to-low plasticity such as OL, OI, MH, and OH. No bead of water is formed in soils types such as inorganic clays or silty clay with medium-to-high plasticity, that is, CI and CH. The toughness of the clay is roughly estimated by taking a small quantity of wet soil and using hand pressure. Soil with substantial clay, good plasticity, and no dilatancy possesses high toughness (that is, CH), but fine sandy and silty soil with no plasticity but showing quick dilatancy has no toughness, that is, ML.

5.8.2 Additional Information Needed for Design Purposes

For engineering design purposes, the description of coarse-grained soils provided under different divisions (Table 5.2) is substantiated by further information on stratification, degree of compaction, cementation, moisture content, and drainage characteristics from study of soils in situ and in undisturbed conditions. Information on the percentage of gravel and sand, their angularity, surface condition, hardness of coarse particles, and geological name of soils are also obtained. The additional information required for fine-grained soils include structure, stratification, moisture content, consistency limit, and drainage conditions in undisturbed or remoulded soils. From the studies of disturbed soils, information on colour in wet condition, plasticity index, percentage of particles in different size grades, and geological names may also be obtained.

5.9 CLAY MINERALS IN SOILS AND THEIR ENGINEERING SIGNIFICANCE

Clay minerals are an assemblage of crystalline minerals that provide plasticity and cohesion to soil. A clayey soil, in addition to non-clay minerals such as quartz, feldspar, mica, and calcite, may contain one or two clay minerals made of microscopically thin sheets. Chemically, the clay minerals are composed of aluminium silicate or a combination of iron and magnesium silicate, but some contain alkaline earth. Structurally, the clay minerals have two building blocks, one having the configuration of a tetrahedron and the other of an octahedron. The tetrahedral unit contains a silica atom with four hydroxyls and the octahedral unit comprises an aluminium atom and an iron, or a magnesium atom enclosed in six hydroxyls. The structural frameworks of both the units are made up of densely packed sheets or layers.

5.9.1 Types of Clay Minerals

The three main groups of clay minerals commonly present in soil are *montmorillonite*, *kaolinite*, and *illite*.

Montmorillonite

Montmorillonite is an expansive clay mineral of soil characterized by a sheet-like structure. An octahedral gibbsite sheet packed between two silica sheets forms the structure of montmorillonite. The swelling of this clay mineral is caused due to the bond of water inside the sheets. Each thin sheet or platelet can attract to its flat surface a layer of adsorbed water nearly 20 times larger than the thickness of a sheet. Soil containing montmorillonite also shows high shrinkage in addition to excessive swelling. This clay mineral may sometimes split into individual layers in the presence of abundant water. The well-known *bentonite* usually originated from volcanic ash is a montmorillonite clay utilized as drilling mud, which due to its expansive character prevents the collapse of bore hole walls, but its presence in the foundation will have a very hazardous effect.

Kaolinite

Kaolinite is a non-swelling clay mineral formed from kaoline, the altered product of feldspar, which is a basic mineral component of many rocks such as granite and pegmatite. Kaolinite contains gibbsite and silicate sheets or layers in its structure. The successive layers are held together firmly by hydrogen bonds providing a stable structure. Water cannot penetrate into the layered structure. Hence, kaolinite when mixed with water does not show any appreciable swelling. The fine sheets or platelets in its structure carry negative electromagnetic charges on their surfaces that attract thick layers of adsorbed water producing good plasticity under wet conditions. China clay is the common name of kaolinite, which is widely used in the pottery industry.

Illite

Illite is a non-expansive clay mineral. Its structure is similar to that of montmorillonite. However, in illite, the silicates are replaced to a certain extent by aluminium in the tetrahedral layers and potassium intrudes into the replaced part to tie the sheet-units together. Unlike kaolinite, the illite crystals of a soil break easily into platelets comprising a gibbsite layer packed between two silicate layers.

5.9.2 Sensitive Clay—Characteristics and Rectification Measures

When a certain type of clay is remoulded, some disturbances are created in the structure of the clay particles, which are then reoriented to nearly parallel arrangements. This causes reduction in shearing resistance. Such type of clay sensitive to disturbances is called *sensitive clay*. This clay acts normally when partly saturated but it yields to pressure forming a viscous flow. When the clay is of highly sensitive type, it is known as *quick clay*. The degree of sensitivity (S_t) of a type of clay is expressed by the ratio between the non-compacted compressive strength of an undisturbed specimen and the strength of the same specimen at the same water content but in a remoulded state. The value of S_t in sensitive clay is generally in the range 4–8 but may be 8–16 in clay of high sensitivity. In quick clay, natural water content is well above the liquid limit and sensitivity S_t often exceeds 40 (Terzaghi Karl, 1968).

Quick clay is abundantly present in Scandinavian countries and in parts of Canada. If disturbed by an earthquake tremor or a landslide, quick clay causes viscous flow destructing houses and properties of the area containing such sensitive clay. The author of this book had the experience of viewing landslide disturbances in parts of Norway where large segments of land with quick clay layers move like viscous flow, carrying collapsed houses, inclined trees, and so on far away from their original position. Research undertaken by Norwegian Geotechnical Institute (NGI) has indicated that leaching of sodium salts from the soil due to glacial movement is responsible for the formation of quick clay. The soil scientist in NGI demonstrates this with the help of this simple experiment: Take a small lump of soil (quick clay) in a bowl and stir it with a spatula until the soil becomes viscous and eventually appears like clayey water. Now spray some table salt (sodium chloride) to this clayey water; it becomes solid and is not liquefied by disturbance. This suggests that if sodium salt is taken out from the natural clay, it forms quick clay, and the re-entry of the salt brings back its original form.

Earlier, any structure built in soil having a quick clay layer faced the danger of collapse due to the viscous flow of the clay layer caused by even small a disturbance, However, later it became possible to construct multi-storied buildings in such soil by stabilizing the quick clay soil. This is done by introducing sodium salt (common salt) into the soil through drill holes and then dispersing it, following the method of electro-osmosis (Fig. 5.13) developed by Casagrande (1949). The method introduces electric current through a metal tube acting as

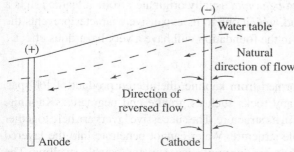

Fig. 5.13 Arrangement of electrodes for electro-osmosis process of stabilization of soil

cathode when positive ions move towards the cathode taking with them molecules of water. This enforces inter-pore water to collect in the cathode areas and they are dragged by ion charges. Thus, the process causes reduction in water content of the clay materials. In addition, by suitable choice of the anodic electrode, a structural transformation of the clay can be induced triggering a series of physicochemical reactions towards improving the stability of the quick clay soil.

5.10 SOILS OF INDIA—CHARACTERISTICS AND OCCURRENCE

Large parts of India are covered by various types of soils found in the plains, valleys, and hill slopes and even in the high-altitude mountainous terrains. The major soil types of India may be classified into the following groups:

- Alluvial soil
- Black soil
- Lateritic soil
- Red soil

In addition, considerable parts of the country are occupied by desert soil, forest soil, and mountain soil. In certain parts, there are peaty, marshy, and saline types of soils too. A description of all these soil types with respect to their formation, characteristics, and occurrence is provided in the following subsections.

5.10.1 Alluvial Soil

Alluvial soil covers nearly 40 per cent of the land area of the country. The economy of the country is dependent to a large extent on its alluvial soil. A large tract of India washed by the rivers Indus, Ganga, and Brahmaputra extending from Assam to Punjab is covered by alluvial soil and this is the richest source of fertile soil in the world. This soil is riverine soil transported and deposited by the numerous rivers of the country. In addition to the alluvial tract of North India, considerable parts of peninsular India including the coastal plains and river deltas also contain alluvial soil. The Krishna, Cauvery, Godavari, and Narmada–Tapti river systems washing a large landmass of South India have formed a very fertile alluvial soil.

Alluvial soils are also found nearly 30 m above the flood level of the streams belonging to Sutlej, Yamuna, Ghagra, and other rivers of Indo-Gangetic plains. The northern alluvial tract extends over many states—Punjab, Haryana, Uttar Pradesh, Bihar, Orissa, West Bengal, and Assam. In the peninsular region, the alluvial soil occurs in many of the river valleys and in the east coast. The alluvial area also includes parts of Gujarat and deltaic soil in peninsular India.

Alluvial soil contains fine sediments of silt and clay. This soil also contains very small humus and is deficient in nitrogen but rich in potash. If the alluvial soil is of recent deposit, it is called *Khaddar*; if it is of old alluvium, it is called *Bhangar*. If it contains sandy and gravelly matters in large proportions, it is termed *kankary*. The *Khaddar* soils of newer alluvium are pale brown in colour and non-porous, clayey, and loamy in composition. This soil is found in the lower areas of valleys, being enriched every year by flood sediments. The *Bhangar* soils of older alluvium are dark in colour containing a good percentage of nodules and shingles.

Though the alluvial soils differ in texture in different places, they are overall very fertile and respond well to manures and irrigation for both *rabi* and *kharif* crops. These soils are well suited for jute cultivation and also for the cultivation of rice, wheat, sugarcane, and oilseeds.

5.10.2 Black Soil

Black soil, also called *black-cotton soil*, is a typical Indian soil. This soil is locally known as *regur* (equivalent to Russian *Chernozem*). The black-cotton soil is mainly the product of in situ decomposition of the Deccan trap basalts. The colour of the soil varies from black to chestnut brown. Its black colour is due to its iron content derived from the constituent of basalt.

Black soil is mainly found over the Deccan trap areas including Maharashtra, Madhya Pradesh, Gujarat, Andhra Pradesh, and parts of Tamil Nadu. The river valley areas of Narmada, Tapti, Godavari, and Krishna also contain this soil. The total area covered by black soil is about 5.4 lakh km², that is, about 16.6 per cent of the total land cover of the country.

Black soil occurs in areas of low (45–75 cm) rainfall. The soil develops conspicuous cracks on drying. It is highly porous and swells enormously after wetting. Laboratory tests of black-cotton soil indicated abnormal increase in expansive pressure when allowed to absorb water continuously.

The following are the characteristics of black soil:

- It swells and becomes sticky when wet and shrinks when dried.
- During dry season, it develops wide cracks.
- The soil consists of high alumina, lime, and magnesia.
- It contains small potash, low nitrogen, and phosphorous and lacks organic matter.
- This soil is very good for cotton cultivation and hence the name.
- In addition to cotton, it is suitable for the growth of tobacco, sugarcane, and oil-seeds.
- Cereals such as rice, wheat, and *jowar*, citrus fruits, and vegetables are also grown in this soil.
- The moisture-retaining character of the soil makes it suitable for dry farming.

5.10.3 Lateritic Soil

Lateritic soil, another typical Indian soil, occurs as in situ deposits occupying the upper surface of laterite. The soil is the decomposed product of laterite and is course textured, soft, and friable. It is a porous soil because its silica content is removed from it by chemical action. Lateritic soil is poor in lime and magnesia and deficient in nitrogen.

The soil is formed under typical monsoon climate experiencing heavy rainfall and high temperature with alternate dry and wet periods that allow leaching of silica and lime of the parent rock. It is deep brown to red in colour due to the presence of iron oxide, which is formed by the process of leaching. Due to leaching, the nutrients are percolated below the soil by heavy rains, leaving the top soil impermeable. The soluble plant foods such as potash are also removed from the top soil by leaching leaving alumina and iron oxide. This makes the soil less fertile.

There are two types of lateritic soil, namely upland laterite and lowland laterite. Upland laterites are found over hills and upland areas. From there, they are transported by stream towards the lowlands. Such transferred soils form lowland laterite, which is acidic in nature as alkalis are leached away. Lowland laterite soil does not retain moisture and hence it is not fertile. It is suitable only for special crops such as tapioca and cashew nuts. In general, lateritic soil is poor in nitrogen minerals and is good for plantation of tea, coffee, coconut, rubber, and even millet with the use of fertilizers.

This soil covers an area of nearly 2.4 lakh km² of the country. The soil occurs in many parts of India, especially in the Deccan traps of the peninsula and Khondalites of Orissa. It also occupies large areas of Western Ghats, Eastern Ghats, Rajmahal Hills, Maharashtra, Karnataka, Kerala, and West Bengal.

5.10.4 Red Soil

Red soil is another important soil type of India. Large parts of the peninsular region, especially the areas represented by gneisses and granites, are covered by red soil, which is produced by in situ disintegration of these rocks. The soil is found as transported soil but occurs to very deeper parts from the surface. The soil is coarse in the deeper part, medium in middle part, and fine at the upper part in the delta regions.

The red colour is due to the wide diffusion of iron oxide through the soil material. This red colour of the soil varies from very pale to deep depending upon the iron minerals present in the parent rocks. The soil is deficient in nitrogen, humus, and phosphorous but rich in potash. It is a fertile soil and varies considerably in composition and characteristics depending upon the constituent minerals of the parent rocks. This soil is suitable for cultivation of pulses, millets, linseeds, tobacco, and so on.

This soil occupies an area of 3.5 lakh km^2 area, which is about 10 per cent area of the country. Red soil is prevalent in areas of Maharashtra, Gujarat, Tamil Nadu, Andhra Pradesh, Karnataka, Orissa, Bundelkhand, and Chota Nagpur plateau.

5.10.5 Other Soil Types

Other than the ones described in the previous subsections, there are some more soil types in India.

Desert soil This type of soil prevails in the arid regions of Rajasthan, Haryana, and Punjab where rainfall is less than 50 cm causing desert-like conditions. The soil is rich in sand particles and contains high phosphorous but is low in nitrogen. Desert soil contains 90 per cent sand and 5 per cent clay. It also contains a rich percentage of soluble salts but lacks in organic matter. This soil type covers approximately 2 lakh km^2 of the dry areas of Punjab and Haryana. It is made fertile by adding the deficient minerals as fertilizers for cultivation purposes. *Jowar*, *bajra*, maize, and cotton are mainly cultivated in desert soil.

Forest soil This type of soil is good for luxurious growth of several types of plants and trees. It is found in mountain areas. Forest soil contains high organic content derived from the decomposition of forest products and is very rich in humus but poor in potash, lime, and phosphorous.

Mountain soil The high-altitude areas including the Himalayan terrain contain mountain soils. This soil occurs as a thin deposit over the bedrock and is mostly infertile. It is quite prone to soil erosion as it occurs on steep slopes and gets heavy rainfall.

Peaty and marshy soil This soil occurs in the humid climatic regions of Kerala, Uttar Pradesh, Orissa, Tamil Nadu, and West Bengal containing large amounts of soluble salts and a high percentage (10–49%) of organic matter.

Saline and alkali soils These soils contain high alkaline content impregnated by salt and are available in parts of Punjab, Haryana, and Bihar. These soils are known as *reh* and *kallar* in different local areas.

SUMMARY

- Geological processes are responsible for changing landforms that control the construction of engineering structures in various ways. Actions of rivers, ice, and wind shape the earth's surface and create various types of soil deposits.
- Weathering processes are responsible for the disintegration of rocks, the ultimate product being residual soil. There are six stages of rock weathering that converts fresh rock (Grade I) to residual soil (Grade VI). The intermediate grades (Grades II to V) include slightly weathered to extremely weathered rocks.
- Soil that remains in the place of its origin overlying the bedrock from which it has been formed is called residual soil. Soil that originates from transportation and deposition of particles in a place different from its parent body is known as transported soil.
- Geological processes are responsible for the denudation of landmass and formation of soil under different depositional environments. Natural hazards such as earthquakes, floods, and landslides accentuate the process of soil formation bringing about rapid changes in the landform.
- Based on the mode of transportation, place of deposition, and organic content, soils are designated as alluvial soil (river deposit), glacial soil (glacial drift), Aeolian soil (windborne dune sand, loess), and organic soil (swamp deposit).

- Glacial soil is highly permeable and requires compaction and stabilization measures prior to its use. Sand dunes, loess, and organic soils present specific problems that need to be addressed before any engineering construction work is taken up in such soils.
- The engineering classification of soil outlined by the Bureau of Indian Standards is broadly divided into coarse-grained soil and fine-grained soil depending upon the percentage of gravels and sands. Each of these divisions is then subdivided into different categories taking into account the percentage of sand, clay, and organic contents and properties such as liquid limit, plasticity, and grading.
- Clayey soil contains clay minerals such as montmorillonite, kaolinite, and illite. The montmorillonite-bearing soil causes swelling and hence is avoided in engineering use or its expansive character is retarded by mixing with non-expansive clays.
- Soil containing certain types of sensitive clay such as quick clay when disturbed by an earthquake or other tremors may cause devastating slides behaving like viscous flow. Leaching of sodium salt from the soil constituent is responsible for such behaviour. Quick clay soil is stabilized by the reintroduction of sodium salt into the soil by electro-osmosis process.

EXERCISES

Multiple Choice Questions

Choose the correct answer from the choices given:

1. Residual soil is formed:
   (a) at a lake floor or river bed by deposition of particles
   (b) on the bedrock by decomposition of rock constituents
   (c) on mountains with accumulation of debris
   (d) in coastal belts due to alluvial deposits

2. Out of the six grades of weathering, residual soil is formed after weathering of rock:
   (a) in Grade VI   (b) in Grade I
   (c) in Grade III   (d) all of the above

3. Residual soil may occur as a:
   (a) very thick deposit of above 100 m

   (b) deposit of limited thickness about 15 m to 50 m
   (c) few millimetres from discontinuity surface
   (d) surface dust

4. An example of residual soil is:
   (a) *regur* of peninsular India
   (b) *khaddar* of Indo-Gangetic plains
   (c) *Bhangar* or old alluvium
   (d) *kankary* containing sandy and gravelly matters

5. Aeolian soil is formed of:
   (a) river deposits
   (b) windborne materials
   (c) artificial filling
   (d) decomposition of vegetation

6. Soil derived from decomposed vegetable matters is known as:
   (a) black-cotton soil   (b) organic soil
   (c) lateritic soil   (d) red soil

7. Slow deposition of particles in calm lake water is called:
   (a) varves
   (b) loess
   (c) berms
   (d) talus
8. The correct statement among the following is:
   (a) Residual soil is harmful for construction of roads, rails, and runways.
   (b) Alluvial soil can be safely used in founding buildings or dam structures.
   (c) Both residual and alluvial soils can be used for construction of roads, rails, runways, and dams after adopting suitable measures as required by engineering design.
   (d) A deposit of residual soil can never remain in the same position since its origin.

Review Questions

1. List the geological processes responsible for producing soil.
2. Tabulate the different grades of weathering that form soil from rocks. State the role played by each grade of weathering.
3. Describe the geological processes—erosion, deposition, and transportation—in relation to the formation of soil deposits. What are the other geological and natural agencies responsible for soil formation?
4. Give a brief account of soils derived from the various types of depositional regimes.
5. What is glacial activity? Explain the terms moraines. drumlins, eskers, and kames.
6. Write short notes on fluvioglacial deposit, aeolian deposit, colluvial deposit, and terrace deposit.
7. How is organic soil formed and what are its characteristics? What is the action to be taken if such soil is present in a highway section or a building site?
8. Give an outline of the engineering classification of soils. Write the characteristics of soils represented by the group symbols GW, GM, SW, SM, ML, MI, MH, and PT.
9. What are clay minerals? How are they formed? Describe the structures and properties of the three main groups of clay minerals commonly present in soil.
10. What is sensitive clay? State how its presence in a soil deposit creates instability. Add a note on the measures required to stabilize the soil mass.
11. Give a short account of the different types of soils available in India including their mode of formation and occurrence.

Answers to Multiple Choice Questions

1. (b)   2. (a)   3. (b)   4. (a)   5. (b)   6. (b)   7. (a)   8. (c)

6

Fundamentals of Soil Mechanics

LEARNING OBJECTIVES

After studying this chapter, the reader will be familiar with the following:

- Laboratory determination of index properties and Atterberg limits of soil
- Field methods of density measurement and laboratory analysis of soil
- Determination of soil permeability by laboratory and field pumping tests
- Consolidation and compactness tests of soils in the laboratory
- Earth pressure and the different types of retaining structures to contain soil slopes
- Shallow and deep foundations in soil for construction buildings

6.1 INTRODUCTION

Soil mechanics can be stated as the study of the properties, behaviour, and application of soil consisting of an assemblage of fine rock particles and clay minerals. This chapter discusses the various aspects of soil mechanics, the knowledge of which is essential for geological as well as engineering works. It describes the procedures for determining the index properties and Atterberg limits of soil by laboratory analyses. The chapter also explains the method of determination of soil permeability by pumping, which is the practical way of understanding the nature of water movement through the subsurface soil mass and fractured rocks. It also elucidates with illustrations the testing procedures of consolidation and compactness of soil and further describes the different types of retaining structures used to arrest disruption of soil slopes caused by earth pressure. The other aspects that are discussed in this chapter are the methods of field testing to measure the load-bearing capacity of soil for evaluating designs for building foundation.

6.2 SOIL PHASES, COMPONENTS OF SOIL, AND SIZE OF SOIL PARTICLES

The topmost part of the earth's crust is composed of different types of soil. Geological processes are responsible for the formation of these soil types as described in Chapter 5. A mass of soil consists of solid particles of soil and void spaces that occur between the soil particles generally filled with water, air, or gas. Naturally occurring soil mass always consists of solid particles with water and air in the pore

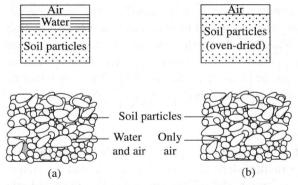

Fig. 6.1 Soil mass: (a) three-phase system; and (b) two-phase system

spaces and this soil is known as three-phase soil system, see Fig. 6.1(a). Soil mass with dry soil particles and water, or oven-dried soil particles and air are two-phase soils, see Fig. 6.1(b).

Soil is defined differently according to the field of its use. It is considered by geologists as the unconsolidated earth material that exists in the relatively thin portion of the earth's crust within which roots occur. The rest of the crust consists of rock. The rock occurring immediately below the soil zone is termed *bedrock*. Agriculturists consider soil as the material covering relatively thin parts of the surface where plants can grow.

To engineers, soils include all unconsolidated deposits with inorganic and/or organic material consisting of grains and rock fragments of varying sizes. Evaluation of the properties of soil in quantitative terms is required for the design of various engineering structures that are made of soil or constructed on soil foundation and use soil as a building material.

6.2.1 Size Fractions of Soils

In order of decreasing size, the components of soils include boulder, cobble, gravel, sand, silt, and clay. Table 6.1 provides a list of soil components classified according to Indian Standards (IS). The clay component in soil includes *colloids*. Colloids may be like jelly, called *gel*, or like liquid, called *sol*. If some types of clay are subjected to vibration or sudden shocks, the gel portion gets converted into sol, but after few hours it regains its original form of gel. This type of clay is known as *thixotropic clay*.

Table 6.1 Size of soil components (IS 460–1962)

Component	Passing through IS sieve	Retained on IS sieve
Boulder (rock fragments)	–	300 mm
Cobble (rock fragments)	300 mm	80 mm
Gravel (coarse grained)	80 mm	20 mm
Gravel (fine grained)	20 mm	4.75 mm
Sand (coarse grained)	4.75 mm	2.00 mm
Sand (medium grained)	2.00 mm	425 microns
Sand (fine grained)	425 microns	75 microns
Silt	75 microns	2 microns
Clay	2 microns	–

The soils may be dry or saturated fully or partially with water. The voids in soils are completely or partially filled with water and air. Soils consisting of gravels, coarse sands, and medium dry sands called *bulky particles* are cohesionless, but silt- and clay-type soils are cohesive. Wet sandy soils—especially with some clays or silts—are also cohesive. Being derived from rock destruction, the bulky particles in soil are initially angular in shape, but the geological process of erosion makes the sharp edges of the grains rounded and spherical shaped. The variation in shapes of the particles also depends on soil minerals and their transportation and depositional characteristics. Soils with fine-grained particles such as silt and clay are found to have a dominant content of crystalline minerals. When the surface activity of the crystalline minerals develops plasticity and cohesion, they are known as *clay minerals*. The clay minerals in soil are found as an assemblage of book-like sheets or tubes.

6.2.2 Structure of Soils

The structure of a soil is the arrangement of the constituent particles in the soil matrix that contains voids, fissures, and cracks. Soil structure depends on many factors such as shape, size, mineral composition, orientation of the grains, the relation of soil water in their ionic components, and the interactive forces among the particles. Since coarse-grained particles when clean and dry do not possess plasticity and cohesion, soil containing these particles cannot form stable structures. In moist sands, however, the contact pressure developed by capillary water ring may provide the necessary bonding to create stable structures. However, when the grains become dry, the structure again breaks down. Gravity and surface forces are the main forces that play an active part on soil particles in developing the soil structure. Gravitational force is operative only in soil containing coarse-grained particles. In small particles, surface forces are effective in most of the soil types. These attractive forces are electrostatic in nature and are popularly known as *van der Waals'* forces. Repulsive surface forces also interact among the particles to form layered structures. The different types of structures (Fig. 6.2) mainly observed in soil developed by the depositional character and interactive forces of particles are described in the following paragraphs.

Single-grain structure   Grain-to-grain contact during deposition of predominantly coarse particles (>20 microns) in stream or lake water produces the single-grain structure. The suspended grains settle in water under their own weight by gravitational force. The surface forces of the particles do not play any role in attracting the grains. The deposition with particle-to-particle contact may take place either in a loose state or in a dense state, and accordingly, the void ratio of the soil with single-grain structure becomes high or low.

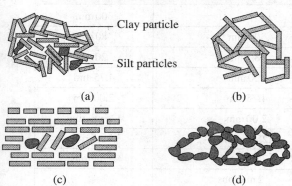

— Clay particle

— Silt particles

(a)   (b)

(c)   (d)

Fig. 6.2   Structure of soils: (a) single-grain structure; (b) flocculent structure; (c) dispersed structure; and (d) honeycomb structure

Flocculated and dispersed structures   These structures of soil are observed in clay consisting of colloids. In flocculent type, the structural arrangements are made of an assemblage of flaky inorganic colloid particles instead of individual soil particles. During deposition of the fine soil particles in lake or river water, if the net electrical force between adjacent colloidal particles in clays is sufficient to attract, flocculation will take place resulting in the flocculent structure. The chances of developing the flocculated structure are more if the dissolved materials remain in high concentration in water as in marine water. Lambe (1953) is of the opinion that flocculent structure is developed by edge-to-edge contact of the colloidal platelets. An *oriented* or *dispersed* structure will be formed if platelets come in contact with one another at the end faces.

Honeycomb structure   Several single mineral grains in a soil bonded together by contact forces form a honeycomb structure. When silt and clay particles (varying in size range between 0.2 microns and 20 microns) settle in water under gravity, the surface forces at the contact areas of the particles may overcome the gravitational force to prevent their immediate settlement over the already-deposited particles. These particles under settlement then come in close contact and are held in position until miniature arches are formed over the void spaces to make honeycomb structures. The types of soil that exhibit such structures possess high load-bearing capacity.

6.3 INDEX PROPERTIES OF SOILS AND THEIR LABORATORY DETERMINATION

The index properties of soils are those used in their proper identification. They are divided into two categories namely *soil aggregate* and *soil grain* properties. The soil aggregate properties are related to the properties of the soil as a whole and not to its constituent grains. The determination of the properties of soil as aggregate involves laboratory tests relating to its water content and specific gravity, unit weight and density, porosity and void ratio, and consistency limits. In addition, the field test method of density is also included in the discussion here.

The soil grain properties include the size and shape of the grains in soil. The size of soil grains is measured by sieve analysis (also known as *mechanical analysis*) of dry soil. However, when the grains are fine, the size is measured in a wet condition by sedimentation method after dispersing the soil particles in water. The shape of the soil grains is known from their characteristics such as angularity and roundness from visual observation of coarse grains or microscopic examination of fine particles.

Test procedures for determination of index properties of soil are mentioned briefly in the following subsections conforming to the IS.

6.3.1 Water Content

The water content of a soil is the ratio of the weight of water to the weight of solids. The common method of measuring the water content in a soil sample involves the following steps. First, a small quantity (say, 10 g to 20 g) of moist soil is taken in a clean container after accurate measurement (say, W). The soil sample is then dried in an oven for 24 hours maintaining a temperature of 100°C to 110°C and thereafter cooled in a desiccator. Then, the weight of the cooled dry sample (W_d) is taken. The difference of these two weights ($W - W_d$) gives the weight of the water of the soil sample. The percentage of water content (w) of the sample is calculated as follows:

$$w = \frac{W - W_d}{W_d} \times 100 \tag{6.1}$$

6.3.2 Specific Gravity

Specific gravity of a soil is the ratio of the weight of dry soil (γ_s) to the weight of water (γ_w) of equal volume of soil, both measured vat the same temperature (standard temperature 27°C). A density bottle is generally used in the determination of specific gravity of soil. The weight of the density bottle (W_1) is first taken. Then, the weight of the oven-dried sample (W_2) is measured after cooling in a desiccator. The difference of these two weights ($W_2 - W_1$) gives the weight of the dry sample. The bottle with soil sample is then filled up with distilled water and its weight (W_3) taken. Hence, the weight of water in the bottle with the soil sample is ($W_3 - W_2$). The material is taken out of the bottle and cleaned. The bottle is then filled with distilled water and its weight (W_4) taken. The weight of water in the bottle without soil sample is ($W_4 - W_1$). Therefore, the specific gravity (G) of soil is obtained from the following relation:

$$G = \frac{\gamma_s}{\gamma_w} = \frac{W_2 - W_1}{(W_4 - W_1)(W_3 - W_2)} \qquad (6.2)$$

6.3.3 Unit Weight

The unit weight of a soil mass is defined as its weight per unit volume. It is expressed as kilonewtons per cubic metre (kN/m³) in the metric system. Unit weights are reported as dry unit weight γ_{dry} or wet unit weight γ_{wet}.

$$\gamma_{dry} = W_d/V \cdot (\text{kN/m}^3)$$
$$\gamma_{wet} = W_s/V \cdot (\text{kN/m}^3)$$

Determination of unit weight in the laboratory involves taking the weight as well as the volume of the sample. The volume of soil solids is obtained from the weight of the solids divided by the specific gravity (G) of the solids (particles). To determine the volume of a cohesive type of soil in the laboratory, the soil sample is cut into a regular shape such as cube, then coated with paraffin, and finally the volume is taken by the water replacement method. The unit weight of soil solids (γ_s) is the weight of the soil solids (W_d) per unit volume of the solids (V_s). Thus,

$$\gamma_s = \frac{W_d}{V_s} \qquad (6.3)$$

6.3.4 Density of Soil

The density of a soil is defined as the mass of the soil per unit volume. Density may be of different types such as bulk density, dry density, saturated density, and submerged density. Determination of density requires measurement of the weights and volumes of the soil sample in the laboratory in different states such as dry, saturated, or submerged conditions depending upon the type of the density to be determined. In metric system, density is expressed in g/cm³. For calculation purposes, the unit weight of water is taken as 1 g/cm³.

The relation between unit weight (γ_s) and density (γ) is expressed as γ_s kN/m³ = 9.81 γ g/cm³.

Bulk density The bulk density (γ) of soil is the weight per unit volume. If W is the weight of a soil sample and V is the volume of the soil, the bulk density is expressed as

$$\gamma = \frac{W}{V} \qquad (6.4)$$

Dry density The dry density of soil (γ_d) is the ratio of the weight (W_d) of the soil mass in dry condition (total weight of the particles) to its volume (V) before drying. Thus,

$$\gamma_d = \frac{W_d}{V} \qquad (6.5)$$

Saturated density   The saturated density of a soil sample (γ_{sat}) is the ratio of the weight of the saturated soil sample (W_{sat}) to the volume of the saturated sample (V_{sat}). The saturated density is expressed as

$$\gamma_{sat} = \frac{W_{sat}}{V_{sat}} \tag{6.6}$$

Submerged density   The submerged density (γ') is the ratio of submerged weight of the soil solids (W_d)$_{sub}$ to the volume (V) of the soil mass and is given by the expression

$$\gamma' = \frac{(W_d)_{sub}}{V} \tag{6.7}$$

6.3.5 Porosity

Porosity of the soil is the ratio of volume of voids (V_v) to the total volume (V) of the soil mass. Similar to rocks, it is also designated by the symbol n and is expressed as

$$n = \frac{V_v}{V} \tag{6.8}$$

6.3.6 Void Ratio

Void ratio (e) of a soil sample is the ratio of the volume of voids (V_v) to the volume of the solid mater (V_s) in that soil mass and given by the expression

$$e = \frac{V_v}{V_s} \tag{6.9}$$

6.3.7 Relation Between Porosity and Void Ratio

If porosity is expressed as a percentage, it is referred to as the percentage of voids. According to Eq. (6.8), if the volume of voids in a soil mass is taken as n, then the volume of soil mass becomes equal to 1. Then we have $1 - n$ as the total volume of the soil solids. So, the void ratio of Eq. (6.9) can be expressed as

$$e = \frac{V_v}{V_s} = n\left(\frac{n}{1-n}\right) \tag{6.10}$$

Similarly, if the volume of the voids is taken as e, the volume of the soil solids according to Eq. (6.9) becomes 1 and the total volume of soil mass would be $1 + e$. So, the porosity, given by Eq. (6.8), can be expressed as

$$n = \frac{V_v}{V} = \frac{e}{1+e} \tag{6.11}$$

The size, shape, and nature of packing of the particles decide the void spaces in the soil mass. The *degree of saturation* of the soil voids depends upon the ratio of voids filled with water to the total volume of voids. A soil mass is said to be fully saturated if all the voids are filled with water. In a partially saturated soil voids, the rest of the voids are filled with air.

6.4 CONSISTENCY LIMITS (ATTERBERG LIMITS) AND THEIR DETERMINATION

The consistency of a soil refers to its *stiffness* or *firmness* that depends upon the water content of the soil. The stiffness decreases and soil loses its cohesion with increasing water. With gradual increase in water, the cohesion is reduced to such an extent that the soil mass no longer retains its shape and flows as a liquid. However, if the soil is allowed to dry, it regains its shearing strength and stiffness. While drying up, the soil mass passes through four states of consistency namely liquid, plastic, semi-solid, and solid states. The arbitrary limits set for these states in terms of water content at which the soil mass changes from one state to the next are called the consistency limits or Atterberg limits. These limits include *liquid limit* (L_w), *plastic limit* (P_w), and *shrinkage limit* (S_w) and are of significance in the engineering use of soils. Plastic limit is used in soil classification. Soils of particle size less than 425 microns are taken for conducting all tests for laboratory determination of Atterberg limits.

Liquid limit (L_w)

At the liquid limit state, the water content is such that it still possesses a small shearing strength against flowing. The liquid limit is measured in a standard liquid testing apparatus that consists of a brass cup seated on a rubber base and provided with a handle that can be raised or dropped from a standard height, see Fig. 6.3(a). About 120 g of soil is mixed with water to make a cake. The cake is kept in the brass cup making a 'V'-shaped separation such that the gap between them is 2 mm at the bottom and 10 mm at the top and 8 mm in depth, as shown in Fig. 6.3(b). Strokes are then given to the cup by rotating the handle until the two parts touch each other, see Fig. 6.3(c). At this stage, the number of strokes used is noted and a soil sample wis taken from the cake and its water content measured. Tests are repeated with different water contents such that the number of strokes required remains between 15 and 50. At least four such tests are done, each time recording the number of strokes and the corresponding water content of the soil mass. A graph prepared from the four records for water content versus number of strokes gives the L_w value, which is equal to the water content for 25 strokes, see Fig. 6.3(d).

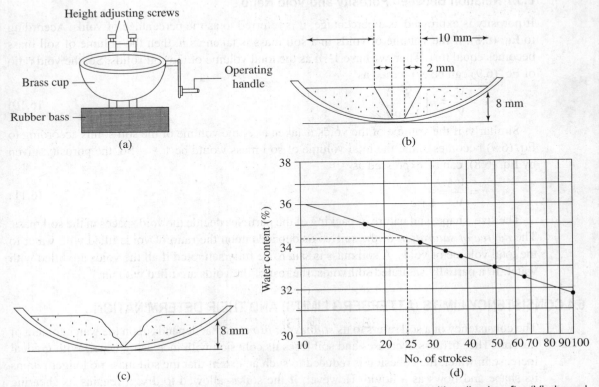

Fig. 6.3 Liquid limit test: (a) Apparatus for standard liquid test; (b) Brass cup with soil cake after 'V'-shaped separation; (c) Separated soil cake touching together after certain strokes; (d) Curve showing plots of water content versus number of strokes

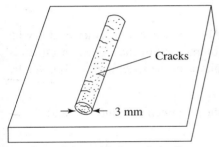

Fig. 6.4 Determination of plastic limit (P_w)

Plastic limit (P_w)

It is the minimum water content at which the soil when rolled into threads of 3 mm diameter just starts crumbling. Nearly 20 g of air-dried soil is mixed with water until it becomes plastic in nature. A small quantity (say, 8 g) of this soil is rolled on a glass plate to make a thread of uniform diameter of about 3 mm for the entire length. If no cracks are developed in it, the thread is kneaded together and the soil is rolled again. The process is continued until fine cracks just start developing on the surface of the threads (Fig. 6.4). At this point, the water content of the soil is measured which gives the plastic limit of the soil. The average result of at least three tests should be taken. *Plasticity index* is deduced from the difference in values of liquid limit and plastic limit. Thus, plasticity index $I_w = L_w - P_w$. In case of sandy soil, which is not plastic in nature, plasticity index cannot be deduced.

Shrinkage limit (S_w)

If a saturated soil sample with water content more than the shrinkage limit is allowed to dry up slowly, its volume will be gradually reduced. However, at a certain stage, further reduction in water will not change the volume of the soil sample, the pore spaces being filled with air. The water content corresponding to this stage is the measure of the shrinkage limit. To determine the shrinkage limit, a wet soil sample (Fig. 6.5) is taken in a dish and its weight W_1 is measured by subtracting the weight of empty dish from that of the dish plus soil. The volume of the sample is obtained by mercury replacement method, which involves immersing the dry sample in a container filled with mercury and then measuring the overflow portion of mercury. The volume V_1 is computed by dividing the weight (g) of displaced mercury by the density of mercury (13.6 g/cm³). Let at the state of shrinkage limit the volume attained by the soil sample be V_2 (Fig. 6.5). The dry weight (W_d) and the corresponding dry volume (V_d) of the soil sample are determined.

Soil before shrinking (wet soil) Soil after shrinking (dry soil)

Fig. 6.5 Determination of shrinkage limit (S_w)

Then, according to the definition $V_2 = V_d$.
The weight of water in original sample $= W_1 - W_d$.
Reduction (by weight) of water due to shrinkage $= (V_1 - V_2)\gamma_w$.
Here, γ_w is the density of water, that is, 1 g/cm³.
So, weight of water at shrinkage in soil $= (W_1 - W_d) - (V_1 - V_2)\gamma_w$.

Shrinkage limit (percentage) $S_w = \dfrac{(W_1 - W_d) - (V_1 - V_2)\gamma_w}{W_d} \times 100$ (6.12)

6.5 FIELD METHOD OF DENSITY DETERMINATION

The field measurement of fill or natural soil is important in engineering constructions. The test of density on fill indicates whether it has attained compaction to the extent required by the design. Field test of density on soil is conducted in borrow areas to determine its suitability for construction of earthen structures. The methods commonly followed in the field density test include water displacement method, sand replacement method, core cutting method, and rubber balloon method. These methods are described in the following subsections.

6.5.1 Water Displacement Method

A soil specimen of regular shape is collected from the field by a core cutter. It is then coated with paraffin wax to make it impervious to water. The total volume (V_t) of the waxed specimen is found by determining the volume of the specimen. The volume of the specimen (V) is given by

$$V = V_1 - (M_t - M)/P_p$$

Here, M_t is the mass of waxed solid, M is the mass of the specimen without wax, and P_p is the density of paraffin.

Dry density of specimen $= (M/V)/(1+w)$

Here, w is the water content.

6.5.2 Sand Replacement Method

The apparatus needed for the sand replacement method consists of a calibrated container, soil cutter chisel, balance, tray, and a sand pouring cylinder fitted above a sand pouring cone with a shutter. Dry and clean sand particles of size ranging between 600 micron and 300 micron are used in the cylinder. First, a flat ground of soil is chosen for conducting the test. Soil is then excavated from the place with the help of the chisel by making a test hole about the size of the calibrated container. The weight of the excavated soil is taken (W). Next, the cylinder is filled with sand and the weight of the cylinder plus sand is measured (W_1). The sand from the cylinder is then run into the calibrated container so that the volume of the sand is equal to that of the container (V). The sand-filled cylinder is then placed over the calibrated cone and the sand is allowed to pour, until it stops falling when the shutter is closed and the sand in the cone is collected and its weight (W_2) taken. The cylinder is then placed over the container and the sand is allowed to pour until no further sand runs down. The weight (W_3) of the cylinder with sand is taken. The cylinder filled with sand is then placed over the excavated hole until it fills the hole. The cylinder with the remaining sand is weighed (W_4). Thus, we have the volume of the calibrating container V.

If W' is the weight of sand needed to fill the calibrated container, then $W' = W_1 - W_2 - W_3$.

Therefore, bulk density of sand $\gamma_s = \dfrac{W'}{V}$ g/cm^3.

Weight of sand in the hole $W'' = W_1 - W_4 - W_2$

Therefore, the bulk density of soil $\gamma = \dfrac{W\gamma_s}{W''}$ g/cm^3 (6.13)

Dry density of soil $\gamma_d = \dfrac{\gamma}{1+w}$ g/cm^3 (6.14)

6.5.3 Core Cutter Method

In core cutter method of density determination of soil in the field, a cylindrical core cutter is inserted into a clean soil surface and a cylindrical soil sample having the same volume (V in cm^3) as the cylindrical soil cutter is obtained. First, the weight of the cutter plus soil is taken. Second, the weight of only the cutter is noted. The difference of the two readings gives the weight of the soil sample (W in g). Next, the water content (w) of the excavated soil is determined in the laboratory.

Thus, the bulk density of soil $\gamma = \dfrac{W}{V}\gamma$ g/cm^3 and the dry density $\gamma_d = \dfrac{\gamma}{1+W}\gamma$ g/cm^3.

6.5.4 Rubber Balloon Method

The apparatus of the balloon cone method of testing of soil includes a water-filled vertical cylinder having a bottom opening over which a rubber membrane or balloon is stretched. The cylinder is so graduated that water levels are visible to read the volumes. A small hand pump is attached so that air can be forced into the top of the cylinder. When a specimen is placed over a test hole (for test measurement), the pumped air forces the balloon and water into the hole. A typical hole is 100 mm in diameter and is 100 m deep.

The volume is determined directly by noting the water level in the cylinder before (initial) and after (final) the balloon is forced into the test hole. This difference in volume (final − initial volumes) is the volume of the hole. The water and balloon are retracted from the test hole by reversing the air pump and evacuating air from the cylinder. The outside atmospheric pressure forces the water and balloon back into the cylinder. The soil is removed from the test hole and its wet and dry weights are recorded. Then, similar to the sand cone method, the volume of the test hole and the weights (wet and dry) of the soil being known, the in situ bulk and dry densities of the soil can be calculated (see Section 6.5.2).

6.6 SIZE AND SHAPE OF SOIL PARTICLES

The size of the soil particles is determined by mechanical analysis, that is, the method of separation of particles into different size fractions by means of standard sieves. The IS sieve numbers are designated in millimetres and microns and the corresponding size of the openings (apertures) in millimetres. The American Society for Testing Materials (ASTM) sieve numbers are based on the number of meshes (openings) per inch at the sieve bottom and the apertures are in millimetres (Table 6.2).

Table 6.2 Sieve numbers and corresponding apertures of IS and ASTM sieves

IS sieve number	Aperture (mm)	ASTM sieve number	Aperture (mm)
5.60 mm	5.60	3½	5.66
4.75 mm	4.75	4	4.76
4.00 mm	4.00	5	4.00
2.80 mm	2.80	7	2.83
2.36 mm	2.36	8	2.38
2.00 mm	2.00	10	2.00
1.40 mm	1.40	14	1.41
1.18 mm	1.18	16	1.19
1.00 mm	1.00	18	1.00
710 microns	0.710	25	0.707
600 microns	0.600	30	0.595
500 microns	0.500	35	0.500
425 microns	0.425	40	0.420
355 microns	0.355	45	0.354
300 microns	0.300	50	0.297
250 microns	0.250	60	0.250
212 microns	0.212	70	0.210
180 microns	0.180	80	0.177
150 microns	0.150	100	0.149
125 microns	0.125	120	0.125

(Contd)

Table 6.2 (Contd)

IS sieve number	Aperture (mm)	ASTM sieve number	Aperture (mm)
90 microns	0.090	170	0.088
75 microns	0.075	200	0.074
63 microns	0.063	210	0.063
45 microns	0.045	230	0.044

6.6.1 Mechanical Analysis for Size Fractions

Mechanical analysis of soil sample is conducted in two parts. First, the particles of coarse-grained soil (above 75 microns) are separated into size fractions using standard sieves. Next, the particles of fine-grained soil (less than 75 microns) are separated into size fraction following sedimentation method (wet analysis). For sieve analysis, a set of sieves is arranged with the 75-micron sieve at the bottom and larger aperture sieves above it one after another in order of increasing mesh openings. A cover is always kept over the largest aperture sieve and a pan (receiver) under the bottom sieve to collect the particles passing through it. An electric shaker is used to shake the set of sieves for 15 minutes.

If the soil contains gravel-size particles, the IS 100 mm sieve is placed at the top followed by, say, 63 mm, 50 mm, 20 mm, 10 mm, and 4.75 mm sieves towards the bottom. For sand-size particles, the 4.75 mm sieve is placed at the top followed by the other sieves towards the bottom in the following order: sieve numbers 4.00 mm, 2.00 mm, 1.00 mm, 500 microns, 300 microns, 180 microns, 125 microns, and 75 microns. The weights of the individual fractions indicated by the 'passing through' sieves are noted. The percentage of each fraction is then calculated from the total weight of all the fractions passing through the set of sieves. The weight of the fraction in the pan is also to be noted and added to get the total weight of the sieved sample.

6.6.2 Wet Analysis (Sedimentation Method)

Today, there are sophisticated automatic instruments for measuring the size of fine-grained particles under wet condition. Computerized automatic size measuring instruments for particles based on laser beam techniques are used in the geotechnical laboratories of Geological Survey of India. The sedimentation method is undertaken in particles of fine-grained soil (< 75 microns). It requires a graduated cylinder and a pipette or a hydrometer. The method involves mixing suitable amount of the soil (W g) with a liquid (generally water) to make a known volume of the mix (say, V) in a graduated cylinder. The mix (soil and liquid) is agitated so that the particles remain in suspension. With the help of a pipette, a sample of the mix is drawn from a fixed height (h cm) after any time interval (t minutes). If the diameter D of the particles drawn have weight w, then the weight percentage (n) is given by

$$n = \frac{w}{W/V} \times 100 \tag{6.15}$$

For different diameter (D) values, there will be different weight percentage (n) values. The D values are determined using Stokes' law. The settling velocity (v) of the material obeys Stokes' law, which is based on the shape, weight, and size of the particles. According to this law, the relatively coarser particles will settle first followed by relatively finer grains. In Eq. (6.16) derived from Stokes' law, it is assumed that the particles are spherical in shape and all have the same specific gravity (G). Using water (viscosity η) as the liquid medium for suspension, it can be deduced that the settling velocity (v) and particle diameter (D) falling from a height of h cm in t minutes have the following relationship:

$$v = \frac{h}{60t} \text{ cm/s} \tag{6.16}$$

$$D = \sqrt{\frac{1800\eta v}{G-1}} \text{mm} \tag{6.17}$$

Combining Eqs (6.16) and (6.17), we have

$$D = 10^{-5} M \sqrt{\frac{h}{t}} \tag{6.18}$$

where the factor $M = 10^5 \sqrt{\dfrac{30\eta}{G-1}}$

The values of M for different values of specific gravity (G) with viscosity of water (η) at different temperatures (°C) are obtained from the standard table. So, the different size (D) values and the corresponding weight per cent (n) values are obtained from Eqs (6.18) and (6.15), respectively.

6.6.3 Size Distribution Curve of Soil Particles

The data obtained from the size analyses of soil particles is plotted on a logarithmic graph paper to obtain the *size distribution curves*, which are also known as *cumulative curves*. The particle size (in millimetres) is plotted as the abscissa and the percentage of particles 'finer than' by weight as the ordinate (Fig. 6.6). The nature of the curves clearly brings out the soil characteristics including the coarseness or fineness of the soil particles, uniformity of the grain size, and the nature of gradation. For example, a soil becomes *well graded* (curve c, Fig. 6.6) when fine particles fill the voids among the coarse particles. As such, well-graded soil will have a mixture of fine and coarse particles and minimum void space. In *poorly graded* soil or *uniformly graded* soil (curve e, Fig. 6.6), the particles of the same size occur in dominance over the other or grains of certain range may be nearly or completely absent.

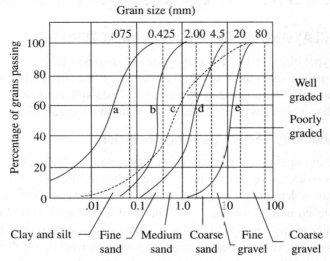

Fig. 6.6 Cumulative curves showing distribution of size and nature of gradation of soil particles

Some values of size distribution curves such as D_{10}, D_{30}, and D_{60} are used to show the soil characteristics by means of their ratios. *Uniformity coefficient* (C_u) of the soil is given by the ratio of size D_{60} and the effective size D_{10}. Thus,

$$C_u = \frac{D_{60}}{D_{10}} \qquad\qquad (6.19)$$

The value of C_u is nearly unity in very poorly graded soil. It is less than six for poorly graded sandy soil and less than four for poorly graded gravelly soil. The gravelly soil estimated to have a C_u value of 3.1 (curve a, Fig. 6.6) is poorly graded. The pattern of the size distribution curve is expressed as the *coefficient of gradation* or *coefficient of curvature* (C_g) and is given by

$$C_g = \frac{(D_{30})^2}{D_{10} D_{60}} \qquad\qquad (6.20)$$

The value of C_g lies between one and three in well-graded soil. In addition, C_u must be more than six in well-graded sands and more than four in well-graded gravels. The estimated C_g and C_u values of the sandy soil (curve c, Fig. 6.6) are 1.8 and 20, respectively, and as such it is a well-graded sand or sandy soil.

6.6.4 Shape (Angularity and Roundness) of Soil Particles

Angularity and roundness of coarse soil particles (gravel and sand) depend upon the nature of the soil deposit as well as the mineral constituents. Residual soil deposits consisting of rock fragments, gravels, and coarse sand are mostly angular in shape. However, in transported soil, when water acts as the transporting agent (i.e., alluvial deposits), the grains become rounded due to abrasion and rolling. The particles of transported soil of river origin may be subrounded to angular if the place of origin is nearby as the degree of wear is less. Beach sands are also rounded due to the abrading action of the particles under movement by waves. Marine sand (soil formed of deposits in marine water) is, however, mostly angular. The ice-worn coarse sands are mostly flat and generally angular. The Aeolian (wind-blown) sands are fine grained but rounded. In the fine-grained soil, there are abundant scale-like and flake-shaped minerals such as micas. These fine-grained minerals, if soft in nature, are subjected to further breakage and become finer but retain their flaky shape.

6.7 SWELLING CLAY AND ITS EXPANSIVE CHARACTERISTIC

Certain types of soil, especially clayey soil, when soaked with water may increase in volume. This characteristic is essentially due to the presence of swelling type of clay-like montmorillonite, which can hold a large quantity of water and then swell many times its volume. In fact, this clay mineral has a high absorbing capacity of water that increases its volume.

6.7.1 Laboratory Test for Swelling Coefficient

Swelling or expansion characteristic of clayey soil or clay is related to its plasticity. Clay with plasticity index above 35 is found to have high swelling. Whether a clay sample is of swelling type or not is determined by a simple test in the laboratory. The clayey soil collected from the field is sieved in a 45 micron sieve. The portion of fines (clay) that remains in the pan below the 45 micron sieve is collected. Then, a measured quantity (say, 10 ml to 20 ml) of this clay sample is taken in a graduated cylinder and dispersed in 1,000 ml water. The water-mixed clay sample in the cylinder is kept undisturbed for 24 hours to find the change in volume. Any observed increase in volume indicates that the clay is of swelling type. The percentage increase in volume with respect to the original volume is the measure of swelling.

If L is the original volume of the sample inside the cylinder and L' is the volume of the sample after swelling, the swelling coefficient is expressed as

$$S = (L' - L)/L \times 100$$

6.7.2 Measurement of Expansion Pressure

The pressure exerted by expansion of swelling type of clayey soil or clay can be measured in a consolidometer or proving ring apparatus. The specimen of soil is compacted in a Proctor mould and then cut into the size of the cell in which it is to be placed for the test. The cell of the fixed type of consolidometer contains two porous stones, the lower one being fixed and the upper one movable (see Section 6.8). The specimen is allowed to soak in water. As the specimen starts to swell, a very small increase in expansion in volume is subjected to load until it comes to its original position. The load is applied by operating the spindle of the proving ring apparatus or the trolley head of the consolidometer. A curve is plotted for the swelling percentage versus applied load (in tonnes) from the results of three such tests. This helps to understand the soil behaviour under different loads. The moisture content and the density of the specimen are measured before and after the test. This helps to determine the moisture content and density at which the expansion will be minimum. In practice, the expansion pressure can be reduced by the use of a layer of non-expansive material overlying the swelling clay in foundations.

6.8 SOIL PERMEABILITY

Permeability of a soil is defined as the capacity of the soil to allow flow of water through its voids. When a porous media such as sand permits water to pass through the interconnected voids or pore spaces, it is called *permeable*. However, clay, though highly porous, is said to be *impermeable* as the voids are not interconnected and as such water cannot pass through it. Movement of water through soil exerts pressure known as *seepage pressure*. Permeability of soil is of importance in the study of groundwater flow through wells, seepage through the body of earth dams, piping of hydraulic structures and safety measures, and drainage and associated problems related to the stability of soil slopes.

6.8.1 Flow Characteristics and Darcy's Law

The flow characteristic of water through a soil mass is illustrated in Fig. 6.7. Water entering in pervious soil travels along smooth curves known as *flow line*. If the lines are parallel, the

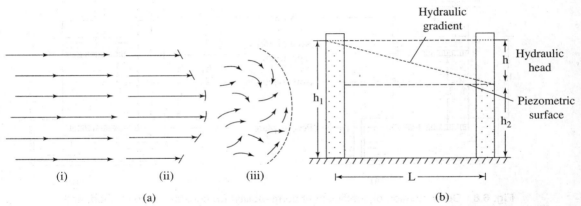

Fig. 6.7 Flow characteristic of water through soil: (a) flows of different types (i) linear, (ii) laminar, and (iii) turbulent; and (b) hydraulic gradient, $i = h/L$

flow is said to be *linear flow*. If the flow follows a definite path, it is called *laminar flow*, but if the flow paths are irregular and cross at random, it is known as *turbulent flow*, see Fig. 6.7(a). The water levels in the two tubes inserted in a water-bearing soil rise to a height of h_1 and h_2 known as *piezometric heads*. The difference in the piezometric heads $(h_1 - h_2)$ between the two is known as the *hydraulic head* (h). The ratio of hydraulic heads to the distance (L) between the two points is known as the *hydraulic gradient, i*. Thus, $i = h/L$, see Fig. 6.7(b).

Permeability of a soil is measured by the *coefficient of permeability*, which varies depending on the type of soil. The greater the coefficient of permeability, the more permeable is the material. In laminar or non-turbulent flow conditions, the permeability of soil obeys Darcy's law, which states that the rate of flow is proportional to the hydraulic gradient. Thus,

$$v = ki \tag{6.21}$$

Darcy's law can also be expressed as $q = kia = va$, or $v = \dfrac{q}{a}$ (6.22)

where v is the velocity of flow, i is the hydraulic gradient, k is Darcy's coefficient of permeability, q is the discharge per unit time, and a is the cross-sectional area of soil mass normal to flow direction.

Thus, coefficient of permeability (or simply, permeability) is the average velocity of flow through the total cross-sectional area of soil mass under unit hydraulic gradient. Permeability is dependent on various factors such as the size and shape of the soil particles, nature of fluids that flow through the pores, void ratio of soil mass, adsorbed water in soil, structural arrangement of the particles, and presence of organic matter and other admixtures.

6.8.2 Laboratory Determination of Soil Permeability

Coefficient of permeability is determined in the laboratory by constant head method for coarse-grained soil, which is generally very permeable, and by falling head method for fine-grained soil, which is less permeable. The apparatus used in the permeability test is known as permeameter. In both the methods, the permeameters used allow flow of water through the soil and establish a hydraulic gradient within the sample. In the constant head method, see Fig. 6.8 (a) when the steady state of flow is obtained, the quantity of flow (Q in ml) is recorded from

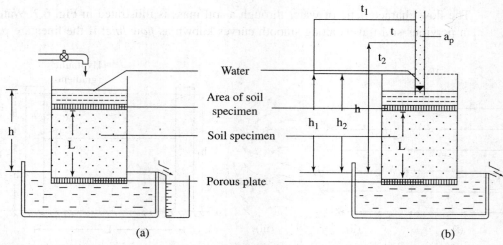

Fig. 6.8 Determination of coefficient of permeability: (a) constant head method; and (b) falling head method

the measurement of the water collected in the bottom vessel and the time (t in seconds) taken for this quantity of water flow in the vessel is noted. If L is the height of the specimen, gradient $i = h/L$. The values of specimen height L (cm), its cross-sectional area a (cm^2), and the head of water h (cm) being known, the coefficient of permeability k (cm/s) can be computed. According to Darcy's law, we have

$$q = Q/t = kia$$

or

$$k = \frac{Q/t}{ia} = \frac{Q/t}{ha/L} \tag{6.23}$$

The apparatus for measuring permeability by falling head method consists of a standpipe (cross-sectional area a_p cm^2) through which water can be allowed to flow down to the soil sample, see Fig. 6.8(b). The head of water is measured from the difference of water levels between the standpipe and the bottom tank. When the steady state of flow is attained, the head h_1 at the time instant t_1 is recorded. After an interval of time, height h_2 and the corresponding time t_2 are again noted (when $t_1 < t_2$). Now, consider the head h at any intermediate time interval t. Then, the coefficient of permeability k is computed following Darcy's law on the basis of the change in head (in cm) in a small time (in seconds) interval. The height L (cm) and cross-sectional area of the sample a (cm^2) being known, the coefficient of permeability k cm/s is computed as follows:

$$k = \frac{a_p L}{at} \log_e \frac{h_1}{h_2} = 2.3 \frac{a_p L}{at} \log_{10} \frac{h_1}{h_2} \tag{6.24}$$

Source of errors in laboratory test for permeability   Several types of errors could affect the determination of permeability by laboratory test. The air trapped in the sample may prevent it from getting fully saturated. The soil sample is also disturbed by flowing water, a portion of soil being washed away from the sample during the test. Formation of filter skin of fine materials on the surface of the sample and segregation of air in the form of bubbles within the sample are other sources of error. The most important factor is that the soil sample used for the laboratory test is not truly representative of the field condition, and as such, determination of coefficient of permeability of soil by field test (pumping test) is favoured.

6.8.3 Determination of Field Permeability of Soil by Pumping Test

Unlike the laboratory test, the result of the field permeability test of soil represents the overall permeability of the soil covering a large area. According to Terzaghi and Peck (1967, a reliable value of permeability can be obtained by *pumping test* through a test well and selected number of observation wells, see Figs 6.9(a) and (b). Permeability is determined by pumping water from the test well. Initial ground water levels in the wells are observed in all the holes for a sufficient time to find the nature of fluctuation. Pumping is started at a uniform rate of discharge (Q) until equilibrium position is reached. Initially, the ground water level remains horizontal, but on continuing the pumping, the water level descends creating a *cone of depression*. The rate of descend in levels (drawdown) corresponding to a steady discharge is recorded. For the *confined aquifer* shown in Fig. 6.9(a) the value of permeability k is calculated from the following equation:

$$k = \frac{Q}{2\pi H_0 (h_2 - h_1)} \log_e \frac{r_1}{r_2} \tag{6.25}$$

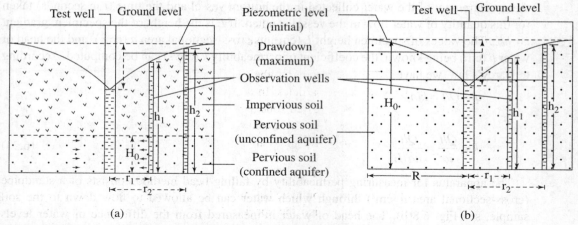

Fig. 6.9 Field permeability test: (a) confined aquifer; and (b) unconfined aquifer

Here, h_1 and h_2 are the depths of water levels in the observation wells for the corresponding radial distance r_1 and r_2 $(r_2 > r_1)$ and H_0 is the thickness of the confined aquifer, see Fig. 6.9(a).

In *unconfined aquifer* shown in Fig. 6.9(b) the value of k can be calculated from the following relation:

$$Q = \frac{\pi k(h_2^2 - h_1^2)}{\log_e \dfrac{r_2}{r_1}} = \frac{1.36k(h_2^2 - h_1^2)}{\log_{10} \dfrac{r_2}{r_1}} \tag{6.26}$$

Here, as before, h_1 and h_2 are depths of water level in the observation wells and r_1 and r_2 are the corresponding radial distances from the test well. It has been assumed that the flow is laminar, the aquifer homogeneous, and flow of water uniform in nature throughout the aquifer. It is also assumed that the velocity of flow of water is proportional to the hydraulic gradient and the wells penetrate to the entire depths of pervious strata through which water flows to the wells. The degree of permeability varies widely depending upon the type of soil or the unconsolidated material through which water flows. It is high in clean gravel, medium in coarse sand, low in fine sand or silt, and very low to practically nil in clay.

6.9 CONSOLIDATION OF SOIL

A thick deposit of clay compressed by the load of superincumbent materials through geologic time gets hardened with reduction in volume and becomes shale. In a similar way, when a saturated and compressible soil (such as clay) is kept under static load for a long time, its volume decreases due to *compression*. The compressive load squeezes the pore fluids and reduces the pore spaces causing an overall reduction of the volume of the soil mass. Such process of compression resulted from long-time loading to cause decrease in soil volume is termed *consolidation*.

Compression of clay under increase in pressure proceeds very slowly as the permeability of clay is very low. In consolidation at constant load, the decrease in water content is caused by the gradual adjustment of soil particles and thus reduction in volume is slow. After the initial or primary stage of consolidation, increment in load causes excess *pore water pressure* or *hydrostatic pressure* because of which water gradually escapes from the voids. The increment of pressure causes increase in effective pressure on soil solids and the volume of the soil mass gradually decreases. At the end of the primary consolidation stage, the effective pressure (also called *consolidation pressure*) in the soil mass ceases and no further drain of water occurs and a condition of equilibrium comes into existence. The void ratio changes according to the varied consolidation pressures.

The primary consolidation that decreases the soil volume is attributed principally to the decrease in void spaces between soil particles. In primary consolidation, compression refers to the processes where particles in the stressed zone are rearranged into a more compact or tighter configuration as void water is squeezed out. Secondary compression is the additional compression that occurs at a constant value of effective stress after excess pore water has been dissipated.

6.9.1 Laboratory Test of Consolidation

In the laboratory, consolidation test is conducted using a *consolidometer* that consists of a loading frame and a consolidation cell in which the specimen is placed between two porous stones (Fig. 6.10). The porous stones enable the squeezing out of water from the sample during testing. Both the porous stones placed at the two ends of the specimen take part in the compression of the specimen. (In the fixed ring type cell, only the top porous stone moves down when the specimen is subjected to vertical pressure [compression].) To start with, an initial pressure, say, 0.1 kg/cm², is applied, which is gradually increased to 0.2, 0.5, 1, 2, 4, 5, 8, and finally 10 kg/cm². For each increase in pressure, the compression is read out from the dial gauge fitted with the machine at intervals of 0.25, 1.0, 2.5, 4, 6, 9, 12, 20, 25, 36, 49, and 60 minutes, and then 2, 4, 8, and 24 hours. After the completion of consolidation by applying the maximum pressure, the specimen is removed and the weights of the soil solids and the water content are determined. The dial gauge reading for rebound after removal of the pressure (after at least 24 hours) is also recorded. The test data is used to calculate the void ratio and the coefficient of volume change.

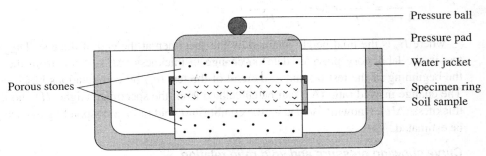

Fig. 6.10 Consolidation test apparatus (consolidometer)

6.9.2 Calculation of Void Ratio and Coefficient of Volume Change

Void ratio can be calculated by two methods namely the height of solids method and the change in void ratio method.

Height of solids method

Let the diameter of the soil specimen be a, the weight of the dry specimen W_d, and the specific gravity of the soil G. Then, the height (H_s) of the soil solids is calculated from the following expression:

$$H_s = \frac{W_d}{Ga}$$

(6.27)

The void ratio is calculated from the relation:

$$e = \frac{H - H_s}{H_s} \qquad (6.28)$$

where, H is the specimen height at equilibrium under various applied pressures. This is given by the relation

$$H = H_0 + \Sigma \Delta H = H_1 + \Delta H$$

Here, H_0 is the initial height of the specimen, H_1 is the height of the specimen at the beginning of the load increment, and ΔH is the change in the thickness of the specimen for each pressure increment.

The value (H_s) of dry soil specimen is obtained from Eq. (6.27) using the specific gravity, dry weight, and diameter of the soil specimen measured in the laboratory. The pressure increment and the corresponding change in height are noted in a tabulated form from the dial gauge readings of the consolidation test, which will give the value of H for different pressure increments ($H = H_1 \pm \Delta H$). As the values of H and H_s are known, the void ratio, e, can be computed from Eq. (6.28).

Change in void ratio method

In the case of completely saturated soil, the final void ratio (e_f) at the end of the test is estimated from the relation:

$$e_f = w_f G \qquad (6.29)$$

Here, G is the specific gravity and w_f is the final water content at the end of the test. The change in void ratio Δe under each pressure increment can be computed from the following relationship:

$$\frac{\Delta_e}{1 + e} = \frac{\Delta H}{H} \text{ or } \Delta e = \frac{\Delta H (1 + e_f)}{H_f} \qquad (6.30)$$

where H_f is the final height attained by the specimen at the end of the test. The consolidation test in the laboratory provides data on change in thickness (ΔH) of soil from the void ratio at the beginning of the test (e) and at the end of the test (e_f) by computations based on Eq. (6.30). The change in void ratio (Δe) can be calculated from the specimen height (H) and the change in thickness (ΔH). Knowing Δe and e_f, the equilibrium void ratio corresponding to each pressure can be estimated.

Curve showing pressure and void ratio relation

The data on consolidation test is plotted on a graph paper taking the increment of pressure (p) as abscissa and the corresponding void ratio (e) as ordinate to obtain the *virgin compressive curve*, shown in Fig. 6.11(a). When plotted on a semi-logarithmic paper, the virgin compression curve is obtained as a straight line, see Fig. 6.11(b) which can be expressed by the empirical relationship given by Terzaghi 1967 as follows:

$$e = e_0 - C_c \log_{10} \delta'/\delta_0'$$

Here, e_0 is the initial void ratio at initial pressure δ_0', e is the void ratio at increased pressure δ', and C_c is the corresponding index.

The theory of consolidation is complex in nature and a textbook on soil mechanics (*Soil Mechanics in Engineering Practice*, Karl Terzaghi, Ralph Brazelton Peck Wiley, 1967 on this aspect may be thoroughly studied before using it to solve practical problems.

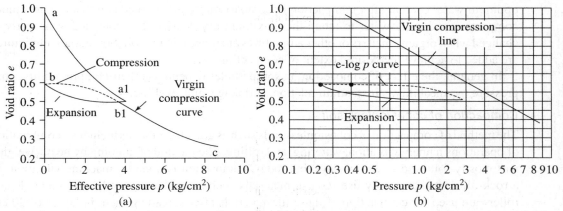

Fig. 6.11 Consolidation test result: (a) curve showing void ratio–pressure relationship of remoulded specimen; and (b) curve showing void ratio–pressure relationship in semi-logarithmic plot

6.10 SOIL COMPACTION

Compression of soil accomplished by loading for a short period is known as *compaction*. It is a quick process of densification and stabilization of soil. A soil mass loses its compactness when it is excavated from its natural source and deposited elsewhere. As such, the soil of borrow areas brought for construction of earth-fill structures requires to be compacted. Compaction is a process by which the soil particles are rearranged and closely packed, thus resulting in a decrease in the void ratio and an increase in density. The extent of densification of soil is dependent on the amount of compaction. Increase of compaction leads to an increase in dry density and a decrease in water content.

6.10.1 Process of Compaction

Soil from the borrow areas is deposited at the earthwork site where it is compacted by rollers or other machineries until the desired result with respect to density and water content is achieved. In general, if the soil is very dry it is sprinkled with water before compaction to make it wet to obtain the desired compaction. Alternatively, if the soil is wet, it is piled near the construction site to allow it to dry and reduce the water content before its placement at the site for compaction. Laboratory test is conducted on the compacted soil of the work site to check whether the water content and density have achieved the compaction as per design requirement. It has been found that soil achieves the highest compaction and maximum density with a specific water content termed as *optimum water content*.

6.10.2 Compaction Machineries and Their Performance

Compaction of soil in earthwork sites utilizes mainly three types of compaction equipment, namely *rollers*, *rammers*, and *vibrators*. Rollers may be of several varieties such as the pneumatic tyred rollers, smooth wheel rollers, and sheep-foot rollers. Rammers used in field compaction of soil may be pneumatic or internal combustion types. Vibrators comprise vibrating units of pulsating hydraulic types moulded on rollers or plates, or out-of-balance weight types.

Performances of the equipment as regards compaction depends on several factors such as the nature of soil including the grain size, water content, nature of admixture, and the degree of compaction. Compaction increases the dry density but the extent of increase depends upon the type of soil. Coarse-grained and well-graded soils attain maximum dry density with lower optimum water content after compaction in comparison to fine-grained soil that requires more compaction and more water to attain the maximum dry density. The compaction property is changed when organic matters or other admixtures are present in the soil. Sometimes admixtures are added to achieve the desired extent of compaction.

In general, the effect of compaction on soil particles depends on their compressibility, pore pressure, permeability, stress–strain relation, shrinkage, and swelling.

Compaction of cohesionless soil

The method of compaction of cohesionless soils such as gravels and sands includes a combination of operations namely vibration, watering, and rolling. The method of tamping by heavy weights was mostly followed in earlier days, but today, compaction is done by machine vibration. In a rock-fill dam that mostly uses coarse materials such as cobbles, gravels, and sands, heavy rollers are used for compaction of these fill materials after spreading them in layers of 20 cm to 30 cm thickness. Most fills are compacted by pneumatic tyred rollers. It may be required to add water during compaction to eliminate the capillary forces in soil grains. In clear sand, water drains out easily and compaction may not be effective to the desired extent.

Compaction of cohesive soil

In cohesive soils such as clayey silt or in admixture of gravel–sand–silt–clay, compaction can be well effected by pneumatic-tyred rollers. The cohesive fills for buildings are compacted in 15 cm-to 30 cm-thick layers by heavy rollers of about 25-tonne capacity, whereas for embankments and dams, heavier rollers of capacity up to 50 tonnes are used. The degree of compaction is mainly dependent on the nature of moisture content. The effectiveness of compaction of soil under varying moisture contents is known from the changing density. The relationship between water content and density of a cohesive soil guides the engineers to decide the thickness of layers and the types of rollers to be used to obtain 100 per cent compaction at optimum moisture content.

Compaction of in-situ soil

The in-situ soil is compacted by piles. Subsidence of loose sand associated with vibration at considerable depth may take place due to pile driving. When piles are driven in such soil consisting of loose sand from the ground to considerable depth, the soil gets compacted causing reduction in volume and decrease in porosity to a great extent (30–40 per cent). Sandy soil of small to medium cohesion can also be compacted by pile driving. In this soil, the compaction is achieved not by vibration but by static pressure. Some types of vibrators combined with water jet are also effective in bringing about compaction of sandy deposits in natural state. Pile driving below water table may result in some viscous mass instead of compaction. In such cases, the method of compaction by preloading may be adopted.

Compaction of organic soil, silt, and clay

Organic soil, silt, and clay that are not easily compressible can be compacted by preloading. The method includes covering the area of these soils by a fill having higher weight per unit area sufficient to compress the soil to the required extent. Silty soil with lenses of sands gets easily compacted but more pressure is required for organic soil without sand lens due to its impermeable nature. Some drains are provided before preloading in organic and clayey soils to quicken the process of compacting. A pervious drainage blanket is also laid to allow escape of water from the soil to the blanket. Though water percentage is high in clay, permeability is very low to practically nil. Clay is excavated in chunks to a place on the site of the earth structure for use as a construction material. Neither vibration nor pressure of short duration can change the water content.

6.10.3 Standard and Modified Proctor Test

Proctor test is conducted in the laboratory to decide the relation of density with water content. The main purpose of the test is to measure the optimum water content at which the soil is to be compacted to attain maximum dry density. The equipment recommended by BIS for Standard Proctor test includes a cylindrical metal mould of 10 cm diameter, height 12.7 cm, and volume 1,000 ml, a detachable base plate, a collar of 6 cm effective height, and a rammer of 2.6 kg to fall from a standard height of 31 cm. The test is conducted in samples of dry soil of less than 4.75 mm size. The test sample of soil with certain water content is compacted in the mould in three equal layers with 25 blows by the standard rammer. The dry density of the compacted soil is determined for the water content and weight of the sample. Several tests are conducted on the samples with varying water contents and the corresponding dry densities are determined. The bulk density γ for the corresponding dry density γ_d of each sample is then computed from the following relation:

$$\gamma = \frac{W}{V} \text{ g/cm}^3 \text{ and } \gamma_d = \frac{\gamma}{1+w} \text{ g/cm}^3 \tag{6.31}$$

Here, V is the volume of the mould (1,000 ml), w is the water content, and W is the weight (g) of the wet compacted soil.

A *compaction curve* for a soil is drawn from the plots of test data with water content percentage as abscissa and dry density as ordinate, as shown in Fig. 6.12(a). Dry density increases with decrease in water content until the maximum dry density (γ_d) is attained at a point 'X' in the ordinate corresponding to *optimum water content* (w_o) in the abscissa. The maximum compaction is reached under a specific water content when the air voids will be nil. A line *called zero air void line* showing the dry density as a function of water content for soil with no air voids is established by the following relation:

$$\gamma_d = \frac{G\gamma_w}{1+wG} \text{ g/cm}^3 \tag{6.32}$$

where w is the water content corresponding to specific gravity G of the soil.

The modified Proctor test is based on compaction that is about 4.5 times higher than the Standard Proctor test. This test is carried out to simulate field conditions for soil in areas where heavy machineries are used in the construction such as earth dams or where there is a movement of heavy vehicular transport as in an airport. According to IS, in this test, instead of three layers, five layers are compacted with 25 blows from a 4.89 kg rammer dropped from a height of 45 cm. Due to such higher compaction of soil in the modified Proctor test, the water content decreases and the soil density increases. As can be seen from the two curves given in Fig. 6.12(a) maximum dry density is higher in modified Proctor test with less water content for the same soil. The maximum water content corresponding to maximum dry density is known as optimum water content.

6.10.4 Proctor Needle and Its Use

Proctor needle is a handy instrument used for rapid measurement of the water content in a field. The instrument consists of a spring-loaded plunger that holds a graduated needle shank attached to a needle tip at its lower end. The other end contains a calibrated stem with a handle for operating the instrument, see Fig. 6.12(b). When the needle tip of certain cross-sectional area penetrates the soil, a resistance is offered by the soil that is measured by a gauge attached to the handle. The depths of penetration of the Proctor needle of different cross-sectional areas are calibrated with respect to soils with different water contents.

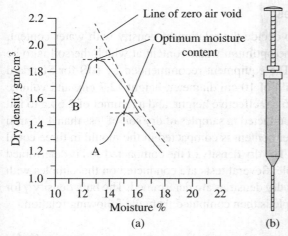

(a)

(b)

Fig. 6.12 (a) Compaction curves for Standard Proctor test (curve A) and modified Proctor test (curve B); and (b) Proctor needle

A curve is drawn on the result of such tests. In the field, the penetration resistance of the compacted soil is measured by inserting the Proctor needle and the reading of the water content is recorded from the calibration curve. The field density of the compacted soil is determined following the method stated in Section 6.5. The extent of compaction attained in the field is measured as per cent compaction, which is the ratio of field dry density to that obtained by the laboratory compaction test. At present, battery-operated soil 'compaction meters' are available, which directly measure and record the resistance data as they are inserted into the ground. Data is recorded and displayed for the depth of penetration of the plunger or needle at 1 inch (2.5 cm) intervals in psi or kPa.

6.11 EARTH PRESSURE AND RETAINING STRUCTURES

If a cut slope in a soil is very steep, it will not stand unsupported. A retaining structure is, therefore, required to resist the lateral thrust of the soil mass behind the structure. This soil mass is called *backfill* and the thrust is known as *lateral earth pressure*. In the case of *active* earth pressure, the retaining structure moves away from the backfill and a portion of the backfill may break apart keeping the remaining backfill portion at rest, see Fig. 6.13(a). The backfill portion that tends to move apart is wedge shaped and is called *failure wedge*. The portion at the top of backfill above a horizontal plane is called the *surcharge* and the slope angle with respect to horizontal is the *surcharge angle*. At the active state of earth pressure, the moving force is at the minimum. However, if maximum force is mobilized to bring failure of the soil mass or backfill, it is called *passive* earth pressure, see Fig. 6.13(b).

While on the verge of failure, the soil mass is in a plastic state of equilibrium. Rankine's theory (1857) deals with this state of plastic deformation considering the soil to be non-cohesive homogeneous in nature. The theory assumes that there is no shearing stress between the soil and the wall. However, concrete and masonry structures acting as retaining walls have rough surfaces that develop frictional forces, which are non-existent in Rankin's assumption. The existing friction makes the resultant pressure act inclined to the normal to the wall at an angle between the wall and the soil. This angle is known as the *friction angle*.

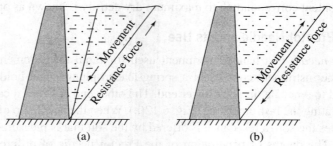

(a)

(b)

Fig. 6.13 Earth pressure: (a) active earth pressure; and (b) passive earth pressure

6.11.1 Retaining Structures

The retaining structure namely retaining wall, is constructed to save a vertical or very steeply excavated soil mass. It is mainly used against steeply inclined soil cuts made for highways and railways. It is also used as the side wall of embankment fills and bridge abutments and for providing support to the steep sides of industrial structures. The retaining wall may be of several types such as gravity, cantilever, buttress, and crib walls. The gravity retaining wall may be plain concrete, masonry, or timber having no tensile stress in any portion of the wall, see Fig. 6.14(a). The cantilever wall generally made of reinforced concrete appears like an inverted 'T', each projecting part acting as a cantilever, see Fig. 6.14(b). This type of structure is suitable for medium height (6–7m) retaining walls. There is a subtle difference between buttress wall and counterforts. When the counterforts are provided in the toe side, the retaining wall is called a buttress wall. So, in the counterfort type, which is suitable for high walls (>6 m), both the walls span horizontally between vertical brackets, see Fig. 6.14(c). Sometimes cantilevered walls are buttressed on the front, or include a counterfort on the back (soil side), to improve their strength resisting high loads. The crib walls, suitable for medium high structures, are made of timber or pre-cast concrete or pre-fabricated steel frames and filled with granular soil.

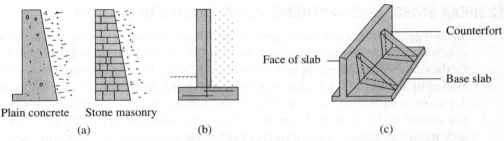

Fig. 6.14 Retaining structures: (a) gravity walls; (b) cantilever wall of reinforced concrete; and (c) counterfort wall

6.11.2 Backfill—Materials Used and Drainage

The materials used for backfill behind the retaining wall are pervious and semi-pervious types of soil including clean sand or crushed rock (Fig. 6.15). The GW and SW types of soil (see Section 5.7, Table 5.2) are considered good and the GP and SP types of soil satisfactory for backfill. The granular material, however, requires compaction to reduce the lateral pressure and increase the angle of internal friction. The compaction is done by vibration. Compaction done in 15 cm to 20 cm layers is sufficiently dense. In areas free from frost action, GM, GC, SM, and SC types of soil are also used as backfill behind retaining walls, but require elaborate drainage arrangements. In general, soil containing more than five per cent fine particles is not desirable as this prevents good drainage.

The *weep holes* are provided to the retaining walls to drain out the seepage of water from the backfill using perforated pipes which are inserted into the retaining walls, see Fig. 6.15(a). When the backfill material is pervious, weep holes of 15 cm in diameter are placed horizontally at intervals of 2 m to 5 m. A filter of granular materials is provided at the entrance of each weep hole to avoid clogging by fines. Water seeping through the weep holes may fall on the toe causing weakness of the retaining wall. To prevent this, a pipe drain is provided along the base wall in addition to the filter material, as shown in Fig. 6.15(b). If the backfill is of semi-pervious type, a vertical strip of filter material is provided midway between the weep holes in conjunction with the horizontal strip of filter material, see Fig. 6.15(c). When the backfill is fine-grained, nearly 30 cm-thick blanket of pervious material is used at the top, as shown in Fig. 6.15(d).

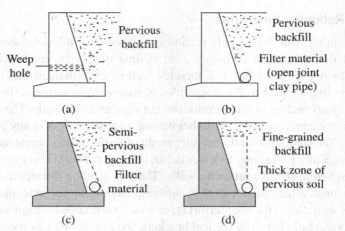

Fig. 6.15 Different types of backfill: (a) pervious backfill; (b) pervious backfill with filter material; (c) semi-pervious backfill; and(d) fine-grained backfill

6.12 SHEAR STRESS, SHEAR STRENGTH, AND FAILURE MECHANISM

Stress is generated in a soil mass when it is loaded by a structure. The weight of the structure exerts a vertical force, but another force called *shear stress* is also developed in the soil and is responsible for the failure of structures. For example, a retaining wall to a soil slope may fail due to the pressure exerted by the backfill. A loaded earth mass such as an earth embankment may slide due to the stress within the soil that forces a wedge of soil to move out. In all these cases, it is the shear stress that acts on soil and tends to deform the soil mass eventually bringing about its failure. A soil body is on the verge of failure when it reaches the *limiting shear stress* that equals the *shear strength* of the soil. The shear stress at this stage is in a *state of equilibrium* with the shear strength. Any further increase in shear stress, or decrease in shearing resistance, will disturb the state of equilibrium and bring about the failure (called *shear failure*) of the soil body. Failure of a soil mass may also take place under differential settlement if loads are not uniformly distributed on the surface.

The maximum shear stress that can be sustained by a soil mass before failure is the shear strength. It is also stated as the *shearing resistance* of the soil body to deformation or to movement. The shear strength of soil mass is constituted mainly of *cohesion* and *angle of internal friction* of soil particles. Cohesion is a measure of the forces that cement the particles of soil. Angle of internal friction is the measure of the shear strength of soil due to friction. In cohesive soil, both cohesion and friction act as shear resistance, but in non-cohesive soil, only friction provides shear resistance.

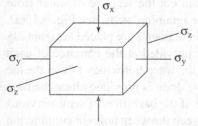

Fig. 6.16 Principal stresses in a soil mass

6.12.1 Mohr–Coulomb Shear Failure Criterion

In a loaded soil mass, there are innumerable planes along which stresses occur. Of these, there are three typical planes mutually perpendicular to each other on which the stress is wholly normal and no shear stress acts. These are the three principal planes and the stresses that act on these three planes are designated as *major principal stress* (σ_x), *intermediate principal stress* (σ_y), and *minor principal stress* (σ_z). Of these, σ_x is the vertical stress and the other two are horizontal stresses (Fig. 6.16).

The Mohr–Coulomb shear failure criterion is useful to understand the principles of failure of soil mass. The essential points to be remembered while using Mohr–Coulomb theory are that a soil material fails due to normal stress on the material as well as shear stress on the failure plane and that the stresses on the potential failure plane determine the ultimate strength of the material. The theory is expressed by the following equation:

$$\tau = c + \sigma \tan \phi$$

Here, τ is the shear strength, σ is the normal stress, c is the cohesion, and ϕ is the angle of internal friction.

6.12.2 Method to Draw Mohr's Circle

To start with, the normal stress is plotted along the abscissa (OX) and the shear stress results of a triaxial test along the ordinate (OY). In Fig. 6.17, the distances Oa and Oa' plotted along OX represent the results of two consecutive lateral pressures of triaxial tests on a soil sample. The distances ab and a'b' are the axial loads required to break the samples in the test. Then, circles are drawn with ab and a'b' as diameters to represent Mohr's circles (Fig. 6.17). In practice, instead of full circles, half circles are commonly drawn. A common tangent DE is then drawn such that a straight edge on the circumference of one of the circles just touches the other circle. The tangent DE meets the ordinate OY at the point D. The magnitude of cohesion (c) is given by OD and the angle of internal friction (ϕ) is obtained from the angle between DE and D'E', the latter being parallel to OX.

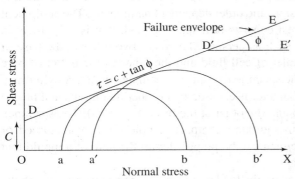

Fig. 6.17 Mohr's circle showing failure envelop, cohesion (c), and angle of internal friction (ϕ)

Determination of shear strength   Shear strength or resistance to shear of soil is determined by laboratory tests. The test result is used in the design of a structure in soil mass against possible failure. The tests that are generally conducted in the geotechnical laboratory to measure the shear strength of soil are direct shear test, triaxial compression test (drained and undrained), unconfined compression test, and Vane shear test. The procedures of these tests are briefly described in the following subsections.

6.12.3 Direct Shear Test

The principle involved in the direct shear test is illustrated in the simplified drawing, Fig. 6.18(a). In the laboratory, the test is conducted in a shear box apparatus consisting of a box in two separable parts each provided with a porous stone, as shown in Fig. 6.18(b). One part is kept rigid and the other part movable. The soil specimen is placed between the two porous plates such that the central surface of the specimen remains at the joint of the shear box. The soil specimen can be compacted in the shear box by clamping both the parts by means of two screws. Normal load (σ_n) is applied to the specimen through a pressure pad at the top that tends to keep the specimen in place, whereas shearing force applied to the lower box by jack tends to separate the specimen into two parts. The shear stress (σ_s) is measured with the help of a proving ring fitted at the upper part of the shear box. A number of tests are conducted under increasing normal loads. The applied loads and the shear forces required for failures of the soil specimens are recorded.

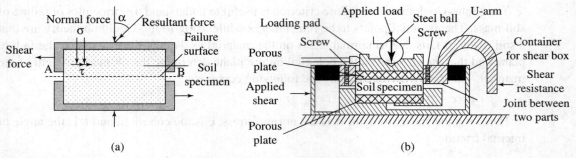

Fig. 6.18 Direct shear test: (a) principle involved; and (b) shear box apparatus

6.12.4 Triaxial Compression Test

Triaxial compression test is conducted on a variety of soil types for obtaining the shear strength parameters namely friction angle and cohesion, and other dependent parameters. The equipment used in conducting triaxial compression test consists of a specimen cell that is cylindrical in shape and made of a transparent material such as perspex. There are two tubes at the base of the testing equipment, one acting as the inlet of cell fluid and the other as the outlet of pore water. A third tube attached to the top of the specimen acts as the drainage outlet. A cylindrical soil specimen of length 2 to 2½ times its diameter is encased in a rubber membrane and placed between two porous discs—one at the top and the other at the base (Fig. 6.19). The specimen of cohesive soil is prepared directly from saturated compact sample either undisturbed or remoulded. For cohesionless soil, the specimen can be prepared with the help of a mould that maintains the required shape of the specimen.

During testing, vertical stress σ_x is applied by the load on the top of the specimen through the proving ring. The stresses σ_y and σ_z ($\sigma_y = \sigma_z$) act in the horizontal direction through the pressure of fluid that surrounds the specimen cell. Suppose in one test the fluid pressure applied is σ_z and the corresponding value of vertical stress obtained at failure is σ_x. Several such tests are conducted and several sets of values are obtained, which are used to draw Mohr's circles. The failure envelop at shear is provided by the common tangent to these circles (Fig. 6.20). A Mohr's diagram can be drawn from a minimum of two sets of test results to determine c and ϕ values.

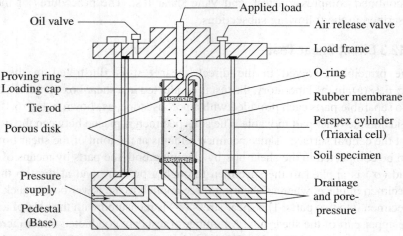

Fig. 6.19 Triaxial compression testing machine for soil

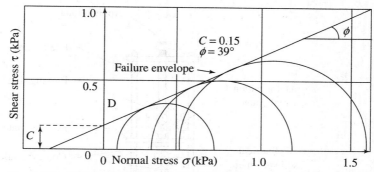

Fig. 6.20 Mohr's circles showing failure envelop, values of cohesion (c), and angle of internal friction (φ) obtained from the plots of triaxial test result

There are several variations of triaxial compression tests. The test is done in both consolidated and unconsolidated specimens in drained and undrained conditions. In the *drained test*, porous discs are kept in place and the pore water is allowed to drain out through the outlet. In the *undrained test*, instead of porous discs, solid caps are placed at the two ends so that no water is drained out. In a *consolidated drained (CD) test*, the specimen is consolidated and sheared in compression allowing drainage and keeping the rate of axial deformation constant. In the case of *consolidated undrained (CU) test,* the specimen is allowed to consolidate with porous caps until consolidation is complete at the desired confining pressure. The drain water outlet is then closed and the specimen is sheared and the shear characteristics are measured. In the *unconsolidated undrained (UU) test*, the specimen is not allowed to drain. The specimen is compressed at a constant rate (i.e., strain controlled).

6.12.5 Unconfined Compressive Test

Unconfined compressive test is a simple laboratory testing method to quickly determine the unconfined compressive strength of a soil and rock. In the case of soil, this test can be done if the soil is fine grained possessing sufficient cohesion. The test result is used to calculate the undrained compression of the soil under unconfined conditions.

The unconfined compressive strength (δ_1) is a measure of the load per unit area at which the cylindrical specimen of a cohesive soil fails in compression and this is expressed by the simple relation:

$$\delta_1 = q_u / A$$

where q_u is the axial load at failure in unconfined compressive test and A is the corrected cross-sectional area of the specimen $= A_0/(1 - \varepsilon)$.

Here A_0 is the initial area of the specimen and ε is the axial strain (change in length/original length).

The apparatus for unconfined compressive test consists of a load frame fitted with a proving ring attached to a dial gauge (Fig. 6.21). The test is conducted on a cylindrical specimen (of length 2 to 2½ times of diameter) placed on hollowed cones to keep the end restraints at the minimum until the soil specimen fails due to shearing along a critical plane. The normal load (δ_1) is applied from the load frame through the proving to measure the deformation of the soil specimen, but there is no confined horizontal pressure applied on this specimen cell unlike that in a triaxial specimen. Hence, the test is unconfined and uniaxial in which $\delta_2 = \delta_3 = 0$. The loading is continued until the load values decrease or remain constant with increasing strain, or until they reach 20 per cent (sometimes 15 per cent) axial strain. At this state, the soil specimen is considered to be at failure

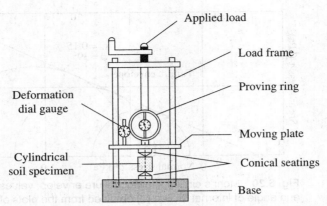

Fig. 6.21 Apparatus for testing unconfined compressive strength of soil

and the undrained shear strength of soil (S_u) is equal to one half of the unconfined compressive strength. Thus, $S_u = \delta_1/2$ where δ_1 is the unconfined compressive strength.

6.12.6 Vane Shear Test

Vane shear test of soil is done to determine the undrained shear strength of cohesive soil both in the laboratory and in the field by a simple instrument shown in Fig. 6.22. The vane shear tester for laboratory use consists of four steel blades, each of 1 mm thickness, 1.2 cm diameter, and 2 cm height. The blades are welded at the bottom of a rod, the top being attached to a torsion wheel having a calibrated spring arrangement to measure the torque. The vane is pushed inside the soil sample and the torque rod is rotated slowly at the rate of 1° per minute resulting in the shearing of the soil along a cylindrical surface. After the shearing is complete, the dial reading is noted. The spring constant multiplied by the dial reading gives the torque T in kg-cm.

The vane shear tester is the most convenient instrument to carry to the field and conduct field tests on shear strength of cohesive soil. It can also be attached to the bottom of a borehole in which case the vane must penetrate at least five times the hole diameter into the soil for testing. It is to be ensured that the soil at the testing spot is in undisturbed state. The vane size of the field instrument is generally 5 cm to 10 cm in diameter and 10 cm to 20 cm in height and it is made of about 2.5 mm thick rectangular metal plates welded orthogonally with a torque rod and torque applicator assembly. A pit of one or two metres deep is dug for a fresh surface for the test. To proceed with the test, the torque rod carrying the vane is gently pushed into the soil. Then, by the gearing arrangement, a torque is applied to the rod passing through a stiff torque spring. Assuming that both the top and bottom ends of the vane take part in the shearing of the soil, the torque T required to twist the vane in the soil gives the measure of the shear resistance of the soil distributed uniformly over the cylinder rupture surface. From mathematical computation, the following relation can be derived to calculate the maximum shear strength (τ in kg/cm^2) of the soil:

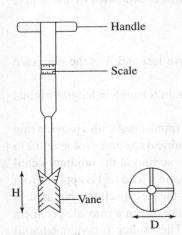

Fig. 6.22 Field vane shear testing instrument with vane height H and vane diameter D

$$\tau = T / \left\{ \pi D^2 \left(\frac{H}{2} + \frac{D}{6} \right) \right\} \qquad (6.33)$$

where T is the torque in kg-cm, D is the overall diameter of vane in cm, and H is the height of the vane penetration in cm.

6.13 BUILDING SITE GEOTECHNICAL INVESTIGATION

In the construction of a building, geotechnical investigation is necessary to evaluate the site condition with respect to the load-bearing capacity of the foundation materials. The investigation includes site exploration to find the depth and nature of the materials by laboratory and field testing. In a light structure, the exploration is restricted to few metres of top layers, but deep drilling is necessary for heavy structures so that the load of the building can be transmitted to a deep layer. The geology of the area such as the nature of subsoil condition (bore hole log of subsoil) and the groundwater level is generally known from the constructed buildings of the neighbouring area. The foundation load of a building depends mainly on two factors namely the weight of the building or the dead load and the life load due to the people who will use the building.

6.13.1 Loads of a Building and Foundation Exploration

In fact, the dead load and the life load will vary depending on the type of the building and its purpose of use. There may be residential buildings of limited dead load and live load. However, a school building will have high live load. Similarly, a shopping centre building may have lot of merchandise increasing the dead load. A building may be constructed for a monument or a temple of heavy loads depending on use of rock slabs or similar construction materials. A power plant has the impact of regular vibration. Thus, the load factor is decided first from the type of building to be constructed including the purpose of its use before taking up exploratory work at the site and foundation of the building. In general, the dead load that acts vertically down is calculated from the weights of building components and the life load is considered to be half of dead load or a fraction of it. The other factor is the vibration as in a powerhouse foundation, which is taken care of in the design according to the type of building.

The methods applied for exploration at the foundation site of building include some pitting and trenching followed by drilling with a sampler to collect undisturbed samples for laboratory tests. The programme of drilling depends upon the size and layout of the building or structure. One hole is to be drilled at the centre and for a large building four more in the four corners. Depending upon the local geology, generally a hole may be drilled down to the bedrock to get a complete picture of the subsurface materials and others by wash boring. In alluvial areas with homogeneous soil type, the holes may penetrate until firm soil is available.

The type of laboratory or in-situ testing of the foundation soil will depend upon the soil condition. In case of any chance of shear failure, the shear testing in laboratory and in-situ condition is to be done. Consolidation test is performed if settlement of foundation soil is anticipated in addition to routine laboratory tests on undisturbed soil samples. In addition to exploration, geotechnical investigation of the building site includes excavation of the site to fix the foundation grade. In some uneven hilly terrain, there may be sloping surfaces, which need to be flattened to obtain foundation grade.

6.13.2 Bearing Capacity of Soil for Building Foundation

The *bearing capacity of a soil mass* is its ability to support the load of structures for the different types of buildings. It is related to the extent of load the soil can safely support without causing failure or damage of structure. This allowable load is also called *allowable soil pressure*. The failure of foundation soil takes place when the *ultimate bearing capacity* exceeds the load of the structure placed on it. The ultimate bearing capacity (q_f) is the gross pressure at the base of the foundation.

The *safe bearing capacity* (q_s) of a soil is the load per unit area that can be safely supported by the soil mass. The ultimate bearing capacity is divided by the factor of safety, F (resistance force/driving force), with values such as 2 or 3 or 4, etc., to give the safe bearing capacity. Thus, we have

$$q_s = q_f / F$$

The ultimate bearing capacity of soil for building foundation depends not only on the nature of the soil, but also on the method of application of the load. The failure of a structure may be caused by its sinking down or toppling of its one side and bulging of the underlying soil. Several layers of the soil around the structure are affected due to the failure of structure, generally showing bulging (Fig. 6.23).

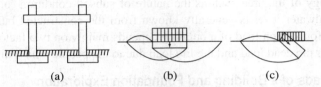

(a)   (b)   (c)

Fig. 6.23   Failure of the structure: (a) structure in intact position;
(b) bulging of soil under structural load on foundation;
(c) bulging of soil and sinking down (failure) of foundation

6.14 DETERMINATION OF BEARING CAPACITY OF SOIL FOR BUILDING SITE

Geotechnical investigations related to determination of ultimate bearing capacity of soil in a building site or other sites of engineering structures in the field condition can be done using some simple instruments. The following subsections describe two such instrumental tests.

6.14.1 Static Cone Penetration Test for Building Foundation

Static cone penetration test is generally conducted in a shallow pit to get the in-situ bearing capacity of the soil. In its simple form, the penetrometer consists of a cone screwed at the end of a steel rod and a calibrated meter (dial gauge) at the upper end. Depending upon the angle of the cone, the base area may differ. Figure 6.24 shows a static cone penetrometer with a cone of 30°, base area 5cm², and height 4.6 cm. The bearing capacity of the soil is measured by the resistance offered by the soil to the penetration of the cone neglecting skin friction. The cone is gently pushed into the soil by applying body weight to the cross handle and load (P) is recorded from the calibration of dial gauge. The ultimate bearing capacity is obtained by dividing the load by the base area of the cone.

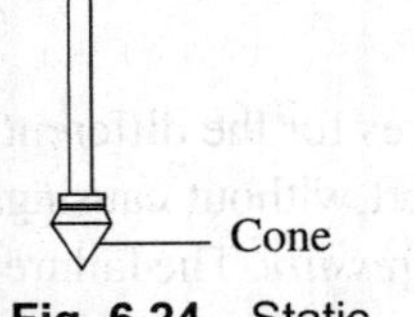

— Dial guage

— Steel rod

— Cone

Fig. 6.24 Static cone penetrometer

6.14.2 Plate Load Test

The bearing capacity of a soil is determined in the field by the plate load test. This test provides data on the rate of settlement of the test plate under gradually increased loading and ultimate bearing load at which the plate starts settlement at a rapid rate. The plate load test involves the digging of soil up to a depth of foundation level to make a flat ground covering sufficient area to place a square-shaped (50 cm² or 75 cm²) rigid plate having a thickness of about 25 mm. A layer of sand may need to be spread below the plate to level the uneven ground. The loading to test plate is generally provided by weighed sand bags or steel beams kept on a platform that connects a calibrated jack as shown in Fig. 6.25(a). The reaction of the pressure (load) applied through the jack is taken by the load on the platform and transmitted to the ground.

As per IS recommendation (IS codes 1888 and 1982), the test plate is to be provided with a seating load of 70 g/cm^2 before the start of the actual test. The loads are then placed in the order of one-fifth of the expected safe bearing capacity or one-tenth of the ultimate bearing capacity of the soil. Settlement is observed at certain intervals (1, 5, 10, and 20 minutes until one hour) by means of two dial gauges of 0.02 sensitivity placed on the two sides of the test plate. The test with one load is continued until the settlement is reduced to less than 0.02 mm per hour. Then, the next load increment is made and the test is continued as before. In this way, the maximum load applied is three times the proposed allowable pressure. After the test, a load increment versus settlement curve is plotted on log-log paper to obtain the failure stress, see Fig. 6.25(b). The safe bearing capacity is taken with safety factor 2 on ultimate bearing capacity. The test is good for homogeneous and uniform nature of soil throughout the foundation. Since the plate is small, the test gives a conservative value of the bearing capacity of the foundation soil.

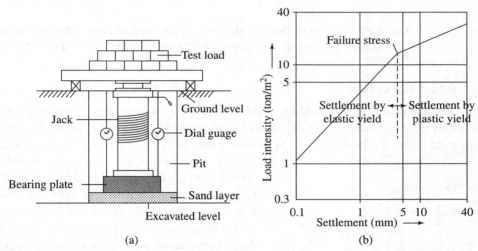

Fig. 6.25 Plate load test: (a) plate loading set-up; and (b) curve showing relation between pressure increment and settlement rate till failure

6.15 SHALLOW BUILDING FOUNDATIONS

Soil foundations may be described under two groups, namely *shallow* and *deep*. A shallow foundation is located on the surface of the ground or slightly below it. Common examples of shallow foundations are spread footings and mat or raft, which is suitable for a light structure or a building, the total load of which is very limited. The foundation of a structure is considered to be deep if its depth is equal to or greater than its width. Deep foundation transfers the load to the underlying stratum. Even if the subsoil is soft or compressible to a sufficient extent, deep foundation is taken up to provide adequate support to the structure. The common types of deep foundations are pile foundation, pier foundation, and caisson or well foundation.

6.15.1 Spread Footing

Spread footing, also known simply as *footing,* is a type of shallow foundation used to support columns or walls in the construction of small or medium-size buildings. The footing denotes the part of the foundation, generally square or rectangular in shape, that remains in contact with the soil. The footing transmits the load of a *column* or a wall directly into the soil. The base of the footing is kept large enough to safely support the imposed load. The common types of footing include *single footing, stepped footing,* and *slope footing* or *battered footing, see* Fig. 6.26. The pressure distribution below the footing is dependent on the rigidity, type, and characteristics of the soil.

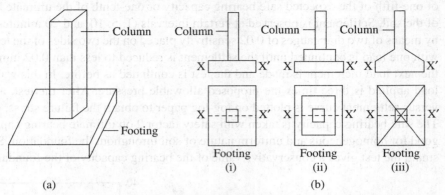

Fig. 6.26 Spread footing: (a) 3D view; (b) plan and section of different types of footing (i) single footing, (ii) stepped footing, and (iii) slope footing (or battered footing)

6.15.2 Mat or Raft

The *mat* or *raft* type of shallow foundation occupies a large area generally covering the entire base of the structure (Fig. 6.27). Rafts are generally made of reinforced concrete slabs and used where the bearing capacity of soil is low. In case the structure (i.e., a building) is heavy and the soil is soft, the use of the mat or raft foundation will be more economical than the spread footings as the total load being spread over a very large area on mat, the actual load per unit area (m²), will be low in the underneath soil. If the foundation area consists of soil of heterogeneous nature containing soft clay lenses, the raft foundation provides better safety without involving differential settlement. However, mat being lightweight may be uplifted by hydrostatic pressure. Hence, the position of the groundwater table in relation to the level of mat foundation is to be carefully studied before opting for mat foundation. The sectional-cut is a–b. The cross-sectional cut a–b is shown on the right diagram.

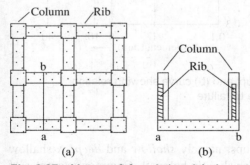

Fig. 6.27 Mat or raft foundation: (a) plan; and (b) section of mat with ribs

6.15.3 On-grade Mat Foundation

The *on-grate mat* foundation is extensively used in house building in the US, which has been proved to have high load-bearing capacity in expansive or 'hydro collapsible' (water-bearing) soils (Fig. 6.28). The foundation (called *wafflemat*) is created by connecting a series of 20 cm- to 30 cm- high, 45 cm × 45 cm, thermal-grade heat-resistant plastic forms, 'waffleboxes', set directly on grade and then monolithically pouring a post-tension, re-enforced concrete slab (10–20 cm thick). The completed slab then sits on the ground like a raft, the void areas underneath the slab formed by the placement of the forms allowing for expansive soil movement. The on-grade mat foundation possesses great

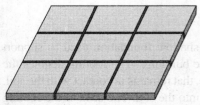

Fig. 6.28 On-grade mat foundation

stiffness, with the strength to resist differential swelling resulting from landscaping practices, surface drainage, or flooding from any source.

6.16 DEEP FOUNDATIONS FOR BUILDINGS

The methods followed in deep foundation include pile foundation, pier foundation, and caisson and are described with illustrations in the following subsections.

6.16.1 Pile Foundation

The widely used type of deep foundation is the *pile foundation* that supports various types of heavy structures, especially high-rise buildings. The piles are made of timber, concrete, or steel. The average diameter of a timber pile is 25 cm to 35 cm. Timber piles are prone to easy damage and they have small bearing capacity. The concrete piles may be *pre-cast* or *cast-in-situ* types. The pre-cast concrete piles, generally reinforced and square or octagonal in cross section, are capable of withstanding load up to 80 tonnes. The cast-in-situ types are designed for a maximum load of 75 tonnes. Prior excavation or boring is required for cast-in-situ piles. A steel sheet with steel core is driven first into the soil to make a bore. Then, the steel core is taken out and the annular space is filled with concrete.

Different types of piles are used depending upon the subsoil condition as illustrated in Fig. 6.29. If bedrock is available at a shallow depth, it is preferable to use end-bearing pile that transfers the load through soft soil to the hard strata below. When the subsoil is of homogenous nature capable of friction load-carrying, the friction pile is used to transfer the load. In case of granular sandy soil, compaction pile is used to compact the loose sandy materials and increase the bearing capacity of the soil. Other types of piles such as tension pile, anchor pile, and fender pile or their combinations are also used depending upon the subsoil condition.

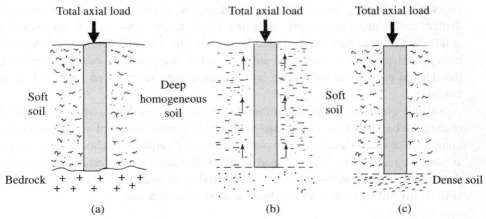

Fig. 6.29   Different types of piles: (a) end-bearing pile; (b) friction pile; (c) compaction pile

The *steel piles* may be H-piles, pipe piles, or sheet piles. The measurement of 'H' in a steel H-pile is 25 cm, 30 cm, and 35 cm in each of its three directions. The H-piles are used as long piles to withstand the load of heavy structures, especially in the construction of high-rise buildings. They are used when the bottom stratum is very compact having high supporting capacity. A combination of concrete and timber, or steel and concrete piles is also used. The piles are forced into the soil mass to bear the load of the structure. The forcing of piles into the soil is done by means of a hammer supported by a crane or pile driver. The 'resistance' offered by a pile is measured by the number of blows needed for one centimetre penetration.

Engineering geological investigation of a site comprising soil where the structure is to be founded on piles is needed to provide data on the nature of the foundation material and other relevant aspects. This will help to select the type of the pile to be used according to the soil type of the site. For example, if the ground contains water or soft soil overlying hard soil or rock, the decision will be in favour of using end-bearing piles. However, if the structural site is alternated with layers of clay and sand that provide frictional forces, the deciding pile type will be frictional pile. Where bedrock or hard and compact soil occurs at a shallow depth, the piles may be driven up to this firm foundation.

An exploratory drilling is to be carried out at the site and samples are to be collected from the entire depths by use of soil samplers. For building constructions in cities and towns, the data on foundation soil is available from work already done, though an exploratory work should be taken up for design requirement. In addition, pile load test is essential to understand the functions of pile after it is driven into the soil stratum. Observation in this respect is to be made for a period of nearly one month to find how the pile behaves within the soil mass.

6.16.2 Pier Foundation

Cylindrical and prismatic columns that serve as supports of structures transferring the load to soil or bedrock are called *piers*. If rock stratum is available within a shallow depth, it is always better to extend the pier so that the base rests on the rock. The piers of the extended base are sunk by open excavation or machine-drilled. These cannot be driven in subsoil containing rock fragments or boulders unless these are removed by excavation. Piers are of varying dimensions. Large cylindrical piers are usually 3 m or more in diameter. Construction of a shallow pier in dry soil is done by open excavation keeping the wall protected by timber support. However, piers up to sufficient depth in subsoil are constructed by using machine-drilled shafts providing casing to protect soil slides from side walls. When piers are of small dimensions, they are installed by cast-in-situ method. The procedure includes driving heavy steel pipes or cylinders to subsoil having sharp cutting edges. After the pipe reaches the desired part of subsoil at the bottom, the pipes are removed and the hole is filled with concrete.

Subsoil condition is an important factor to be considered while selecting the sites for resting the base of piers. In a bridge involving the construction of several piers, core drilling is conducted at each of the proposed pier position. It is required to choose dense soil or rock in the subsurface region on which the load of the bridge can be transferred by the piers that support the bridge. In case subsoil consists of layered sand and clay deposits, the base must rest on sufficiently stiff clay or thick sandy deposit after compaction.

6.16.3 Caisson

Caisson is the French word for 'box'. A caisson is generally applied to concrete box-like structures made of timber, masonry, or reinforced concrete. Caissons are taken down from ground surface to subsurface through water or soft loose soil to meet thick sand, stiff clay layers, or bedrock for foundation of building and bridge structures. There are three types of caissons, namely box type, well type, and pneumatic type. A box caisson is open at the top but closed at the bottom. This type of caisson is used when the load of the structure is relatively small. In pneumatic caissons, there is a working chamber for men at the bottom part where entry of water is prevented by using compressed air. Pneumatic caissons facilitate work in subwater conditions such as the construction of a bridge pier below river water. The caisson becomes an integral part of the bridge construction for its entire stretch from ground surface to foundation at depth in dense soil or rock. Caissons that open at both ends are called *wells*. Well foundations are commonly used in bridge foundations. A detailed discussion on well foundation is provided in Chapter 18 (see Section 18.4).

SUMMARY

- Soils are unconsolidated materials that occur on the earth's surface overlying bedrock. The grain arrangements of soils show different types of structures such as single-grain structure, flocculated structure, dispersed structure, and honeycomb structure. The load-bearing capacity of soil is guided by the nature of these structures.
- Quantitative evaluation of index properties of soil such as water content, specific gravity, density, plasticity, porosity, grain size, and consistency limits by laboratory tests helps in distinguishing the soil types and the engineering classification of soil. These properties also serve as valuable criteria in the design of structures in soil.
- Seepage through soil creates problems of piping in earth dams and stability problems in excavated soil slopes. Permeability values of soil determined both by laboratory and field tests help in assessing the seepage problem and designing safety measures.
- Soil is tested for the probable presence of swelling type of clay, which is harmful if used in the foundation of an earth structure such as a dam, as after saturation with water it develops expansive pressure damaging the foundation.
- In the construction of earth-fill structures, tamping, rolling, and vibration are followed to compress the soil. Moist soil layers placed on an earth dam are compressed by vibratory loading for rapid expulsion of water from the pore spaces that increase the density of the soil. This process of rapid compression is known as compaction.
- Long-time compression by static loading is essential in lowly permeable material for its densification and the process is called consolidation. In the design of earth dams and embankments, the optimum water content at which maximum density of soil can be achieved on compression is determined by Proctor test.
- A retaining structure is constructed to arrest the earth pressure that comes as lateral pressure from the backfill behind a cut slope. An active earth pressure may dislodge the retaining wall when a portion of the backfill breaks apart. If the failure of the backfill generates from mobilization of maximum force, it is called passive earth pressure. Depending upon the nature of soil and slope inclination, retaining walls of different types such as gravity, cantilever, buttress, and crib are constructed of concrete, masonry, timber, or pre-fabricated steel frames.
- Shearing resistance develops in a soil mass when it is subjected to stress by an engineering structure. The shear strength or resistance to shear of a soil is measured in the laboratory by direct shear or triaxial shear tests.
- The shear test data and failure envelop when plotted in a stress circle provide the values of cohesion and angle of internal friction of the soil necessary for design of earth structures. The knowledge of stress distribution below loaded area in soil foundation is needed to avoid settlement of structure.
- The safe bearing capacity of a building foundation depends on its dead load and live load. In case of shallow foundation, the structure is laid on spread footing, raft, or mat. However, different types of piles such as end-bearing pile, friction pile, or tension pile are driven in soil mass for deep foundation for the safety of a heavy building. In loose materials such as river deposits, cylindrical and prismatic columns called piers are sunk by excavation down to underlying firm soil or bedrock to transfer the entire load of the structures such as a bridge.

Multiple Choice Questions

Choose the correct answer from the choices given:

1. Index properties of soil determined in the laboratory include:
   (a) water content, unit weight, and specific gravity
   (b) density, porosity, and void ratio
   (c) both (a) and (b)

2. Consistency limits of soil determined in the laboratory include:
   (a) liquid limit, plastic limit, and shrinkage limit of soil
   (b) density and permeability
   (c) size and shape of particles

3. Field density of soil can be measured in the field by:
   (a) sand replacement method
   (b) rubber balloon method
   (c) core cutter method

4. Size of the coarse-grained soil is determined in the laboratory by:
   (a) sieving using standard sieves
   (b) wet analysis
   (c) both (a) and (b)

5. The type of water flow through soil is:
   (a) linear
   (b) laminar
   (c) turbulent

6. During pumping test in field permeability test:
   (a) the water level descends
   (b) a cone of depression is formed
   (c) both (a) and (b)

7. Consolidation test in the laboratory is conducted by using a:
   (a) consolidometer
   (b) rammer
   (c) both (a) and (b)

8. Proctor needle test is conducted to know about soil:
   (a) consolidation
   (b) compaction
   (c) both (a) and (b)

9. Vane shear test apparatus is used to determine:
   (a) bearing capacity of soil
   (b) shearing strength of soil
   (c) both (a) and (b)

10. Plate load test is conducted in the field for determining the:
    (a) compression strength of soil
    (b) bearing capacity of soil
    (c) none of (a) and (b)

11. Retaining walls are constructed for the protection of:
    (a) unstable hill slopes
    (b) backfill
    (c) both (a) and (b)

12. Raft or footing is provided in constructing houses or like structures for:
    (a) shallow foundation
    (b) deep foundation
    (c) both (a) and (b)

13. Caissons are taken down from ground to subsurface through soft soil to meet thick sand or stiff clay layers for foundation of a:
    (a) bridge
    (b) building
    (c) both (a) and (b)

Review Questions

1. State briefly the index properties of soil giving expressions for their calculation. Write the equation showing the relation between porosity and void ratio.

2. Describe Atterberg limits and the testing procedures of soil to determine these limits in the laboratory.

3. Define bulk density of soil and describe the procedure of in-situ bulk density measurement of soil by sand replacement method.

4. Describe in brief with a diagram the method of determination of permeability coefficient of soil by falling head method.

5. Explain the method of pumping test for the determination of field permeability of a soil.

6. Explain the procedure of consolidation test in the laboratory. Give your observation on void ratio–pressure relation in consolidation test from the study of the given curve.

7. Describe the procedure of soil compaction test by compaction machine for cohesive soil. Explain the method of standard and modified Proctor tests.

8. Give a brief account of the different types of retaining structures.

9. Describe Mohr–Coulomb failure criterion of a soil. Explain the principle of triaxial compression strength test.

10. Draw a sketch of vane shear testing instrument and write the procedures for using this instrument for in-situ shear test of soil.

11. State briefly the procedure for determining the bearing capacity of a soil in the field plate test.

Explain the extent to which the test result is related to plate dimensions.

12. Describe with sketches the types of spread footings used for shallow foundations and the types of piles used in deep foundations of a building.

NUMERICAL EXERCISES

1. In a laboratory testing, while determining the specific gravity, the weight of the density bottle was noted as 19.25 g. The soil sample after drying in the oven and cooling was weighed as 35.38 g. The bottle with sample was then filled with distilled water and weighed to be 120.21 g and the bottle with water to be 112.61 g. Find the specific gravity (G) of the soil sample.

[Answer: $G = 2.12$]

2. A vane testing instrument with vane height 12 cm and diameter 6 cm was deployed for field testing of cohesive soil in a pit in undisturbed soil. Torque was slowly applied to the two ends of the vane to a maximum of 300 kg-cm to bring about failure of the soil. Calculate the shear strength of the soil.

[Answer: Shear strength = 0.379 kg/cm2]

3. In an analysis of 200 g of soil particles, the following result was recorded with respect to their size distribution: coarse gravel 80 g, fine gravel 50 g, coarse sand 30 g, medium sand 20 g, fine sand 15 g, and silt with clay 5 g. Draw a cumulative curve showing the distribution of particles.

Hint: Refer Table 5.1 for IS size gradation and proceed to draw in a graph paper following the illustration given in Fig. 6.6.

Answers to Multiple Choice Questions

1. (c)	2. (a)	3. (all a, b, c)	4. (a)	5. (all a, b, c)	6. (c)	7. (a)
8. (b)	9 (b)	10. (b)	11. (c)	12. (a)	13. (c)	

7 Hydrology and Geological Works of Rivers

●●●●●●● **LEARNING OBJECTIVES** ●●●●●●●●

After studying this chapter, the reader will be familiar with the following:

- Hydraulic parameters of a river
- Sculpturing processes of a river valley
- Erosional activities of a river
- River action in the youth stage
- Mature stage river action and sediment load

- Old stage river action and formation of peneplain
- Formation of delta and features of a deltaic region

7.1 INTRODUCTION

This chapter discusses the hydraulic parameters of a river followed by its activities. Rivers originate from upland areas, especially from regions with hills and ridges. At the initial stage, the water flow in a river is very limited, but fed by rainwater, snow, and ice-melt water, the flow increases. As it moves downward, the river erodes the rocks and soils and forms a deep, V-shaped valley. In the middle or mature stage, the eroding action of the river is more on the banks and the valley widens. The river at this stage meanders and at places forms ox-bow lakes. The zigzag movement of the mature stage river causes bank cutting and smoothening of the valley floor and increases deposition of the eroded sediments, which are carried by the river as traction load. In the old stage, the river carries lots of fine sediments by suspension in addition to coarse sediments along the floor, which are all deposited near the sea. This is the peneplain stage of the river when it tends to cut through the deposited sediments to form deltas. The deposited sediments of the river form alluvial soil, which is very fertile and good for agriculture and is also used in dam construction and pottery industry. This chapter discusses all these activities of a river and the utility of the river deposits.

7.2 RIVER HYDROLOGY

River hydrology is the study of a river including its flow pattern. A river is constituted of its *channel*, *bed*, and *banks*. The landform or the valley through which a river flows is its channel, the bottom part of the channel is the bed, and its two sides are the banks. *Hydraulic geometry* is the collective term used to describe a river channel and the flow of its water. The river channel is simply a sloping trough created by the flow of water in the most effective manner for movement of water with sediments. Strictly speaking, a river channel need not necessarily always contain flowing water. Many channels of desert streams are dry most of the time. In River Falgu of Bihar, the water flows below the sandy bed, the upper surface remaining dry most of the time.

7.2.1 Hydraulic Parameters

River flow occurs in a long trough-like depression bounded by banks or valley walls that slope towards the channel. The depth of a river channel is measured at any desired point. The width is the distance across the river from bank to bank. The channel width is initially very narrow, but with the movement, the trench of the channel becomes wider and in case of large rivers may be hundreds of metres.

The *hydraulic radius* (r) of a channel is measured as

$$r = A/P$$

where A is the cross-sectional area, given as the vertical cross section at right angles to the direction of flow, and P is the wetted perimeter, the length of the line of contact of water with the channel measured along the cross section. The hydraulic radius is essentially equivalent to the average depth of the river.

The *discharge* of a river is measured as

$$Q = a/v$$

where a is the cross-sectional area of the river and v is the mean velocity. The *velocity* of a river depends on the channel gradient, the volume of water flow, and the configuration of the channel.

The *discharge* or the *volumetric flow rate* of a river is measured as the volume of water that passes through a given cross-sectional area of the river channel per unit time. In metric system, it is recorded in cubic metres per second (cumec). Sediments and soluble materials carried and deposited by a river may also be referred to as system discharge. The volume of water discharge is expressed by the following equation:

$$Q \text{ (cumec)} = AV$$

where A is the cross-sectional area (m^2) and V is the mean velocity (m/sec). The discharge variation at a single station may be caused due to the seasonal distribution of rain, seasonal melting of snow and ice, and water release from upstream dam.

Drainage density The drainage density of a river is given by the total length of the stream channel in kilometres divided by the basin area.

Sediment yield It is the total quantity of suspended sediments and bed load reaching the outlet of a drainage basin over a fixed period of time. Yield is usually expressed as kilograms per square kilometre per year. It depends on various factors such as the size of drainage area, basin slope, diameter, sediment type (lithology), vegetable cover, and human land-use practice.

7.3 EROSIONAL PROCESSES OF A RIVER

The total process of sculpturing by a river involves weathering, mass wasting, and overland flow called fluvial denudation. The geological work is performed by a river in three integrated processes, namely erosion, transportation, and deposition. These phases of geological works cannot be separated from one another, as where erosion occurs, there must be some transportation and eventually the particles must come to rest.

When water flows over the soil surface, it exerts a shear stress or drag upon the grains. If the stress is sufficient to overcome the resistance of cohesive forces binding the grains to the parent mass, the grains enter into the flowing water by rolling and dragging down the slope. The progressive removal of the grains in this manner is called *soil erosion*. On barren soil surface, the dropping rainwater dislodges soil particles called *splash erosion*. On barren land due to deforestation, the soil is removed by *sheet erosion* by sheet flow, which gradually takes away the upper fertile layer of soil when subsoil and bedrock remains.

7.3.1 Types of River Erosion

A flowing river can erode both the banks and the bottom by different processes. In general, erosion by river water takes place where the river forces exceed the resistance offered by the materials over which it runs. The various types of river erosion are *hydraulic action*, *attrition*, *abrasion*, and *corrosion*.

Hydraulic action is the force of water on the grains projecting from the bed and banks. Attrition is the disintegration by the collision of the particles that are suspended in water. Abrasion or corrasion is the mechanical wear caused by the impact of rock particles carried in the current striking against the rock exposed at the channel surface. It produces solid particles or sediments that are carried away by the water as suspended material.

Corrosion is the process of dissolving the materials that remain with the water of the river. Chemical reactions between the ions carried in the river water and the exposed mineral surfaces result in a form of erosion called corrosion. It is similar to chemical weathering of rocks (see Section 4.4). The erosion cycle is considered to begin with the uplift of the land mass and ends when the uplifted land mass is eroded and all that is left is the plain known as peneplain at the base level.

7.3.2 Factors Deciding the Rate of Erosion

The rate of erosion by a river is dependent on the following four factors:

  (i) Volume of water and velocity of flow
 (ii) Character and size of the load
(iii) Rock type and geological structure
(iv) Vegetation that affects stability and permeability

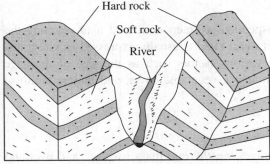

Fig. 7.1 Alternative beds of hard and soft rocks

At the place of its origin in a hilly or mountainous terrain, the river gets water from seepage, springs, or snow melt, which increases on its way downwards being fed by rains and small tributary streams. Broad river valleys are formed when rivers cut down sideways into rocks, wearing the slopes on either side. Alternative beds of hard and soft rocks are easily eroded resulting in a river valley with stepped cross profile (Fig. 7.1). The soft rock layers are easily scoured out while the hard layers crumple down forming a valley.

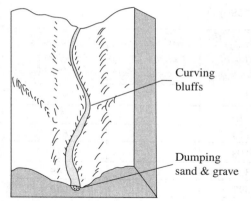

Curving bluffs

Dumping sand & grave

Fig. 7.2 Widening of valley floor by river erosion resulting in a V-shaped valley

This forms a very steep-sided valley initially, but produces a broad, flat-bottom bed valley on further erosion (Fig. 7.2)

7.4 DEPOSITIONAL AND TRANSPORTATION CHARACTERISTICS OF A RIVER

When the river moves at a high velocity, the solid rocks by the bank walls or other parts may collapse. This generally happens when the water moves as rapids or waterfalls. The load of the river is the sum of all the dissolved and suspended solid materials that flow by forward movement, vertical movement, and lateral movement. The force of water erodes the rock and soil on the way, and the sediments with coarse particles and even broken rock fragments are deposited on the bed and banks.

7.4.1 Movement of Sediment Load

The bed load or traction load moves along the river channel. The load increases with the bank erosion of soil and weathered rocks and corrosion of hard and fractured rocks that may come on the path of river flow. The increase in load also checks the velocity of the water. At places, by scouring or cavity formation at the bank, the river water may bring bank failures causing increase in its bed load. The flow of a river is generally turbulent in nature. Hence, if there are any obstacles on its way, eddies are formed in the river causing more erosion and consequent increase in its load. Eddies take away most of the suspended load.

Both rock and soil mass take part in the process of erosion and deposition in the river bed. The eroding power of river water depends not only on the softness or weathering condition of the rocks, but also on the presence of fissures and looseness of grains in the rocks. The river carries a part of the bed load, the coarser and heavier portion by traction and the lighter and finer parts by suspension. Some portions of coarse sediments including fragmented rocks, cobbles, and pebbles that are formed by water action are also deposited in the two banks.

Thus, the denudation processes of erosion, transportation, and deposition take place during the different stages of the river flow from its origin in the upland hilly terrain to its meeting with a sea or lake. The bed load or bank deposit of a river contains boulders, shingles, gravel, and sand and the river deposits produce alluvial soils, all of which are of great importance in engineering constructions.

7.4.2 Sediment Load of Indian Rivers

From the earlier discussion, it is apparent that a river degrades or erodes some stretches of the bed and aggrades or builds other parts, thus keeping a concave profile. Even after these activities, it has enough velocity to carry the load so that just above the sea level it forms a peneplain. An estimate shows that the rivers of the globe discharge about $15–16 \times 10^{16}$ tonnes of sediments per year (Walling and Webb 1983) into the oceans. The erosion and deposition of an individual river depends largely on its discharge—the product of its volume and velocity. For example, the discharge of the river Amazon in South America is 180,000 cubic metres per second, which is ten times more than the discharge of the river Mississippi. The sediment loads carried by the major Indian rivers are shown in Table 7.1.

Table 7.1 Sediment load in major Indian rivers (Subramanian 1996)

River	Discharge $m^3 \times 108/year$	Drainage area $km^2 \times 10^3$	Load—chemical M tonnes/year	Load—sediment M tonnes/year	Erosion rate—chemical tonnes km^2/year	Erosion rate—sediment tonnes km^2/year
Mahanadi	67	142	9.6	1.9	67.6	80.9
Krishna	30	251	10.4	4	41	57
Godavari	92	310	17.0	170	55	610
Cauvery	21	88	3.5	0.04	40	40.5
Ganges	493	750	84.0	329	111	549
Brahmaputra	510	580	51	597	88	953

For large rivers of India such as the Ganges and the Brahmaputra, the flow pattern is affected to a large extent by the wet and dry seasons. The stream velocity also depends to a large extent on the particle size. In other words, the stream velocity affects differently sized particles in different ways. Figure 7.3 shows the sediment load and the suspended particles of the Ganges in different places of flow.

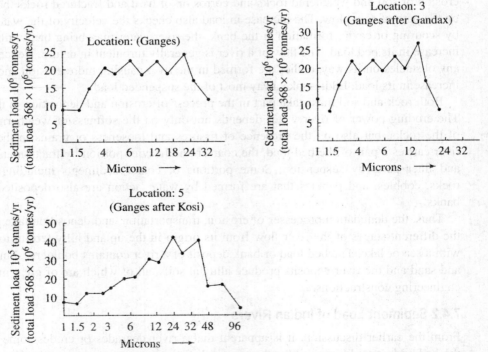

Fig. 7.3 Sediment load carried by the Ganges at different places (Subramanian 1996)

7.4.3 Different Ways of Transporting Load by a River

The transportation of the load is executed by a river in two contrasting ways, that is, within the current and along the base of the current. In fact, transportation takes place by four different ways—*traction*, *saltation*, *suspension*, and *solution*.

The rolling of the coarse particles along the river bed is called traction. In the process of saltation, a particle may rise as much as 30 cm above the floor and travel down some distance and then be pulled back by gravity, which is in fact a jumping motion. Suspension is the process of carrying fine sediments such as silt and clay with water flow. The materials that are dissolved by water come from rocks such as limestone and some salts of potash and other compounds present in the soil. These dissolved materials in the sediments are carried by flowing water as solution.

7.5 WORK-ACTIVITIES OF A RIVER IN DIFFERENT STAGES

William Morris David (1850–1934) first gave the idea of the erosion cycle, which is known as David erosion cycle. It includes three stages of river activity beginning from its origin in a hilly region until its flow into a sea, in many cases forming a delta (Figs 7.4 and 7.5). These three stages are as follows:

(i) Youth stage
(ii) Mature stage
(iii) Old stage

In the youth stage of the cycle, the river erodes headward actively and the valley is deep and V-shaped due to downcutting, see Fig. 7.6(a). The stage of maturity is activated when the valley is cut into uplifted blocks, new network of tributaries are formed, and all valleys show flattened V-shaped profiles and large deposition Fig. 7.6(b). In the old stage, the river has low relief across which it flows with huge deposits of fine sediments, see Fig. 7.6(c).

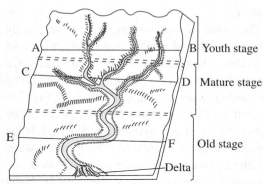

Fig. 7.4 River flow with different configurations during its journey through youth, mature, and old stages

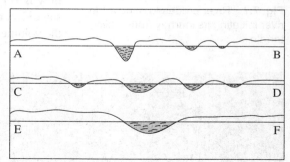

Fig. 7.5 Sectional view (not to scale) along A–B, C–D, and E–F of Fig. 7.4, showing basin characteristics in different stages: A–B, youth stage with V-shaped valley; C–D, mature stage with widening valley; and E–F, old stage with large shallow basin

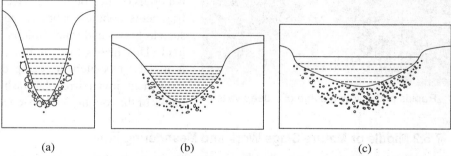

Fig. 7.6 Configuration and sediment depositions of a river valley: (a) V-shaped at young stage with very coarse rock fragments; (b) flattened V at mature stage with pebbles and gravels; and (c) flat profile at old stage with sand, silt, and mud

7.5.1 Work-activities of a Young Stage River

At the initial or young stage, the river valley starts to form from high up on hill slopes where springs erupt or rain water causes splash and then rill erosions. In its upper course, a river may be a small, fast-flowing torrent cutting down into its bed and forms rapids and waterfalls. Floods accelerate the process, transforming rocks that create rock pool. Several features are commonly seen in the young stage when the river moves through the hills and mountains. Of these features, the most prominent is the headwater erosion when undercutting, rain wash, and soil creep help to slice the hillside and lengthens the river. At places, circular holes (potholes) are formed in stream bed by whirling water with pebbles and rock fragments.

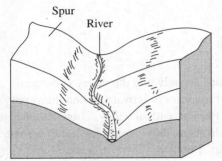

Fig. 7.7 River erodes forming a the river initiating its journey from a hilly terrain

During the zigzag movement of the river in the outside of bends, the current flow is vigorous. This develops a deep and steep cross section with downward erosion through interlocking spurs and ridges projecting from the sides of the valley (Fig. 7.7). During its forward movement, the young river erodes the spurs and widens the valley floor. A young stage river flows through the uneven hilly track of its origin carrying little water but the volume of water increases with rains. At this stage, the gradient being steep, stream flow is very fast and results in high erosion of rock. The young stage river contains many rapids (Fig. 7.8) on its way that increase the volume of its water.

The force of the flowing river water easily erodes the soft and soluble rocks, but if the rocks on its flow path are hard alternated with soft beds, a waterfall may be formed by differential erosion by the running water as clearly illustrated in Fig. 7.9. As seen in this figure, the soft rocks are eroded by undercutting and the overhanging hard rock ledge is broken and taken away by the flush of water forming a plunge pool. In case the rock is weak due to joints, fractures, and faults, they may be easily dislodged forming very deep waterfalls (Fig. 7.10).

As a result of this widespread erosion of the river, the valley becomes wider. The eroded materials of the youth stage river are full of rock fragments, cobbles, pebbles, and coarse particles such as gravel and sand, which are carried as bed load. This heavy load of the eroded materials is mostly deposited in course of its downward movement in an irregular fashion on the two banks of the roughly V-shaped valley (Fig. 7.11).

Fig. 7.8 Rapids at the young stage of a deep valley

7.5.2 Middle or Mature Stage Work and Meandering Flow

During the course of its flow down from the hilly terrains, the gradient of the river flattens and the volume of water further increases with rains and the water of the tributary streams (Fig. 7.6b). This is the middle or mature stage of the river. The length of the river at this stage is increased

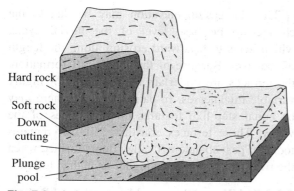

Fig. 7.9 A river at young stage flowing through alternated soft and hard layered rocks forms waterfalls

Fig. 7.10 Erosion along a fault in limestone resulted in the deep waterfall

Fig. 7.11 A young stage V-shaped valley with huge deposits of rock fragments, cobbles, and pebbles along its two banks

by head erosion and widened by bank erosion and backward cutting. It changes course frequently and flows in a meandering pattern causing deposition of sediments along and inside of the curve. In fact, this is where river currents move around a bend and hit the concave bank. The river deposits load on the convex bend and form a pile-up of sediments and at places form *ox-bow lakes* (Fig. 7.12). When the erosion causes the bottom of the river to reach the level of the water table, it becomes a permanent river with water remaining round the year. The lighter particles of the river remain suspended, whereas the heavier particles including pebbles and cobbles are bounced or dragged at the river bottom and partly taken to the sides.

With more downward flow at this mature stage, the river is further enriched by the groundwater drainage from other rainwater catchments. As the river flows down to its middle section, it shows more mature features than the upper course. The gradient of the river becomes gentler but the current flow is sufficient to transport the load of sediments from its upper course to the downward path. The downcutting at this stage is minimal and the river cuts more to the sides taking the load downward. In the process, the valley floor broadens. The eroded materials from downcutting and bank erosion still include pebbles and rock fragments but are mostly gravels, sand, and silt, which fill the wide valley (Fig. 7.13). The river at this stage during rains inundates the areas on both banks by flood and damages villages but enriches the areas with new soil deposits.

River terrace

At the mature stage, a river may start vigorous cutting of its sides boosted by heavy rainfall. This causes formation of a bench or step that extends along the sides of a valley and represents a former level of the valley floor. The step generally has a flat top made up of sedimentary deposits and a steep fore edge. These step-like features called river terrace (Fig. 7.14) may result from a change in ocean level or by tectonic uplift of sediments carried by the river. Normally, these are the remnants of former *floodplains*, cut through by the river and left standing above the present floodplain level.

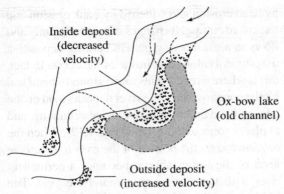

Fig. 7.12 Meandering river forming an ox-bow lake

The changes in elevation may be due to the change in the base level of the fluvial cycle, which leads to headward erosion along the length of the river. Because of the manner of formation, the terraces are underlain by sediments of highly variable thicknesses. In paired terraces, there may be equal step-like features on both sides of the valley. These terraces constitute all that remains today of past abandoned valley floors formed when the river was flowing at a higher level than today.

7.5.3 Old Stage Work and Formation of Deltas

In course of further downstream movement towards its final destination to a lake or sea, the river flow is relatively slow and its gradient becomes almost flat (the peneplain stage) and the load is mainly composed of sand and finer particles such as clay and silt (Fig. 7.15). Vast quantities of these sediments built the floodplains. When its river floor attains the level of the sea or the lake into which it finally flows, the river reaches its *base level of erosion* and then the erosion of bottom part or downcutting is stopped. At this old stage, huge piles of sediments consisting mainly of silt and clay-size particles are deposited in the sea or lake. At the confluence of a river with the sea, the reduction in flow rate causes large deposition of the sediments into the river, thereby raising the level of the deposit close to its water surface.

At the old stage, as the river proceeds to meet the sea, its flow cuts through the deposited materials in different branches forming swamps and *deltas* (named so as they look like the Greek word delta Δ) roughly of triangular shape (Fig. 7.16). Deltas are formed when a river flows into a body of standing water, such as a sea or lake, and deposits large quantities of sediment. The deposition rate is faster than the rate at which the tides and currents can carry the load away. The triangular mass of sediment consists mainly of silt and sands deposited at the mouth of the river. The environment of a large delta building into the sea is transitional between the normal marine and continental, and the deposits of each may be expected to interlens with the other. They are usually crossed by numerous stream channels and have exposed as well as submerged areas. The Ganges delta formed when the Ganges meets the Bay of Bengal is one of the largest deltas of the world.

A delta, in the geological sense, is defined as a deposit of sediment partly subaerial and

Fig. 7.13 Mature stage river with coarse sediments (front side) and a sand bar (distant side)

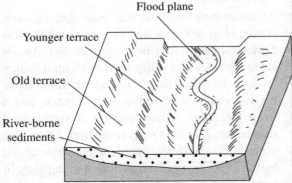

Fig. 7.14 River terraces formed at different stages

Fig. 7.15 Peneplain stage of a river with extensive deposits of sand and silt covering the floodplain

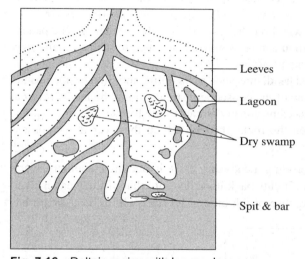

— Leeves
— Lagoon
— Dry swamp
— Spit & bar

Fig. 7.16 Deltaic region with leeves, lagoons, swamps, spit, and bar

made by a stream at the place of entrance into a permanent body of water (Twenhofel 1950). The major portion of the deltaic sediments is deposited subaqueously in the permanent body of water where waves and currents aid in the transportation and deposition. A delta is formed primarily by the deposition of river sediments in both the subaerial and submerged bodies of water (ocean or lake). The deposition of sediments at deltaic regions result in characteristic structures such as *topset* deposits, *foreset* deposits, and *bottomset* deposits. These bedded structures can be described as follows:

The topset bed of a deltaic region is made up of marsh deposits with silt and sand along with levee deposits and crevasee splay deposits. Muddy deposits with shell layers and tidal current deposits are also mixed with the sediments of the topset bed.

The foreset bed is made up of silty and clayey materials and rather coarse sand, silt, and clay deposits formed off the major deltaic distributaries. Delta front gullies also distribute their depositional materials to the foreset bed.

The bottomset bed consists of offshore clays obtained under the influence of active deltas. It also has marginal deposits obtained from subsurface materials. A part of the continental shelf at the slope of the topmost part sometimes have the materials of bottomset delta deposits.

7.6 CHARACTERISTIC DEPOSITION OF DELTAIC ENVIRONMENT

The product of the sediments carried by rivers and deposited in lake water is the *lacustrine soil* composed mostly of fine sand, silt, and clay. Since the lake water is still, graded bedding with well-sorted grains is formed in lacustrine deposits. The lacustrine soils are frequently found to contain large quantities of organic matters. In shallow lakes, the deposition of vegetable matters even forms *peat* from the assemblage of vegetable decay, and as such this organic soil is avoided in engineering constructions.

The rivers while being geological agents of erosion and transportation also act as environments of sediments deposits. The deposit of sediments by rivers is known as *alluvium*. When a river leaves a bedrock valley in a highland and enters open lowland where it can change directions, it deposits *fans*. There are low, fan-shaped cones of alluvial sand and rock fragments ranging from a few metres to several kilometres in extent.

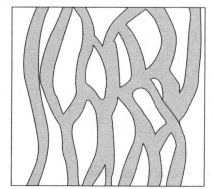

Fig. 7.17 Braided channel

Braided channels (Fig. 7.17) that are formed near a delta are typically broad and shallow. They are formed where the river has variable discharge and easily erodible banks. The building of braided channel is interspersed with bars and islands. The Ganges is a good example of a river having braided channels close to the deltaic region.

7.7 RIVER DRAINAGE PATTERNS AND RIVER CAPTURE

The land drained by a river is eroded to form a drainage basin of different patterns depending on the nature of the bedrock, the rock structure, and the erosional effects of the river. If a river drains through a sequence of sedimentary rocks, these will be worn away rapidly compared to other rocks due to their softness. In a soluble rock such as limestone, the river may create cavities by solution action and may drain through the cavities. If the rocks possess faults and fissures, the river will easily widen the fissures or faulted rocks. Thus, the type of the rock and the rock structures will decide the drainage pattern to a great extent. Of the various drainage patterns generally seen, the common ones are *dendritic*, *radial*, *trellis*, and *rectangular* (Fig. 7.18).

The dendritic pattern appears like a tree with branches provided by the tributaries. The pattern is formed when the bedrock is fairly uniform in nature. A geological structure such as a fault may cause certain changes as seen in the northern part of Fig. 7.18(b).

The radial drainage develops over central headlands such as an uplift, anticline fold, or dome.

The trellis pattern is produced by the flow of the consequent stream along the land's original slope and later by the flow of subsequent streams at right angles to the former.

The rectangular pattern is formed when the rocks through which the river flows possess joints intersecting at right angles or soft bedrock.

River capture This involves a river eroding headward until it acquires another river's headwaters at a so-called *elbow of capture*. It cuts back its valley into that of the weaker river, thus enlarging its drainage basin at the expense of the other. The river whose headstream has

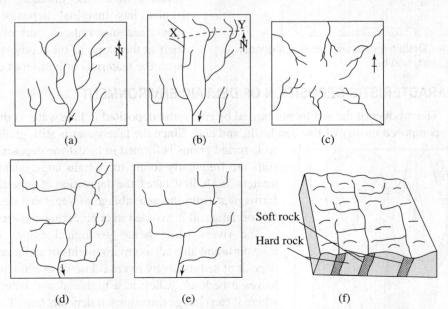

Fig. 7.18 Different drainage patterns: (a) dendritic; (b) dendritic pattern with a fault passing at the upper part; (c) radial; (d) trellis; (e) rectangular; and (f) geological structure resulting in a rectangular pattern

been captured often dries up or flows as a *misfit river* in a valley that is too large to have been eroded by the present capacity of the river.

In Fig. 7.19(a), river A, having a good discharge, captures the headwater of another river B flowing at a distance by elbow near C. This results in the downward portion of the captured river B being completely dried up (Fig. 7.19b).

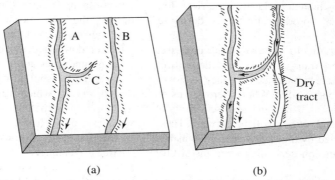

(a)   (b)

Fig. 7.19 River capture: (a) two rivers A and B flow independently with good discharge; (b) river B dries up being the victim of river capture by river A

An instance of river capture is an ancient river of India. 'It transpires that the Aryans explored the river Ganges and its tributaries thoroughly during the Vedic period. The river that *Saraswati* flowed during the Vedic time was later lost in the Bikanir desert due to the river-capture by the Yamuna which now flows through north India and meets with the Ganges. The river Saraswati, now completely dried up, was the holiest of all the rivers of India during the Vedic time and its name has been mentioned in many of the *slokas* of *Rig Veda*. The river Saraswati, a victim of river-capture, is specially significance in Indian history as it nourished a glorious civilisation of India along it large valley area nearly 5000 years ago' (Gangopadhyay, 2002a).

7.8 ENGINEERING USES OF RIVER DEPOSITS/ALLUVIAL SOIL

Rivers in different stages of their activities create different kinds of deposits depending on the nature and size of the materials. The materials of these deposits have manifold uses. The homogeneous types of riverine deposits containing mostly gravel or coarse-grained sand are widely used as construction materials for concrete aggregates. Since the finer river deposits (alluvial soil) are mostly porous and permeable in nature, construction of earthen structures on this soil requires an impervious core in addition to drainage arrangements. Alluvial deposit with sufficient clay is used for earthen dams. The clayey and silty type of alluvial soil with good plasticity is used for the impervious core of embankments and rock-fill dams.

Alluvial soil is widely used for making bricks and in pottery industry. Many of the large cities and towns of India are located on alluvial soil where numerous tall buildings are constructed on pile foundation because of the low shear strength of the materials. Many highways and railways are also constructed on the alluvial soil after proper compaction and adequate drainage arrangements.

Expansive soils or organic matters also sometimes occur in alluvial soil along road alignments when these require replacement by non-expansive and non-organic soil followed by compaction. The piers of the bridges located on alluvial soil require founding on hard strata at sufficient depth or on thick deposit of sandy clay of adequate bearing capacity to prevent settlements (see Section 18.7).

- Hydraulic geometry is the collective term used to describe a river channel and the flow of water. The important hydraulic parameters include volumetric flow rate, hydraulic radius, drainage density, and sediment yield, which are related to the river flow and the sediments it produces by erosion during its flow.

- Landform sculpturing by a river involves processes such as erosion, transportation, and deposition. These processes are so integrated that they cannot be separated from one another. Where there is erosion of landform such as banks and bed, sediments of different sizes and shapes will be formed. These will be transported by the river water either as bed load or in suspension depending upon the size of the sediments.

- Hydraulic action, abrasion, and attrition are the processes that are actively associated with erosion by the rivers. Both soil and rocks take part in the process of erosion. The bed load including bank deposits of the river contain mainly boulder, shingle, and gravels, whereas the finer particles such as sand, silt, and clay are taken into suspension. The erosion cycle is considered to begin with the uplift of land mass and end with peneplain at the base level when the rivers meet the ocean or sea after completing their long journey with transportation and deposition of sediments on their way.

- The geological work activities of a river take place in three stages, namely youth, mature, and old stages. In the youth or initial stage, a river channel is narrow and is mainly fed by rainwater and snow-melt water. Rapids and waterfalls are formed in this stage as the river moves downwards. The most prominent features of this young stage are the headwater erosion and undercutting, which form a V-shaped valley. The product of erosion is mainly boulder, shingles, and gravels, which are deposited irregularly in both the banks and on the bed.

- In the mature stage, the valley is widened and its bottom flattened. In addition to rainwater, water from small streams flows into the river. At this stage, the river meanders with frequent change of course and at places forms ox-bow lakes. With further downward movement, the river is fed by the groundwater and the water table rises. At this stage, the transported materials are mostly gravel, sand, and silt, which are deposited on wide valleys. The river is sometimes flooded due to heavy rains, which deposits fertile soil on land and at the same time inundates houses and properties causing extensive damage.

- In the old stage, the river meets a lake or ocean when it reaches the base level of erosion. Many of the rivers form deltas where fine sediments mainly of silt and clay cover large areas forming levees and marshy land with peats. At the delta deposits, there are characteristic structures such as topset, foreset, and bottomset deposits, which are constituted of sand, silt, and clay.

- In a map, a river appears in several distinct patterns such as dendritic, radial, trellis, and rectangular, which help to recognize the geology and rock structure of the area drained by the river. River capture is another phenomenon related with river flow. A river with good discharge can capture the discharge of another river, making it completely dry. The ancient Indian river Saraswati's flow was captured by the river Yamuna, thus causing it to completely dry up.

- The alluvial soil produced by the river deposits is fertile and is very good for irrigation purpose. This soil is also used in various engineering constructions such as dams, roads, highways, and railways and in pottery industries.

Multiple Choice Questions

Choose the correct answer from the choices given:

1. A river begins to widen and its average slope declines at the:
   - (a) youth stage
   - (b) mature stage
   - (c) old stage
   - (d) both (a) and (b)

2. The relief of a river channel is steep and it has rapids at the:
   - (a) youth stage
   - (b) mature stage
   - (c) old stage
   - (d) both (a) and (c)

3. A river divides into several channels to deposit mainly silt and mud at the:
   - (a) youth stage
   - (b) mature stage
   - (c) old stage
   - (d) both (b) and (c)

4. A river will meander and forms ox-bow lake in the:
   (a) youth stage   (b) mature stage
   (c) old stage   (d) both (b) and (c)
5. The type and form of deposit in the old age of a river is:
   (a) silt, mud, and organic matter
   (b) deltaic
   (c) peneplain
   (d) all of the above
6. State which one of the following statements is correct:
   (a) Attrition is the disintegration of suspended particles in a river by collision.
   (b) Corrossion is the mechanical wear due to the impact of rock particles in a river against a bedrock.
   (c) Corrosion is the solvent action of river water similar to rock weathering.
   (d) All of the above.
7. State which one of the following statements is wrong:
   (a) Traction is the rolling of coarse fragments along the river bed.
   (b) Saltation is the jumping motion of particles along the river bed.
   (c) Suspension is the carrying of fine sediments with the river flow.
   (d) None of the above.
8. The types of geological work mainly performed by a river are:
   (a) erosion of land by its channel
   (b) transportation of sediments along its flow
   (c) deposition of sediments carried on its way and at the end
   (d) all of the above
9. The sediment transportation by a river occurs:
   (a) within the current of the river
   (b) along the base of the current
   (c) both (a) and (b)
   (d) none of the above
10. A river in its old stage deposits its sediment mass in the:
   (a) ocean water
   (b) lake water
   (c) both (a) and (b)
   (d) none of the above

Review Questions

1. What do you understand by the term hydraulic geometry? What is a river channel? Is it necessary for a river channel to contain water all the time?
2. What are corrasion and corrosion? Name the factors on which the velocity of river flow is dependent.
3. From Table 7.1, find the total sediment load of Rivers Ganges and Brahmaputra. Which river has more total load and how is it related to the erosional rate of the rivers?
4. Describe the main work activities of rivers in the young stage.
5. State the main works of the rivers in the middle stage.
6. What are the main works of rivers in the old stage?
7. What causes flood in a river? What is its harmful effect and how is it beneficial to agricultural people?
8. How and at what stage of a river is an ox-bow lake formed?
9. Write notes on the following: discharge, drainage density, sediment yield, and peneplain.
10. What is the reason of a V-shaped valley during the youth stage of a river compared to its old stage?

Answers to Multiple Choice Questions

1. (b)   2. (a)   3. (c)   4. (b)   5. (d)   6. (d)   7. (d)   8. (d)
9. (c)   10. (c)

Geological Works of Oceans and Coastal Management

⬤ ⬤ ⬤ ⬤ ⬤ ⬤ ⬤ **LEARNING OBJECTIVES** ⬤ ⬤ ⬤ ⬤ ⬤ ⬤ ⬤

After studying this chapter, the reader will be familiar with the following:

- Ocean features and divisions of ocean floor
- Functions of waves, tides, and currents of oceans
- Activities of oceans in creating different landforms

- Coastal erosion by agents of oceans
- Littoral processes in eroding coastal landform
- Various means of coastal management
- Case study on protection of coastal area

8.1 INTRODUCTION

Geological works of oceans are responsible in many ways for changing the earth's landform, especially the coastal areas. The chief agents of an ocean that take part in the changing processes are waves, tides, and currents. The action of waves depends on the strength of the winds. Tides are caused by the gravitational pull of the moon and the sun. Waves produce currents that cause undercutting of the cliff rocks, resulting in the fragmentation of rocks. The seaward movement of the fragments of rocks causes further disintegration and deposition along the coastal land, forming beaches full of sand, gravel, and shingles.

Widespread erosion takes place along the coastal areas especially by littoral current, which may destroy beaches. The erosional effect including subsidence of coastal landform hinders navigation of ships to and from the coastal harbours and causes distress to the inhabitants of the coastal areas. This chapter describes with illustrations the processes responsible for the erosion of wide areas along the coast and the coastal management required to protect the coastal landform.

8.2 OCEAN FEATURES AND DIVISIONS OF OCEAN FLOOR

The oceans of our planet span across several kilometres and they cover 70 per cent of the earth's surface area. However, they are not only large water bodies but they also shape the inland surfaces of the earth including their floors to a great extent. The study of the ocean requires knowledge of different zones of oceans and ocean floor features. The zone where the ocean meets the land is the *intertidal zone*. This zone is divided into high-tide and low-tide zones depending on the depths at which these tides occur. Above the tidal zone is the *beach*. The area termed *neritic zone* is next to the tidal zone. It is rather a narrow shallow region along the continental shelf. This zone always remains under water as it stretches from the low tide line to the edge of the continental shelf. Below the neritic zone is the oceanic zone (Fig. 8.1).

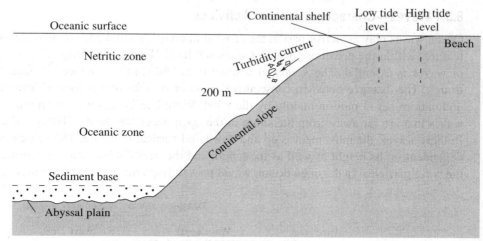

Fig. 8.1   Major features of ocean, continental floor, and coastal area

The ocean floor down to a depth of about 200 m is termed the *continental shelf*. The floor steepens further down along the continental slope to several kilometres until it meets the deepest part of the ocean called *abyssal zone*. This is separated from the continental slope by an abrupt marginal slope at the end of which there are sediment deposits mainly of clay and silts. The abyssal zone of the ocean floor varies in depth from 2.2 km to 5.5 km and this zone lies between the continental rise and the mid-ocean ridges. The abyssal plains are very flat because of the deposition of piles of fine sediments at the ocean floor.

The sediments of the abyssal depth are mainly transported by *turbidity current*. Transporting of the sediments off the edges of the continental shelf and onto the ocean floor including the abyssal plains form *turbidity current* and the thick deposits of sediments brought by this current are termed *turbidites*. A submarine landslide stirs up lots of sediments into suspension. This mixture is considerably dense than the surrounding ocean floor and the turbidity mass of sediments of the entrained water which comes down the slope.

The ocean floor also contains several ridges and volcanic peaks, predominantly in the Pacific Ocean. Most of these volcanic peaks that have released lava in the ocean floor have now risen above the sea as in the case of the Hawaiian Islands. Several large trenches are also present in the ocean near the Hawaiian Islands. The mid-Atlantic ridges are a ridge system present at the middle of the ocean. In general, these ridges are associated with the plate divergence and the production of new crustal material (see Section 20.4).

8.3 AGENTS OF OCEAN ACTIVITY

The most active as well as obvious work of the oceans is the sculpturing of the coasts, the zones where the land meets the ocean water. Coastal lands include cliffs and shores, which are areas between low water and highest storm water waves, and also beaches, the place of shore deposits. When ocean water is in violent motion, it scours coasts around the world. The eroded materials are deposited on the lands at the fringe of the ocean water, that is, the coasts. Along rocky shores, the water laden with rock fragments batters the land away. However, fragments torn from the cliff-fringed coasts are deposited gently on shores to raise beaches with sand, gravel, and shingles. Hence, most coasts advance or retreat. The chief agents in the process of ocean activities are the waves and currents, but tides also contribute to a great extent.

8.3.1 Waves—Characteristics and Activities

Ocean waves play an active part in the erosion of coastlines and can create many types of land features within the ocean or close to the coastal land. Hence, knowledge of the characteristics of waves is essential. The *height* of a wave is the maximum distance between its crest and trough. The distance between consecutive crests or troughs of a is its wavelength. Waves are undulations set in motion mainly by the wind. Wave energy increases with wind energy. The waves that are far away from the shore in the open water are the oscillatory waves (Fig. 8.2). In these waves, the movement is up and down and vertical in nature. The velocity of a wave is dependent on its height as well as the length and the depth of water and the orbital motion of the wave particles. In the open ocean, waves pass through the water without moving it forward.

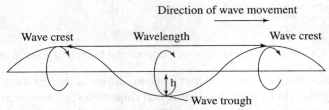

Fig. 8.2 Wave parameters and oscillatory movement of water at each point of wave surface

When waves approach towards the coastline, their bases drag on the ocean floor, whereas the crests continue to move towards the land. Eventually, the waves topple over themselves. During storms, waves behave in two distinct ways as an agent of erosion:

First, the high waves dashing against the rocky shore with sufficient force loosen the materials from the rock body and cause its fragmentation. If the area contains loose sandy materials, the erosion will be of serious nature. The wave acts at an angle on the shore and not normal to the shoreline. This creates a net flow of water called longshore currents flowing along the coastline.

Second, as the waves loosen the materials and carry them towards the offshore, the longshore currents transport them over a distance of several kilometres. Thus, a beach of sand may lose most of its materials and eventually a good beach may lose its existence. Coastal management aims to check such situations and save the beach.

8.3.2 Current—Types and Behaviour

The two common types of current are undertow or *underwater current* and the longshore or *littoral current*. When ocean waves hit the shoreline, there will be currents that move back towards the ocean. Such retreating current below the oncoming waves is known as the underwater current. When a wave attacks a cliff in the shore, it will batter the rock and causes its fragmentation which will move forward and backward. In the process, the size of the fragments will be reduced and they are transferred to the shallower to deeper parts. *Density currents* are suspension currents that transport sediments by suspension. These currents may carry particles over a distance of several kilometres from the mouth of large rivers flowing into the ocean. Such kinds of current can be seen even in the reservoir water of dam projects.

8.3.3 Tides—Patterns and Effects

Tides are the rhythmic rise and fall of ocean water caused by the gravitational pull of the moon and the sun, noticeable along the coastline. Coinciding with storm waves, the highest tides affect the highest level of the ocean shore. The difference between high tide and low tide is the *tidal range*. Since the moon is nearly 400 times nearer to the earth than the sun, it is the main celestial body that produces tides. A tidal bulge is produced on two sides of the earth, the side facing the moon and the one opposite to the moon. This results in two high tides and two low tides every day as the earth rotates through the bulges.

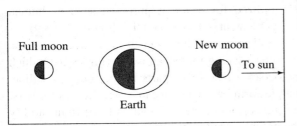

Fig. 8.3 Spring tide (maximum tidal range)

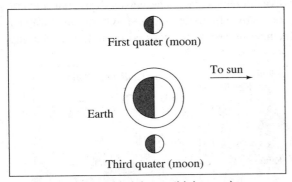

Fig. 8.4 Neap tide (minimum tidal range)

Tidal heights vary throughout the month due to the changes in the positions of the sun, moon, and the earth. The two extreme conditions are the spring tides and neap tides. *Spring tides* occur when the earth, the moon, and the sun are all in alignment, that is, under the greatest gravitational pull during the full moon and the new moon phases (Fig. 8.3). Tidal ranges are at a maximum and the tides at their highest. *Neap tides* occur when the sun, moon, and earth are at right angles (Fig. 8.4). The tidal range is minimal at neap tide since the gravitational stresses are not aligned. In a diurnal or daily tidal pattern, there will be one high tide and one low tide each day. The semidiurnal or semi-daily pattern exhibits two high tides and two low tides within a 24-hour period having approximately equal heights each tidal day, whereas in the mixed type there are two high tides and two low tides each day with markedly different heights. So, in contrast to the semidiurnal type, the heights of the high and low tides of a mixed tidal pattern vary significantly.

8.4 LANDFORMS CREATED BY OCEAN EROSION

In shallow water, the forward movement of waves causes undercutting of landscape and formation of cliff until its unsupported top collapses. If the cliff rock is hard, the cliff slope formed is generally steep except splitting of jointed rocks (Fig. 8.5). If the beds or bands in the cliff rock slope towards the ocean, overhanging cliffs may be formed, but if the beds are landwards, the slope will be gentle in nature (Fig. 8.6).

Where the cliff contains soft rocks dipping landward, the cliff is liable to be undermined by ocean water. In such circumstances, it is common to see caves at the cliff foot. Geological structures

Fig. 8.5 Hard rock creating steep slopes of cliff rocks with a sand beach in front, the bed slope being landwards

Fig. 8.6 Cliff rock with gentle seaward slope

such as faults, shear zones, or joints often control the erosion to form headlands and bays. Figure 8.7 shows how geological structures control the formation of headlands and bays. Ocean water easily erodes the crushed rocks from the fault zones keeping the resistant rocks intact, thereby producing the bay and headland. The block diagram in Fig. 8.8 gives a clear view of mainland and well-developed headland and bay, which is utilized for different purposes such as human settlement, sea port, and lighthouse (Fig. 8.9).

There may be long cliffs in areas near the coast facing the ocean water (Fig. 8.10), which by erosion will form different features. The ocean waves scour parts of the rocky coastline, leaving a vertical face above the water and a platform below the ocean water. It does this first by cutting a notch out of the rocky land near the bottom of the cliff, which leads to instability above the notch. Then, the portion above the notch is eroded by the process of mass wasting, while the portion below the notch is left intact as rocky cliff as explained by Fig. 8.11.

Further battering of the coastline by the waves creates caves. The waves throw particles of rock at the land, which causes abrasion, the process by which the land is slowly rubbed away by friction between the coastline sediment and the headline rock. At weak points in the cliffs along

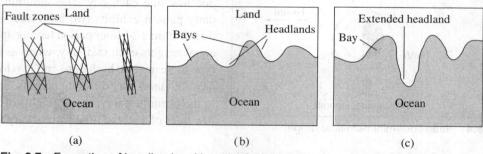

Fig. 8.7 Formation of headland and bay: (a) faulted rock in land; (b) bay and headland formed by erosion along faulted rocks by ocean water; and (c) the headland and bay enlarged by further erosion

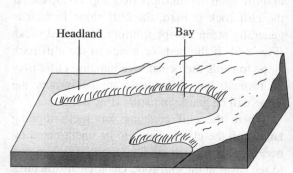

Fig. 8.8 An extended headland

Fig. 8.9 Headland used as a port

Fig. 8.10 High cliff extended along coastal land facing the ocean

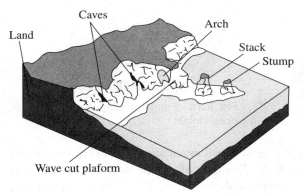

Fig. 8.11 Transformation of coastal landform into different features (as shown in Fig. 8.12)

a coastline, the abrasion rounds out a bowl-shaped feature at the bottom of the cliff called an ocean cave. The ocean arches are the next step and are formed when the caves continue to be eroded by the waves.

An arch is formed when two caves on each side of a headland join in the middle, creating a mini-tunnel through the cliffs, which on further erosion takes on the shape of an arch. Over a period of time, the top of the arch collapses creating stacks. These are features of coastal erosion that occur when the land that connects the pillar in the ocean to the headland collapses leaving a tower of rock in the ocean. Figure 8.12 clearly brings out the creation of these features such as formation of caves, then subsequent erosion to arch, and further removal of a slice of rock from the head of the arch by natural process (mechanical weathering/hydraulic action) creating stack and stump that stand as an upright steep-sided rocky island. See Fig. 8.11 and compare how this fits with the actual situation shown in Fig. 8.12.

Fig. 8.12 Formation of cave, arch, stack, and stump in the coastal rocks

8.5 SOME TYPICAL OCEANIC LANDFORMS

If ocean level rises or coastal land subsides, ocean water may enter the low-lying coastal areas to form different features as observed in different parts of the world. Some of these oceanic features such as fjords, fjards, dalmatian coastlines, and estuaries are described in this section.

Fjords are lowland ocean inlets along the coast with high sides and U-shaped cross profile (Fig. 8.13). These are deepened by glaciers that deposit the end moraines. Such coastal features are found in Norway, Alaska, New Zealand, Greenland, and some other parts of the world.

Fjards are also lowland ocean inlets deepened by glaciers, but are V-shaped in surface configuration. They are similar to fjords but are often with small islands at the ocean end (Fig. 8.14). These are found mostly in the southern part of Sweden and also in Nova Scotia.

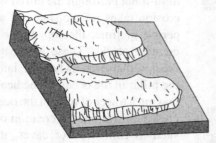 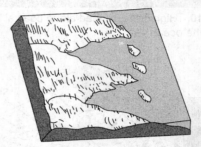

Fig. 8.13   Fjords: U-shaped ocean inlets

Fig. 8.14   Fjards: V-shaped ocean inlets

Dalmatian coastlines are features that appear as drowned mountain ridges in parts of pacific coastlines occurring parallel to ocean (Fig. 8.15). Flooding along the coastal land has also formed valleys and small isolated ridges as narrow offshore islands. The name has been taken from croatia's dalmatian coast where these features are very prominently present.

Estuaries are parts of the lowland tidal river mouth flanked by mudflats (Fig. 8.16). Many of the estuaries occur as submerged river valleys owing their existence to the post-glacial rise in sea level and hence they usually contain lots of deposited sediments. Some estuaries show intricate patterns of channels, largely the results of erosion by both incoming and outgoing tidal rivers. These characteristic features are seen in many river mouths such as the Thames, St Lawrence, and the Ganges.

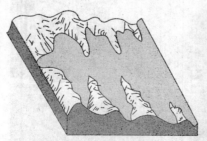 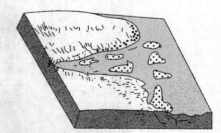

Fig. 8.15   Dalmatian coastlines

Fig. 8.16   Estuaries

8.6 COASTAL LANDFORMS OF VARIED PATTERNS AND DEPOSITS

Oceans may give rise to different types of landforms along the coasts by the deposition of eroded materials. Waves especially during storms act as the main architect of sculpturing the coast by means of deposition of eroded materials. In fact, storm waves deposit a large quantity of pebbles, sand, silt, and clay in the coastal bed to give rise to landforms of different patterns.

As explained in Section 8.8, littoral current produced by the action of waves moving at an angle to the coast carry huge loads of sand, silt, and pebbles. Extensive beaches are formed when the coast land slopes gently towards the ocean. However, when the current is drifted back towards the ocean, materials are also carried away from the coast as offshore deposits even on steeply sloping coastal floor. The following are a few illustrative examples of beaches of varied materials and characteristic deposition:

Crescent-shaped beaches   These are formed at the bay lying between two rocky lands. The beach area is small in extension and is made mainly of sand (Fig. 8.17).

Fig. 8.17   A distant view of several small beaches between two rocky areas (see Fig. 8.5 for a closer view of a part)

Longshore beaches   These are present in a narrow belt extending for a long distance in coastal areas. Such beaches are covered mainly by thick deposits of sand (Fig. 8.18).

Fig. 8.18   Longshore beach of sand

Lowland beach   This is formed in the upper shore region occupying broad areas consisting mainly of sand with a fringe of pebbles and rock fragments in the upper parts of the shore. This is often backed by sand dunes blown inland by the onshore winds (Fig. 8.19).

Fig. 8.19   Lowland sandy beach

Mudflats and salt marshes   Deposition of silt and clay by tides that are drowned at high tide forms mudflats and salt marshes in estuaries and bay areas. In the Bay of Bengal of eastern India, swamps of mangrove trees grow in plenty in such mudflats and marshy lands.

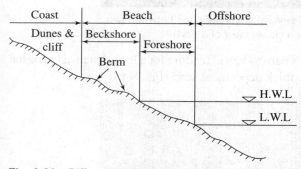

Fig. 8.20   Different parts of coastal land

8.7 EROSION OF SHALLOW COASTAL LAND

Large-scale erosion can be observed over long extents of the coast including a part of the shallow water of the ocean. The terms *foreshore* and *backshore* are used to designate the coast inclusive of a small part of water. Foreshore is the portion extended towards the sea and the backward part is the backshore (Fig. 8.20).

The foreshore is affected by the waves and currents of ocean water, whereas the backshore having several berms is affected by erosion and subsidence, mostly by high tides and swelling water. However, tsunamis have a devastating impact on both the foreshore and the backshore.

8.7.1 Basic Aspects

Coastal erosion causes long-term losses of rocks and sediments from the areas near the sea. It may also cause redistribution of coastal sediments. The erosion may be caused by hydraulic action, abrasion, impact, and corrosion. The waves generated by the wind move forward with oscillatory movements of high amplitudes. During the wave movement, the sands also move to and fro. The rocks forming the coastal landforms are also subjected to erosion. The eroded materials are taken away with the water when the rocks chip and the denser particles are taken back and deposited again at the edge. This forward and backward movement of the particles is a continuous process.

Two types of currents flow along the coast, namely *undertow* and *longshore* or littoral currents. The undertow current is formed below the wave and usually flows away from the shore. The longshore current is of great significance to engineers as this plays an important role in the erosional activity of the seashore. Another important agent of sea erosion is the tide, which has the maximum impact when the sun, moon, and earth are in a line. Sometimes, tidal rivers flow into the sea and influenced by the tidal effect cause deep erosion of the confluence area.

Sands of varying thicknesses in the coastal areas may form a large beach but the seashore remains submerged under water. Beaches may be of varying types according to their shape. A foreshore beach may be almost flat (slope 1:120), gentle (slope 1:30), or steep (slope 1:15). The sand particles of a beach are finer towards the steeper slope and coarser towards the gentler slope. In other words, the increase in coarseness of sand depends on the increase in steepness of the beach. The morphology of a foreshore varies to a great extent having troughs at some places and terraces in some areas. The littoral zone together with some berms and the gentle slope of the sea extending down to about 200 m is the *continental shelf* beyond which is the neritic zone followed by the oceanic zone and the abyssal plain (Fig. 8.1).

8.7.2 Harmful Effects of Coastal Erosion

Coastal erosion is one of the most significant hazards associated with coasts. It is responsible for the loss of subaerial landmass to sea due to natural processes such as waves, currents, and tides. The force of the waves and currents tends to move the sand from the shore and thus the landform slowly gets eroded bringing about a change in both the size and the shape of the coast. However, the erosional effect does not cause instantaneous destruction like a tsunami. Tsunami is associated with earthquake occurring in the seabed creating a massive release of energy and materials and causing a storm-like effect, whereas coastal erosion is a slow process of wave, current, and tide actions that extend over several months and even years. Other causes of coastal erosion include the change in sea level and the morphology of the shore.

8.7.3 Erosion of Beaches and Dunes

Large-scale erosion of beaches may be induced by storms. This phenomenon is associated with the natural evolutionary process of the sediments in sloping beaches. A storm increases the wave energy, which can easily remove the sediments from the berms and dunes. The removed sediments are then deposited as near-shore bars. Sand dunes can also act as stores of sediments used to carry out the coastal processes. This bulk removal of sand from the beaches and dunes is significant from hazard perspective.

Human interference is also responsible to some extent for the erosional processes. Construction of building and other engineering structures, dredging of sea shore, mining of beach sand, and even the construction of a dam across a river with its tailwater flowing into the sea add to seashore erosion. The presence of humans and their developmental activities in the coastal areas without considering or rather neglecting the coastal environment has had harmful effects.

8.7.4 Erosion of Rocky Coast or Cliff Areas

Though coastal erosion is more prominent along the plain landform, the rocky cliff areas present at the coast are also affected to a great extent by erosion. These areas are exposed to storms. Hence, the waves act as a very powerful erosive agent on the coastline and cut the coastline back into a cliff. The erosion is strongest near the level where the waves break and as such it does not cut down much below the intertidal level. A platform created as a result of cutting by the waves may be seen during low tide, extending towards the ocean water for 100 m or more from the foot of the cliff.

Coastal hill slopes covered by soil mass are subjected to easy erosion and slides after saturation with rainwater, which hastens the process of erosion. If the soil mass or rock of the cliff face is porous, it can absorb more water and the cliff face undergoes slope failure, causing extensive loss of coastal land after the slid materials are taken to the foreshore sea. The rate at which the cliff fall debris is removed from the foreshore depends on the power of the waves moving towards the coast.

The erodibility of the ocean-facing rocks is controlled by the strength and presence of fissures and beds of non-cohesion materials such as sand and silt. Rocky coastal areas having layers of fractured rocks produce varying amounts of resistance to erosion. Fractured rocks are eroded fast and change the coastal landform forming pillars, columns, tunnels or bridge-like structures in the rocky cliff. The rocky cliff of the shore comprising hard rocks may avoid erosion to some extent, but in course of time—over hundreds and thousands of years—the process of erosion by high tidal water may eventually result in isolated upright pinnacles of rock ledge in place of massive rocks as may be seen in Fig. 8.21.

(a)

(b)

(c)

Fig. 8.21 Erosional features of coastal areas: (a) hard rock body partly disintegrated; (b), upright smaller rock bodies formed; and (c), upright pillar-like rock body standing isolated

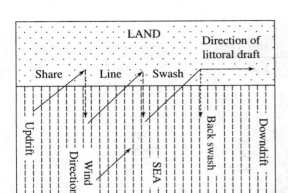

Fig. 8.22 Coastal area near ocean under the impact of littoral drift

8.8 LITTORAL DRIFTS

Longshore drift or littoral drift can be defined in terms of the systems within the surf zone (Fig. 8.22). The direction of the littoral drift generally becomes obvious on observation. Fluorescent die or radioactive tracer may also be used to find the direction and quantification of drift material. The deposition of the drift materials is influenced by the processes occurring in the surf zone that also largely control the erosion of the sediments. The longshore currents generate waves oblique to shore-break waves, which results in the transportation of materials as littoral drift.

Figure 8.22 shows that sediments transported along the shore and surf zone are influenced by the swash in the direction of the prevailing wind, which moves the shingles up the beach at an angle of wind direction, and also the backwash, which moves the shingles back down the beach. Longshore drift commonly transports sediments of sizes varying from shingles to gravel. Sand is affected by the oscillatory forces of breaking waves, and hence shingle beaches are generally much steeper than sandy beaches.

Creation of ports and harbours throughout the world can have a serious impact on the natural course of littoral drift. The major impact is the alteration of the sedimentation pattern, which may lead to erosion of a beach or coastal system. Waves usually surge onto a beach at an oblique angle and their swash takes sediments up and along the beach. The backwash usually drains back down the beach at an angle to the coast. The movement of the sediments depends upon the wind direction in the area close to the surf area of the littoral zone.

8.9 COASTAL MANAGEMENT

Coastal management includes both restoration and protection of the shore. In general, people living close to the seashore have a profound interest in maintaining their beach area. In fact, certain protective measures are taken to prevent the erosion of the beach and the coastal land. Such measures include the construction of groynes, concrete or masonry walls, and dunes and involve certain restoration works for the beaches as explained in the following subsections.

8.9.1 Groynes, Gabions, Concrete Walls, and Sand Dunes

Groynes One method of maintaining the shore involves the construction of groynes (Fig. 8.23), which can be made of wood, cement, or rock. A groyne is a littoral barrier placed generally at right angles to the shore. It involves building a wall of height generally 1 m to 2 m above the high-tide line, extending nearly 50 m from the beach. In fact, the length of the groyne should be 50 per cent longer than that of the beach. The purpose is to capture the sand, which will otherwise be taken away by the longshore current. The sands that are carried away from the beach area by the longshore current are stopped on the up-current side. Groynes are very cost-effective for beach protection as they do not need any maintenance. However, they are considered detrimental to the aesthetics of the coastline. Moreover, they do not protect the beach against the storm-driven waves.

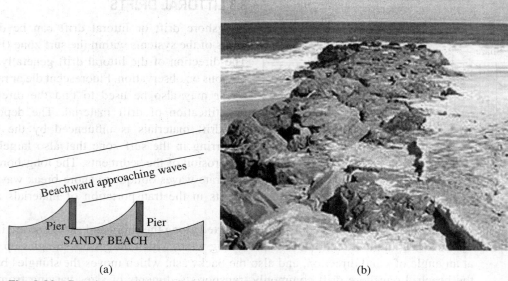

Fig. 8.23 Beach management by groynes: (a) concrete pier and (b) piles of rock (Wikipedia 2011)

Gabion   These are boulders and rocks wired into mesh cages. These are usually placed in front of areas vulnerable to heavy erosion such as cliff edges. Gabions are placed at right angles to the beach similar to a large groyne of rocks. When the ocean water breaks on the gabion, the water drains through it leaving the sediments. The rocks and boulders also absorb a moderate amount of wave energy. Gabions need to be tied properly to prevent abrasion of wire by rocks or detachment by stretching. Hexagonal mesh wires are better than the rectangular mesh type.

Breakwater structures   Enormous concrete blocks and natural boulders are sunk offshore to alter wave direction and to filter the energy of waves and tides. The waves break further offshore and therefore reduce their erosive power. This leads to wider beaches, which absorb the reduced wave energy, protecting the cliff and the settlement behind. The use of enormous concrete blocks has of late been replaced by several alternative structures that are also resistant to wave action and require less concrete to produce a superior result.

Building of concrete walls   Massive walls of concrete or masonry structures are built in the shore area to prevent erosion of cliff rocks and sand transportation from the beach and at places prevent flooding of the human settlement from tide during storms. Erection of the ocean wall normally prevents the destructive action of tidal water and separates the beach from property and population. However, it may also bring about destruction of the beach. If a series of storm waves with greater height and more aggressive force than that of normal waves crashes on the beach, the high walls may be broken down to smaller size. This requires a new approach for coastal management. Creation of sand dunes and beach restoration are two new approaches.

Sand dunes   These are a natural and effective way of creating a barrier to protect the lives and property along the shorelines. However, in recent years, coastal communities plant new dunes hoping to preserve their beaches. There are two prominent ways by which dunes protect the beaches. First, as the waves batter shores and roll up beaches, the dunes provide a sloping surface for dissipating the wave energy thereby halting the erosion to a great extent. Second, the dune grass by means of their spread-up roots holds the sand particles together and prevents erosion. Overall, dunes are relatively inexpensive and are effective in preserving a beach.

Beach replenishment One of the most popular soft engineering techniques of coastal management is beach replenishment or nourishment. This involves importing sand off the beach and piling it on top of the existing sands. The imported sand must be of a similar quality to the existing beach material so that it can integrate with the natural processes occurring there, without causing any adverse effects. Beach nourishment can be used alongside the groyne structures. However, this is a slow and expensive process as sand is dredged from a near-shore location, transported to the beach to be restored, and pumped into the beach.

8.9.2 Considerations for Effective Measures of Coastal Protection

As a means of coastal management, after consideration of the different ways of protecting the coast, it is argued that the following measures will be effective not only in controlling coastal erosion, but also in reducing the distress of the people living close to coastal landform:

- Coastal areas in many places are populated by human settlement very close to the shore. Such human settlements need to be prevented. Planting of trees close to the human habitation in coastal areas protects from degradation of the sandy soil and contains erosion. It is necessary to grow saline-resistant plants along the coastlines.
- An engineer is to consider the extent of damage that may be caused by littoral drift to design for remedial measures. There is sufficient scope for using the natural beach materials obtained from the littoral drifts for continuous coastal protection along the long extent of the coastline.
- Use of natural heaps of sand and shingles to make walls is an economical means of absorbing the energy of breaking waves and thus protecting an erodible coastline (Wood and Muir 1969). Hence, this is a useful option to be considered for the protection of coastal areas. Easily available materials such as clay, sand, and shingles are also worth considering for use as construction materials for walls. However, concrete is the most constructive material for use in the retaining walls for coastal protection.
- Growing of mangroves and building concrete sea walls or walls of dense and durable rock blocks are other methods of saving the coastal landform from the effects of erosion. The sea walls should be designed based on the maximum probable wave height.
- Construction of walls of rocks or wave breakers not only saves the coast but also serves as a means of protection frowwm the impacts of tsunami. Depending upon the morphology of the coast, rock walls can be built as close to the seashore as practicable.
- The modern practice of preventing coastal erosion by the use of groynes should be followed. They are to be built transverse to the breaking crest of storm waves and anchored at the base of a suitable bedrock. The beach can be protected by arresting the littoral drift with the use of groynes or by artificially changing the beach. Groynes require anchoring with bedrock for placing at the required angle to the coastline.

- Although coastal protection can be achieved by placing groynes to arrest longshore drifts and retention of shingles at the cliff toe, the most important measure of protection is providing concrete walls.
- Construction of houses or human habitation by encroachment of the shore should be prevented in the areas vulnerable to inundation or areas affected by tidal action, if required, by law enforcement.
- In recent times, coastal erosion is prevented by use of *geotextile* tubes (Stabiplage unit) as is being done effectively in France for the last two decades (UNESCO 2002). If located along the coastline, these are very useful in protecting the coast from the impacts of tsunami.

8.10 CASE STUDY—COASTAL EROSION AT UPPADA ALONG KAKINADA COAST, ANDHRA PRADESH (PAL AND RAO 2009)

Large-scale coastal erosion is prevalent over the 30-km long coast between Kakinada and Antarvedi. The impact of the sea erosion at Uppada is so intense that the sea walls and groynes erected south of it have proved ineffective in protecting the beach road, which is considerably damaged at several places. More than half of the village, predominantly a fishermen's colony, has gone into the sea. Evidence shows loss of more than 600 acres of coastal land in less than two months.

Major geomorphic units found in the area include beach ridge, younger beach–dune complex, older beach ridge, tidal mud flat, and tidal creek. The most vulnerable geomorphic units are the long and extensive beach ridge and the younger beach–dune complex because of their proximity to shore and weak geological formation. The longshore current has caused maximum erosion to the stable beach bridge and at the same time has widened the mouth of small rivers draining them in the Bay of Bengal, north of Uppada.

The following are the major causes of coastal hazards.

- Natural causes: Combined action of waves, winds, and tides near shore currents, storms, slope process, and sea-level rise
- Anthropogenic causes: Dredging of tidal entrance, construction of harbours near shore, construction of groynes, sea walls, and jetties, and destruction of mangroves and other natural buffers such as beach barrier dunes

The following two preventive and mitigation measures have been proposed.

(a) Structural measures: Construction of sea walls and groynes
(b) Non-structural measures: Vegetation cover and dune stabilization

It was concluded that comprehensive multi-hazard and risk assessment studies with the participation of local communities and scientists need to be initiated. Sea erosion causing rapid migration of shoreline towards land and inundation of low-lying flats need greater attention in view of the potential rise of sea level and its impact on communities.

SUMMARY

- Oceans occupy 70 per cent of the earth's surface. The coastal strip down to a depth of about 200 m from the coast is the continental shelf. This is followed by the neritic zone, oceanic zone, and abyssal zone that is several kilometres deep.
- The main agents of ocean activity are waves, currents, and tides. Waves cause the maximum destruction and erosion of the lands. The currents, namely the underwater and littoral currents, erode the coastal lands, take away the sediments, and deposit them in offshore parts. In addition, density currents carry sediments by suspension.
- Oceans form varied types of landforms near the coasts such as headlands and bays depending on the rock types and rock structures. Of these, cavity formation, arching, and finally separation of the rock mass into stalks and stumps are features formed by the erosion of cliffs.
- Ocean activity is also responsible for creation of features such as fjords, fjards, and Dalmatian coastlines. These features can be found in the lowland areas near the coastal areas.

- Coastal erosion is a natural hazard to the people living close to the coastal lands. The longshore current that travels along the coast covering extensive distance plays the most damaging role in coastal erosion.
- The soft materials along the coast are easily eroded by the tides, but the cliffs that occupy a substantial part of the coastal area comprising hard rocks may be protective to erosion. However, with time, these rocks are also eroded retaining only the remnants.
- Human activities such as mining of beach sand, dredging of shore areas, and also construction of building close to the coast makes the coastal land vulnerable to easy erosion.
- The principal protective measures for coasts close to ocean end, especially the beaches, include providing walls to save the sand of coastal zone from wave and current actions by means of groynes, gabions, or construction of concrete walls.
- Areas near coasts that are subject to easy erosion should be barred from human habitation. Use of geotextile tube is a modern method of stabilizing coastal erosion.

EXERCISES

Multiple Choice Questions

Choose the correct answer from the choices given:

1. Neritic zone is present along:
   - (a) continental shelf
   - (b) abyssal plain
   - (c) tidal zone
   - (d) all of the above
2. Intertidal zone is present between the:
   - (a) neritic zone and continental shelf
   - (b) beach and continental shelf
   - (c) high tide and low tide zone
   - (d) (a) and (b)
3. Waves are formed by the:
   - (a) intensity of wind
   - (b) duration of wind
   - (c) gravitational pull
   - (d) both (a) and (b)
4. Tides are primarily caused by the:
   - (a) sun
   - (b) moon
   - (c) wind
   - (d) all of the above
5. Spring tides occur during:
   - (a) full moon
   - (b) quarter moon
   - (c) new moon
   - (d) both (a) and (c)
6. Tides of great height occur during:
   - (a) full moon
   - (b) neap tides
   - (c) spring tide
   - (d) both (a) and (c)

7. A lowland beach is formed:
   - (a) in the upper shore region
   - (b) with fringes of rock fragments and pebbles in the higher part
   - (c) often joined by sand dunes in the upper part
   - (d) all of the above
8. Fjords are found in the coast of:
   - (a) Alaska
   - (b) Norway
   - (c) New Zealand
   - (d) all of the above
9. Estuaries are found in the river mouth of:
   - (a) Thames of the UK
   - (b) Ganges of Bay of Bengal in India
   - (c) Dalmatian coast of Croatia
   - (d) both (a) and (b)
10. A beach can be best protected by providing a natural and effective barrier by a:
   - (a) groyne
   - (b) concrete wall
   - (c) sand dune
   - (d) gabion

Review Questions

1. What are the main features of ocean floor? What is the approximate depth of the continental shelf?

2. Name the agents of ocean activity. How are spring and neap tides formed?

3. How are headlands and bays formed in an ocean?

4. How are the features such as caves, arches, stacks, and stumps created close to ocean shore?

5. What is a fjord? Add a note on its formation.

6. Give a short account of the processes responsible for coastal erosion.

7. Describe the principal ways of coastal management related to the prevention of coastal erosion.

8. What is beach reconstruction? What are its merits and demerits?

9. In what way is the littoral current associated with coastal erosion?

10. Write notes on continental shelf, abyssal plain, littoral current, headland, groyne, and gabion.

Answers to Multiple Choice Questions

1. (a)   2. (c)   3. (d)   4. (b)   5. (d)   6. (d)   7. (d)   8. (d)
9. (d)   10. (c)

9 Underground Water in Relation to Engineering Works

● ● ● ● ● ● ● ● **LEARNING OBJECTIVES** ● ● ● ● ● ● ● ●

After studying this chapter, the reader will be familiar with the following:

- Hydrological cycle and source of groundwater
- Water table configuration in relation to landform and materials
- Different ways by which water remains in the voids of soil
- Confined and unconfined aquifers and formation of artesian flow and springs
- Porosity and permeability of soils and fractured rocks
- Utility, yield, and investigation methods of groundwater
- Hazards posed by underground water in engineering projects and their remedy

9.1 INTRODUCTION

This chapter deals with the various aspects of subsurface water in relation to their engineering significance. It discusses the processes involved in hydrological cycle to maintain the balance of the earth's water. It describes with illustration the water table configuration and its relation to topography and the materials through which underground water flows. The chapter further explains the utility of groundwater and the method of its withdrawal by pumping water from dug wells in urban and rural areas in relation to its yield and highlights the hazards of its excessive withdrawal. The groundwater investigation in alluvial terrain and hard rock areas includes preparation of groundwater maps and measurement of yields, which are very helpful in deciphering the groundwater resource of an area for its exploitation. The chapter also elaborates the harmful effects caused by the presence of underground water in engineering structures such as dams, tunnels, and underground power houses.

9.2 HYDROLOGICAL CYCLE

Water covers nearly 75 per cent of the earth's surface. The radiation from the sun changes the surface water of rivers, lakes, and oceans from liquid to vapour state. The water vapour rises in the atmosphere and forms clouds and then falls back on the earth as precipitation in the form of rains, hails, or snow. In this way, the earth's water is under continuous circulation between land, atmosphere, and ocean. This process of movement, exchange, and storage of earth's water is known as the *hydrological cycle* (Fig. 9.1). When the precipitation falls on the land surface, a part

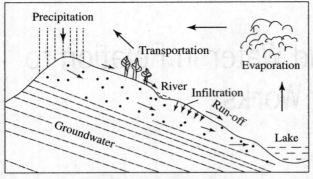

Fig. 9.1 Hydrological cycle of earth's water

of the water moves down under gravity into the ground as *infiltration*. This is the major source of groundwater and is commonly referred to as recharge as it brings replenishment or *recharge* of groundwater resources.

The precipitation that occurs on the land surface represents the early stage of the hydrological cycle. The portion of water that percolates below the ground moves slowly through the subsurface and ultimately escapes into streams, lakes, and oceans where evaporation from these water bodies completes the hydrological cycle. Continuous exchange takes place between the groundwater and atmosphere during the downward movement of the precipitated water. Transpiration of plants and respiration of animals also contribute to this process. Plant roots in the soil adsorb a portion of the water that again enters into the atmosphere by a process called *evapotranspiration*. Diagenesis of sediments also has a distinct role to play in the hydrological cycle.

9.3 MODES OF OCCURENCE AND SOURCE OF UNDERGROUND WATER

Water that flows along the land surface remains manifested in rivers, lakes, and marshes as *surface water*. The water that percolates down and occurs beneath the ground surface occupying the pores of soil and open fractures of rocks is known as subsurface water or *underground water*. The *meteoric water*, such as rains and that derived from melting of ice or snow, is the main source from which underground water originates and gets replenished. Water from streams, canals, lakes, reservoirs, and dams also seeps into the ground and contributes to underground water.

9.3.1 Groundwater and Vadose Water

Surface water percolating down through the top unsaturated zone fills the pore spaces of soil and openings of fractured rocks of the underground region. Thus, a *zone of saturation* is created below the ground surface where all pores and voids are completely filled with water. The water that remains in the zone of saturation is termed *groundwater*. The top surface of the zone of saturation is called the *water table*. Above the zone of saturation is the *unsaturated zone* or *vadose zone*. The part of water that remains in the unsaturated zone is called *vadose water* (Fig. 9.2).

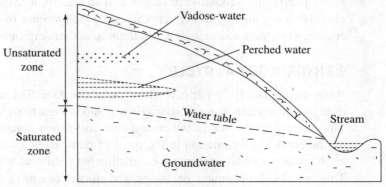

Fig. 9.2 Modes of occurrence of underground water

9.3.2 Juvenile Water and Connate Water

The underground water that originates from molten magma is called *juvenile water*. The water that coexists with igneous and sedimentary rocks, since its origin is the *connate water*. Both these types of water have their source wholly in deep subterranean region. As such, if the water table is very deep where the hydrostatic pressure is high, the meteoric water may mix with juvenile or connate water expelled during compaction. In the hydrological cycle, the sea that receives water from surface and rains loses part of it not only by evaporation to atmosphere, but also as connate water to the lithosphere. Finally, the expulsion of the connate water by diagenesis of sediments completes the hydrological cycle. The gain in the water in the lower crust as juvenile water is balanced to some extent by loss through the migration of water to a deeper level in the *subduction* zone (see Section 20.4).

9.4 FACTORS DECIDING THE CONFIGURATION OF WATER TABLE

The water table is not a flat surface. It roughly takes the configuration of the surface relief (Fig. 9.3). The depth to water table depends upon the nature of the constituent materials, land slope pattern, and density of precipitation. In general, there is an even water table in homogeneous, porous, and permeable materials, but it shifts its position as well as level depending upon the conditions of infiltration from source water. The configuration of the water table can be arrived at by measuring and plotting the water levels in several wells or boreholes of an area. It will be observed that the highest point in the water level occurs below the peak of a hill. The lowermost point is generally found to be intersecting the surface near a stream or at a marshy land. The difference in level between the highest point below a peak and the lowest point near the stream valley is the *hydraulic head*.

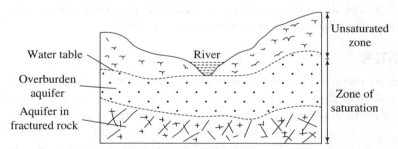

Fig. 9.3 Configuration of water table and groundwater aquifer in overburden and fractured rocks

9.4.1 Fluctuation of Water Table

The water table fluctuates with the change of seasons. During the rainy season, the water table of high rainfall areas shows an upward shift due to high infiltration of surface water. During dry periods such as summer months, the water table tends to move downward; however, in hilly terrains where a large quantity of water enters into the ground during summer months from melting of ice and snow, there is a consequent rise in the water table. In winter months too, the water table goes down because of less inflow of water. In plains, however, during drought, the water table falls appreciably due to discharge of groundwater through springs, wells, and rivers.

The position of the water table can be measured by taking the level of the standing water level of a well or a borehole that penetrates down to the zone of saturation. The down-gradient direction of the water table nearly follows the ground slope, being high beneath the topographic high and low towards the stream valley. Under the available head, water commonly flows at a slow rate but the rate of flow becomes faster near the area of escape to a stream.

The nature of subsurface material or rock strata controls the position of the water table. In cavernous limestone terrain called *karst* area, the water table is commonly an irregular surface. The piezometric head having equal water pressure is measured while investigating the groundwater behaviour of karstic terrain to plot the position of the water table including the flow pattern of subterranean water. The pattern of the piezometric surface (generally undulating in nature) in a karst country can be obtained from the measurement of water levels in a number of boreholes covering a large area. This surface shows an overall down-gradient direction of underground water flow but it does not necessarily follow the topographic slope.

9.4.2 Perched Water Table

In general, in an area, there is only one water table under the ground; however, in some places, a portion of the meteoric water while moving through the unsaturated zone may remain trapped as a local body of water above the water table, being surrounded by impervious layers. Such body of water in an isolated zone of saturation within the unsaturated zone is called the *perched water* (Fig. 9.2). The formation of a perched water body depends on the geological conditions favourable for it. A clay or shale bed is porous but impermeable to water. As such, a pervious deposit such as sandy soil or sandstone confined within shale or clay beds may create a *perched water table*. This means there will be one more water table besides the main one. In fact, the nature of the rock with respect to its permeable property is the main factor in forming the water table. Thus, several perched water tables may be formed in a thick soil deposit having alternate layers of impervious and pervious materials obstructing continuous downward movement of water. A well dug into the zone of perched water may not yield water for a continuous period as the well can be drained of its water in a short time.

9.5 SOIL WATER

In an area with a thick soil cover, the water that remains inside the voids of the soil is called the *soil water* (Fig. 9.4). There are two types of soil water, namely *free water* and *held water*. In the zone of saturation where soil water moves freely under the influence of gravity, it is known as free water. A part of the water while moving through the unsaturated zone is held in soil pores by attractive forces existing within the pores. This water called held water occurs in soil pores in three distinct ways as stated in the following subsections.

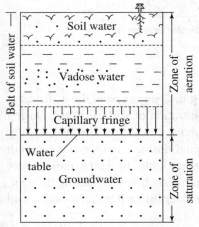

Fig. 9.4 Soil water and capillary fringe

9.5.1 Structural Water

Structural water remains in the crystal structure of soil minerals and can be separated only if the structure is broken. However, under loading due to an engineering structure on the soil, structural water cannot be removed. As such, structural water can be said to be inseparable from the soil particles.

9.5.2 Hygroscopic Water

Soil can freely adsorb a part of water from atmospheric moisture by physical forces of attraction. The water that remains in soil as tightly bound water by force of adhesion is known as *adsorbed water* or *hygroscopic water*. The attractive forces within the soil influence the behaviour of the adsorbed water but pore water is free of strong attractive forces. The quantity of adsorbed or hygroscopic water in soil depends on the nature of the soil particles, topographic configuration, temperature, and humidity. The hygroscopic capacity of a soil for a given temperature is expressed by the ratio of adsorbed moisture from saturated atmosphere to the weight of the oven-dried sample. The coarser the particles, the lower is the hygroscopic capacity, which is found to be about 1 per cent in sand, 7 per cent in silt, and 17 per cent in clays. Soil possessing adsorbed water has only a limited effect on engineering structures. The adsorbed water may get accumulated even under pavements and it may also spread as ice (as fine grains) into soil formed by freezing of moist air in frost and permafrost zones.

9.5.3 Capillary Water

The attractive force of the soil in the unsaturated zone close to the water table causes the upward movement of water. The pores or interstices of soil act as minute capillary tubes through which moisture (in the form of water) rises above the water table and fills the pore spaces due to capillary force. This moisture held thus in the soil is called the *capillary water* and the phenomenon is known as *capillary action*. The zone directly above the water table in which water is held by capillary action is the *capillary fringe*.

The capillary forces that create the zone of capillary water depend on factors such as the surface tension of water, pressure in water in relation to atmospheric pressure, and size of the soil pores. In soil mostly with coarse sand, the capillary water rises only a few centimetres above the water table, but in fine-grained (clayey) soil, the rise may be as high as 10 m. The rise is rapid in coarse-grained soil compared to fine-grained soil, where it may take several months or more than a year to attain maximum rise of capillary water. The capillary water can fully saturate the ground immediately above the water table, but further up, the soil may be partially filled with vadose water. However, if the water table is shallow and the soil is fine grained facilitating maximum rise of water, the zone above the groundwater table may be completely filled with capillary water.

9.6 CONFINED AND UNCONFINED AQUIFERS

Layers of sediment or rock below the surface that have high porosity and permeability are called aquifers. Layers of clay or bedrock that are highly impermeable to groundwater are commonly referred to as aquicludes.

There are mainly two types of aquifers—confined and unconfined aquifers. In a confined aquifer the saturated zone is capped by a confining layer, whereas in the unconfined aquifer there is no confining layer on top of the saturated zone. The confinement or un-confinement of the aquifer is based on porosity and permeability—examples being gravel, coarse-to-fine sand, fractured shale, silty and clayey soils. Some good aquicludes are unfractured shales, granite, and marble.

9.6.1 Aquifer and Aquiclude

An *aquifer* is a water-bearing body formed in the underground by infiltration of surface water. It may be found in overburden material or in bedrock that can hold and transmit large quantities of groundwater. In *overburden aquifer*, water fills the voids of the grains through which it moves.

In *bedrock aquifer*, water fills and flows through the interconnected open spaces due to joints, faults, fissures, and solution cavities in rocks (Fig. 9.3). An overburden aquifer constituted of gravel and sand can yield a large quantity of water, as the void spaces are large enough to allow flow of large volumes of water. However, when silty and clayey types of soils occur in the overburden aquifer, the yield of water is poor due to low permeability of the soils. In general, good water-bearing aquifer occurs in the interface of overburden aquifer and bedrock aquifer.

The yield of water in a bedrock aquifer depends on the availability of the interconnected open spaces for movement of underground water. If the bedrock is open jointed, porous, and permeable, appreciable quantity of water is expected from the aquifer. Sandstones having voids among the grains and also fracture openings are good aquifers as these hold and transmit a large volume of water. If an aquifer has its upper limit as the top of the saturated zone, it is known as an *unconfined aquifer*. In some geological set-ups, an aquifer may be overlain by clayey material or rocks such as shale through which water cannot move easily. Such a deposit or layer of rock through which water cannot move easily is called *aquiclude* and the underlying aquifer is a *confined aquifer*. A thick clay layer or a shale bed that is impervious that occurs overlying a porous sandstone aquifer will form an aquiclude.

9.6.2 Hydrogeological Criteria to Identify Aquifers

Groundwater aquifers occur in both soil (overburden) and rock. In general, overburden aquifers are the main source of groundwater and they provide a better yield than the bedrock aquifers. The aquifer sites are identified from the measurements of water table and the recharge of water in wells in both soils and rocks. The main hydrogeological indicators of the presence or absence of an aquifer of a place is the nature of the soil and rock types.

An alluvial country is the best producer of groundwater. Alluvium consists of loosely compacted sand grains with high porosity and permeability. These sand beds produce the maximum yield. At places, the sand and silt layers may occur with impervious clay. The sand layers confined within clay layers at places produce perched water bodies that also supply sufficient quantity of water. If a well is extended down to the perched water body, the supply may be for a short period only as the recharge takes time once the trapped water is exhausted.

Among the rock types, the clastic sedimentary rocks consisting of permeable strata such as sandstone and conglomerate are good sources of groundwater. Soluble sedimentary rocks such as limestone, if cavernous, provide a high yield of water. Igneous and metamorphic rocks are, in general, poor aquifers as the rocks are not capable of storing water, but if the rocks are intensely joined with interconnected open spaces, they may serve as good aquifers. The presence of springs, soil moisture, dense vegetation, wetlands, and streams acts as a surface indicator of the presence of groundwater at a shallow depth. The indirect evidence of the presence or absence of groundwater in the overburden is provided by the grain size of the soil, the degree of sorting or gradation, and the percentage of fines, whereas the nature of rock, fracture density, intensity of weathering, and the degree of solubility of rocks decide the availability of groundwater in rocks. Field measurement of permeability and water level through wells provide direct evidence about the quantum of yield of groundwater aquifers.

9.6.3 Artesian Flow and Natural Fountain

Water in a confined aquifer is under pressure. The high pressure head is caused by the restricted movement of water and also due to the weight of the overlying deposits. If a well is dug, water rises up to a point above the top of the aquifer called the *piezometric surface*. The water comes out of the well at the ground surface under pressure. This type of groundwater flow is known as *artesian flow* and the well dug is the *artesian well* (Fig. 9.5).

The water pressure (*piezometric head*) of a confined aquifer where artesian condition has developed can be known by measuring the water level using a borehole acting as a piezometer. The pressure on the surface may be higher than that on the ground surface. This will cause the

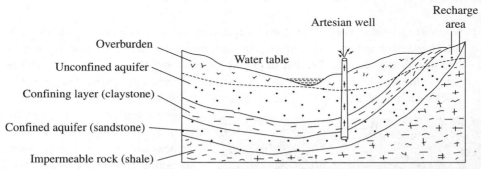

Fig. 9.5 Flow of groundwater from an artesian well penetrating confined and unconfined aquifers

water to flow out making a *natural fountain*. The following conditions are necessary for the existence of an artesian well:

- A confined aquifer (say, in sandstone bed) in a dipping or synclinal disposition should be present.
- The sandstone must receive discharge of water from precipitation in the exposed part that may be a place far away from the well site.
- The water must move down through the dipping aquifer and accumulate at the bottom so that hydrostatic pressure is produced as a result of being confined between two aquicludes such as shale beds.

For an aquifer to continuously produce sufficient quantity of water, it must be replenished after withdrawal. If the water table occurs at a shallow depth, the unconfined aquifer is recharged from local precipitation. However, in case of confined aquifer having artesian condition, the aquifer is charged with water if there is rainfall in the area where the strata meet with the surface. The rate of flow of underground water is generally slow—only a metre or so in a day. In case of movement through limestone cavities, the flow may take place rapidly like a surface channel. The duration of water residing underground (*resident time*) at a place may be a few days to several months and even hundreds of years. In an unconfined aquifer, where groundwater is available at a shallow depth close to ground surface, a dug well will supply water, but in confined aquifer the well may need to penetrate very deep to withdraw water.

9.6.4 Springs and Their Origin

Underground water naturally emerging at the surface with some force is called a *spring*. Springs appear when the water table intersects the surface with development of swamps or lakes.

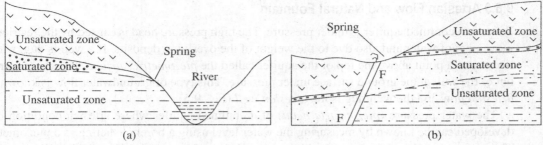

Fig. 9.6 Formation of a spring: (a) at the interface of saturated and unsaturated zones; and (b) along a fault (F-F) separating saturated and unsaturated zones

The course and flow rate of a spring depend on the permeability and structure of the materials through which water moves. Springs are commonly formed from groundwater issuing out from the interface of saturated and unsaturated zones meeting the ground slope. The captive water in sandstone is generally found to flow down to surface as spring where it is capped by impermeable rocks such as shale, see Fig. 9.6(a). Such a condition may also develop along a fault cutting a bedrock aquifer where groundwater will flow out as spring from contact between the saturated bedrock and the fault zone with impermeable clay gouge, Fig. 9.6(b). Water moving down from a perched water table towards a valley slope may also form springs. Springs are most common in sandstone, karstic limestone, vesicular lava, and highly fissured or fragmented rocks (Fig. 9.7). When the temperature of the water of a spring is more than the yearly average temperature of the given place, the spring is called a *thermal spring* or *hot spring*. The actual temperature of a hot spring may even be very close to the boiling point of water.

Fig. 9.7 Springs in weathered and fragmented rocks above a tunnel drift indicating seepage problem during tunnelling along the drift level

9.7 WATER RETAINING AND TRANSMITTING CAPACITY OF SOIL AND ROCK

Whether a rock mass or soil mass can retain as well as transmit water through it is dependent on its properties such as porosity and permeability. These two properties of soil and rocks discussed in the following subsections are important in the study of groundwater resources of an area.

9.7.1 Porosity of Soil

Porosity is a measure of the capacity of a rock or soil mass to store fluid in it. It is expressed as the percentage of the total volume of voids contained in the gross volume of a specimen of soil or rock. Groundwater saturates porous unconsolidated deposit and fractured bedrock. River sand and gravel possess high porosity, but when mixed with silt and clay, the porosity is reduced as these fill the spaces between the coarser particles. Graded sands with coarse and rounded grains are more porous than the poorly sorted sediments with angular grains, see Fig. 9.8(a) and Fig. 9.8(b).

Voids of varying dimensions are present in scoriaceous lavas formed from escape of gases, Fig. 9.8(c) and also in soluble rocks such as limestone by formation of caverns, Fig. 9.8(d). In some igneous and metamorphic rocks free from weathering and fractures, porosity is almost nil because of the tightly woven minerals. In sedimentary rocks, the pore spaces are dependent on compaction or packing during formation of the strata and also on cementing materials and grain size. Clastic sedimentary rocks such as conglomerate and coarse-grained sandstone may have very high porosity due to large intergranular voids. Most significant open spaces are associated with rock structures such as folds, faults, and joints. A weathered product of rock, called *regolith*, possesses high porosity.

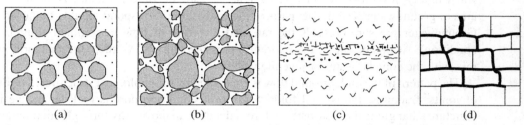

Fig. 9.8 Porosity in different types of soils and rocks: (a) high porosity in well-sorted sand; (b) poor porosity in poorly sorted sand; (c) porosity due to voids in scoriaceous lava; and (d) porosity in cavernous limestone

The average values of porosity and permeability of unconsolidated materials are shown in Table 9.1.

Table 9.1 Porosity and permeability of unconsolidated material (Watson 1983)

Medium	Porosity (percentage)	Permeability (cm/sec)
Gravel	25–45	More than 10^{-1}
Coarse sand	30–50	10^{-1}–10^{-3}
Fine sand and silt	40–50	10^{-3}–10^{-5}
Clay and silty clay	35–55	10^{-5}–10^{-9}

9.7.2 Permeability and Its Relation to Porosity

Permeability is the transmitting capacity of the water. The rate of groundwater movement and quantity of water that can be pumped out from underground through wells are known from the determination of permeability of soil or rock through which water flows. Permeability of unconsolidated gravels and sand deposits is very high. Clay, though has high porosity, is low in permeability. In bauxite (a rock used as aluminium ore), there are numerous void spaces, but they are not interconnected and as such the rock is impermeable. Rocks subjected to faulting and intense jointing provide sufficient open spaces for water to flow and as such these rocks are permeable to a great extent. A non-porous and less-fractured rock has low porosity and low permeability. Limestone without solution effect and crystalline (igneous) rocks are of this category. However, soluble limestone with interconnected cavities exhibits high permeability and allows flow of large quantities of water. Coarse-grained sandstone, pebble stone, and conglomerate generally exhibit high permeability, but if the pore spaces are filled with mineral matter, the permeability is reduced. Pumping tests are conducted through boreholes to measure the rate of groundwater flow through subsoil and subsurface rock mass (see Section 11.8).

9.8 GROUNDWATER MOVEMENT

Groundwater flows from higher elevation to lower elevation due to potential energy derived from the gravity. During its downward flow, groundwater roughly follows the slope of the water table. If the water table of a region is represented by a contour map, the movement of groundwater will be from higher to lower contour direction until its escape to a lower valley. The water finally flows towards the discharge area. A stream valley or a lake is the ideal discharge area of groundwater. The deeper parts of the water table correspond roughly to the stream level. Groundwater pressure controls the rate and direction of flow in confined aquifers (artesian condition). At higher depth, the hydraulic pressure is high. As such, the meteoric water may mix with the connate water or juvenile water.

The pattern of the underground flow is controlled by the geological structures. The orientation of joints, fractures, faults, and folds with porous and non-porous rock layers are important structures that guide the flow pattern. Groundwater can move only through interconnected open spaces. The permeability of a sedimentary rock may be affected if interrupted by other non-porous and impermeable rocks. In accordance with Darcy's law, the rate of flow of water in an aquifer depends on both the hydraulic head and the capacity of the aquifer to transmit by any mechanism. The rate of groundwater flow (*hydraulic conductivity*) is very high through the interconnected open spaces of the fractured and highly porous rocks such as sandstone and the voids in gravel and sand deposits. Conversely, rocks such as shale and clay deposits have very poor water-transmitting capacity.

Tracers or dyes of various types are used to measure the direction and gradient of groundwater flow. From the middle of the last century, tritium in groundwater derived from rain water is being used as a tracer. Nuclear devices (e.g., Hiroshima bombing) in the middle of the last century increased the tritium content in water. Tritium concentrates have been used to investigate the rate of recharge of aquifer, where recharge is due to the entry of rain water from adjacent strata. Tracer study using environmental tritium was carried out in karstic limestone of Kopili project in Meghalaya (see Section 22.5).

9.9 SYSTEMATIC GROUNDWATER INVESTIGATION

Geological Survey of India first started the systematic groundwater investigation covering large parts of the country. Currently, Central Ground Water Board conducts systematic survey of groundwater throughout India. Such activities are oriented towards a comprehensive study of the groundwater regimes in different geohydrological zones controlled by geomorphic, geological, and tectonic environments.

9.9.1 Basin-wise Study

The main objectives of groundwater study include the following:

- Identification of aquifers and preparation of water table maps for the various seasons, assessment of the groundwater recharge, and determination of the chemical characteristics of the groundwater
- Study of the relation between surface water and groundwater

- Determination of the hydraulic characteristics of aquifers
- Study of the problem of saline water invasion into aquifers in the coastal areas

The investigation of the groundwater resources was initially taken up in alluvial areas such as all the major river basins and sub-basins, but now the study has been extended covering the entire country including the hard rock regions. A drilling unit is attached to help exploratory work and several regional chemical laboratories take up analysis of water samples. Several hydrograph stations are set up to measure the seasonal water level fluctuations throughout the country.

9.9.2 Groundwater in Hard Rock Areas

A large part of India is covered by hard rocks in which the occurrence of groundwater is controlled by factors different from those of the alluvial and unconsolidated sediments. In hard rocks, groundwater occurs mainly in the top mantle of weathered rocks and also in the open spaces of interconnected joints, fractures, and solution cavities. All igneous and metamorphic rocks and compacted sedimentary formations are included in hard rock terrains. Based on the lithology and structure, hard rocks can be divided into the following four categories in which occurrence of groundwater is controlled by distinctive features:

- (i) Crystalline and intrusive rocks such as granite, gneiss, schist, and quartzite
- (ii) Effusive lava flows (the trap rocks)
- (iii) Compact sedimentary formations such as the Cuddapah and Vindhyan rocks and compact sedimentaries of the Himalayas
- (iv) Carbonate rocks (e.g., limestone) that often undergo solution to form cavities

In all the four types, groundwater is present in different quantities in the zones of weathering and fracturing. However, the effusive rocks with 'red boles' and inter-trappean beds (under category (ii)) and the carbonate rocks such as limestone and dolomite with solution channels (under category (iv)) have the largest groundwater resources among hard rocks. Groundwater studies for the hard rock terrains include the understanding of the tectonic framework of the region, the inherent physical features of the rocks controlling weathering, depth and lateral extension of weathering, and geochemical characteristics of groundwater available from the zones of weathering and fracturing.

9.9.3 Groundwater Maps

Groundwater maps are prepared showing many features such as depth to water table, yield, and quality of water in aquifers. In systematic groundwater survey, the depth to water table of a site with respect to surface (parapet level) is measured from dug wells or boreholes. If several such wells or boreholes are available covering a large area, the measurement of water levels throughout the year will show the fluctuation in the groundwater level. In boreholes, the water table is measured by inserting piezometric tubes when the nature of water bearing strata and the existence of artesian condition are also recorded. The water table is generally found to be at a shallow depth during rainy season and at a comparatively deeper level in the summer and winter months.

The data collected by measuring the water table from wells or boreholes helps to prepare the water table contour maps of the terrain showing seasonal variations. The contours plotted for water levels will represent a curved surface and may be similar to the overlying ground configuration. Water flows from the higher contour level towards down-gradient direction until it escapes to a stream, lake, or swamp. As the water table fluctuates with seasonal changes in precipitation, such maps are prepared for the dry period as well as for the period of maximum rainfall. The measurement of groundwater is, therefore, continued round the year. Such maps and yield measurements are of immense help in the planning and management of groundwater for the purpose of drinking and irrigation. These maps also help in deciphering problems at construction sites and adopting remedial measures.

9.10 UTILITY, YIELD, WITHDRAWAL, AND ARTIFICIAL RECHARGE OF GROUNDWATER

Man cannot live without water. Groundwater is stated to be the gift of nature to man. Over 90 per cent of world's fresh water is available from groundwater that occurs in most areas in overburden or bedrock aquifers. Groundwater is used for many purposes such as drinking and other domestic purposes and irrigation of agricultural lands, as well as in the industrial units. As the water moves through the soil, it gets filtered and purified. Surface water available in streams and lakes may be easily contaminated and requires treatment and long transportation before being used for drinking purposes.

Groundwater is not easily contaminated and can be used safely in most places as drinking water just by tapping through wells. The quality of the groundwater is, however, dependent on the nature of material or bedrock through which it flows in the underground region and also on the pH value (hydrogen ion content) and Eh (oxidation–reduction) condition. During its passage through bedrocks or overburden aquifers, groundwater is enriched with minerals such as calcium, magnesium, and iron by solution from aquifer materials. When groundwater comes from a deeper part, sometimes by influx of juvenile water, the water contains more of the dissolved minerals good for health.

9.10.1 Uses in Urban and Rural Areas

Groundwater is used in both urban and rural areas. The most common method of groundwater development in India is the dug well and to a lesser extent dug-cum-bored well. Many cities and small towns meet their need of water supply from underground water combined with surface water. Groundwater is withdrawn by pumping through deep wells. Withdrawal and supply of groundwater do not need costly treatment as in surface water. In plain areas, groundwater is used for irrigation purpose as it is easy to withdraw from ground by simply pumping from the area of irrigation without involving any transportation cost. Groundwater is also used in some industrial units for cooling purposes of machinery and industrial processes. Groundwater is a better choice for industrial use as it maintains a uniform temperature and uniform quality when obtained from an area. Though all variants of groundwater contain mineral matters dissolved from soil or rock through which it flows, its quality and temperature, which remain constant at a given source, make it more desirable for industrial development.

Groundwater resources are of enormous value to the people of India where large sections of the population live in villages in the plains. They get their supply of drinking water from dug wells or tube wells. These water wells are simply tubes or shafts dug to a depth below the water table. The dug wells may range from about a metre to as much as 10 m in diameter. In general, shallow wells restricted to 5–30 m in depth obtain groundwater, but in some dry areas the wells may have to penetrate deeper to meet the water table. These water wells are influenced by gravity flow in which the discharge comes from groundwater zone surrounding the well. Even after withdrawing water, the wells tend to maintain the level with the surrounding areas. All gravity wells, however, show seasonal fluctuations in water level and may dry up during summer.

9.10.2 Yield in Different Parts of India

In rocky areas, wells in weathered and jointed rocks may yield from about 22,500 litres to as much as 450,000 litres per day. In general, the yield is restricted to 100,000–150,000 litres per day. These wells may irrigate from a hectare to ten hectares of land. Many towns in hard rock areas also receive water supply from dug wells or dug-cum-bored wells of rocky terrains. Alluvial terrains are, however, the largest suppliers of groundwater. In the Ganges basin of West Bengal, aquifers at a depth of 150 m from surface yield groundwater to an extent of 100,000–200,000 litres per hour. Aquifers occurring even at a depth around 300 m may supply a large quantity of groundwater. Tube wells tapping artesian aquifers recorded in parts of Kashmir valley may yield around 22,500 litres per hour. In Punjab within the Sindh basin, the yield from 100 m deep tube wells is up to about 180,000 litres per hour (Niyogi and Seth. 1972).

9.10.3 Cone of Depression

A *cone of depression* is formed as a result of withdrawal of water by pumping from an aquifer through a well (Fig. 9.9). The pumping rate and the permeability of the aquifer material decide the shape of the cone. Continuous drawing of water by pumping from the well causes progressive drawdown and increase in the cone of depression. The difference in height between the normal water table and the water surface is the measure of the drawdown. If the withdrawal is reduced, the drawdown decreases when the water surface moves up with recharge. However, after a critical limit of drawdown due to excessive withdrawal, the water level remains standing without further rise in level. Heavy pumping of underground water may result in such a condition, and as such pumping from wells should be limited so that recharge does not suffer.

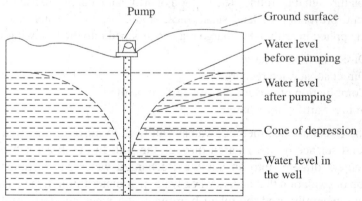

Fig. 9.9   Cone of depression formed in groundwater level

While using underground water from a well for drinking purpose, care should be taken that the water is not polluted. Shallow dug wells may get polluted from surface pollutants trickling in during rains. Waste products dumped in the areas surrounding wells are also potential source of danger of polluting the water. The most prolific effect of *chemical pollution* due to the presence of arsenic in water has been recorded from various parts of West Bengal including the capital city of Kolkata and its neighbouring regions. Serious skin diseases have been reported from the use of groundwater where arsenic is present even in traces beyond the threshold value. It is not easy to remove this chemical from the water, and as such arsenic-contaminated water should not be used for drinking or domestic purposes. Of late, water treatment plants have been installed in some selected places to remove arsenic from water and make it suitable for drinking purposes.

9.10.4 Environmental Degradation

Excessive withdrawal of groundwater by pumping through wells may create problems of land subsidence and other environmental degradation. Several instances of this problem can be found in different parts of India. In Farrukhabad district of Uttar Pradesh, excessive withdrawal of groundwater from shallow aquifers of the Gangetic alluvium has led to the development of piezometric pressure and reduction in volume of subsoil causing ground subsidence and cracking of ground and even minor damage of the dwellings (Pande 1999).

Over the years, expansion of cities with construction of multi-storeyed building complexes has generated a greater demand of water that is being met through overdrawing of groundwater by digging deep wells. This has caused serious drawdown of the water table endangering the stability of buildings. Sandy and silty soil, wherever it becomes unsaturated by the drawdown, is likely to get compressed under the load of overlying buildings causing their deformation. Kolkata, located on the Gangetic alluvium, has faced such problems of deformation and tilting of some buildings from excessive use of groundwater by sinking deep tube wells. This is, however, not a groundwater problem but a man-made problem that can be solved by a balanced use of groundwater.

9.10.5 Artificial Recharge

Artificial recharge is the planned effort of augmenting the depleting groundwater resources through works designed to increase natural replenishment or ensure percolation of surface water into the groundwater aquifers. The beneficial purposes include rainwater harvesting and conservation or disposal of floodwater, control of saltwater intrusion, storage of water to reduce pumping cost, and improvement of water quality by removing suspended solids by the process of filtration through the ground. The method of artificial recharge also has application in wastewater treatment, prevention of land subsidence, storage of fresh water, and crop development. The method applied in artificial recharge may be grouped into the following four categories:

(i) Direct surface recharge
(ii) Direct subsurface recharge
(iii) Combination of surface and subsurface method
(iv) Indirect artificial recharge

In the direct surface method, water is allowed to move to the groundwater aquifer directly from land surface by percolation through the soil. In the direct subsurface recharge method, water enters directly into an aquifer. Recharged water passes into the aquifer without the filtration or oxidation that occurs when water percolates through the unsaturated zone. Recharge wells are generally used to replenish groundwater when aquifers are deep and separated from land surface by materials of low permeability. Recharge wells are used to dispose of industrial wastewater and to add freshwater to coastal aquifers.

In the combination of surface and subsurface method, the two methods are applied in conjunction with each other to meet specific needs of recharge. In the indirect method of artificial recharge, infiltration gallery pumping stations are installed near the water bodies such as streams or lakes. The process involves lowering of groundwater level and allowing infiltration into the drainage basin to enhance groundwater reserve. Groundwater barriers or dams are also built within river beds in many places to obstruct and detain groundwater flows so as to sustain the storage capacity of the aquifer and to meet the water demand when needed.

This method, however, has less control on the quantity and quality of the water than the direct method. If an aquifer contains poor quality water, this can be cleansed by means of artificial recharge. This is done by increasing the oxygen and reducing the carbon dioxide content. The quality of water in the groundwater aquifer can also be improved by allowing infiltration of accumulating rainwater after heavy downpour or flood water into the aquifer. Groundwater aquifer is also cleansed by means of transferring surface water through natural sand infiltration to re-originate the water. In India, the method of artificial recharge, irrespective of the quality, provides reliable irrigation facility for cultivators in rural areas.

9.10.6 Saltwater Intrusion in Groundwater

Sea water moving inland is called *saltwater intrusion*. Alternatively, mineral-bearing beds containing salt may leach into groundwater to create salinity of fresh underground water. In fact, mixing of freshwater and saline water may take place in different ways. In areas close to sea, salt water may intrude on fresh water causing contamination. It may also happen that part of salt water in certain parts of subsurface regions is enriched by fresh water, which overrides and displaces the salt water. In coastal areas, it is also observed that a water body is in equilibrium with fresh water that floats over the salt water. This feature is known as Ghyben–Herzberg balance (Krynine and Judd, 1957). The equilibrium condition is lost due to pumping while drawing groundwater from this part resulting in contamination of groundwater by intermixing of fresh and salt water.

9.11 GROUNDWATER PROVINCE AND HAZARDS OF USING ITS WATER

A groundwater province or basin is the underground water reservoir or aquifer extending over large areas having distinct hydrological and geological boundaries. It is generally difficult to determine where the exact boundary of a groundwater province lies. In certain cases, however, there is a definite geological boundary such as a deposit of sedimentary formation, which gives a clear understanding of the aquifer with its ends. A groundwater province may be very large like the alluvium of the Ganges or it may occur as long but narrow underground channel or buried channel having sufficient flow of water. In a groundwater basin, water may flow like a surface stream. The slope of such flow can be known from measuring the water table at different places of the basin through boreholes and dug wells and then preparing a contour map of water table (see Section 9.13) such as a structure contour map covering the entire basin area. The direction of surface of higher contour to lower contour indicates the direction of groundwater movement.

The aquifer systems that have formed large and distinct groundwater provinces of India include extensive alluvium, alluvium and sandstone combined, limestone with solution cavities, crystalline rocks having joints and fractures, and trap basalts. According to Taylor (1959), India has eight groundwater provinces on hydrological evidence as follows:

  (i) Precambrian crystalline province
 (ii) Precambrian sedimentary province
(iii) Gondwana sedimentary province
 (iv) Deccan trap province
  (v) Cenozoic sedimentary province
 (vi) Cenozoic fault province
(vii) Ganges–Brahmaputra–Indus alluvial province
(viii) Himalayan highland province

India's main groundwater provinces cover the Ganges–Brahmaputra and Himalayan region, each constituting a single province. The rocks of the Himalayan high mountain region are, however, not conducive to creating good groundwater aquifers. The *Babars* and *Trais* of the Himalayas act as a potential groundwater recharge zone for the aquifer systems downhill.

The deeper confined aquifers in these formations show flowing artesian conditions. The Indo-Gangetic alluvium occurring in the Himalayan fore-deep forms the most productive and extensive multi-aquifer system of India. The sedimentary formations of Gondwana and tertiary sedimentary deposits are also productive groundwater aquifers. The porosity and permeability characteristics provided by open joints and fissures in the Deccan basalt and Precambrian crystalline rocks have created groundwater aquifer systems in these hard rocks.

There are chances of contamination of water in a groundwater province. While investigating an aquifer for groundwater, it will be known whether the water is contaminated or it is safe to use the water. Contaminated surface water may infiltrate into groundwater and pollute it. Industrial waste products (even nuclear waste) may percolate and contaminate groundwater. Pollutants released from the ground can go down into the groundwater thus adding the harmful pollutants into the aquifer.

In many cases, the hazard is dependent of the geology of the basin. Stratigraphy of an area plays an important part in transporting pollutants. Cracks are developed in the ground due to earthquakes that create avenues for the entrance of surface water with pollutants and contamination with groundwater. Careful consideration is necessary before supplying water to people for drinking, agriculture, and irrigation purposes. In case the water is contaminated, it is necessary to process the water for purification. However, in some cases, the water may go beyond the usable limit when alternative water supply is needed for the people.

If there is substantial bacteria content in the surface material (e.g., where a sewage disposal problem exists), the percolated water from the surface will lead to change in the components of groundwater. Microbial dissolution of iron oxide from the mineral constituent of groundwater basin rock may cause release of arsenic from the strata, which remains dissolved in water and causes arsenic pollution of water. Humans may develop serious skin and other diseases if such water is used for drinking or other purposes. Such arsenic contamination in groundwater aquifers and people suffering from its use are very widespread in rural areas of West Bengal and parts of Bangladesh.

Overutilization of water from aquifers of groundwater basins beyond the capacity of recharge is another concern. People consume more quantity of water than the aquifer can recharge, and as a result a water shortage may develop. As such, in the investigation of a project for groundwater aquifer, it is necessary to find how much water is available and how much can be safely used from the aquifer and suggest the project authority accordingly.

9.12 INFLUENCE OF UNDERGROUND WATER IN ENGINEERING CONSTRUCTIONS

Problems created by groundwater in areas of dams, reservoirs, and other engineering structures are numerous. The problems become acute if the rocks are fractured and soluble in nature. The following subsections give a picture of groundwater effect in the sites of major engineering structures.

9.12.1 Dams and Reservoirs

In the case of an earth embankment, capillary rise in water may saturate the soil below the earthen structure, thus reducing its bearing strength. In addition, seepage pressure of groundwater flow may take place below an earthen structure. The combined effect may even cause failure of the structure. As a remedy, the earthen structure may have to be placed below the zone of capillary action. The soil below the structure is required to be compressed to an extent as though an equivalent external load is placed on the surface of the soil mass.

Fractured rocks, faults, and shear zones and folded strata with tension joints are the structural features in rocks that provide path to movement of groundwater, which emerges as seepage through areas near dams and reservoirs. The configuration of the water table changes with the gradual filling of a reservoir. Under full reservoir condition, the water table rises to the upper surfaces where the material may be highly porous and permeable or bedrock may be faulted. The new condition may result in an excessive seepage or leakage of impounded water. Rise in groundwater level exerts more pressure on the pores or spaces within an already saturated rock, and this may create a new situation wherein leakage may take place. The sides of the reservoir under the new condition become vulnerable to slide.

If porous sandstones alternating with shales occur by the side of a reservoir, it may cause a problem of slide under the new condition when bedrock becomes saturated. Seepage takes place through the interface of sandstone and shale. The condition becomes more prolific if an unsaturated material exists along the side overlying a crystalline rock. The percolating water reduces the strength of the shale beds that become unstable. Such unstable beds create a danger of sliding. The hydraulic gradient of the reservoir changes and poses leakage problem of stored water to adjoining valleys.

Underground water circulating in soluble rocks creates solution cavities of different dimensions. When such a feature of cavity formation in limestone is very prolific, it results in karst condition in underground limestone through which a large volume of water may escape from a reservoir after it is filled with water. The cavities previously filled with soil or other materials are washed away making paths for the flow of water through the solution cavities. In some places, the flow of water through the underground cavities is like a channel when the entire reservoir may become dry. Such leakage may also take place below a dam or its abutments if fissures and solution cavities occur. Occurrence of widespread solution cavities may create a problem of leakage in storage dams in a limestone terrain and its remedy lies in the proper identification of the cavities and then resorting to commensurate treatment.

9.12.2 Road Pavements and Soil Slopes

Movement of water through the soil mass may cause a pavement's failure. Water may be drawn in a pavement by capillary action or infiltrated through cracks into the subgrade soil beneath a road pavement or runway. The accumulated water reduces the bearing capacity of the soil. As a result, the pavement may subside or get deformed due to wheel movement (Fig. 9.10). One simple solution is to remove the material prone to capillary action and backfill the portion with granular material. Another approach is to provide a layer of soil to the pavement structure unaffected by capillary action. Seepage of water through soil mass or fractured rock of a land slope or open-cut may cause erosion or its failure. To control the effect, the design of the pavement requires provision of adequate drainage of the subsoil so that seepage pressure does not build up.

Fig. 9.10 Subsidence and deformation of road by seepage water beneath the road pavement and later vehicular movement in a highway in Darjeeling, West Bengal

Retaining structures with drainage by weep holes are provided to protect the slope from groundwater hazards. Highways or runways constructed in areas with water table close to the ground surface face problems from capillary water or seepage of water that reduces the strength of the structure. The control measure lies in lowering the water table by excavating side trenches. The problem is also tackled by building the road or runway on a shallow fill (sand blanket) about a metre above the ground surface that prevents capillary action (see Section 19.2.2).

9.12.3 Problems in Tunnelling

Groundwater brings about acute problems in tunnelling. A dry tunnel is relatively safe, but when the tunnel passes through zones saturated with water the condition aggravates. Folds, faults, and joints are common structural problems. The problem is intensified many times when the divisional planes are filled with water, as saturation with water weakens the shear strength of the rock. Such water-bearing zones in structurally weak rocks, if present in a tunnel, may cause collapse of the tunnel.

The presence of underground water can be known by measuring the depth of water table along the tunnel alignment through boreholes keeping piezometers at least for a year during the investigation stage. The groundwater study helps to decipher the nature of probable water flow into the tunnel under the tunnelling condition. It is not only the water but also the loose materials that may rush into the tunnel that are of concern. In Barapani tunnel in Meghalaya, the inflow of groundwater along with sand rush had completely choked the tunnel. As a result, tunnelling work remained suspended for several months until the dry season when the sandy material was removed and the tunnel was thickly lined before resuming the work (Gangopadhyay 1971). It was necessary to use high-power pump for dewatering during tunnelling. The fractured rocks through which water flows were required to be grouted and lining provided to stop tunnel collapse. Grouting reduced the inflow even though initially it could not be stopped completely, increasing further workability.

9.12.4 Water Retaining Structures

The configuration of the water table and its relationship to the ground surface are very significant aspects in the construction of water retaining structures. The surface water (e.g., rains) percolates through the unsaturated zones and reaches the zone of groundwater, but at places it comes out as seepage water or gush out as *spring* if it gets an avenue through the ground. The interface of the ground and the spring marks the position (level) of the water table. With the construction of a reservoir, there will be a tendency of the stored water to seep out if the bedrock material of the reservoir is porous and permeable. In such a case, the land where seepage takes place becomes generally swampy. In rainy season, the swampy land becomes filled with water, but in dry periods, when the water table goes down, these swamps get dried up.

9.13 CASE STUDY ON GROUNDWATER PROBLEMS AND THEIR SOLUTION

The knowledge of groundwater and water table maps are helpful in solving engineering and mining problems. During the construction of Bansagar dam in Madhya Pradesh, it was observed that the proposed reservoir in karstic terrain was likely to create flooding of the open-cast mine of steel grade limestone used in the Bhilai Plant. The Central Ground Water Board carried out a detailed investigation of the groundwater condition in limestone of the area; this included the preparation of a hydrogeological map of the limestone terrain to decide the suitability of locating the reservoir. Groundwater levels were measured through several dug wells and drill holes using them as piezometers.

Figure 9.11(a) represents the simplified hydrogeological map of the investigated area. As may be seen in the map, the study indicated two distinct groundwater systems in the limestone (Kajrahat limestone) occurring in the right bank of the dam. System A exhibited deep karstification and copious flow of groundwater. System B showed the presence of massive limestone with general absence of solution cavities down to 80 m with meagre discharge, Fig. 9.11(b).

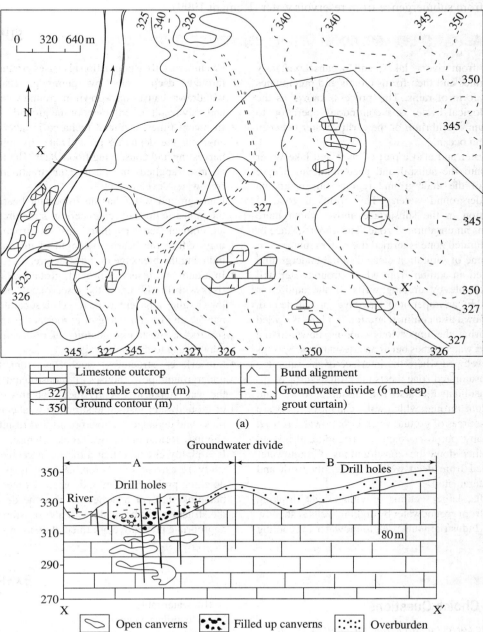

	Limestone outcrop		Bund alignment
327	Water table contour (m)	⊧ = =	Groundwater divide (6 m-deep
~ 350	Ground contour (m)		grout curtain)

(a)

Groundwater divide

Open canverns   Filled up canverns   Overburden

(b)

Fig. 9.11 Bansagar project, Madhya Pradesh: (a) hydrogeological map of reservoir limestone; and (b) geological section along X–X' showing groundwater divide with subsurface cavities (zone A) and massive limestone (zone B)

After consideration of the limestone condition of the reservoir area, it was planned to take up corrective measures. Accordingly, along the dividing ground of this zone, a 6 m-deep and 1.5 km-long grout curtain was provided as a precautionary measure against submergence by underground flow. In addition, a contour *bund* was raised around the mining area. Thus, knowledge of the groundwater condition helped in taking protective measures to save the mine from submergence from reservoir water (Romani 1999).

SUMMARY

- Water from rivers, lakes, and oceans constantly evaporates and then forms clouds and precipitates in the form of rains. This process, known as the hydrological cycle, is a continuous operation to maintain equilibrium of the earth's water between land and ocean.

- Rainwater and also a part of river and lake water seep into the subsoil and accumulate in deeper parts of the underground region; this is known as underground water. This forms a zone of saturation in the subsurface above which there remains an unsaturated zone. The upper surface of the saturated zone is termed the water table.

- The zone of saturation created in the underground is called an aquifer from where groundwater will be available by digging wells. If the aquifer is capped by an impervious clay layer or a shale bed, it is known as a confined aquifer. If a hole is drilled puncturing the impervious capping down to the aquifer, water comes out as an artesian flow. Springs are developed where the interface of the saturated and unsaturated zone meets the ground surface.

- Unconsolidated coarse soil, fissured rocks, and sandstone through which water can move freely are good sources of groundwater. Groundwater is used for many purposes such as irrigation, drinking, and other domestic consumptions. Groundwater obtained from deep wells is generally fresh and uncontaminated.

- In India, large sections of villages get drinking water from groundwater by digging wells at shallow depth. Industrial units use groundwater for cooling machinery. Urban population also uses groundwater through deep wells by pumping. Excessive withdrawal causes drawdown in groundwater level and may result in settlement of the ground.

- In recent times, artificial recharge is taken up to augment the depleting groundwater by rainwater harvesting and conserving floodwater. The method is also applied in wastewater treatment and freshwater development.

- Underground water has an important bearing on the construction of engineering structures. In an embankment, capillary rise in groundwater may reduce the bearing strength of fill materials inducing the breakage of the structure.

- In a dam and reservoir project, groundwater level rises with the filling of the reservoir. If there is a lower valley adjacent to the reservoir separated by a narrow divide, consisting of porous and permeable soil/rocks, substantial quantity of reservoir water may escape to the lower valley.

- Subsurface structures such as tunnels or underground powerhouses, if constructed below the water table, frequently face problems of rush of groundwater inside the structure weakening the rocks and endangering overbreaks and resulting in chimney formation and wall rock collapse.

- It is highly essential that a thorough geotechnical study is carried out beforehand to assess such potential problems from underground water in the planning stage of a project so that the design of the structure can be made with a cost estimate for appropriate preventive or remedial measures

EXERCISES

Multiple Choice Questions

Choose the correct answer from the choices given:

1. The top surface of the zone of saturation of underground water is called:
   (a) water surface
   (b) water table
   (c) capillary water

2. The underground water that originates from molten magma is called:
   (a) juvenile water
   (b) connate water
   (c) hygroscopic water

3. The water that remains in an unsaturated zone above a saturated zone in the underground is:
   (a) free water
   (b) perched water
   (c) vadose water

4. Water that remains in igneous and sedimentary rocks since its origin is known as:
   (a) free water
   (b) connate water
   (c) groundwater

5. Aquiclude is formed in an impervious shale bed occurring in the subsurface if:
   (a) a porous sandstone aquifer underlies this shale bed
   (b) the sandstone aquifer is capped by an impervious shale bed
   (c) both (a) and (b)

6. Contamination of groundwater happens by:
   (a) contaminated surface water entering into groundwater
   (b) mixing of industrial waste with groundwater
   (c) both (a) and (b)

7. Withdrawal of groundwater by pumping creates:
   (a) upheaval of underground strata
   (b) collapse of wall rocks in wells
   (c) formation of depression in water level like a cone

8. Permeability of gravel deposit is:
   (a) less than 10 m/sec
   (b) more than 10 m/sec
   (c) almost nil

9. The yield for dug well in an aquifer of jointed and fractured rock is generally:
   (a) less than 20,500 litres per day
   (b) more than one million litres per hour
   (c) one million litres per day

10. In a karstic terrain, construction of reservoir poses the problem of widespread leakage. Under this condition, it will be better to:
    (a) abandon the reservoir project
    (b) investigate and find a solution to stop the leakage at an expense
    (c) keep the reservoir for years allowing the cavities to be sealed by natural siltation

Review Questions

1. Explain the hydrological cycle of earth's water. What is evapotranspiration?
2. What is the source of groundwater? Write short notes on perched water, vadose water, juvenile water, and connate water.
3. What is water table? Give a short account of the configuration of water table in relation to topography and rain water infiltration to ground.
4. Describe soil water, hygroscopic water, and capillary water. Distinguish between aquifer and aquiclude.
5. Where do groundwater aquifers occur? Name the type of geological formation that is a good source of groundwater.
6. Discuss and illustrate geological conditions that give rise to artisan flow of groundwater.
7. How are springs formed? Explain the type of sratigraphic and structural set-up suitable for formation of springs.
8. Explain porosity and permeability of unconsolidated materials and rocks.
9. What are the objectives of groundwater investigation? Describe the method of systematic groundwater survey in alluvial areas as well as hard rock terrains and preparation of groundwater maps.
10. Give an account of the development and use of groundwater in urban and rural areas of India. What are the methods applied in artificial recharge of groundwater?
11. What is a cone of depression and how is it formed? Explain the result of excessive withdrawal of groundwater by pumping in rural areas as well as in cities. What may be the effect of chemical pollution of water? Cite examples.
12. Describe briefly the problems of underground water in engineering constructions such as dams and reservoirs, road pavements, tunnels, and water retaining structures.

Answers to Multiple Choice Questions

1. (b) 2. (a) 3. (b) 4. (b) 5. (c) 6. (c) 7. (c) 8. (b)
9. (a) 10. (b)

10 Applications of Rock Mechanics in Engineering Geology

● ● ● ● ● ● ● ● **LEARNING OBJECTIVES** ● ● ● ● ● ● ● ●

After studying this chapter, the reader will be familiar with the following:

- Importance of rock mechanics in engineering geology and civil engineering works
- Laboratory testing of common properties of rocks
- Instrumental measurement of in-situ stress of rock mass

- Methods to determine shear strength and compressive strength
- Rock mass classification of Norwegian Geotechnical Institute and geomechanics classification
- Solving practical problems of support requirement for underground structures

10.1 INTRODUCTION

This chapter deals with the methods of quantitative evaluation of rock properties and the elastic and plastic behaviour of rocks under stress. It explains with illustration the various types of instruments used for determining the strength properties of intact rocks and rock mass. The chapter elucidates different approaches for estimating the rock quality designation. It also explains the rock mass classification according to the Norwegian Geotechnical Institute (NGI), as well as the geomechanics classification giving examples of their uses to identify the support requirement of underground engineering structures. It further describes the method of calculation of geological strength index (GSI) of tectonically disturbed rocks such as the Himalayan terrain using the two tables provided.

10.2 RELEVANCE OF ROCK MECHANICS IN EVALUATING ROCK AND ROCK MASS PROPERTIES

The study of the physical characteristics and mechanical behaviour of rocks in response to the forces imposed on them comes under the purview of rock mechanics. Application of the principles of rock mechanics is necessary in engineering geological works related to civil engineering structures, for example, concrete and masonry dams, tunnels, and underground powerhouses that are built in or of rocks.

'The engineering geologist must recognize that for his work to be of maximum benefit to the design engineer, it will be necessary for the engineering geologist to quantify his findings. That is, engineering geologic results must not only be presented in a descriptive form that is understandable by the engineer, but the engineer now is requesting numerically defined limits for engineering geologic description', was the message of Judd (1969) in the inaugural ceremony of the symposium on the role of rock mechanics in engineering geology held in India in 1968. In fact, knowledge of rock mechanics is necessary in engineering geological works to understand the behaviour of rocks under force field and to estimate the rock properties in quantitative terms for engineering design.

The science of rock mechanics is based on the engineering principles used in the analysis of rock and rock mass for engineering purposes. Rocks are used for engineering purposes mainly in two ways—one as construction material involving only intact rocks and, two, as foundation to engineering structures on rock mass. Road metals, railway ballasts, concrete aggregates, cut-stones, masonry works, support columns, and beams are only few of many instances of engineering use of varied sizes of intact rocks as construction materials.

Construction of heavy structures such as high-rise buildings and concrete dams needs a foundation of firm in-situ rocks. The quantitative values of rock properties derived by conducting various tests on intact rocks and measurement of relevant features in rock mass find importance in the engineering analysis of foundation condition, design of slopes, and underground excavation in rocks.

In case the rocks are used as construction materials, intact rock specimens are taken from the field site and tested for determination of rock properties in a laboratory set-up. Several instruments are also available to measure the strength properties of intact rocks in the field itself (Fig. 10.1). However, it is the rock mass properties that are important to engineers and involve the determination of the bulk strength properties for foundation of engineering structures and excavation purposes.

The rock mass properties are in a way controlled by planes of discontinuities and weak structural features of rocks including faults, fractures, joints, bedding planes, foliation planes, and clay seam. Field measurements of the geological parameters related to these planes of discontinues and other weak features are the main considerations along with the analytical data of laboratory tests on rock specimens in estimating rock mass properties. Wherever needed for design purposes, rock mass properties such as bearing strength of foundation rock and stress condition of rocks at depths are measured by in-situ instrumental tests.

Fig. 10.1 Determining the strength of intact rock by point load testing machine

This chapter deals with the basic aspects of rock mechanics such as quantitative evaluation of the properties of intact rocks as well as in-situ rocks or rock mass having relevance in the construction of surface and subsurface engineering structures with which engineering geological works are closely associated. Brief descriptions of such instruments including their functions are also provided in this chapter.

10.3 DETERMINATION OF COMMON PROPERTIES OF ROCKS

The common rock properties such as specific gravity, porosity, void ratio, and absorption can be determined in a laboratory set-up with an oven, an accurate balance, and some glassware. Brief descriptions of the testing procedures for determining these rock properties in intact specimens are given in the following subsections.

10.3.1 Specific Gravity

Specific gravity of a rock specimen is defined as the ratio of the weight of the specimen at a given temperature to the weight of an equal volume of water (that weighs 1 gm/cm³). The procedure to determine the specific gravity in the laboratory is as follows:

The specimen is oven-dried for 24 hours and cooled, and its weight (W_0) is taken. It is then soaked in distilled water for 24 hours and its weight (W_w) is noted. Finally, the specimen is immersed in water and its weight (W_s) is taken under suspended condition.

The specific gravity (G) of the rock specimen is then given by

$$G = \frac{W_0}{W_w - W_s} \qquad (10.1)$$

The specific gravity thus obtained is the apparent specific gravity of the rock. In igneous and metamorphic rocks—in which the pore spaces are negligible—the apparent specific gravity is almost the same as the true specific gravity. However, in a porous sedimentary rock, the apparent specific gravity will vary to a certain extent depending upon the volume of pore spaces. The true specific gravity of a rock specimen can be obtained by powdering the specimen and then following the method described under 'specific gravity' for soil (see Section 6.3).

10.3.2 Density

Density is defined as the mass per unit volume. The density (ρ) of a rock specimen is derived by dividing the weight of the specimen by its volume. Density is determined in the same way as specific gravity, that is, by measuring the dry weight (W_0), water-saturated weight (W_w), and water-suspended weight (W_s). However, unlike the specific gravity, which is a dimensionless number, density has a unit and can be expressed as follows:

$$\rho = \frac{W_0}{W_w - W_s} \text{ gm/cm}^2 \qquad (10.2)$$

Thus, the density of a material is the same as the specific gravity when expressed in metric units. The specific gravity of granite is 2.68, its density is 2.68 gm/cm³ = 2680 kg/m³ = 2.68 tonnes/m³.

Density may be of the following different types:

- Dry density, ρ_d is the weight of dry specimen with pores free of water/unit volume.
- Saturated density, ρ_s is the weight of the specimen soaked in water/unit volume.

- Grain density, ρ_g is the weight of the powdered sample/unit volume.
- Bulk density, ρ_b is the weight of the specimen with pores partially filled/unit volume.

In a porous sedimentary rock, saturated density varies to a great extent from dry density due to pore spaces. The strength of a rock also reduces when its density is reduced due to the presence of void spaces.

10.3.3 Unit Weight

In civil engineering works, it is desirable to use the term 'unit weight', which is the same as density when expressed in the unit of metric system. Thus, unit weight of basalt ($G = 2.65$ or $\rho = 2.65$ gm/cm^3) is 2.65 gm/cm^3 or 2650 kg/m^3.

In English system of measurement as used in the US, unit weight is expressed as pound (lb) per cubic feet (ft^3). The unit weight of a rock sample is derived by multiplying the specific gravity of the sample by the density of water, that is, 62.4lb/ft^3. Thus, the unit weight of basalt ($G = 2.65$) in English system of measurement will be (2.65×62.4lb/ft^3) = 165 lb/ft^3.

10.3.4 Porosity

Porosity (η) of a rock specimen is the volume of voids contained in it and is expressed as the percentage of the gross volume (V) of the specimen. In the determination of porosity, if a rock specimen of regular shape is used, the volume (V) can be directly measured. A rock cube or a rock core with parallel cutting of two ends is generally used to facilitate direct measurement of volume. The specimen is first oven-dried for 24 hours at a temperature of 105°C and then its weight (W_0) is taken. It is then kept immersed in distilled water for 24 hours and the weight (W_w) is noted. Porosity is given by the following equation:

$$\eta = \frac{W_w - W_0}{V} \times 100 \tag{10.3}$$

A rock with low porosity has high density and possesses high strength. For example, granite, dolerite, charnockite, gneiss, and massive quartzite that have negligible porosity are dense rocks possessing high strength.

10.3.5 Absorption

Absorption is the ratio of the weight of water filling the pores in a rock specimen to the weight of the dry specimen expressed as percentage weight. To measure the absorption of a rock specimen, it is oven-dried for 24 hours and its weight W_0 is taken. The specimen is then kept immersed in water for a period of 72 hours and the weight W_1 is noted. The absorption of the specimen is estimated by the following expression:

$$\frac{W_1 - W_0}{W_0} \times 100 \tag{10.4}$$

A rock specimen kept immersed in water for sufficient time may not absorb water to fill all the pore spaces of the rock due to air in the pores. Some clay present in the pore spaces may swell when the specimen is immersed in water and hinder entry of water into the pores. Thus, absorption is not essentially dependant on the total volume of pore spaces in a rock specimen but depends on its capacity to absorb water. Rocks with high absorption values are generally low in strength (see the tables given in Section 10.4.6).

10.4 MEASUREMENT OF STRENGTH OF INTACT ROCK

Rock strength is the most important parameter in the design of a structure. The stability of a structure depends upon the strength of the foundation rock and its behaviour under stress. The strength of a rock can be assessed by subjecting it to any of the three stresses, namely *compressive stress*, *shear stress,* and *tensile stress* and studying the resistivity of the rock as follows:

Compressive strength   Compressive stresses comprising two opposite forces applied on a rock specimen act to decrease the volume of the rock specimen. Compressive strength is the maximum stress that is necessary to break a loaded specimen of rock. It is measured as the total load applied per unit area in kg/cm².

Shear strength   Shearing action is caused by two forces acting in opposite directions along a plane of weakness (e.g., fracture, fault, bedding plane) inclined at an angle to the forces. It tends to move one part separated from the other part with respect to each other.

Tensile strength   When a rock specimen is placed under tensile stress, its volume decreases due to the forces directed outwards, opposite in action. The stresses tend to produce cracks in the rock. Tensile strength is lower than compressive strength.

The laboratory measurement of the strength of a rock is taken in intact specimen in the form of a cube, cylinder (generally in rock core), or disc. The laboratory equipped with rock cutting and rock drilling machines facilitates preparation of the specimen for the test. A laboratory type drilling machine can drill cylindrical specimens generally of diameters 25 mm, 38 mm, and 63 mm. The rock cutting machine has an arrangement to cut the two opposite sides of the cylindrical rock specimens perfectly parallel by means of its diamond saw. The instruments used and the methods followed in the determination of strength properties of intact rocks in the laboratory are described in the following subsections.

10.4.1 Rebound Hammer Test

Rebound hammer is a handy instrument consisting of a spring-loaded steel hammer (Fig. 10.2). It is used to measure the approximate value of compressive strength of rock in hand specimen and also to assess the uniformity characteristics of rock strength in outcrops. When the plunger housing with the hammer is firmly compressed against a rock surface, the spring-loaded hammer gets released automatically and rebounds after giving an impact of energy, the exact amount can be read out from the scale given in the graph card of the rebound hammer. Usually, an energy impact of 0.075 m-kg is satisfactory.

The rebound of the hammer is related to the ultimate compressive strength and modulus of elasticity of rock and is indicated in a scale as rebound number. The rebound distance of the hammer with respect to the plunger is measured from the scale attached to the frame of the hammer. The ultimate compressive strength of the rock is thus obtained within 75 per cent confidence limit. For example, rebound numbers 49, 61, and 65 in the standard model of *Schmidt hammer* are indicative of compressive strengths

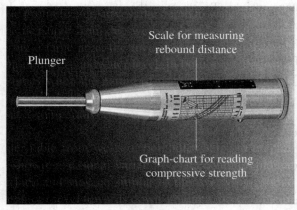

Plunger

Scale for measuring rebound distance

Graph-chart for reading compressive strength

Fig. 10.2   Standard rebound hammer

633 kg/cm² 857 kg/cm², and 935 kg/cm², respectively. The digital model of rebound hammer automatically calculates the rebound numbers and the compressive strength.

10.4.2 Point Load Test

Point load testing machine, shown in Fig. 10.3(a), is a portable instrument used to measure the strength of rock specimens in the laboratory as well as in the field. The instrument consists of a rigid frame, two point load platens that are conical in shape, a hydraulically activated ram with pressure gauges, and a device to facilitate measurement of the distance between the loading points. The pressure gauges of different ranges are given and the one to be fitted with the machine during testing should be of the type that can record the failure pressure. Several sophisticated instruments with digital device for measuring pressure are also available. The rock specimens tested by point load method fail under tension, developing cracks parallel to the loading direction. The point load provides data as point index that can be converted into compressive strength with reasonable accuracy by applying the empirical formula depending upon the nature of the samples. The advantage of this instrument is that it can be used to test irregular samples in the field in addition to testing core and block specimens in the laboratory. Ideally, the samples should be 50 mm in thickness.

In general, point load tests are conducted on the following four possible sample types shown in Fig. 10.3(b):

- Type 1: Core samples for diametrical test
- Type 2: Core samples for axial test
- Type 3: Irregular lump sample
- Type 4: Block sample

Point load index (Is) can be calculated using the following relation:

$$Is = \frac{P}{D^2} \, \text{kg/cm}^2 \qquad (10.5)$$

where P is the failure load and D is the distance between the two platens.

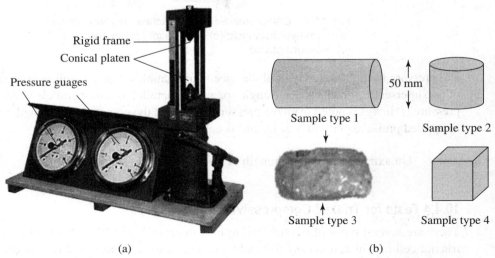

Rigid frame
Conical platen
Pressure guages

50 mm

Sample type 1

Sample type 2

Sample type 3

Sample type 4

(a)   (b)

Fig. 10.3   Point load testing: (a) instrument; and (b) different types of samples

The point load test involves placing the rock sample tightly fixed between the two conical platens and compressing until failure occurs. In selecting samples for testing of blocks and irregular samples, it is to be seen that the value of D remains at 50 mm or very close to it. Otherwise, a correction is needed. The correction is done by using the size correction chart given with the machine. For computation of uniaxial compressive strength parameters (Ic) from the point load index (Is), the following relation is used:

$$Ic = (14 + 0.175D) \times Is \qquad (10.6)$$

where D is the diameter or distance in millimetres. It has been proved that uniaxial compressive strength calculated from point load index gives reasonably reliable values.

10.4.3 Test for Uniaxial Compressive Strength

Uniaxial unconfined compressive strength is the common test conducted in rock specimens in the shape of a cylinder or cube. In general, NX-size (54 mm in diameter) cores obtained from drilling are used for testing. The length of the core specimen is kept at 2 to 2.5 times the diameter by cutting and grinding the two end faces parallel to each other. In its simple form, the uniaxial compressive strength testing machine consists of a hydraulic jack for applying pressure with increasing order. The machine is kept on a solid base and is fitted with pressure gauges to measure the applied pressure (Fig. 10.4). The specimen is placed between two platens

Fig. 10.4 Compressive testing machine: (a) Nx-size rock core; (b) hydraulic jack; (c) solid base; (d) platens; and (e) pressure gauge

and pressure is applied slowly until the specimen crumbles. The crumbling may be in the form of two intersecting cracks at acute angles or a set of parallel to semi-parallel cracks. The applied pressure (P in kg) is noted from the pressure gauge and the compressive strength (σ) expressed as applied pressure per unit area (A cm) is calculated as follows:

$$\text{Uniaxial compressive strength } \sigma = \frac{P}{A} \text{ kg/cm}^2 \qquad (10.7)$$

10.4.4 Tests for Triaxial Compressive Strength

There are several types of triaxial testing instruments. Figure 10.5 illustrates a simple type of triaxial cell in which it is easy to conduct several tests without draining between tests (Hoek 2007). It consists of a steel cylinder with two platens for placing the rock specimen kept in a

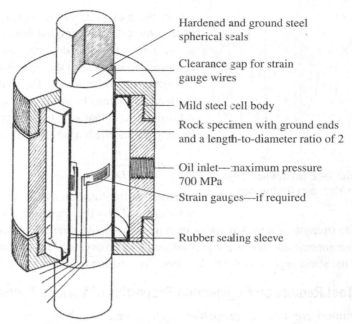

Fig. 10.5 Triaxial cell (Hoek 2007)

- Hardened and ground steel spherical seals
- Clearance gap for strain gauge wires
- Mild steel cell body
- Rock specimen with ground ends and a length-to-diameter ratio of 2
- Oil inlet—maximum pressure 700 MPa
- Strain gauges—if required
- Rubber sealing sleeve

rubber jacket, and the space of the cylinder surrounding the specimen is filled with oil that acts as lateral pressure. The height of the cylindrical rock (generally NX size rock core) specimen that is to be tested should be 2 to 2.5 times its diameter. Hoek's cells of other sizes such as AX and BX are also available. The oil that surrounds the rock specimen acts as the confining pressure. Thus, in triaxial test, unlike uniaxial test, the rock specimen is subjected to lateral pressure in addition to vertical (axial) stress. The axial pressure is exerted through the top platen. The axial and confining loads can be increased simultaneously, and then keeping the confining stress constant, the axial stress is increased until the failure of the specimen occurs.

For rock specimens having high moisture content (75 per cent or more), it is necessary to measure the pore pressure. The strain gauge fitted with the load cell measures the ultimate strength (the ability of rock without yielding to break). This strength of the rock depends upon various factors such as the nature of rock including its composition, grain size, and angularity and is also related to the increased loading and fluid content in rock pores. Using the test results on various axial loads and lateral pressures until failure, Mohr's diagram and failure envelop for the tested rock specimen are plotted to obtain the values of cohesion c and internal friction angle ϕ as done in triaxial soil tests (see Section 6.12, Fig. 6.20).

- Platen
- Core specimen
- Platen

Fig. 10.6 Instrument (HEICO) for Brazilian test for tensile strength showing core specimen of diameter D and length L ($L/D < 1$) fixed between two platens

10.4.5 Brazilian Test for Tensile Strength

Brazilian test for tensile strength is conducted by applying diametrical compression to induce tensile stress in a thin disc of rock core (Fig. 10.6). The ratio between the length (L) and diameter

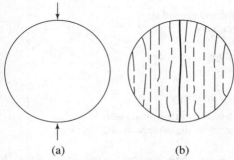

(a)   (b)

Fig. 10.7 Brazilian test: (a) applied load on specimen diameter; and (b) tension cracks

(D) of the rock core test specimen should be less than one (thus $L/D < 1$). The applied load creates a uniform tensile stress across the disc diameter accompanied by a vertical compressive stress as in the Brazilian test (Fig. 10.7). Special precautions must be taken at the contact between the rock and the loading platens to avoid crushing. If the failure is a true splitting failure, that is, a tension crack running between the loading points, the tensile strength is calculated as follows:

$$\sigma_1 = \frac{2P}{\pi DL} \tag{10.8}$$

where σ_1 is the Brazilian tensile strength (MPa), P is the load at failure, and D is the diameter (mm).

Tensile strength of a rock is lower than its compressive strength. A rock under tension will fail earlier to compression. Hence, in the construction of an engineering structure on rocks, it is important to know the stress regime of the rock, especially whether tension will be the primary force.

10.4.6 Test Results on Engineering Properties of Various Types of Rocks

The common engineering properties such as density, absorption, and uniaxial compressive strength of major rock types are presented in Tables 10.1–10.3. The results are based on large numbers of specimens of each rock type that were collected in different times from various project sites and tested in the geotechnical laboratories of Geological Survey of India (Gangopadhyay 1990). It may be seen that, in general, igneous and metamorphic rocks are denser and possess higher strengths compared to the sedimentary rocks, except the Vindhyan sandstone and limestone (Table 10.2).

Table 10.1 Engineering properties of igneous rocks

Rock type	Density (kg/cm³)	Absorption* (%)	Compressive strength (kg/cm²)**
Medium-grained granite	2.66–2.72	Negligible	1850–4150
Basalt (massive)	2.69–2.92	0.03–1.19	1410–4550
Basalt (amygdaloidal)	2.47–2.87	0.68–4.49	1135–2109
Dolerite	2.78–3.03	Negligible	2550–5400
Gabbro	2.81–2.95	Negligible	2160–5250
Charnockite	2.68–3.01	Negligible	1870–3850

* 24 hours of saturation   ** Dry and fresh rock

Table 10.2 Engineering properties of sedimentary rocks

Rock type	Density (kg/cm³)	Absorption* (%)	Compressive strength (kg/cm²) **
Sandstone (Vindhyan)	2.54–2.60	0.85–1.94	550–1975
Sandy shale (Vindhyan)	2.56–2.65	0.30–1.34	415–682
Limestone (Vindhyan)	2.75–2.76	0.25–0.26	766–1399

(Contd)

Table 10.2 *(Contd)*

Rock type	Density (kg/cm³)	Absorption* (%)	Compressive strength (kg/cm²) **
Sandstone (Gondwana)	2.41–2.59	1.15–5.77	90–525
Sandstone (Lower Siwalik)	2.42–2.68	1.88–2.90	450–1060
Sandstone(Middle Siwalik)	2.40–2.47	2.50–2.93	70–250
Sand rock (Upper Siwalik)	2.31–2.41	2.42–3.75	26–240
Shale/claystone (Siwalik)	2.66–2 .77	7.59–9.38	89–123
Sandstone (Tertiary)	2.37–2.59	2.50–5.53	50–160
Limestone (Tertiary)	2.66–2.76	0.15–3.99	261–479
Limestone (Cuddapah)	2.67–2.70	0.02–0.42	335–1350
Shale (Cuddapah)	2.56–2.70	0.41–2.57	144–342
Sandstone (Cretaceous)	2.10–2.32	6.60–7.10	65–350
Tuff breccia	2.71	3.9–18.0	24–290

* 24 hours saturation   ** Dry and fresh rock

Table 10.3 Engineering properties of metamorphic rocks

Rock type	Density (kg/cm³)	Absorption* (%)	Compressive strength (kg/cm²)**
Granite gneiss (medium grained)	2.64–2.68	Negligible	1715–3580
Hornblende–mica gneiss	2.68–2.70	0.50	346–1338
Quartzite (massive)	2.66–2.67	Negligible	550–2850
Quartzitic phyllite	2.63–2.66	Negligible	505–2250
Shaley phyllite	2.32–2.62	0.46–5.32	193–441
Meta-dolerite	2.78–3.10	0.61–1.42	695–2280
Marble	2.66–2.73	Negligible	440–930
Mica schist	2.59–2.60	3.0	176–626
Khondalite	2.60–2.67	0.72–0.95	456–603

* 24 hours saturation   ** Dry and fresh rock

10.5 ELASTIC PROPERTIES OF ROCKS

In case of deformation under continuously increasing stress on a rock body, it may be found that at a certain stage, the body returns to its original shape if the stress is removed. In this stage, the body is said to have elasticity and the strain is proportional to stress. In addition, a strained elastic material stores the energy used to deform it, and the energy is recoverable. The body is said to have reached its *elastic limit* at the stage when the magnitude of strain begins to exceed the magnitude of stress permanently. When the strain goes beyond the elastic limit of the rock, *plastic flow* takes place. If the rock is constituted of brittle materials, the plastic flow will be small in extent, but in case of the rock that is ductile in nature, the plastic flow will be large.

This elastic behaviour of a rock is related to stress (σ) and strain (ε). Stress is measured by the relation $\sigma = P/A$, where P is the force exerted in intact rock in an area A. Strain is given by the expression $\varepsilon = \Delta L/L$, where L is the length and (ΔL) is the change in length of rock specimen.

The ratio between stress and strain is known as the modulus of elasticity or *Young's modulus* (*E*) and is expressed as

$$\frac{\sigma}{\varepsilon} = \frac{P/A}{\Delta L/L} = E \tag{10.9}$$

If the lateral stress in a rock is given by *B*, the ratio between strain of a material in lateral extension (lateral strain) to the strain under vertical extension (axial strain) designated as *Poisson's ratio* is given by the following relation:

$$\frac{Lateral\ strain}{Axial\ straim} = \frac{\Delta B/B}{\Delta L/L} = \mu\ (\text{Poisson's ratio}) \tag{10.10}$$

Poisson's ratio (μ) in a rock varies between 0.1 and 0.5. During earthquake, the waves move through the rock guided by the elastic properties of the rock. The velocity of wave propagation depends on Poisson's ratio, which is variable in different rocks. Thus, depending upon the nature of rock, the wave velocity generated by earthquake will change. In the same rock too, the property may vary depending upon the interlocking nature of the mineral grains. Rocks with rigid interlocking grains will have high modulus of elasticity, but with a large content of moisture as in the case of an immersed body of rock, the modulus of elasticity will be reduced. This aspect of elastic behaviour is significant in the design of engineering structure to be founded on rock.

Modulus of elasticity or Young's modulus is a measure of the rock property that resists deformation. When a cylindrical specimen of rock is subjected to stress parallel to its long axis, it will lengthen and the diameter will be under tension. Poisson's ratio, that is, the ratio of lateral strain to axial stress is measured when a cylindrical rock specimen is subjected to compression parallel to the axis of the rock specimen; the rock shortens along its axis while its diameter increases.

10.6 MEASUREMENT OF STRESS IN UNDERGROUND ROCKS

The stability of an underground opening depends on the rock mass strength and the stress that existed in the rock before the excavation. Tectonic events during geological time are responsible for inducing this in-situ stress in rock. The magnitude of this pre-existing stress varies widely depending upon the nature of geological history of the rock formation in which the in-situ stress exists. The measurement of this in-situ stress in underground rocks is necessary before excavation is taken up for constructing an underground structure.

The test results are utilized in the design of the underground structures and finding remedial measures against any deformation of strata that may be caused by the release of stress under the new condition. In fact, excavation design of all large underground openings such as tunnels, railways, powerhouses, and storages for oil or nuclear waste disposal requires measurement of the in-situ stress in rock mass. The methods of measuring in-situ stresses are varied but two significant methods include the flat jack test and the borehole deformation over-coring method.

10.6.1 Flat Jack Test (Direct Stress Measurement)

Flat jack test is the most convenient method for measuring in-situ stress in rocks of underground openings. The flat jack is a thin envelop-like bladder made of stainless steel with inlet and outlet portals that can be pressurized with hydraulic oil. It measures the pressure required to restore the set of measuring pins fixed in rock wall on either side of a slot. Flat jacks are manufactured in different configurations such as square, rectangular, or with curved edges and their sizes also

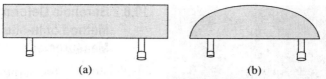

Fig. 10.8 Flat jack with inlet and outlet portals:
(a) rectangular; and (b) curved edged

vary to a great extent, generally between 30 cm × 30 cm and 60 cm × 60 cm, but may even be a square metre depending on the application and slot preparation.

Flat jacks may be of different shapes (Fig. 10.8) such as square or rectangular but a curved flat jack is generally used to fit a slot cut by a circular saw. The test is based on the principles of stress release phenomenon and elimination of local stresses followed by controlled stress compensation. The test is conducted in three phases (Fig. 10.9) as follows:

(i) A small portion of the wall is selected for the test where the rock is not affected by any joint or facture and is also free from any external effects such as cracks due to blasting.

(ii) Two sets of stainless steel pins (reference pins) are then fixed in the wall rock keeping a space between them for inserting the flat jack. After setting the pins and measuring the distance (say, s), a slot is cut by means of a diamond saw so that the pins on either side tend to move their positions due to stress relief phenomenon. The process relieves the rock surface of the stress that originally existed across it. This is measured between the slots by means of the pins fixed prior to cutting. If the distance between the pins is s_1, then $s_1 < s$, where s is the distance in the undamaged state.

(iii) The flat jack is then inserted into the slot and inflated by applying hydraulic pressure (p_f) to restore the original position of reference pins when $s = s_1$ and p (original) $= p_f$ (cancelling pressure). Assuming that the rock mass is elastic within the range of working stress, the measured cancellation pressure (p_f) is very nearly equal to the virgin stress that existed in the rock normal to the plane of slot before the cutting of the slot. The virgin stress is thus estimated by means of the flat jack test.

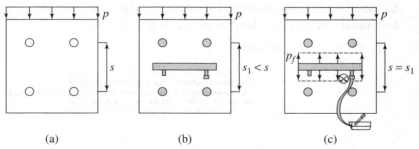

Fig. 10.9 Different phases of conducting flat jack tests; (a) before cutting slot; (b) after cutting slot; and (c) after inserting flat jack and applying pressure

The complete set of equipment necessary to conduct this test is shown in Fig. 10.10. For conducting the flat jack test in underground openings, it is necessary that an exploration adit or pilot tunnel is driven up to the exact position or place of the proposed structural cavern. The test is then taken up in a limited area of the excavated rock wall. The virgin stress remains unaffected even after excavation, but if the wall rock is extremely fractured, either by tectonic deformation or by blasting effect, the test cannot be effectively done.

Fig. 10.10 A set of flat jack testing equipment including a flat jack, a hydraulic pump (used for inflating flat jack) with pressure gauge, hose pipe, and guide frame for positioning of reference pins and slot (HEICO)

10.6.2 Borehole Deformation Over-coring Method of In-situ Stress Measurement

In-situ stress in rock proposed for construction of an underground structure can be measured as follows: A small borehole is drilled to the desired depth. Then, the instrument containing strain gauges is inserted through the borehole and the initial reading is taken. Then, another large hole is drilled outside the small borehole to relieve the stress around this drilled part and the strain gauge reading is taken again. The difference in readings provides the measure of the borehole deformation. The techniques used are explained with illustrations (Figs 10.11–10.13) in the following steps:

(i) To start with, a large diameter (NX size) diamond drill hole is drilled to the depth zone at which stress measurement is desired. After recovery of the drill cores, the bottom of the hole is flattened by a special drill bit as shown in Figs 10.11(a) and (b). A smaller diameter (EX size) borehole is then drilled further down from the end of the large diameter hole, see Fig. 10.11(c).

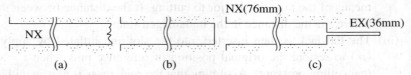

Fig. 10.11 Borehole over-coring method: drilling of holes

(ii) The measuring cell containing a number of strain gauges is then inserted with a special installing tool having an oriented device and a cable to read out unit. Compressed air is used to expand the cell in the hole and the strain gauges are cemented to the wall rock of the small borehole as shown in Fig. 10.12(a). The measuring cell now is fixed to the hole and initial reading (0 reading) is taken, see Fig. 10.12(b).

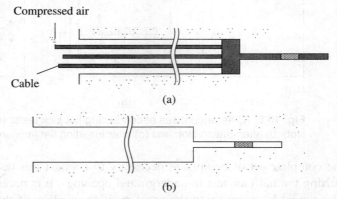

Fig. 10.12 Borehole over-coring method: insertion of measuring cell

(iii) The installing tool is then removed and the small hole is over-cored by a large diameter thin-walled diamond bit, thus relieving the stress at the core. The corresponding strains are recorded by the strain gauge rosettes, see Fig. 10.13(a). The core is then recovered with a special core catcher, see Fig.10.13(b), and immediately after removal, the second reading is taken. From the recorded strain, the stress is computed from laboratory determination of the elastic moduli of rock cores (see Section 10.5).

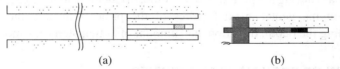

(a)   (b)

Fig. 10.13   Borehole over-coring method: Stress measurement

This method may not be suitable if the boreholes are very deep as it is difficult to handle the instrument at great depths. In boreholes deeper than 50 m, the *hydraulic pressure technique* is generally followed for estimating in-situ stress in rocks. The principle of this method is to determine the magnitude of the in-situ stress by the fracture pressure and the sealing pressure.

The fracture pressure is needed to produce cracks in the borehole walls and the sealing pressures are needed to maintain the cracks when the pump is stopped. In this method, at the required depth, a section of the borehole is sealed by two rubber packers. Hydraulic pressure is applied to the internal walls between the two packers. When the breakdown pressure is reached, the rock surrounding the borehole fails in tension and develops cracks. This fracture is extended away from the boreholes by continuous pumping. When the pumps are shut off with the hydraulic circuit kept closed, a shutdown pressure is recorded. This pressure is necessary to keep the cracks open. The breakdown and shutdown pressures are related to the virgin stress in the site. A borehole camera can be used to measure the direction of the cracks. Thus, both the magnitude and the direction of the principal stress can be estimated.

10.6.3 Borehole Extensometer Test for Measuring Rock Movement

The rocks in an excavated underground cavern tend to move towards the centre of the opening caused by induced stress exceeding the uniaxial compressive strength of the surrounding rocks. The rate of such movement of an underground structure can be measured by a simple instrument named *borehole extensometer* (Fig. 10.14). The instrument consists of a single rod or wire called single position extensometer that extends between the anchor and the reference head. Extensometer with more than one or two rods (up to a maximum of eight) is known as the multiple position (or multipoint) extensometer (Fig. 10.15).

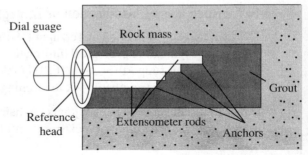

Fig. 10.14   Borehole extensometer   **Fig. 10.15**   A sketch of rod type extensometer

Both wire type and rod type extensometers may work on mechanical or electrical devices. In practice, the instrument is grouted into the borehole keeping the reference head on the rock wall of the excavated opening. Periodic readings are taken by the sensor on all the points in the reference head. The difference between the initial and the final reading indictes the movement of rock mass during the period. The location, orientation, and length of a borehole extensometer depend on the geotechnical features of the project site and the construction method for determining the depth, direction, and amount of anticipated rock movement.

10.7 ESTIMATION OF ROCK MASS PROPERTIES

In the laboratory, only a small portion of the volume of rock mass is considered for testing the rock properties. Hence, it is not representative of the rock mass or in-situ rocks where an engineering structure will be built. Excavation of an underground cavity also involves a large volume of rock mass, and test data on intact rock specimens cannot provide necessary information for design purposes. It is, therefore, necessary to find some procedures to estimate the rock mass properties for design of engineering structures, excavation, and other purposes. Several rock mass classifications are now available, which have been developed by several authors to solve the practical problems related to rock mass properties that satisfy the needs of engineering design. The various geological properties have been considered in devising these classifications.

10.7.1 Rock Mass Classifications

It is to be remembered that rock mass classification is essential at the early stage of project planning. As such, engineering geological works are aimed at collecting the field data related to rock properties by measurement on in-situ rocks. This data accompanied by the laboratory test data is fed into the classification system for obtaining the quantitative values for the design. The engineering requirement and the engineering geological information pertain to evaluation of the in-situ characteristics of rocks especially for rock deformation and support purposes and other follow-up measures for the design. Information on water condition and in-situ stress are also necessary.

In general, the main purpose of the classifications is the safe design of tunnel and underground structures and support system, but they are also applicable for other underground excavations including mining. Several classifications were devised by a number of authors that include Deere et al. (1967), Proctor (1971), Barton, Lien, and Lunde (1974 and 1975) Barton et al. (1980), and Bieniawski (1974, 1975, 1979, and 1989). Earlier to that, Terzaghi (1946) developed a rock mass classification for the purpose of design of tunnel support in which loads are estimated on the basis of descriptive geology.

10.7.2 Rock Mass Classification of Terzaghi

The rock mass classification of Terzaghi (1946) is based on the rock condition in the tunnel. It bears good engineering geological information pertaining to rock quality meant for engineering design of tunnel support (Section 16.10, Table 16.2). Tunnel rocks have been grouped under different categories from hard rock to swelling rock giving the weak features as follows:

Intact rock   It contains neither joints nor hair cracks and it breaks across sound rock. Blasting in rock may cause *spalling* of rock from roof. Hard intact rock may face *popping* conditions of sudden violent outbursts.

Stratified rock It consists of individual stratum with little or no resistance to separation along boundaries of strata. Transverse joints may or may not weaken the strata and spalling condition may be common in this rock.

Moderately jointed rock It contains joints and hair cracks but blocks created by joints will not require support. Both spalling and popping conditions may be encountered.

Blocky and seamy rock It consists of rock fragments nearly or entirely separated from each other and as such vertical wall may need lateral support.

Crushed but chemically intact rock In this type of rock, the fragments are like sand grains. When crushed rock occurs below water table, it will behave as water-bearing sand.

Squeezing rock It advances slowly into the tunnel without any perceptible change in volume. Small micaceous materials or clay minerals of low swelling capacity remain in the rock.

Swelling rock It advances into the tunnel because of expansion caused by clay minerals such as montmorillonite having high swelling capacity.

10.7.3 Rock Quality Designation Index

The rock quality designation (RQD) index developed by Deere (Deere et al. 1967; D.U. Deere and D.W. Deere 1988) is extremely useful in engineering geological description with quantitative estimate of rock cores obtained from subsurface drilling. In fact, most of the authors have used RQD as one of the important geological parameters in their rock mass classifications.

RQD is a measure of lengths of core pieces separated due to joints with respect to total length of rock cores. It is the percentage of core recovery obtained in core pieces of 10 cm or more in length out of the total length of the drill hole run in rock. Thus,

$$RQD(\%) = \frac{\text{(Total length of cores of 10 cm or more)}}{\text{Total length of core run in cm}} \times 100$$

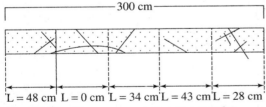

Total length of rock core run

L = 48 cm L = 0 cm L = 34 cm L = 43 cm L = 28 cm

Fig. 10.16 Measurement of rock cores for determination of RQD

Consider that in a site, rock mass has been drilled for a length of 300 cm in five runs. Rock cores obtained from the five runs are of lengths 28 cm, 43 cm, 34 cm, 0 cm, and 48 cm (Fig. 10.16). Thus, the total length of cores obtained is 153 cm.

Therefore,

$$RQD = \frac{(28 + 43 + 34 + 0 + 48) \times 100}{300} = 51\%$$

This falls in the category of *fair* according to Deere's classification (Table 10.4).

Table 10.4 Deere's RQD classification and description

Rock quality designation (%)	Description of rock quality
0–25	Very poor
25–50	Poor
50–75	Fair
75–90	Good
90–100	Excellent

Since RQD represents rock mass quality, the pieces of rock cores measured for RQD must be representative of the in-situ condition. The orientation of the drill hole with respect to rock attitudes may cause some splitting of cores and affect the value of RQD. Faulty drilling operation may also result in more fragmentation of rock cores. It is therefore necessary that utmost care is taken to obtain core pieces that resulted from geological causes such as joints and not by artificial reasons. The drill hole should be made to obtain NX size cores using double tube core barrel and diamond bits ensuring that fracturing of cores does not take place by drilling process.

When drill hole core is not available, RQD can be computed from the measurement of number of joints in one cubic metre of rock in tunnel or drift by applying the following relation:

$$RQD = 115 - 3.3J_v$$

where J_v is the sum of the number of joints per unit length for all joint sets. This is known as volumetric joint count.

10.8 NGI ROCK MASS CLASSIFICATION TO ESTIMATE TUNNELLING QUALITY INDEX

The rock mass classification (Table 10.5) developed by Barton, Lien, and Lunde (1974) of NGI is a widely used classification in engineering geological works for measuring the rock quality of tunnel and other underground excavations in rocks. The NGI classification estimates the tunnelling quality index (Q) from various geological parameters of tunnel rocks that can be used for the design of support systems.

10.8.1 Empirical Equation Used in NGI Classification

The empirical equation adopted in NGI classification to derive rock tunnelling quality index Q the numerical value of which varies from 0.001 to 1000 in a logarithmic scale is as follows:

$$Q = \frac{RQD}{Jn} \times \frac{Jr}{Ja} \times \frac{Jw}{SRF} \tag{10.11}$$

where RQD is the rock quality designation,
Jn is the joint set number,
Jr is the joint roughness number,
Ja is the joint alteration number,
Jw is the joint water reduction factor, and
SRF is the stress reduction factor.

The rock tunnelling quality Q is considered to be a function of the following three measures:

(i) Block shear RQD/Jn
(ii) Inter-block shear strength Jr/Ja
(iii) Active stress Jw/SRF

Explaining these three quotients, the authors (Barton, Lien, and Lunde 1974) offered the following comments:

'*The first quotient (RQD/Jn)*, representing the structure of the rock mass, is a crude measure of the block or particle size, with the two extreme values (100/0.5 and 10/20) differing by a factor of 400. If the quotient is interpreted in units of centimetres, the extreme 'particle sizes' of 200 to 0.5 cm are seen to be crude but fairly realistic approximations. Probably the largest blocks should be several times this size and the smallest fragments less than half the size. (Clay particles are of course excluded.)

'The second quotient (Jr/Ja) represents the roughness and frictional characteristics of the joint walls or filling materials. This quotient is weighted in favour of rough, unaltered joints in direct contact. It is to be expected that such surfaces will be close to peak strength, that they will dilate strongly when sheared, and they will therefore be especially favourable to tunnel stability. When rock joints have thin clay mineral coatings and fillings, the strength is reduced significantly. Nevertheless, rock wall contact after small shear displacements have occurred may be a very important factor for preserving the excavation from ultimate failure. Where no rock wall contact exists, the conditions are extremely unfavourable to tunnel stability. The 'friction angles' (given in Table 10.5) are a little below the residual strength values for most clays, and are possibly down-graded by the fact that these clay bands or fillings may tend to consolidate during shear, at least if normal consolidation or if softening and swelling has occurred. The swelling pressure of montmorillonite may also be a factor here.

'The third quotient (Jw/SRF) consists of two stress parameters. SRF is a measure of (i) loosening load in the case of an excavation through shear zones and clay bearing rock, (ii) rock stress in competent rock, and (iii) squeezing loads in plastic incompetent rocks. It can be regarded as a total stress parameter. The parameter Jw is a measure of water pressure, which has an adverse effect on the shear strength of joints due to a reduction in effective normal stress. Water may, in addition, cause softening and possible outwash in the case of clay-filled joints. It has proved impossible to combine these two parameters in terms of inter-block effective stress, because paradoxically a high value of effective normal stress may sometimes signify less stable conditions than a low value, despite the higher shear strength. The quotient (Jw/SRF) is a complicated empirical factor describing the 'active stress.' (Hoek 2007)

The parameters used in the tunnelling quality index of NGI classification are given in Table 10.5.

Table 10.5   Parameters used in tunnelling quality index Q of NGI classification

1.	Rock quality designation	RQD	Notes
	Very poor	0–25	
	Poor	25–50	1. Where $RQD \leq 10$ (including 0), a nominal value of 10 is used to evaluate Q.
	Fair	50–75	
	Good	75–90	2. RQD intervals of 5, that is, 100, 95, etc., are sufficiently accurate.
	Excellent	90–100	
2.	Joint set number	J_n	
A.	Massive, no, or few joints	0.5–1.0	
B.	One joint set	2	
C.	One joint set plus random	3	1. For intersections use $(3 \times J_n)$
D.	Two joint sets	4	2. For portals use $(2 \times J_n)$
E.	Two joint sets plus random	6	
F.	Three joint sets	9	
G.	Three joint sets plus random	12	

(Contd)

Table 10.5 *(Contd)*

H.	Four or more joint sets, random, heavily jointed, 'sugar cubes', etc.	15	
J.	Crushed rock and earth like materials	20	
3.	*Joint roughness number*	J_r	
	(a) Rock wall contact		
	(b) Rock wall contact before 10 cm shear		
A.	Discontinuous joint	4	
B.	Rough or irregular, undulating	3	
C.	Smooth, undulating	2	
D.	Slickensided, undulating	1.5	
E.	Rough or irregular, planar	1.5	1. Add 1.0 if the mean spacing of relevant joint set is > 3 m.
F.	Smooth, planar	1.0	2. $J_r = 0.5$ can be used for planar, slickensided joints having lineation, provided the lineations are orientated for minimum strength.
G.	Slickensided, planar	0.5	
	(c) No rock wall contact when sheared		
H.	Zone containing clay minerals thick enough to prevent rock wall contact	1.0	
J.	Sandy, gravely, and crushed zone thick enough to prevent rock wall contact	1.0	
4.	*Joint alteration number*	J_a	ϕ_r *degrees (approximately)*
	(a) Rock wall contact		
A.	Tightly healed, hard, non-softening impermeable filling	0.75	–
B.	Unaltered joint walls, surface staining	1.0	(25°–35°)
C.	Slightly altered joint walls, non-softening mineral coating, sand particles, clay-free disintegrated rocks, etc.	2.0	(25°–30°)
D.	Silty or sandy clay coatings, small clay fraction (non-softening)	3.0	(20°–25°)
E.	Softening or low friction clay mineral coatings, that is, kaoline, mica, chlorite, talc, gypsum and graphite, and small quantities of swelling clay (discontinuous coatings, 1–2 mm or less in thickness)	4.0	(8°–16°) *Note*: Values of ϕ_r, the residual friction angle, are intended as an approximate guide to the mineralogical properties of the alteration products, if present.
	(b) Rock wall contour before10 cm shear		
F.	Sandy particles, clay-free disintegrated rock, etc.	4.0	(25°–30°)
G.	Strongly over-consolidated, non-softening clay mineral fillings (continuous, < 5 mm thick)	6.0	(16°–24°)
H.	Medium or low over-consolidation, softening clay mineral fillings (continuous, < 5 mm thick)	8.0	(12°–16°)

(Contd)

Table 10.5 (*Contd*)

J.	Swelling clay fillings, that is, montmorillonite (continuous, < 5 mm thick); values of J_a depend on percentage of swelling, clay-size particles, and access to water	8.0–12.0	(6°–12°)
	(c) No rock wall contact when sheared		
K.	Zones or band of disintegrated or	6.0	
L.	crushed rock and clay (see G, H,	8.0	
M.	and J for clay conditions)	8.0–12.0	(6°–24°)
N.	Zones or band of silty or sandy clay, small clay fraction (non-softening)	5.0	
O.	Thick, continuous zones or	10–13	(6°–24°)
P.	bands of clay (see G, H, and J for clay conditions)	13–20	(6°–24°)
5.	*Joint water reduction factor*	J_w	*Water pressure (kgf/cm²)*
A.	Dry excavation or minor inflow, < 5 l/min, locally	1.0	<1.0
B.	Medium inflow or pressure, occasional outwash of joint fillings	0.66	(1.0–2.5) 1. Factors C to F are crude estimates. Increase J_w if
C.	Large inflow or high pressure in competent rock with unfilling joints	0.5	(2.5–10.0) drainage measures are installed.
D.	Large inflow or high pressure, considerable outwash of filling	0.33	(2.5–10.0) 2. Special problems caused by ice formation are not considered.
E.	Exceptionally high inflow or pressure at blasting, decaying with time	0.2–0.1	(>10)
F.	Exceptionally high flow or pressure continuing without decay	0.1–0.05	(>10)
6.	*Stress reduction factor*	*SRF*	
	(a) Weakness zones intersection excavation, which may cause loosening of rock mass when tunnel is excavated		1. Reduce these values of *SRF* by 25–50% if the relevant shear zones only influence but do not intersect the excavation.
A.	Multiple occurrences of weakness zones containing clay or chemically disintegrated rock, very loose surrounding rock (any depth)	10.0	2. For strongly anisotropic virgin stress field (if measured): when $5 \leq \sigma_1/\sigma_3 \leq 10$, reduce σ_c to $0.8\sigma_c$ and σ_t to $0.8\sigma_t$; when $\sigma_1/\sigma_3 > 10$, reduce σ_c and σ_t to $0.6\sigma_c$ and $0.6\sigma_t$, respectively, where σ_c is the unconfined compressive strength, σ_t is the tensile strength (point load), and σ_1 and σ_3 are the major principal stresses.
B.	Single weakness zone containing clay or chemically disintegrated rock (excavated depth < 50 m)	5.0	3. Few case records are available where depth of crown below surface is less than span width. Suggest *SRF* increase from 2.5 to 5 for such cases (see H).
C.	Single weakness zone containing clay or chemically disintegrated rock (excavation depth > 50 m)	2.5	

(Contd)

Table 10.5 (Contd)

		σ_c/σ_1	σ_θ/σ_1	SRF
D.	Multiple shear zone in competent rock (clay free) loose surrounding rock (any depth)	7.5		
E.	Single shear zone in competent rock (clay free) (excavation depth < 50 m)	5.0		
F.	Single shear zone in competent rock (clay free) (excavation depth > 50 m)	2.5		
G.	Loose open joint, heavily jointed, or 'sugar cube' (any depth)	5.0		
	(b) Competent rock, stress problem	σ_c/σ_1	σ_θ/σ_1	SRF
H.	Low stress, near surface	>200	>13	2.5
J.	Medium stress	200–10	13.0–0.66	1.0
K.	High stress, very tight structure (usually favourable to stability, may be unfavourable for wall stability)	10–5	0.66–0.33	0.5–2
L.	Mild rock burst (massive rock)	5–2.5	0.33–0.16	5–10
M.	Heavy rock burst, massive rock	<2.5	<0.16	10–20
	(c) Squeezing rock, plastic flow of incompetent rock under the influence of high rock pressure	SRF		
N.	Mild squeezing rock pressure	5–10		
O.	Heavy squeezing rock pressure	10–20		
	(d) Swelling rock, chemical swelling activity depending on water pressure	SRF		
P.	Mild swelling rock pressure	5–10		
R.	Heavy swelling rock pressure	10–20		

When making estimates of the rock mass quality (Q), the following guidelines should be followed in addition to the notes listed in Table 10.5:

- When borehole core is unavailable, RQD can be estimated from the number of joints per unit volume, in which the number of joints per metre for each joint set is added. A simple relationship can be used to convert this number to RQD for the case of clay-free rock masses: $RQD = 115 - 3.3Jv$ (approx.), where Jv is the total number of joints per cubic metre ($0 < RQD < 100$ for $35 > Jv > 4.5$).
- The parameter Jn representing the number of joint sets will often be affected by foliation, schistosity, and slaty cleavage or bedding. If strongly developed, these parallel 'joints' should obviously be counted as a complete joint set. However, if there are few joints visible, or if only occasional breaks in the core are due to these features, then it will be more appropriate to count them as 'random' joints when evaluating Jn.
- The parameters Jr and Ja (representing shear strength) should be relevant to the weakest significant joint set or clay-filled discontinuity in the given zone. However, if the joint set or discontinuity with the minimum value of Jr/Ja is favourably oriented for stability, then a second less-favourably oriented joint set or discontinuity may sometimes be more significant, and its higher value of Jr/Ja should be used when evaluating Q. The value of Jr/Ja should in fact relate to the surface most likely to allow initiation of failure.
- When a rock mass contains clay, the SRF factor appropriate to loosening loads should be evaluated. In such cases, the strength of the intact rock is of little interest. However, when

jointing is minimal and clay is completely absent, the strength of the intact rock may become the weakest link, and the stability will then depend on the ratio of rock stress to rock strength. A strongly anisotropic stress field is unfavourable for stability and is roughly accounted for as in Note 2 in the table for stress reduction factor evaluation.

- The compressive and tensile strengths of the intact rock should be evaluated in the saturated condition if this is appropriate to the present and future in-situ conditions. A very conservative estimate of the strength should be made for those rocks that deteriorate when exposed to moist or saturated conditions.

10.8.2 Practical Example of Using Tunnelling Quality Index Q

The evaluation of the value of Q with the ultimate aim of estimating the support requirement consulting Table 10.5 may appear to be a complex job, but it is easy to use in practice. The following is an example of its use in North Koel tunnel in Bihar in granite gneiss (Gangopadhyay and Mishra 1991). The values of the various parameters are as follows:

Table 10.6 Estimating support requirement using tunnelling quality index Q

Rock quality	Good to excellent	$RQD = 90$
Joint sets	Two sets	$Jn = 4\text{--}6$ (average 5)
Joint roughness	Rough, undulating	$Jr = 3$
Joint alteration	Tight and hard	$Ja = 0.75$
Joint water	Dry, minor flow	$Jw = 1.0$
Stress reduction	Competent, single shear	$SRF = 2.5$

Thus, substituting the values of RQD, Jn, Jr, Jw, and SRF in Eq. (10.11), we have $Q = (90/5) \times (3/0.75) \times (1.0/2.5) = 28$.

10.9 GEOMECHANICS CLASSIFICATION BASED ON ROCK MASS RATING

The rock mass classification called geomechanics classification (Table 10.7) developed by Bieniawski (1974) is based on rock mass ratings (RMR) fixed for the different characteristics of tunnel rocks. In addition, rating adjustment is needed with respect to orientation of tunnel drive with attitudes of strike and dip of joints.

10.9.1 Parameters Used in Rock Mass Ratings with Tables

The six parameters used for the geomechanics classification are as follows:

  (i) Strength of rock
 (ii) RQD
(iii) Spacing of joints/discontinuities
 (iv) Conditions of joints
  (v) Groundwater condition
 (vi) Strike and dip orientations of joints

Both point load and uniaxial compressive strength of rocks in megapascal units (1 MPa = 10 kg/sq) obtained from laboratory tests are considered for the parameter 'strength of rock'. Other parameters are determined from the study of the drill cores and rocks in the drift. The equivalent ratings for each of these six parameters can be estimated from Table 10.7. The summation of the equivalent ratings for the six parameters gives the final RMR. From the value of RMR, the rock mass class with respect to its quality can be known (as in Table 10.7C). It may be noted that Bieniawski from time to time (1974, 1975, and 1979) refined his original rock mass classification and the one reproduced here (Table 10.7) is from the publication in 1989.

Table 10.7A Classification parameters and their ratings

Uniaxial compressive strength (Mpa)	Point load test index (Mpa)	Rating	Drill core quality (RQD %)	Rating	Spacing of joints	Rating
> 250	> 10	15	90–100	20	> 2 m	20
100–250	4–10	12	75–90	17	0.6–2 m	15
50–100	2–4	7	50–75	13	200–600 mm	10
25–50	1–2	4	25–50	8	60–200 mm	8
5–25		2	< 25	3	< 60 mm	5
1–5		1				
< 1		0				
Conditions of joints (see Table 10.6F)	Rating	Groundwater	Inflow per 10 m tunnel length	Joint water pressure	General condition	Rating
Very rough surface, not continuous, no separation, hard joint wall	30		None	0	Completely dry	15
Slightly rough surface, separation < 1 mm, hard joint wall rock	25		< 10 l/min	> 0.1	Damp	10
Slightly rough surface, separation < 1 mm, soft joint wall	20		10–25 l/min	0.1–0.2	Wet	7
Slickensided surface or gouge < 5 mm thick, joint, open 1–5 mm	10		25–125 l/min	0.2–0.5	Dripping	4
Soft gouge > 5 mm thick, or joint open > 5 mm continuous	0		> 125 l/min	> 0.5	Flowing	0

Table 10.7B Rating adjustment for joint orientations (see Table 10.6E)

Joint strike/dip orientation	Very favourable	Favourable	Fair	Unfavourable	Very unfavourable
Tunnels ratings	0	−2	−5	−10	−12
Foundation ratings	0	−2	−7	−15	−25
Slopes ratings	0	−5	−25	−50	−60

Table 10.7C Rock mass classes determined from total ratings

Rating	100–81	80–61	60–41	40–21	< 20
Class	I	II	III	IV	V
Description	Very good rock	Good rock	Fair rock	Poor rock	Very poor rock

Table 10.7D Meaning of rock mass classes

Class	I	II	III	IV	V
Average stand-up time	10 years for 5 m span	6 months for 4 m span	1 week for 3 m span	5 hours for 1.5 m span	10 hours for 0.5 m span
Cohesion of rock mass	> 300 kPa	200–300 kPa	150–200 kPa	100–250 kPa	< 100 kPa
Friction angle of rock mass	> 45°	45°–90°	35°–40°	30°–35°	< 30°

Table 10.7E Effect of joint strike and dip orientation in tunnels

Strike perpendicular to tunnel axis				Strike parallel to tunnel axis		
Drive with dip 45°–90°	Drive with dip 20°–45°	Drive against dip 45°–90°	Drive against dip 20°–45°	Dip 45°–90°	Dip 20°–45°	Dip 0°–20° irrespective of strike
Very favourable	Favourable	Fair	Unfavourable	Very unfavourable	Fair	Fair

Table 10.7F Guidelines for joint classification

Joint length	Rating	Aperture	Rating	Roughness	Rating	Infilling	Rating	Weathering	Rating
< 1 m	6	None	6	Very rough	6	None	6	Unweathered	6
1–3 m	4	< 0.1mm	5	Rough	5	Hard filling < 5 mm	4	Slightly weathered	5
3–10 m	2	0.1–1 mm	4	Slightly rough	3	Hard filling > 5 mm	2	Moderately weathered	3
10–20 m	1	1–5 mm	1	Smooth	1	Soft filling < 5 mm	2	Highly weathered	1
> 20 m	0	> 5	0	Slickenside	0	Soft filling > 5 mm	0	Decomposed	0

10.9.2 Practical Example of Using Rock Mass Rating

The geomechanics classification for the values of RMR was used for North Koel tunnel project of Bihar. The tunnel was driven in gneiss perpendicular to foliation strike and joint dip 45°. Ratings for the parameters can be obtained from Table 10.7A. The rating adjustment value for joint orientation is taken from Table 10.7B through Table 10.7E. The result of study for equivalent rating for each parameter and the total RMR thus obtained is as follows:

Table 10.8 Application of rock mass rating in North Koel tunnel project

Parameters	Value/description	Rating
Strength (compressive)	60–90 MPa	7
RQD	90%	20
Spacing of joints	0.3–0.8m	20
Condition of joints	Rough surface, separation < 1 mm, hard joint wall rock	20
Groundwater	None	10
Rating adjustment	Very favourable	0
Total RMR (after adjustment) = 77		

The RMR value of 77 falls under category II in the geomechanics classification of rock mass quality, which is graded as 'good rock' (Table 10.7C). As stated before, the estimation of Q values for the same tunnel rock following the NGI rock mass classification comes as 28, which also falls in the category 'good rock'.

Comparing the two systems of rock mass classifications, Hoek and Brown—Emperical Strength Criteria (1980) has concluded that the Q values of the NGI system bears the following relation with the RMR system:

$$RMR = 9 \log_e Q + 44 \tag{10.12}$$

Substituting the value of Q (28) of tunnel rocks estimated for the rocks of North Koel project in Eq. (10.12), we have

$RMR = 9 \log_e 28 + 44 = 9 \times 3.332 + 44 = 74$, which is very close to the measured RMR value of 77, indicating that both the classifications are equally effective in estimating the rock mass quality. In fact, the design of tunnel and the support system made on the basis of the values of both Q and RMR are found to be very useful in safe tunnelling (Section 16.10).

10.10 GEOLOGICAL STRENGTH INDEX FOR BLOCKY AND HETEROGENEOUS ROCK MASS

[Note: The Hoek method of rock mass strength determination briefly discussed in this section (along with tables) will be especially required for professionals working in underground projects comprising tectonically disturbed rock formation, such as the Himalayan terrain where the rock formations are affected by several faults and thrust. The related publications of Hoek (reference given in this section) may be studied thoroughly for the application of the method in project works.]

In a tectonically disturbed terrain such as Himalayan areas, the rock mass may be very blocky containing discrete blocks interlocked by matrix. The strength of such blocky rock mass depends upon the nature of small blocks, interlocking grains, and also their angularity and roughness. These blocks may also be altered to various extents and the interfaces between the blocks may have slickenside or may be filled by clay.

In such highly disturbed rocks, the rock quality cannot be evaluated from the NGI and geomechanics methods. The strength of such in-situ rock mass must be evaluated from geological observations and from the test results on individual rock pieces or rock surfaces that have been removed from rock mass. This problem has been extensively discussed by Hoek and Brown—Empirical Strength Criteria (1980).

To estimate the strength characteristics of such highly disturbed rock mass, later Hoek and Brown (1997) proposed the geological strength index (GSI), which was further refined by Marinos and Hoek (2001) who published two charts for estimating the GSI— one for blocky rock mass and the other for heterogeneous rock mass. These have been reproduced here in Tables 10.9 and 10.10, respectively.

Table 10.9   Characterization of blocky rock masses on the basis of interlocking and joint conditions

GEOLOGICAL STRENGTH INDEX FOR JOINTED ROCKS (Hoek and Marinos, 2000) — From the lithology, structure and surface conditions of the discontinuities, estimate the average value of GSI. Do not try to be too precise. Quoting a range from 33 to 37 is more realistic than stating that GSI = 35. Note that the table does not apply to structurally controlled failures. Where weak planar structural planes are present in an unfavourable orientation with respect to the excavation face, these will dominate the rock mass behaviour. The shear strength of surfaces in rocks that are prone to deterioration as a result of changes in moisture content will be reduced is water is present. When working with rocks in the fair to very poor categories, a shift to the right may be made for wet conditions. Water pressure is dealt with by effective stress analysis.	SURFACE CONDITIONS	VERY GOOD — Very rough, fresh unweathered surfaces	GOOD — Rough, slightly weathered, iron stained surfaces	FAIR — Smooth, moderately weathered and altered surfaces	POOR — Slickensided, highly weathered surfaces with compact coatings or fillings or angular fragments	VERY POOR — Slickensided, highly weathered surfaces with soft clay coatings or fillings
STRUCTURE		DECREASING SURFACE QUALITY ⇨				
INTACT OR MASSIVE - intact rock specimens or massive in situ rock with few widely spaced discontinuities	DECREASING INTERLOCKING OF ROCK PIECES	90 / 80			N/A	N/A
BLOCKY - well interlocked undisturbed rock mass consisting of cubical blocks formed by three intersecting discontinuity sets			70 / 60			
VERY BLOCKY - interlocked, partially disturbed mass with multi-faceted angular blocks formed by 4 or more joint sets				50 / 40		
BLOCKY/DISTURBED/SEAMY - folded with angular blocks formed by many intersecting discontinuity sets. Persistence of bedding planes or schistosity					30	
DISINTEGRATED - poorly interlocked, heavily broken rock mass with mixture of angular and rounded rock pieces					20	
LAMINATED/SHEARED - Lack of blockiness due to close spacing of weak schistosity or shear planes		N/A	N/A			10

Table 10.10 Estimate of GSI for heterogeneous rock masses such as flysch (Marinos and Hoek 2001)

GSI FOR HETEROGENEOUS ROCK MASSES SUCH AS FLYSCH
(Marions. P and Hoek. E, 2000)

From a description of the lithology, structure and surface conditions (particularly of the bedding planes), choose a box in the chart. Locate the position in the box that corresponds to the condition of the discontinuities and estimate the average value of GSI from the contours. Do not attempt to be too precise. Quoting a range from 33 to 37 is more realistic than giving GSI = 35. Note that the Hoek-Brown criterion does not apply to structurally controlled failures. Where unfavourably oriented continuous weak planar discontinuities are present, these will dominate the behaviour of the rock mass. The strength of some rock masses is reduced by the presence of groundwater and this can be allowed for by a slight shift to the right in the columns for fair, poor and very poor conditions. Water pressure does not change the value of GSI and it is dealt with by using effective stress analysis.

COMPOSITION AND STRUCTURE

SURFACE CONDITIONS OF DISCONTINUITIES (Predominantly bedding planes)

	VERY GOOD - Very rough, fresh unweathered surfaces	GOOD - Rough, slightly weathered surfaces	FAIR - Smooth, moderately weathered and altered surfaces	POOR - Very smooth, occasionally slickensided surfaces with compact coatings or fillings with angular fragments	VERY POOR - Very smooth slicken-sided or highly weathered surfaces with soft clay coatings or fillings
A. Thick bedded, very blocky sandstone. The effect of pelitic coatings on the bedding planes is minimized by the confinement of the rock mass. In shallow tunnels or slopes these bedding planes may cause structurally controlled instability.	70	A			
B. Sandstone with thin interlayers of siltstone	60		B		
C. Sandstone and siltstone in similar amounts			C		
D. Siltstone or silty shale with sandstone layers		40	D		
E. Weak siltstone or clayey shale with sandstone layers			30 E		
F. Tectonically deformed intensively folded/faulted, sheared clayey shale or siltstone with broken and deformed sandstone layers forming an almost chaotic structure				F 20	
G. Undisturbed silty or clayey shale with or without a few very thin sandstone layers				G	
H. Tectonically deformed silty or clayey shale forming a chaotic structure with pockets of clay. Thin layers of sandstone are transformed into small rock pieces.					H 10

C, D, E and G - may be more or less folded than illustrated but this does not change the strength. Tectonic deformation, faulting and loss of continuity moves these categories to F and H.

→ : Means deformation after tectonic disturbance

The character of rocks, whether blocky or very blocky, seamy, disintegrated, sheared, or of heterogeneous nature is to be decided first for the application of the GSI system. In addition, the following properties of the rock pieces are required for use in the Hoek–Brown criterion with the ultimate purpose of application in design of underground excavation and support system for highly disturbed as well as heterogeneous types of rock mass.

Uniaxial compressive strength of intact rock mass	σ{ci}
Hoek–Brown constant	m_1
Geological strength index	GSI
Hoek–Brown constant	m_b
Hoek–Brown constant	s
Deformation modulus	E_m

Here σ_1 and σ_3 are the maximum and minimum effective principal stresses at failure, respectively, m_b is the Hoek–Brown constant for the rock mass, s and a are constants that depend upon the rock mass characteristics, and σ{ci} is the uniaxial compressive strength of the intact rock. Wherever possible, the values of these constants are to be determined by laboratory analysis of the results of a set of carefully conducted triaxial tests on core samples.

In case the desired tests cannot be done, the assumed values as given by Hoek (2007) for the section 'Rock properties' can be consulted. The latest version of Hoek–Brown criterion has been published by Hoek, Carranza–Torres, and Corkum (2002). This article together with the computer programme known as RockLab can be downloaded for free from the Internet at **www.rocscience.com** for use in implementing the criteria.

The method of GSI system and the Hoek–Brown criterion have been successfully used in many underground projects in the world, giving reasonable estimates of a wide variety of tectonically disturbed rocks. In India, the Hoek's method was successfully utilized in the design and support selection for the underground powerhouse of Nathpa Jhakri hydroelectric projects in the Himalayan terrain of Himachal Pradesh.

SUMMARY

- Engineering geological works related to quantitative evaluation of properties of intact rock and rock mass are very important in the design of engineering structures, especially underground structures. The intact rock properties such as density, porosity, absorption, and strength (compressive, shear, and tensile) are determined in the laboratory using normal wares and simple instruments.
- A point load testing machine can be taken to the field to measure rock strength even in irregular specimens in the field itself. Sophisticated computerized instruments are available for measuring uniaxial and triaxial compressive strengths of intact rocks.
- Test results conducted on large number of Indian rocks collected from various project sites clearly bring out the fact that igneous and metamorphic rocks are in general dense and possess high strength and are suitable to use as building materials in engineering constructions.
- In-situ rock stress, which is an important engineering property, is measured by various methods including the borehole method by strain technique or deformation technique for shallow depths and hydraulic fracture technique for deeper parts. The fracture technique is based on the application of hydraulic pressure for cracking the rocks. Rocks possess the property of elastic deformation. the study of which has significant bearing in the design of foundation on rock.
- Rock mass behaves differently from that of intact rocks. Estimation of rock mass properties is essential in the design of rock tunnelling and support system. These properties cannot be evaluated by the study of intact rocks alone. However, several methods are now

available to estimate the rock mass quality based on the geological information on rock mass condition and measures of structural attitudes of rock.

• Two of the widely used rock mass classifications that provide rock mass quality in quantitative terms for design purposes have been discussed here. Of these, the NGI classification is used in the estimation of tunnelling quality index (Q) but it is applicable also in other underground works such as mining and underground powerhouse. The other rock classification is the geomechanics classification that is based on RMR.

• The two classifications for estimating the rock quality involve measurements of various geological parameters of in-situ rocks, especially the nature of discontinuities and other weak features and also laboratory test data on rock samples. These measured parameters on rock mass and rock samples when introduced into the given tables provide the values of Q and RMR.

• Engineering geological works on rocks of several tunnels of India were carried out and the values of Q and RMR were estimated for design of tunnel excavation. It has been observed that both the classifications are very useful in designing cost-effective safe tunnelling.

• In estimating the strength characteristics of highly disturbed rocks that include blocky and heterogeneous rock mass, the Q and RMR values may not be suitable; instead, the GSI proposed by Hoek and Brown may be used by consulting the two charts given by the authors.

EXERCISES

Multiple Choice Questions

Choose the correct answer from the choices given:

1. Test is performed in the laboratory for compressive strength by applying compression on:
   (a) a smooth core sample of height that is double of its diameter
   (b) diametrical (curved) side of core
   (c) a sample having rough surfaces on both ends of rock core specimen

2. The type of specimen needed to find tensile strength is:
   (a) rock core with length double its diameter
   (b) thin disc of rock core
   (c) rock core with equal length and breadth

3. The compressive strength of granite is:
   (a) 1500 kg/cm^2
   (b) 4000 kg/cm^2
   (c) 500 kg/cm^2

4. The compressive strength of dolerite is:
   (a) 2000 kg/cm^2
   (b) 4000 kg/cm^2
   (c) 5500 kg/cm^2

5. The compressive strength of charnockite is:
   (a) 2000 kg/cm^2
   (b) 5000 kg/cm^2
   (c) 1500 kg/cm^2

6. The compressive strength of Vindhyan sandstone is:
   (a) 5000 kg/cm^2
   (b) 1000 kg/cm^2
   (c) 200 kg/cm^2

7. The compressive strength of tertiary sandstone is:
   (a) 50 kg/cm^2
   (b) 150 kg/cm^2
   (c) 2000 kg/cm^2

8. The compressive strength of Cretaceous sandstone is:
   (a) 50 kg/cm^2
   (b) 250 kg/cm^2
   (c) 100 kg/cm^2

9. The compressive strength of gneiss is:
   (a) 1000 kg/cm^2
   (b) 3000 kg/cm^2
   (c) 6000 kg/cm^2

10. The compressive strength of marble is:
    (a) 200 kg/cm^2
    (b) 800 kg/cm^2
    (c) 1000 kg/cm^2

11. The compressive strength of mica schist is:
    (a) 200 kg/cm^2
    (b) 800 kg/cm^2
    (c) 1000 kg/cm^2

State whether the following statements are true or false.
Mark (a) if true and (b) if false:

12. Poisson ratio is the ratio of the lateral strain to axial strain.

13. In-situ stress in underground rocks can be measured by using flat jack.

14. Stress in subsurface rock can be measured through a borehole by using a borehole deformation gauge.

15. The movement of rocks in an underground chamber can be measured by a borehole extensometer.

16. In the field, compressive strength of rock samples can be measured by point load tester.

Review Questions

1. What is rock mechanics? Discuss the scope of application of rock mechanics in engineering geology.
2. Define and write the equations for the following properties of rocks: specific gravity, density, unit weight, porosity, and absorption.
3. Explain the method of determining the compressive strength of an intact rock specimen using rebound hammer and point load testing machine.

4. Write a short account on testing of intact rock specimens for the determination of uniaxial compressive strength.
5. What are elastic and plastic deformations of rock? Define Young's modulus and Poisson's ratio and give their equations. How will a rock core specimen act under compression parallel to the axis of the specimen?
6. Write a short account on flat jack test for in-situ stress measurement in massive rock.
7. What do you understand by rock mass properties? What are the purposes of rock mass classification?
8. Name the different groups of tunnel rocks under the rock mass classification of Terzaghi.

NUMERICAL EXERCISES

1. What is RQD? Explain how RQD is calculated from the study of rock cores obtained from rotary drilling. From a 30 m-deep drill hole, the core recovery for each 3 m run of drilling from the top towards the bottom is as follows: 0.2 m, 0.5 m, 1.5 m, 1.4 m, 1.9 m, 2.5 m, 0 m, 2.2 m, 2.6 m, and 2.9 m. Calculate the rock quality of drill cores (use Table 10.4). State the method of measuring RQD of tunnel rock by volumetric joint count.

[Answer: 52.3%, fair]

2. Write the equation for NGI rock mass classification (Q values) used in the measurement of rocks in a tunnel or underground cavity. Calculate the value of Q for the tunnel rocks.
[Hint: Use the formula given in Eq. (10.6) for volumetric joint measurement and calculate RQD. Refer Table 10.5 for different joint parameters and stress reduction factors. Suppose the following are the parameters obtained after measurement in the tunnel rocks as hinted:

RQD measured from volumetric joint count	$115 - 3.3Jv = 115 - 3.3 \times (Jv = 12) = 75.4$
Joint set number: three joint sets	$Jn = 9$
Joint roughness number: smooth, planar	$Jr = 1.0$

Joint alteration number: silty coating	$Ja = 3.0$
Joint water reduction factor: minor inflow	$Jw = 1.0$

Stress reduction factor: Single shear zone with clay $SRF = 2.5$
Now use Eq. (10.11) (Section 10.8) to calculate the value of Q.]

[Answer: 12.1]

3. What is geomechanical classification? Name the person who developed this classification. What does RMR stand for? What are the parameters used to find the value of RMR? Calculate the value of RMR from the following parameters (use Table 10.7):

Compressive strength	50 MPa
RQD	76%
Spacing of joint	0.7–1.0
Condition of joints	(slightly rough, separation < 1 mm)
Groundwater	< 10 l/min (damp)
Rating adjustment	favourable

[Answer: 72]

Answers to Multiple Choice Questions

1. (a)	2. (b)	3. (b)	4. (b)	5. (a)	6. (b)	7. (a)	8.(b)	9. (b)
10. (b)	11. (a)	12. (a)	13. (a)	14. (a)	15. (a)	16. (a)		

11 Site Investigation

•••••••••   **LEARNING OBJECTIVES**   ••••••••••

After studying this chapter, the reader will be familiar with the following:

- Different stages of investigation of engineering project sites
- Study of landform and geology from aerial photography and remote sensing
- Different methods of geophysical explorations including nuclear logging
- Rock drilling and subsoil exploration by pitting, trenching, and drilling
- Logging of drill cores and their recording and schematic presentation
- Water pumping test for permeability of subsurface rocks and its utility

11.1 INTRODUCTION

This chapter discusses the various aspects of engineering geological investigation in project sites. The investigation, which is conducted in different stages, starts with in-house study of the site by aerial photography and remote sensing methods followed by field visits to obtain details of the site condition and record the geological features by plane table mapping. Subsurface condition of the site is then evaluated from geophysical and drilling explorations. Geophysical survey in engineering projects is mostly done by seismic and resistivity methods but wherever required gravity, magnetic, and radioactive methods are also applied. Rock cores collected from drilling are tested in the laboratory for their engineering properties. Pumping test is undertaken to find the permeability characteristics of jointed rocks. After each stage of investigation, a report is submitted with all the field information and the site geological map for design and safe construction of the project.

11.2 DIFFERENT STAGES OF SITE INVESTIGATION

Site investigation of an engineering project is the primary concern of an engineering geologist. The geological information provided to the construction engineers from surface and subsurface investigations is the basic requirement for economic design and safe construction of engineering structures. The investigation begins after finding the suitability of the site selected by the engineers with due consideration of geological aspects. The site investigation also includes subsurface exploration by geophysical methods and by rock core drilling. In all major projects, engineering geological investigation is carried out in three stages of project development coinciding with the three main phases of engineering works as follows:

(i) Initial/preliminary stage (planning phase)
(ii) Feasibility/pre-construction stage (design phase)
(iii) Construction stage (excavation and construction phase)

Post-construction stage investigation is also carried out in some projects.

11.2.1 Initial Stage (Planning Phase)

The initial stage investigation of an engineering project takes account of the regional and local geology including the rock types and structural setting of the project area. This may be done from available maps and publications and reports of earlier studies. It is also essential to visit the site to get first-hand knowledge of the broad geological features of the project area. In the preliminary engineering geological investigations, it is very important that an assessment is made of the potential geological hazards or the weakness of the site and their remedial measures. Of course, these assessed geological defects have to be substantiated by further detailed investigation including exploratory works at a later stage so that the engineers may incorporate appropriate corrective measures in the design of the structure including a cost estimate. The important work involved in this stage includes preparation of a geological map with standard geologic and structural symbols.

Preparation of a geological map The preliminary investigation is conducted with the help of topographic maps, aerial photographs, and satellite imagery. The physiographical features and broad geological structures (three-dimensional or 3D) are better studied by remote sensing. The regional geology is also known from Geological Survey of India's systematic maps (scale 1:100,000). This in-house study is followed by reconnaissance of the area and plotting of the geological features in a contour map, to a scale of preferably 1:50,000, showing the possible site including the alternative ones.

A reconnaissance of the site and geological mapping provide useful information within a short span of time about the rock formations including their structural features and are also cost-effective. A map thus prepared provides the basic geological information that helps in the engineering planning and study of the technical and economic feasibility of a project. Geological maps are prepared by delineating the exact boundaries of the different beds and the structural elements of each bed or rock type as observed at the site. The rock types are shown by colours or symbols as shown in Fig. 11.1.

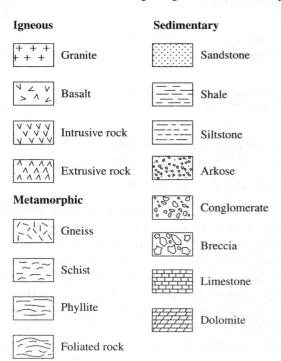

Fig. 11.1 Symbols of common rocks used in a geological map

Dip and strike In a geological map, the attitudes of beds such as *strike* and *dip* of sedimentary rocks, the foliation planes of metamorphic rocks, and the joint planes of any rock type and other structural elements are shown by certain standard symbols (Fig. 11.2). Of these, the strikeand dip of a bed or any planar surface of rock are the most important geological features. Dip is the maximum slope or inclination of a bed or rock facewith respect to horizontal plane. The strike direction is at a right angle to dip direction. Dip is expressed by both amount and direction but strike by only direction. Thus, if a bed dips, say, 40° towards south (S), the strike will be east–west (E–W). A vertical bed dips at an angle of 90°. The dip of a horizontal bed is 0° as shown in Fig. 11.2.

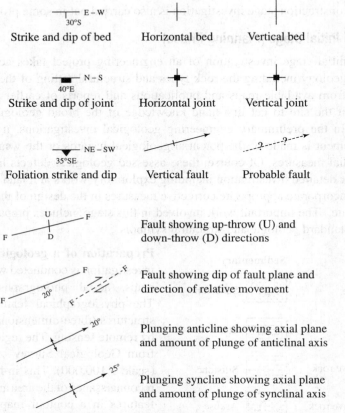

E – W
30°S
Strike and dip of bed   Horizontal bed   Vertical bed

N – S
40°E
Strike and dip of joint   Horizontal joint   Vertical joint

NE – SW
35°SE
Foliation strike and dip   Vertical fault   Probable fault

U   F
F   D
Fault showing up-throw (U) and down-throw (D) directions

F   F
F   20°   F
Fault showing dip of fault plane and direction of relative movement

20°
Plunging anticline showing axial plane and amount of plunge of anticlinal axis

25°
Plunging syncline showing axial plane and amount of plunge of synclinal axis

Fig. 11.2 Standard symbols of structural elements of rocks

Fig. 11.3 Brunton compass

Dip and strike measurements are made with a simple instrument called clinometer or Brunton compass (Fig. 11.3), which measures the strike and dip direction (i.e., the direction of maximum inclination of a rock face). The flat surface of the instrument is placed on the sloping rock face when an indicator measures the inclination swinging between 0° (horizontal) and 90° (vertical). It is to be remembered that any angle other than the maximum slope angle is not the dip but is called *apparent dip*. So, an apparent dip will be always less than the *true dip* and its direction is other than the true dip direction. It may be noted that strike and dip beds, foliation planes, and joint planes are indicated by symbols (Fig. 11.4).

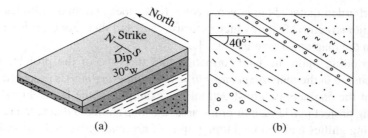

Fig. 11.4   Strike and dip of beds of sedimentary rocks:
(a) dip of beds 30° west and the strike north–south; and
(b) a cross section of beds normal to strike showing beds dip
at 40°

Geological map vs engineering geological map   Engineering geological maps are the same as geological maps with some added features to serve the purpose of engineering in a better way. The salient features of the geological maps and the added features included in the engineering geological maps will be apparent from Table 11.1.

Table 11.1 Salient features of geological maps and engineering geology maps

Main features of geological maps	Added features of engineering geological maps
North line	Project layout such as tunnel alignment
Scale of map	Scale preferably by bar
Section line	Drill hole location, sampling location
Outcrops	Drift location
Rock formations	Isopach for overburden thickness
Lithological contacts	Stratum contour
Unconformity	RQD/Q/RMR values
Bedding, foliation	Joint attitudes—stereo plot of joint data
Attitudes of joint	Compressive strength—Schmidt hammer test
Fault line	Shear zone
Trace of fold axis	Water level and spring

11.2.2 Feasibility Stage (Design Phase)

Feasibility stage includes a detailed investigation to select the best possible site for an engineering project. For example, a number of dam sites or tunnel alignments may be initially marked on the map. Out of these, the most favourable site is to be selected from engineering geological consideration. Feasibility stage investigation involves ground survey for the preparation of a detailed geological map of the selected site. At this stage, mapping is carried out to a scale varying between 1:10,000 and 1:500. The adverse aspects of the foundation or other geological problems of the project site are thoroughly studied in this stage and plotted on a large-scale map, say, 1:1,000. The following two types of maps are of use in this stage:

Special purpose engineering geological maps   The various features plotted in the geological map prepared in an engineering site during the design phase include rock exposures and extent of overburden materials in covered zone, rock types and their attitudes such as strike and dip, and major structural features such as thrust, fault and shear zones, folds, and joints. The symbols used in the map are shown in Fig. 11.2.

Depending upon the land formation and the type of the project, additional features are documented in the map. A *special-purpose engineering map* is prepared in accordance with the need of the engineering construction. For example, in the feasibility stage investigation for a building project in a landslide-prone area, the map should indicate the land slope characteristics including gullies and bad land topography, if present, active and old slides, area of subsidence, drainage pattern, springs, if any, local water table condition, and so on (Fig. 11.5).

In addition to ground survey by surface mapping, feasibility stage study is also implemented by remote sensing and photography interpretation following the techniques described in Section 11.3. Exploratory works such as pitting, trenching, and drilling are taken up after the geological mapping of the site. Exploratory drifts are also necessary in large projects to get the details of subsurface rocks and facilitate in-situ measurement of stress condition. If needed, geophysical studies are carried out to understand the subsurface condition of the rocks including geological structures.

Geophysical survey is conducted through seismic, gravity, magnetic, electric, electromagnetic, and radiometric techniques. Seismic method is generally utilized in many projects to evaluate subsurface structures such as fault and other weak zones and also the thickness of different rock formations. The earthquake intensity of the area is also studied from available records and accordingly reported to the project authority for their inclusion in the design. Selection of quarry sites of required quality and quantity of construction materials for the project is a very important aspect of feasibility stage investigation.

Subsurface maps   In addition to surface mapping, subsurface mapping is also carried out in the feasibility stage. There are several ways to express subsurface geology and structural conditions. Geological sections prepared from the attitudes of rock types are given in a map to show the subsurface condition. Drilling data presented as drill hole logs (see Section 11.7.7) are another way of presenting subsurface information, which helps design of an engineering strata. Data on measurement of lithological contacts of rock formations inside tunnels and study of rock structure are presented in the form of a tunnel log, which portrays subsurface features of a tunnel (see Section 16.6.3). In fact, preparation of a subsurface map depends on the requirement of engineering design. All these details of geological information are necessary for the preparation of project report by the engineers along with design of the structures and cost estimate.

11.2.3 Construction Stage

In many large projects, an engineering geologist's presence is routinely required. The geologist guides the exploratory work required for the site before beginning construction and prepares detailed geological maps of the sites on scales varying between 1:500 and 1:50 when the foundation is exposed after excavation prior to starting the construction. At this stage, plane table survey with an alidade (Fig. 11.6) facilitates the plotting of topographic as well as geological features at the site itself. While surveying, the geologist should be especially vigilant to detect the geological weaknesses of the sites, plot these weak features in the map, and later report to the project engineer suggesting suitable remedial methods for the weaknesses. For example, the foundation rock may be highly fractured requiring treatment. If grouting is suggested to strengthen the foundation rock, a plan for grout holes is to be prepared on the geological map showing the location of the grout holes including their orientation and slope.

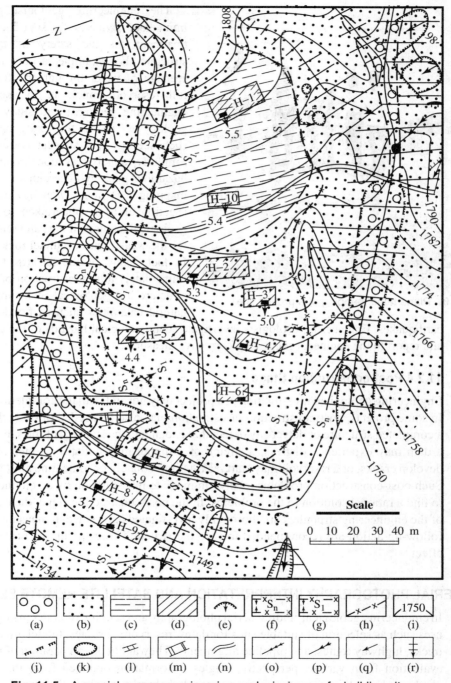

Fig. 11.5 A special-purpose engineering geological map of a building site: (a) bouldery soil; (b) sandy soil; (c) mixed sandy and silty soil; (d) bearing capacity of soil in kg/cm²; (e) slide; (f) highly unstable zone; (g) less unstable zone; (h) separating line of two slide-prone zones; (i) contour; (j) scarp; (k) subsidence; (l) retaining wall; (m) lined drain; (n) metalled road; (o) perennial channel; (p) channel with water flow; (q) dry channel; and (r) limit of active erosion

Fig. 11.6 Plane table mapping of a project site

The engineering geologist gives prior or immediate advice on the spot about the remedial treatment for the safety of the engineering structure. For example, a dam may experience seepage of water through the foundation under full reservoir condition. In such a case, the engineering geologist will study the foundation geology to locate the exact zone or path of such seepage and suggest corrective measures. Similarly, tunnel excavation through rock may face hazardous geology in certain sections. Geological mapping (3D *tunnel logging*) inside the tunnel (Section 16.6) is undertaken to detect adverse geological features, if any, and remedial measures for the weak zones of tunnel rocks are suggested. In fact, many of the hazards faced during or immediately after construction of an engineering structure are geological in nature or created by natural calamities such as floods, landslides, or earthquakes. Engineering geologists investigate such geological hazards of the construction site and report to the engineers suggesting corrective measures.

11.2.4 Post-construction Stage

Engineering geological investigation in a project may not end with the completion of the project. In some projects, post-construction investigation may also be needed to study the behaviour of a completed structure or detect any defect that endangers the stability of the same. For example, a dam may experience distress, a stretch of a tunnel may collapse, a pier of a bridge may develop cracks, or a reservoir may start leaking immediately or some time after its construction. Such post-construction problems involve immediate investigation by an engineering geologist to find a rational solution to the problems. In the post-construction stage, the causes of distress of the engineering structures are thoroughly investigated and remedial measures for immediate follow-up action are recommended to save the structure from failure or further deteriorating effect.

11.3 AERIAL PHOTOGRAPHY INTERPRETATION AND SATELLITE REMOTE SENSING

In engineering geological studies, aerial photography and satellite remote sensing have manifold uses such as selection of suitable sites for dams, reservoirs, bridges, tunnel alignments, railway tracks, highway routes, and pipeline alignments. They aid in the preparation of maps for terrain evaluation from various perspective angles. Information obtained from the study of aerial photographs and satellite imagery is immensely helpful not only in the planning stage, but also during the execution and operation of engineering projects.

Studies of aerial photographs and satellite imagery give valuable information regarding geomorphology, geology, groundwater condition, and surface water distribution. They also help in the evaluation of vegetation cover and ground stability and in the preparation of maps delineating the slide-prone areas showing old slides and recent slope failures. The study also assists in locating construction material and estimating the quantity of gravel deposit or sand in a valley by photogrammetric methods. Fluvial landforms such as flood plains, point bar, channel bar line, and back swamp are other features that can be deciphered from the study. The various aspects of aerial photographic study and its interpretations are described in the following subsections.

11.3.1 Aerial Photography

Indian Air Force is the repository of aerial photographs of the country. Sophisticated cameras fitted in aircraft flying at lower heights take aerial photographs of the earth's surface. The photographs taken are usually vertical, but may also be oblique which are used for special purposes. Vertical photographs are used for the best interpretation of geological details. The photographs taken are mainly of the following types:

Panchromatic photography   This provides black and white photographs and is maximum used.

Infrared photography   This is sensitive to colour and brings out distinctive signatures for water bodies, marshy land, and vegetation.

Colour photography   This is specially prepared for identification of mineral zones with coloured beds.

Spectra zonal photography   This is sensitive to selective rays of the spectrum and is used for special purposes.

An aerial photograph (Fig. 11.7) generally varies in scale from 1:5000 to 1:60,000. For engineering geological purpose, large-scale aerial photographs on scales of 1:5000, 1:15,000, or 1:25,000 are preferred. The photographs can be procured only from Survey of India. An index map (scale 1:250,000) showing latitudes and longitudes helps in precise identification of the required photographs of a desired area. The plan also contains details such as the scale of photographs, aerial camera used and its focal length, the date of photography, and other related aspects.

Fig. 11.7   An aerial photograph of an urban locality with roads and buildings in Gujarat

11.3.2 Applications of Stereoscope

A pair of aerial photographs taken from two camera stations covering some common area (overlap) constitutes a stereo-pair. Stereo-pair photographs with 55 per cent to 65 per cent overlap are used for stereoscopic study. Overlapping photographs, when superimposed by a stereoscope, provide 3D view of the ground surface. The two types of stereoscopes used for 3D views are as follows:

Pocket stereoscope for field use   This consists of convex lenses having focal length of 100 mm and magnification of 2.5 and facilitates viewing only a small portion at a time.

Mirror stereoscope for office use This has two fixed mirrors aligned at an angle of 45°, each reflecting the image to a pair of 45° prisms. The reflected image can be seen through the eyepiece of a binocular. The magnification obtained in this type of stereoscope can be enhanced to 4× and 8×.

Parallax bar is an important tool used in the study of aerial photos with stereoscopes. Parallax is the apparent displacement of position of a body with respect to a reference point, and parallax bar is the instrument that measures the difference in absolute stereoscopic parallaxes of two points imaged on a pair of aerial photographs. It is made up of two glasses connected by a microscopic screw. The glasses are engraved with measuring marks known as 'floating marks'. This is an important method of a preliminary photogrammetric survey of ground details.

11.3.3 Photographic Elements—Tone, Shape, and Texture

Photogeological interpretation of physical features and geology of a terrain is based on the study of *tone*, *shape*, *texture*, and *pattern*, which are the significant photographic elements of aerial photographs. Tone is a measure of the quantity of light reflected by an object and recorded on a black and white film photograph. Tones may have a wide variation usually in shades of white, grey, and black. Shape or forms are the certain distinctive landform features seen in aerial photographs that help in their identification. For example, the shape of a depression resembling a kettle hole indicates a glacial landform. Structures such as folds, domes, and volcanic cones and landform features such as sand dunes, eskers, and river terraces can be recognized from their typical shapes or forms. Texture is an indication of the roughness of a surface and represents a combination of several fundamental characteristics of the photographs such as tone shape and pattern. Texture may be fine, medium, or coarse. Very close spacing of a stream results in fine texture of drainage, whereas wide spacing of the stream causes a coarse texture of drainage.

11.3.4 Ground Pattern

Geological structures, landform features, and vegetation characteristics of a terrain exhibit distinct ground patterns that help in their identification from a study of aerial photographs. For example, structures such as joints, faults, beddings, and dykes are recognized from their characteristic linear patterns. Glacial deposits such as moraines, fluvioglacial deposits such as eskers and kames, alluvial deposits such as a fan, windborne deposits in arid region such as sand dunes, and lacustrine outwash deposits are recognized from their pattern in the 3D view. Ground slope variations are the other features easily distinguishable from the study of the pattern. Plains, plateaus, escarpments, river terraces, and so forth have significant patterns provided by the relief features.

The drainage pattern plays a significant role in the recognition of geological structures such as folds, faults, and joint system. The drainage of an area may be of different types such as dendritic, radial, parallel, annular, rectangular, and trellis (see Fig. 7.18). For example, a radial drainage signifies the presence of a dome or basin. Joints and faults are responsible for the formation of rectangular drainage pattern. Dendritic pattern is indicative of the homogeneity of bedrock having no structural control. Parallel drainage signifies uniform unconsolidated material. Trellis pattern results from lack of preferred orientation of structures or homogeneity of rock types. The drainage pattern is sometimes indicative of the erosion of the rocks. Rapids and waterfalls created by erosion help to identify rocks with alternating soft and hard bands and provide an idea of erosional effect. The distribution of vegetation has a distinct relation with the rocky character of the area including the nature of the soil. For example, a rugged topography is indicative of a hard rock area, a gentle topography is formed by soft rock, and a cultivated area in a flat land suggests floodplain deposits.

11.3.5 Identification of Common Rocks

Experience in inductive evaluation of the photographic elements such as tone, texture, and ground pattern supplemented by the ability to recognize the condition of the terrain help in the identification of the main rock types and broad structural features. The guidelines proposed to identify common rocks are as follows:

Acid igneous rocks such as *granite* are characterized by light tones and basic igneous rocks such as *dolerite* by dark tones. Both these igneous rock types are recognized by granular texture, massive and homogeneous character, resistance to erosion, widely radial drainage, and rounded and hummocky forms. Extrusive rocks such as *rhyolite* are generally of light tone and *andesite* is of grey tone. Volcanic rocks such as *basalt* can be identified by the presence of volcanic cones, flow structures, columnar joints, and poor vegetation cover. Basalts exhibit dark tone, fine texture, drainage, and occasional flow structure.

Metamorphic rocks, especially the high-grade types, show banding, prominent jointing, subdued topography, medium tone, and parallel-to-sub-parallel drainage. *Gneiss* is recognized by high drainage density, light tone, prominent topography, and absence of bedding; it sometimes appears in the form of sharp ridges with an angular or dendritic drainage. *Quartzites* are of light grey tone, medium texture, widely spaced drainage, and sharply crested ridges. *Phyllite, schist*, and *slate* are characterized by grey-to-dark tone, sometimes with mottled appearance, and dendritic drainage pattern with smooth rounded landform. Slate is recognized by its rectangular drainage pattern and usually dark tone.

Sedimentary rocks can be identified from characteristic tone, texture, and topographic expressions. *Sandstone* is of light grey tone, medium texture, pitted appearance, and indistinct bedding. *Limestone* can be recognized from karst topography, fine texture, light tone, internal drainage, and occasional presence of sinkholes. *Shale, silt*, and *clay* are identified from grey to dark tone, fine texture, and closely spaced dendritic drainage. These rocks have intense erosional features including badland topography, gentle gradient of bed, and subdued landform. *Conglomerate* exhibits light tone, medium-to-coarse texture, pitted appearance, and indistinct bedding.

11.3.6 Identification of Large Rock Bodies and Major Rock Structures

A large igneous body has diagnostic pattern of landform and drainage. An intrusive body of dolerite dyke has a distinct relationship with country rock that is known from the study of form or shape. A dyke forms a straight ridge but the rock it has intruded generally forms the lower ground. Metamorphic rocks have a characteristic lithological make-up that is different from stratified rocks or intrusive igneous bodies. Sometimes, igneous and metamorphic rock bodies on a regional scale can be separated from the sedimentary horizons by the zone of tectonic disturbance.

The Precambrian gneissic complex, for example, is separated from the Gondwana sedimentaries by a boundary fault that can be identified from the tonal change and pattern. The regional dip of sedimentary strata is known from the expression of outcrops covering a large area. The individual bedding dip often coincides with the ground slope and the dip direction with the slope direction of the ground. Flat lying or nearly horizontal beds (dip <5°) are easily recognizable from the ground pattern. With some experience, an engineering geologist will be able to estimate the low dip (5°–10°), medium dip (20°–45°), and high dip (>45°) from the shape or disposition of the sloping beds.

Structural features such as a fold can be recognized from the varying dip directions of the strata. Sometimes, the whole of a fold can be viewed in an aerial photo. A plunging fold is recognized from the drainage pattern of the terrain. The major stream generally curves round the nose of such a fold, the convex side of the curve being the direction of the plunge of an anticline. A fault can be identified from features such as abrupt tonal change, repetitive beds, offsetting of lenses and beddings, low relief compared to the adjacent ground, straight depression in the ground or a straight channel for a long distance, waterfalls across a stream, abrupt change in drainage pattern or landform, and rectangular depression. A regional tectonic lineament created by a fault or a thrust can also be recognized from the study of overlapping photographs under a stereoscope.

11.3.7 Satellite Remote Sensing

The nodal agency in India with respect to imagery is the National Remote Sensing Agency, Government of India, Hyderabad. Satellite remote sensing provides data on the broad geological aspects and ideas about topography, landform, drainage pattern, and vegetation cover. The different stages of investigation include the interpretation of a set of actual photographs or satellite imagery to obtain regional level information, detailed photo-interpretation covering engineering project sites, field check, and preparation of maps based on study results. The various remote sensing techniques and their applications are described as follows:

Techniques of producing imagery

Remote sensing is a method of acquiring information about the earth's surface from a distance. The scanning equipment from a space satellite produces the imagery. A satellite sensor records signatures corresponding to a wide range of wavelengths of the sun's rays (or rays from its own source) reflected from the earth. The data referring to different spectral bands (i.e., different wavelengths) can be selected, enhanced, and recombined by computer processing.

At present, in-situ collection of spectral reflectance data of different rock types and structures for computer analysis is used for correlation of the spectral bands for lithological and structural differentiations. The imagery and computer compatible tapes are the data products useful for visual interpretation of geological features. Multi-spectral images and false colour images produced by assigning an arbitrary colour to one or more spectral bands that are specially sensitive to variations in nature are used in ecological, agricultural, and varied geological purposes including environmental study.

Remote sensing, as applied to geological studies in recent time, is based on the precise measurement of the electromagnetic radiation, or spectral signature, from objects on the surface of the earth. The electromagnetic spectrum ranges between very low and very high frequencies, corresponding to very large and extremely tiny wavelengths. Remote sensing instruments, or *sensors*, mounted on board a *platform* (a spacecraft such as a space shuttle or a satellite) measure the spectral signature of the objects on the earth, which are later suitably analysed. However, the signatures corresponding to the zone of visible or infrared rays are most suited for geological remote sensing.

Applications of remote sensing

The imagery obtained by orbiting satellites (Fig. 11.8) provides a good amount of information regarding topography, vegetation, hydrology, and atmosphere of the earth. A number of satellites (Landsat, Geosat, etc.) have been used for collecting data on natural resources including geology. Landsat facilitates digital analysis of remotely sensed data that are useful in studying surface geological features. The geological application of remote sensing techniques has the following advantages over ground reconnaissance. The information on topography and geology can be obtained in a much shorter time compared to ground survey and the regional geological structures are precisely evaluated.

The area covered by a single image is rather large and is of the order of 100×100 km², but an aerial photo taken from a height of a few kilometres covers only a few square kilometres. The scale of the imagery is generally 1:1,000,000 to 1:250,000. Remote sensing techniques facilitate comparative assessment of a changing landform of surface features from survey by flights over the same area conducted in different periods. This helps in identifying the effects of natural calamities such as earthquakes and landslides even on remote localities including those on engineering structures from the study of the changing images of an area before and after e natural hazards. In Figs 11.8(a) and (b) a comparative study of the imagery on May 2005, and that of October 2005, clearly depicts the co-seismic effect of landslide of the earthquake of October 2005 in a remote area of Jammu and Kashmir.

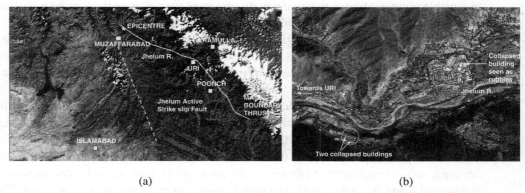

(a)   (b)

Fig. 11.8   Satellite image: (a) pre-earthquake image of a remote locality of Jammu and Kashmir; and (b) post-earthquake image of the locality showing co-seismic effect of landslide

The remote sensing technique of a specific type is used for some specific information. The satellite imagery (in infrared band) provides excellent data on morphological characters, drainage system, seepage, infiltration, and water saturation including the extent of flooding for the preliminary and planning stages of investigation of a project. However, these do not reveal the large scale geological details of an engineering site covering a small area (e.g., a dam site and power plant site). Such details are generally available from the studies of aerial photographs under the stereoscope. Interpretation of satellite imagery and aerial photographs provides a wide range of data useful in engineering geological investigation, and engineering geologists should involve themselves in this interpretation, especially to reveal the following aspects:

(i) Regional structures (fault, fold, etc.), tectonic lineaments, and intrusive bodies such as dykes and the nature of their contact zones with country rock

(ii) Characteristics of soil and rock types including extent of soil cover, outcrop area, attitudes of rocks such as strike and dip of strata, and intensity of fractures and weathering

(iii) Identification of zones of slides and subsidence and slide-prone areas

(iv) Terrain evaluation for railway communication and road alignment

(v) Hydrological aspects such as location of a groundwater basin, the distribution of water saturated areas, and status of infiltration in and around reservoir areas

(vi) Steepness of river gradient, drainage pattern, and swampy areas

(vii) Location of caverns and overall karstic condition, if any, in limestone terrain

A trained eye is, however, necessary for an accurate interpretation. It may be mentioned that remote sensing and photo-interpretation are not the substitute of field investigation. They are tools of quick evaluation of the surface geology and morphology. Ground check-up is absolutely necessary to find the accuracy of the interpreted data.

Applications of geographic information system

Geographic information system (GIS), also known as geospatial information system, uses computers and software to explore the fundamental principles of geography. It is a computer-based tool to enable users to compute, store, analyse, and arrange spatially referenced data. GIS helps geologists to understand questions and interpret and visualize data in many ways that reveal relationship, pattern, and trend in the form of maps, reports, and charts. GIS technology is capable of integrating common database operators such as statistical analysis with the unique visualization and geographical analysis benefits offered by maps.

GIS allows engineering geologists to use existing data in new ways; for example, plotting data on a map helps ask questions such as where, why, and how, with location information in mind. Geographical knowledge helps in making better decisions. GIS is used to put together the vast amount of information available today and provide a meaning to the overall picture. It can be used to prepare maps coded by values from the database to help illustrate patterns. For example, using GIS, the directional signs on a highway are managed. Similarly, the maps on the Internet are made possible with GIS.

GIS can be applied in many fields such as agriculture, business, environment, forestry, military, risk management, hydrology, and geology. In geology, GIS can be used to study geological features, analyse soils and strata, assess seismic information, or create 3D displays of geographical features. Further discussion on this computer-based subject is beyond the scope of this book. Interested readers may study the book by Bruce Davis (2001). This fully updated edition offers comprehensive introduction to the application of GIS concepts. The unique layout provides clear, highly intuitive graphics and corresponding concept descriptions. This reference book is helpful for experienced professionals as well as new readers.

11.4 GEOPHYSICAL EXPLORATION

Geophysical exploration provides information on subsurface rock conditions. The method is used in the investigations of dam sites, tunnel alignments, power plants, bridge sites, and reservoir areas to determine the depth to bedrock, overburden thickness, fault zone, and subterranean cavities. There is some controversy on the authenticity of the data obtained from geophysical exploration. It is, therefore, necessary to substantiate the accuracy of the geophysical data by some boreholes before the data is used for project planning or design. The geophysical methods mainly used in engineering geology are of the following types: seismic, gravity, magnetic, resistivity, and radiometric.

11.4.1 Seismic Survey

The seismic method of survey has wide applications in an engineering project. The seismic wave has characteristic velocities in unconsolidated deposit and in different rock strata that can be utilized to measure the thicknesses of overburden and underlying rock formations (e.g., sedimentary, igneous, and metamorphic rocks). This also enables us to detect unconformity between two rock formations and identify geological structures such as faults and folds and intrusive bodies such as batholith and dyke. To start with, a traverse line with a hole for explosion (*shot* hole) is selected at the site of investigation and geophones are installed along the line at fixed distances related to the shot hole. A small charge of gelatine is exploded from the shot hole. The seismic wave thus generated travels through the media or formation in the subsurface region and is reflected or refracted at a velocity dependent on the nature of media (Fig. 11.9). The denser the media, the faster is the speed of the transmitted wave. In seismic surveys, the fast-moving longitudinal wave is considered rather than the slow-moving shear wave.

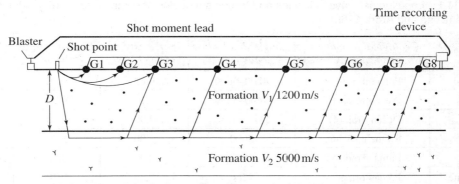

Fig. 11.9 Seismic method of surveying (G, geophone stations)

The initial shot time and the travel time of the impulse of return journey are recorded in the photographic traces in geophones. The difference of initial and return times gives the travel times for the different geophones. The distances of the geophones from the shot hole are known from the field measurement. The times recorded in various geophones are plotted in the vertical line and the corresponding distances along the horizontal line. This gives a straight line for time–distance relation. The average velocity of travel (V) through a rock type, which is mostly used, is obtained by simply dividing the distance by time (Fig 11.9). From the time–distance curve, the distance D which is the intersection point of top formation and underlying formation can be scaled out (Fig. 11.10).

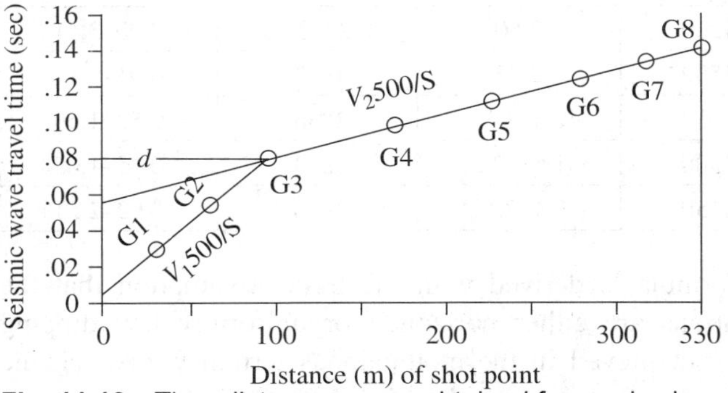

The depth (D) of a rock type or that of overburden is obtained by the following relationship:

$$D = d/2\{(V_2 - V_1)/(V_2 + V_1)\}^{1/2} \qquad (11.1)$$

Here, D is the depth of the topmost formation corresponding to velocity V_1 that overlies the formation corresponding to velocity V_2 and d is the horizontal distance of the intersection of the two zones. The depth D and distance d are expressed in metres and velocities V_1 and V_2 in m/sec.

Fig. 11.10 Time distance curves obtained from seismic survey

The seismic survey also helps to estimate in-situ Young's modulus (E) following the equation:

$$E = V^2 g \gamma \{(1+p)(1-2p)\}/(1-p) \tag{11.2}$$

Velocity V for Eq. (11.2) is known from seismic survey travel time record. The density of several rock specimens is determined in the laboratory and the average value (γ) is taken. The Poisson's ratio (p) of a rock type (core samples) is calculated by determining the compressional and shear wave velocities (Vp and Vs, respectively) of the samples in the laboratory by means of ultrasonic material tester, and then substituting the values in the following equation:

$$Vp/Vs = \{(1-\gamma)/(1/2-\gamma)\}^{1/2} \tag{11.3}$$

From the data available from seismic survey in rock formations of various project sites of India, Bose and Arora (1969) calculated the longitudinal web velocity and the Young's modulus of various rock types as shown in Table 11.2.

Table 11.2 Longitudinal wave velocities and Young's modulus of common rocks of geo-logical formations of India (Bose and Arora 1969)

Rock type	Geological formation/ No. of specimens tested	Longitudinal wave velocity (m/sec)	Density assumed (g/cc)	Poisson's ratio (assumed)	Young's modulus $E \times 10^{-5}$ kg/cm²
Granite	Precambrian (3)	4200–5600	2.65	0.30	3.81–6.78
Gneiss	Precambrian 16	4400–5300	2.70	0.27	4.29–6.07
Basalt	Decan Trap (4)	3940–4150	2.90	0.30	3.71–4.36
Basalt	Rajmahal Trap	5000	2.90	0.30	5.39
Basalt	Punjal Trap	5940	2.90	0.30	7.60
Rhyolite	Precambrian	2250	2.65	0.30	0.99
Khondalite	Precambrian	5400	2.70	0.27	6.30
Schist	Precambrian (2)	3100–3700	265	0.30	1.89–2.75
Quartzite	Precambrian (8)	3500–5500	2.65	0.30	2.15–5.84
Phyllite	Precambrian (5)	2420–3500	2.65	0.30	1.15–2.41
Dolomite	Precambrian (2)	3550–5500	2.6–2.7	0.27–0.33	2.43– 6.54
Limestone	Vindhyan (3)	3550–5500	2.65	0.30	2.48–5.96
Limestone	Tertiary	4500	2.60	0.30	2.43
Limestone	Triassic	2800	2.60	0.33	1.38
Sandstone	Vindhyan (7)	2030–4250	2.65	0.30	1.81–3.56
Sandstone	Gondwana (8)	2400–4200	2.60	0.30–0.33	0.90–3.41
Sandstone	Tertiary (5)	2200–3500	2.55	0.33	0.91–2.11
Sandstone	Siwalik (3)	2400–3050	2.4–2.55	0.36	0.82–1.60
Shale	Tertiary (3)	2200–2500	1.53–2.5	0.33	0.45–1.05
Claystone	Siwalik (3)	2500–4150	2.4–2.45	0.36	0.93–2.51

The seismic depth computation formula is derived with the basic assumption that the interfaces between geological formations are either horizontal or uniformly low dipping (<10°). However, this condition may not prevail in the mountainous terrain where seismic survey result is liable to some error because of very large elevation differences from geophone to geophone along the seismic profile; also, the steep depth of the foundation is a source of error in seismic depth computation. It was found that the values of rock cover and overburden thickness determined by seismic survey in the Himalayan terrain of Jaldhaka project (Banerjee and Rao 1971) vary to an extent of 'nil to +20 per cent' with respect to the values obtained by geological observations and drilling data. Due to the ruggedness of a terrain, various corrections need to be made to get an approximation to the true value.

In engineering geological investigations, seismic survey technique is also applied to identify the weak rock zone and structural features from contrasting velocities in crushed rock due to fault and the intact rocks around. This method is significant in detecting the fault and thrusts in the projects of the Himalayan terrain. Once the fault zone is explored and confirmed by drilling, it is necessary to find if the fault is *active*. This is done by investigating the activity along the fault by seismometers installed along the fault plane or close to it. The instrument uses the pulses transmitted from the ground by means of arms pivoted against a stable base, and the result is recorded in seismograms on a revolving drum. Changes in the magnitude, spacing, and location of minute tremors compared to the stable background provide data on the increasing shocks. A seismometer furnishes data on the activity along the fault for small pulses. In the Thein project of the Himalayan terrain in Punjab, a fault traced close to the dam site by geological and seismic survey was investigated later by installing seismometers to find whether it is active.

For *shallow depth seismic survey*, instead of using explosives, seismic waves are generated by the thumping of a sledge hammer. The technique is the same as the deep subsurface seismic exploration. In this method, the shot points and geophones are arranged in close intervals of a few metres to pick up the return of the wave formed by the striking of the hammer at the shot points. *Hammer seismograph* is the instrument used for such shallow depth seismic survey. This instrument is very handy and provides subsurface information covering a depth of 20–30 m.

11.4.2 Gravity Survey

The aim of a gravity survey is to record the gravity anomaly of a place due to the varying densities of underlying rocks. The instrument used in the measurement of gravity is known as the *gravimeter*. The gravity of an area measured by the instrument is not only the earth's gravity, which depends on the latitude, longitude, topography, and elevation of a place, but also the density of the subsurface rocks. The gravity survey of an area provides data from which *gravity anomalies* associated with subsurface rocks can be evaluated after suitable correction (Bouguer correction) of several variables. The underlying dense rock body shows *high* or positive anomaly and a subsurface light body gives *low* or negative anomaly.

The ground surveys are done at closely spaced intervals to detect the anomalies at a shallow depth of the site. The data obtained is subjected to air, terrain, and Bouguer corrections to obtain the correct value of g of a place. The values of g for different places within a large area are plotted and depicted by contours. The density values of the rocks of a place determined from the laboratory study of rock core specimens help in the calculation of the field value of density anomalies (positive or negative anomalies) of a place. The gravity method can be used for detecting deep-seated cavities. The gravity survey conducted at the Kopili hydroelectric project site in Assam was able to identify caverns at varying depths of limestone formation.

11.4.3 Magnetic Survey

The field operation of magnetic method of survey is aimed at determining the density of rocks in the site of survey. The instrument used in magnetic survey is known as the magnetometer, which, similar to an ordinary compass needle, deflects when magnetic rocks are present in the vicinity. In the place of the compass needle, modern magnetometers consist of a device that can pick up the presence of even a very small trace of magnetic substance. An iron ore deposit or other such metalliferous minerals are detected by the magnetometer. Magnetism is imparted by magnetite (Fe_3O_4), and other magnetic minerals developed in the rocks in the area of survey. The magnetometer records the positive and negative anomalies of magnetism in rocks measured in unit of 'gammas' that range up to ±1000. The data on anomalies is plotted in maps that can be used for making sections and preparing models. The magnetic survey method of subsurface investigation has limited use in engineering geology. Igneous bodies such as domes and dykes covered under overburden in a project site can be detected using this method.

11.4.4 Resistivity Survey

Resistivity survey is also known as the electric or electromagnetic method of survey. It involves the determination of resistance of rock or overburden material to current flow. The method is based on the principle that any change in specific resistance brings about a change in current flow through the media and consequent change in the electric potentials. The porosity of a sand and soil deposit and the presence of fracture space (joint openings) are controlling factors in the resistivity.

To proceed with the resistivity survey in the field, two pairs of electrodes are set on the ground in a line, the distance from one electrode to the other being the same (Fig. 11.11). Current is passed into the ground through the two end electrodes and the reduction in voltage is measured through the central pairs of electrodes. The electrodes are made of porous pots filled with copper sulphate solutions with copper plates immersed in it and passing from the lids of the pots. The sulphate solution seeping from the porous pots makes electrical contacts with the earth. Lateral shifting or increasing the spacing of the central pair of electrodes enables measurement of lateral and vertical variations of voltage by a potentiometer. The following equation is used to measure resistivity (ρ):

$$\rho = 2\pi\alpha\,(v/\tau),$$

Here, α is the spacing between two electrodes; τ is the induced current flow through the two end electrodes measured by a milliammeter; and v is the voltage between the electrodes measured by a potentiometer connected with the inner pair of electrodes. ρ is expressed in ohm/m, α in m, and v/τ in ohms.

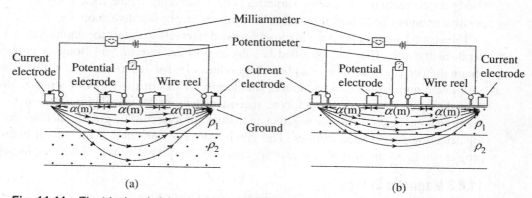

Fig. 11.11 Electrical resistivity survey showing flow of current between electrodes in two layers: (a) top layer resistivity ρ_1 is lower than bottom layer resistivity ρ_2; and (b) top layer resistivity ρ_1 is higher than bottom layer resistivity ρ_2

The resistance values of some soils and rocks are shown in Table 11.3.

Table 11.3 Types of soil/rock and corresponding resistance value

Type of soil/rock	Resistance (ohm/m)
Peat, clay shale	0.5–15
Mudstone	10–200 m
Dry sand, sandstone	200–1000
Saturated sand, sandstone	20–50
Igneous rock	$100–10^7$

Resistivity survey is mainly used in exploring mineral deposits. In engineering geology, this method is applied in determining the depth of porous (water being good conductor of electricity) unconsolidated material and that of rocks at an engineering structural site. It can be seen that low resistivity is indicative of overburden material or soft rocks, whereas hard rocks (igneous or metamorphic) show high resistance values. Thus, the depth of the interface between overburden and hard rock can be known from the resistivity survey.

11.4.5 Radioactivity Logging

Radioactivity logging includes *neutron logging* and *gamma ray logging*. Neutron logging is mostly used for determination of moisture content, porosity, and density of soils and rocks. In addition, radioactive logging can also be utilized to differentiate subsurface formations containing sand, coal, anhydrite, and limestone, which are low in radioactivity compared to bentonite, volcanic ash, and shale, which have higher radioactive values. Nuclear logging involves nuclear reactions between radium or polonium on one side and beryllium on the other to free the fast-travelling neutrons from the nuclei of the beryllium atom.

In practice, a stainless steel tube carrying the radioactive material along with a suitable detector (such as Geiger counter) is placed into the subsurface strata through a bore hole (Fig. 11.12). The scattered radiation is picked up by the detector (Geiger–Muller tube), amplified, and then counted in a scale or rate meter. The neutron scatters in all directions in the earth's mass by colliding with the earth's materials. Collision with hydrogen atom results in decrease in kinetic energy. A change in the count of the slow-moving neutron reflects primarily a change in the water content. The moisture content (quantity per unit volume) can be measured by means of a collision curve. In fact, the technique of *moisture determination* by nuclear method uses the property of elastic back scattering of neutrons. These neutrons collide with the nuclei of the surrounding medium and are slowed down to thermal energy level as the kinetic energy is partially transferred to the target nucleus. Moisture content is determined with the help of calibrated curves drawn between moisture content and count rate.

When a gamma ray photon emitted from a radioactive source enters a material, it may collide and interact with an orbital electron of the material by photoelectric absorption, Compton scattering, or pair production. The technique of *density measurement* by radioactivity method uses the interaction by Compton scattering.

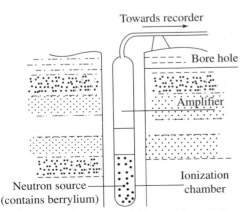

Fig. 11.12 Neutron method of radioactivity logging

Nath and Dhawale (1971), who conducted nuclear logging in the Malabar tunnel project of Maharashtra, have given the details of investigation by radioactivity logging method, by both neutron logging and gamma ray logging, for determining subsoil density and moisture content. In Malabar tunnel, nuclear logging was conducted by using a caesium-137 gamma source and utilizing the phenomenon of elastic back scattering of gamma radiation to obtain the density values of rocks in the borehole. A radium–beryllium neutron source was used for determining porosity using the property of scattering of neutron. The results are given in Fig. 11.13.

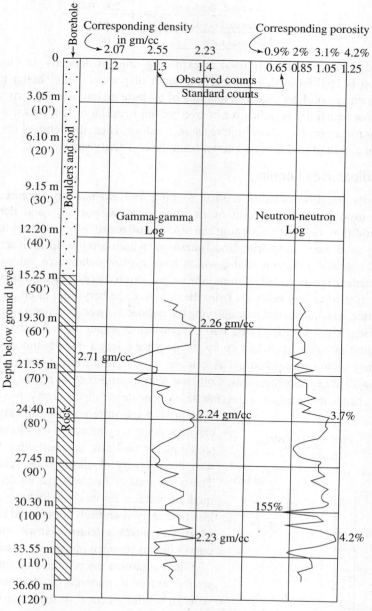

Fig. 11.13 Radioactivity log record at Malabar tunnel project, Mumbai: showing gamma–gamma (density) and neutron–neutron (moisture) logs of bore hole 'A' (Nath R and Dhawale R.A (1971))

11.5 SUBSOIL EXPLORATION AND SAMPLING OF SOILS

The object of subsoil exploration is to obtain detailed information on the nature of the subsurface material for safe and economic design of engineering structures. Exploratory boreholes in thick soil deposit yield data on the layers of changing soil types at depths, the presence of bedrock, if any, within the drilled depth, and the groundwater level. Boreholes also facilitate sampling of soil from subsurface regions for conducting tests on their engineering properties in the laboratory.

11.5.1 Exploration in Sites of Engineering Structures

The depth of exploration of a site made up of unconsolidated materials depends on the type of the construction and the load of the structure to be imposed on the soil. For example, an earth dam site is explored by drill holes to find the depth of the bedrock or an impermeable soil zone suitable for laying the cut-off. It is a general practice to extend the exploratory hole down to the soil deposit where the imposed load of the engineering structure will not create shear failure or any adverse effect on the soil. When a soil deposit is loaded with a structure, a pressure bulb is created in the subsoil region with a gradually diminishing effect from ground surface towards increasing depths. The borehole should at least go down to a depth where the isobar is one-tenth of the surface loading. This depth is generally assumed to be one-and-a-half to two times the width of the loaded structure. The general guidelines for the subsoil exploration of different structural sites are as follows:

(i) At an earth dam site, the exploration should be continued down to a depth equivalent to half the base of the dam, or twice the height of the dam. If the bedrock is available prior to reaching this depth, the exploratory borehole is to be continued for at least 3 m into the bedrock. The boreholes are spaced at every 15 m intervals along the dam axis and also along the upstream and downstream parts. Some widely scattered holes are also taken up depending upon subsoil condition.

(ii) In the case of a retaining wall, spread footings, or pile foundations, the subsurface exploration is to be extended to one-and-a-half times the base width of the structure.

(iii) In road cuts, the exploration should be 1 m to 2 m in subsoil from the base of the cuts. In the case of a fill, the exploration should be down to a depth equivalent to the height of the fill structure, or at least 2 m from the ground surface.

(iv) Borrow areas for earth are also explored to different depths by pits and boreholes to ascertain the quality and estimate the quantity.

11.5.2 Sampling from Pit and Using Soil Sampler with Drill Rod

For collecting an undisturbed sample of soil from close to the ground surface, a pit is dug. A hump is created inside the pit by scooping out soil from all around so that it can exactly fit the inside space of a cylindrical container. The container (top and bottom removed) is then carefully placed over the soil hump and the bottom part is sliced out so as to get an undisturbed sample in the form of a cylinder. The two ends are sealed with paraffin so that the moisture content of the soil remains intact. This type of sampling is possible only in cohesive soil. Alternatively, a large-diameter sampling tube can be driven manually into the bottom of the pit to obtain a tube sample of soil. The tube is capped and carried to the laboratory for testing of soil properties.

Fig. 11.14 Subsoil exploration by borehole with sampler tube

Exploratory drilling is commonly taken up on soil deposit for collecting soil samples from different depths using a soil sampler fitted with the drilling tube (Fig. 11.14). This helps in ascertaining the change in soil characters by laboratory tests of samples collected from different depths. There are two types of soil samples—*disturbed* and *undisturbed*. In disturbed soil samples, the original structures of soil are fully or partially destroyed and these samples are not suitable for determining the strength properties of the soil. However, soil properties such as water content, grain size, consistency limits, and compaction can be determined from disturbed samples if they are *representative* in nature. The undisturbed samples are, however, best suited for the determination of all the geotechnical properties of soil, and they are specially required for conducting tests on shear strength, permeability, and consolidation properties. Several methods of sampling are followed for recovering undisturbed samples to ensure accurate determination of soil properties by laboratory tests. The samplers commonly used for sampling from boreholes are divided into *thick-wall* and *thin-wall* types.

The *open drive sampler* may be of thick-wall or thin-wall type. It is a tube with its lower end open and sharp that facilitates driving into the soil (Fig. 11.15). The tube may be seamless or split into two parts. The upper part of the sampler is provided with a valve to allow escape of water or air when driving into the soil. During the lifting of the sampler, the check valve helps to retain the sample inside the tube. The split-type tube (also called split spoon sampler) is provided with a thin-walled liner inside the tube.

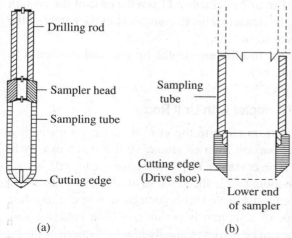

(a)

(b)

Fig. 11.15 Soil sampler used for taking undisturbed soil samples from boreholes: (a) Stationary piston soil sampler; and (b) Rotary sampler

The *stationary piston sampler* consists of a piston attached to a long piston rod for sampling from drill holes. The lower end of the sampler remains closed with the piston during downward movement preventing any entry of water or soil. After reaching the desired depth of sampling, the piston is kept stationary by clamping the piston rod. The sampler is then driven into the soil. When the sampler is full with soil, it is lifted up keeping the piston rod in clamped position. This type of sampler is suitable for collecting moist sand and soft soil.

The *rotary sampler* contains a thin-walled tube inside a core barrel. The tube has a cutting edge and a thin removable wall liner inside. This type of sampler is suitable for recovery of stiff and cohesive type of soil.

11.6 METHODS OF SUBSOIL EXPLORATION

Soil exploration is aimed towards identifying the subsurface soil condition of engineering sites. The common methods used in the exploration of subsoil include pitting and trenching, sounding, standard penetration test, and exploratory drill holes.

11.6.1 Pitting and Trenching

Test pits or trenches are commonly excavated covering the entire area of small structures so that any major change in the soil type within a depth of 3 m to 6 m can be revealed. One pit is commonly dug at each end of the structure. For roads, canals, and similar constructions covering a long distance, one pit or trench is excavated at every 100 m to 500 m intervals covering the entire length along the central line of the cut or fill to find the variation in the nature of soil in its spatial distribution. Penetration resistance test conducted in each pit provides data within a short time on the variations in soil characteristics. Exploration by the method of pitting and trenching serves dual purpose: one, it provides identification with respect to the variations in soil characteristics in spatial and depthwise distribution, and two, it facilitates detailed sampling representative of different soil types covering the entire area of exploration. The samples are tested in the laboratory for their engineering properties as required for design purposes. Wherever required, in-situ testing of soil is also carried out inside the test pits.

11.6.2 Penetration Resistance

Subsoil exploration by penetration resistance (same as subsurface investigation by 'sounding') is a simple technique of driving a rod or a rod with a cone into the soil and measuring the resistance offered by the soil. The penetration resistance of soil is determined by *standard penetration test* and *cone penetration test*. In standard penetration test (Indian Standards 2131–1963), a 3.5 cm-diameter (sampling) tube with a driving tip is used as the 'driving rod'. The tube is encased inside a 5 cm-diameter tube. The driving rod with encasing is forced into soil by blows imparted by a 65 kg drive weight from a 75 cm fall. The rod is allowed to penetrate up to a depth of 45 cm or it is deepened until 100 blows are applied. The 45 cm steady drive into the soil is recorded as penetration resistance 'N'. If the sampler is unable to penetrate even a depth of 2.5 cm with 50 blows, it is considered a 'refusal' of penetration. The equivalent penetration resistance 'Ne' is calculated from the following relation proposed by Terzaghi and Peck (1967):

$$Ne = 15 + 1/2 \, (N - 15)$$

where N is the penetration recorded from observation.

There are standard charts for empirical correlation of the penetration resistance for both cohesionless and cohesive soils that provide data on saturated density and bearing strength (unconfined compressive strength). The *Dutch cone penetration test* is very useful to determine the bearing capacity of the cohesionless soil. The test is done by an instrument having a steel rod (annular pipe/tube) with a 60° steel cone (base area 10 cm^2) at the driving end of the rod. The rod is enclosed in a tube with its outer diameter equal to the base of the cone. The upper end of the rod is fitted with a pressure gauge by a proving ring that measures the resistance. The extent of the steady penetration of the cone under standard blows decides the type of soil with respect to its density and bearing strength. The rod with cone can be forced into the soil with or without the enclosed tube. When it is pushed without the encased tube, it is to overcome the point resistance due to the penetration of cone. The rod with cone alone is pushed down inside the soil by a jack and the value of resistance for 8 cm penetration is recorded, which gives the cone resistance. Then, the rod with cone and the enclosed tube are forced to penetrate further 20 cm to give the value of cone resistance and the frictional resistance along the tube.

11.6.3 Subsoil Exploration by Drill Holes

Drilling for subsoil exploration includes the following:

(i) Auger drilling
(ii) Wash boring
(iii) Percussion drilling (free falling)
(iv) Percussion rotary drilling

Auger drilling

Auger drilling is the simplest method of subsurface exploration and sampling of soil commonly from shallow depths. Auger drills may be *hand operated* or *power driven*. A hand-operated auger is in general operated by two persons by moving or turning a handle as shown in Fig. 11.16(a). The auger is pressed into the soil and at the same time a turning motion is applied to the blades to allow entry of soil inside the auger tip. The size of the hand-operated augers varies from 5 cm to 20 cm in diameter and can drill generally up to a depth of 6 m in cohesive soil above the water table. A tripod stand and a simple hoist are used for the operation of the auger drill for deeper holes. Drill holes of more than 6 m depth can be made by screwing together several of the 1.5-m long extension rods during the progress of drilling. The auger can be used to collect disturbed samples from shallow depth. The sampling ends of the auger that collect subsurface soil by rotational movements may be of various shapes—the most common types being Iwan type, helical (spiral) surface type, and the closed spiral type as shown in Fig. 11.16 (b).

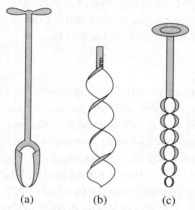

Fig. 11.16 Hand-operated augers for use in soft soil: (a) with a handle and Iwan-type sampling tip; (b) with helical spiral tip; and (c) with closed spiral tip

A power auger, which may be truck mounted, is very useful in boring a large area in a short time. Power augers have continuous flights provided with hard and sharp cutting bits. Continuous flight auger drilling can make holes up to 35 cm in diameter and penetrate up to 50 m or more. In addition to soil, power augers are used for boring and sampling in sand, gravel, and clay deposits. These can also be used in soft rocks such as shale and chalk. In general, no casing is provided in the auger holes except where the drill holes are driven below the water table and through a sandy soil zone. The samples of soil collected from both hand- and machine-operated auger holes are disturbed but they can be used for identification of soil types and their variations at depths. Auger drilling is very helpful in exploring borrow areas for soil of earthen dam and highways but is ineffective in boulder soil.

Wash boring

Wash boring is an inexpensive technique used for getting a general idea of subsurface condition of the soil. The wash boring equipment consists of a drill rod (2 cm diameter) inside a casing (5 cm diameter), which can be raised and dropped alternately into the soil (Fig. 11.17). The lower end of the drilling rod is provided with a sharp chisel or chopping bid to penetrate the soil or cut soft rocks into small pieces. Water is allowed to enter under pressure through the drill rod, which affects recovery of cuttings in the form of soil–water slurry to ground surface through the space between the rod and the casing.

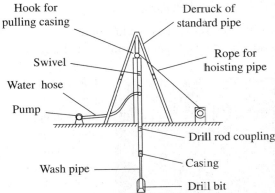

Fig. 11.17 Sketch showing the different parts of the wash boring equipment

The slurry or cuttings preserved as 'wash samples' is not suitable for conducting tests for soil properties, but they give a fair idea about the nature of soil and the type of bedrock. As the wash boring progresses, soil formation of different hardness may be encountered showing the changing soil types at different depths. The presence (and, hence, the depth) of bedrock below the soil deposit can also be deciphered from the feeling of resistance to drilling and identification of cuttings obtained during boring. The groundwater level at the site of exploration is estimated from periodic measurements in wash borings. Undisturbed soil samples can be recovered by inserting appropriate sampling tools after cleaning the bottom of the pipe.

Percussion drilling

The *free falling type percussion drilling* consists of a chisel bit suspended by a cable or attached to a drill rod for boring in soil or soft rock formations by repeated blows against the bottom of the hole. The materials give an approximate idea about the formations at depth, but are not suitable for conducting laboratory tests for their properties. In general, percussion drilling is undertaken in overburden material consisting of sand and clay in varied proportions. To start with, the bit used is 30 cm in diameter which decreases to 7.6 cm at the downward end. If the bedrock is encountered below a soil deposit, the motion imparted by falling type of percussion drilling crushes the rock rather than chipping it. The hole gets saturated with water during drilling below water table. A *bailer* is used to remove the water–soil slurry or pulverized rocks from time to time from inside the hole. The slurry is also bailed out to keep the hole clean before sampling of soil from the bottom of the hole. Separate cables are used for operating the bailer and the bit. A beam moving up and down by a motor acts as a hammer to drive a sampler for collecting soil from depth.

In *percussion rotary drilling,* both percussion and rotary movements act together. This method is rapid in non-coring holes. Percussion rotary drilling is more advanced and gives more progress than the falling type, and as such, this rotary type is mostly used in exploring project sites. The crushed rocks produced by the drilling are taken out by flashing of water or passing air or both together through the drill hole. This type of drilling may be effected by jack hammer or wagon-type drill holes. The percussion rotary drill uses a single-piece drill rod with a tungsten carbide (TC) bit that can penetrate up to 6 m depth. The chisel-shaped bit or four-parts bit is also used for uniform penetration. Extension rods can be added for deeper boring. For rock drilling by percussion rotary type, equipment with extra strong rotation machinery is available, which can drill 4 cm-diameter holes down to a depth of nearly 20 m. This type of drilling equipment is useful for non-coring shallow holes for grouting purpose in tunnels and dam foundations.

11.7 EXPLORATORY DRILLING IN ROCKS

Exploratory drilling in rocks has wide application in engineering projects. Almost all structural sites such as dams, tunnels, bridges, and powerhouses are explored by drilling for rock cores. The machine used in core drilling is known as the drill *rig*. There are several types of rigs operated by motors. Diesel or compressed air is generally used in running the motors, as electricity is seldom available in project sites located in the hilly or remote areas. Figure 11.18 shows a rock drilling operation in a remote area by drilling rig operated by a diesel-run motor, with a *badland* topography at the back. Where good roads are available, truck- or jeep-loaded drill rig can be conveniently used, saving the operation time.

Fig. 11.18   Rock drilling by drill rig

11.7.1 Importance of Rock Drilling in Engineering Geology Works

Subsurface exploration by drilling is a standard practice in engineering geological investigation. Primarily, diamond core drilling is taken up for identifying the nature of rocks at depths and laboratory testing of the core specimens. Though core drilling is an expensive process, the cost is small when compared to the total cost of the project. The utilities of such drilling in rock related to engineering geological investigation are mainly as follows:

(i) In the construction of concrete and masonry dams, the drill holes provide information on the rock type, the depth of bedrock, and foundation grade rock.

(ii) In the investigation for tunnel alignment, the thickness of overburden and rock cover is an important parameter interpreted from drill hole logging.

(iii) The presence of weak zones such as faults, shear zones, joints, solution cavities, and buried channel, if any, in the subsurface region of a project site is assessed from drill hole data followed by pumping tests through the holes.

(iv) In the investigation of construction materials such as stone slabs or stone chips, drilling is done at probable quarry sites to determine the quality and quantity of materials available.

(v) The core specimens of rocks are necessary for quantitative evaluation of their engineering properties required for design purpose.

(vi) Drilling is necessary for grouting purposes. Deep holes are drilled for curtain grouting and also for providing drainage in all large dams.

(vii) Instrumental measurements of rock pressure by inserting extensometers in subsurface rocks require drill holes.

(viii) Drill holes are necessary for permeability testing (pumping test) and measurement of water table.

(ix) Drill holes are also utilized for taking photographs of structural features of subsurface rocks by borehole camera.

11.7.2 Rotary or Core Drilling

In rotary drilling, single or double tube core barrels provided with commercial diamond-studded bits, TC bits, or steel bits with shots are used for obtaining cores of rock or hard soil. Diamond rotary drilling machines for recovery of rock cores are widely deployed in subsurface explorations related to engineering geological investigations. In diamond core drilling, the bits used are the cast set commercial diamonds that are permanently fixed to the cutting end of the rotary tubes. Since diamond is the hardest of all minerals, the diamond drill machine can penetrate all types of rocks. The machine can also drill holes at any angle and at rates of more than 3 m per hour. According to Chugh (1971), 'the essence of diamond drilling continues to be man's ability to interpret from the behaviour of drilling machine as to what is happening in the hole.'

The diamond *core drilling* (*rotary drilling*) uses 3 m-long steel tubes (also called rods or pipes) screwed together as the drilling progresses in cutting rocks in a downward direction. The motor operates a drill head. The tubes are suspended from the tripod. As the tubes are rotated by means of a gear, the bit cuts the rock into circular cores by the pressure of a hydraulic jack. Water is circulated into shallow holes by means of a pump for the purpose of lubricating the passage. Bentonite clay (sometimes added with barites) mixed with water (called *drilling mud*) is pumped into deeper holes to create a bond with the wall rocks for preventing them from caving in. The double-tube core barrel is used for maximum core recovery and less mechanical breakage of cores. The diamond drill machines are of different capacities. A machine weighing 75 kg may drill up to 100 m. Light weight mobile drilling machines shown in Fig. 11.19(a) are also available, which save time in shifting and setting up the machine. Jeep-mounted machines are useful for drilling in rough terrain. The cutting bits, shown in Fig. 11.19(b), are fixed to the outer tube that rotates to cut the rock, while the cores are retained in the inner tube that does not rotate. A sample catching device, see Fig. 11.19(c), at the lowermost drill rod (inner tube) collects the rock core samples and carries them to the surface when the rods are lifted up.

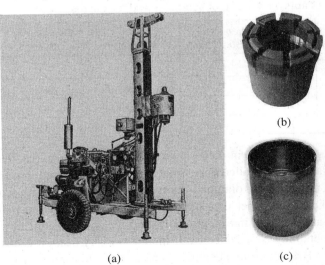

(a)   (b)   (c)

Fig. 11.19 (a) Mobile hydraulic drilling machine suitable for drilling in rock and overburden; (b) diamond impregnated bit (NX size); and (c) core catcher (HEICO)

11.7.3 Depth and Spacing of Drill Holes

The geological programme of drilling with respect to location, depth, and spacing of the holes is based on the nature of the engineering structure and geological condition of the site revealed from surface mapping. In case of a large dam consisting of a uniform rock type, the holes are commonly drilled at least 6 m into the fresh rock and spaced at 15 m intervals along the dam axis and also upstream and downstream parts. Along a tunnel alignment, the drill holes should be spaced initially at 30 m intervals and extending 6 m below the tunnel floor. However, if the bedrock is available at a shallow depth, the hole may be stopped after penetrating 10 m in sound rock. If the dam axis and the tunnel alignment are very long, the spacing may be at 30 m and 60 m intervals, respectively. Depending upon the rock conditions observed in these holes, intermediate holes are planned at places where there are suspected variations in rock textures and structures.

In the case of sites for small structures such as small dams and powerhouses, the holes are to be made at closer intervals, say 6 m to 10 m, and extended at least 6 m into the bedrock. If the site comprises heterogeneous rock types, the location plan of the drill holes should be such that these provide data on rock conditions of each rock type at depth. If weak features are identified from surface investigation of a site, the weak zones are to be probed for their subsurface extension by special holes including inclined holes. Normally, holes are drilled vertically but inclined holes are sometimes planned directing nearly at right angle to dip direction. All holes should be made by diamond drilling, preferably by double-tube core barrel ensuring maximum core recovery.

11.7.4 Selection of Core Bits for Rock Cores

A proper geological study requires continuous cores and maximum core recovery. To achieve this, it is necessary to select the right type of core bits with respect to their size and quality. For example, an NX-size core bit with diamond impregnation provides the best core recovery. In addition, double-tube core barrels are to be used. A casing is first driven using casing core bit. The outer tube or barrel protects the borehole side walls from collapsing and the central tube brings the rock core. The size of the casing or core bits that make the casing hole and core hole is approximately 1.59 to 0.79 mm more than the diameters of the casing and core holes (Table 11.4).

Table 11.4 Size of boreholes and rock cores

Size of hole	Diameter of casing (mm)	Diameter of drill hole (mm)	Diameter of rod (mm)	Diameter of rock core (mm)
NX	89	76	60	54
BX	73	60	48	41
AX	57	48	41	30
EX	46	38	33	22

The bits vary widely in their shape and constituent material. They are made of steel, TC, or industrial diamond called bort. Steel and TC bits are used in soft rock such as sedimentary strata, but in hard rock, diamond-impregnated bits are always used to obtain maximum core recovery. Diamond, other than the bort variety, is also used to make bits, but these are very expensive and are used in exceptionally hard rocks. In geological works, drill holes should be made for obtaining NX-size cores. If the hole is very deep, it is customary to change the hole from NX to BX size from a depth, say, 15 m, and reduce it to AX size at a further depth. This is done to reduce the pressure imposed by the tubes. For geological purposes the EX-size drill holes are seldom planned for recovering rock cores, except where the drilling is limited to about 6 m in rock. Holes of 75 cm diameter and larger dimensions are excavated by calyx drill, which is a non-coring type of shot drilling. These large diameter holes facilitate visual observation of the side walls of the holes.

Vertical holes are generally preferred for easy operations and accurate measurements of the thickness and other parameters of rock formations at depths. Inclined holes can, however, be drilled at any angle and even horizontal holes are drilled to meet specific geological requirements. For example, a vertical hole from the top of a hill is not desirable if the depth of the proposed tunnel is prohibitive, but a horizontal hole driven from the slope can penetrate the desired tunnel level. Inclined holes are taken up to prove a suspected fault or other weak zones. The river bedrocks are also to be explored by angle holes from the banks. In the case of an inclined hole, corrections are to be applied to obtain the exact thickness and attitude of a bed.

11.7.5 Selection of Bits for Drilling in Boulder Deposit

A big boulder trapped in an overburden or a cluster of boulders with cemented matrix may provide cores that may appear like bedrock. The cores are to be carefully studied for smooth and round surface. Knowledge of local geology and the materials below will also help to differentiate the actual bedrock from the overburden (boulder deposit). A loosely compacted boulder bed may cause problems during drilling. It is a common observation in many boreholes that the bit does not cut the rock, but instead the rod rotates only to make the rock surface smooth. The loose boulders keep shifting during drilling and the bits cannot get any grip on them for cutting. This is a condition when a boulder is struck at the lowermost level. This leads to the inference of the presence of a loosely compact bed of boulders at depth. An experienced driller can understand the presence of a boulder or pebble in such a situation.

Rotation of bits instead of cutting also causes considerable wear and tear of the bits including the core barrel and rods. Under this condition, lowering of the pipes is also difficult. To overcome this situation, either the hole is first drilled by a smaller size coring bit and later reamed out with casing shoe bit or if the expected depth of the boulder formation is within 30 m, driving pipes of 10 cm, 7.5 cm, or 5 cm diameters should be used. In deeper parts, it is recommended to start with 15 cm drive pipes if the boulder bed is at a deeper part. Coring is then done with smaller size bits. When core recovery of the overburden material is not required, it is preferable to take up percussion drilling in this part to save the cost. Core drilling may be resorted to once the bedrock is struck. An experienced driller can sense the striking of the bedrock below the overburden by the change in the cutting pressure (Chugh, 1985).

11.7.6 Ways to Improve Core Recovery

The geological interpretation and engineering design to a great extent depend on the percentage of core recovery. Very low core recovery in foundation drilling for rock is interpreted to be due to the soft nature of rock or presence of weak zones when the design of the structure is provided with additional measures, which are not normally required. During drilling in soft rock, there is considerable core loss due to washing of the soft particles by the circulating water used for drilling. This is a normal condition. On the contrary, drilling at a fast rate is also responsible for low recovery of cores caused by grinding. This is an artificial situation. However, normal core recovery is vital for the correct evaluation of rock quality, alteration, joint frequency, and other vital parameters required for judging the stress condition and support requirement for tunnel and underground chambers such as powerhouses. The engineering geologist should assess the drilling and ask the driller to use optimum speed, say, 250 rpm or so, for best possible recovery. To achieve this, drilling must be done with NX-size diamond coring bits and adhere to the following guidelines:

(i) Double-tube core barrel should be used, especially during drilling in soft rock and overburden materials.

(ii) Very hard and massive rock is likely to give 100 per cent core recovery even with single-tube core barrel, but if the rock has separating planes such as joints, double tube core barrel should always be used.

(iii) Best result is obtained by using 0.6 to 1 m-long rods instead of 3 m-long rods.

(iv) Once the inner gauge stones wear out, the bits should be withdrawn (for resetting) and new ones used.

(v) In soft and caving rocks, bentonite slurry should be used to stabilize the rock wall.

(vi) Bits set with small-size diamonds (30–100 numbers per carat) are to be used. The choice for the diamond size and matrix should be as per Table 11.5 (Mustafy 1966).

Table 11.5 Diamond size and matrix for various rock formations

Rock formation	Diamond size (Nos. per carat)	Matrix
Coarse sandstone and clay	8–16	Medium
Hard sandstone and concrete	16–30	Medium
Limestone, schist, and gneiss	30–60	Medium
Granite and quartzite	60–100	Medium
Broken hard quartzite	60–100	Hard
Broken hard quartzite and boulders	Impregnated bit	Medium

11.7.7 Logging of Drill Cores and Diagrammatic Representation

Once the drilling of a hole is completed, the main task of the geologist is the logging of the cores obtained from the hole. The cores are kept in specially-prepared boxes in a systematic way. On the cover of each box, the assigned number of the borehole and its location and ground elevation are to be distinctly indicated. The cores recovered from each 'run' of the drilling (generally of 3 m length) are kept in the *core box* and the depths are written on a wooden piece by non-washable ink (Fig. 11.20). This arrangement of cores is done until all the cores are kept in the box, systematically recording the depths. In general, the driller or engineer in charge of the drilling site keeps the core in the core boxes, recording the relevant information. In case the rock cores are available below the overburden cover (non-coring), some 'wash' of the overburden is kept in the box and the depth at which the overburden ends is noted. This helps to determine the thickness of overburden and the depth to bedrock while studying the core in the core box.

Fig. 11.20 Systematic upkeeping of rock cores in core box

The geological logging of rock cores includes detailed study of the rock cores with respect to percentage core recovery for every run of drilling, measurement of the number of fractures (joints) per run, and calculation of their frequency converting it to per metre of rock (known as fracture frequency), the type of rock and its conditions, the fault or shear zones (from cores of clay, gouge), etc. Figure 11.21 illustrates drill hole log used to record the complete data regarding rock cores and the result of water pumping test (see Section 11.8). The driller also maintains a record of runwise core recovery and other important features that had been encountered during drilling.

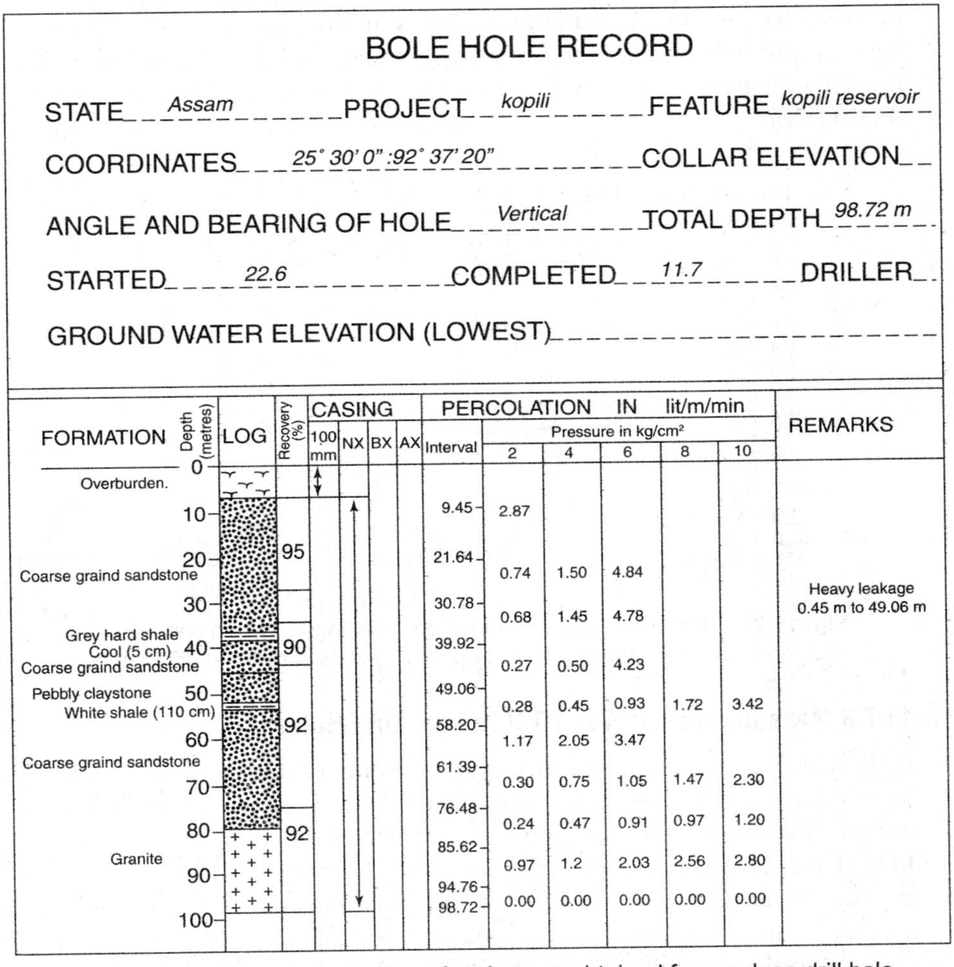

BOLE HOLE RECORD

STATE _ _ _ _ _Assam_ _ _ _ _ _ _PROJECT_ _ _kopili_ _ _ _ _ _FEATURE_ _kopili reservoir_

COORDINATES_ _ _25° 30' 0" :92° 37' 20"_ _ _ _ _ _ _COLLAR ELEVATION_ _

ANGLE AND BEARING OF HOLE_ _ _Vertical_ _ _ _TOTAL DEPTH_98.72 m_

STARTED_ _ _ _22.6_ _ _ _ _ _ _ _COMPLETED_ _11.7_ _ _ _ _DRILLER_

GROUND WATER ELEVATION (LOWEST)_ _ _ _ _ _ _ _ _ _ _ _ _ _ _ _ _

FORMATION	Depth (metres)	LOG	Recovery (%)	CASING 100 mm / NX / BX / AX	PERCOLATION IN lit/m/min						REMARKS
					Interval	Pressure in kg/cm²					
						2	4	6	8	10	
Overburden.	0										
	10				9.45	2.87					
	20		95		21.64	0.74	1.50	4.84			
Coarse graind sandstone											Heavy leakage
	30				30.78	0.68	1.45	4.78			0.45 m to 49.06 m
Grey hard shale					39.92						
Cool (5 cm)	40		90			0.27	0.50	4.23			
Coarse graind sandstone											
Pebbly claystone	50				49.06	0.28	0.45	0.93	1.72	3.42	
White shale (110 cm)			92		58.20	1.17	2.05	3.47			
	60										
Coarse graind sandstone					61.39	0.30	0.75	1.05	1.47	2.30	
	70				76.48						
						0.24	0.47	0.91	0.97	1.20	
	80		92		85.62	0.97	1.2	2.03	2.56	2.80	
Granite	90				94.76						
					98.72	0.00	0.00	0.00	0.00	0.00	
	100										

Fig. 11.21 Example of geological log of rock cores obtained from a deep drill hole

It is important that the logging of the cores is preceded by a reconnaissance of the area covering the drill hole location to obtain a first-hand knowledge of the geological set-up. An engineering geologist working in a project for the first time should never log the rock cores without visiting the drill hole site. The presence of a geologist during drilling helps to understand the significance of certain observations. A fall of drill rod without cutting indicates the presence of fissures. In a karstic terrain, it suggests the existence of a solution cavity at depth. Sudden outflow of water under pressure is indicative of an artesian condition. It is to be ensured by the engineering geologist that the cores are kept systematically in the core boxes and there is no mix-up during the transportation from site to core library.

In addition to the geological logging of rock cores and providing percolation test results as shown in Fig. 11.21, a schematic presentation of the data as in Fig. 11.22 is also necessary. The most common weak features of rocks are the joints, faults, and shearing. In the engineering geological investigation of subsurface rock conditions, such schematic representation of rock core recovery, fracture frequency, and water pumping test results is important for understanding the weakness of rocks including fault zones at depths. High loss through rocks between specific depths indicates that the rock of that section has been affected by excessive jointing or shattering due to fault. For example, the schematic presentation of log of a drill hole of a dam site showing high core loss, excessive fracturing, and very high water loss between dephs 10.5 m and 12.0 m is indicative of a shear zone or a fault zone (Fig. 11.22).

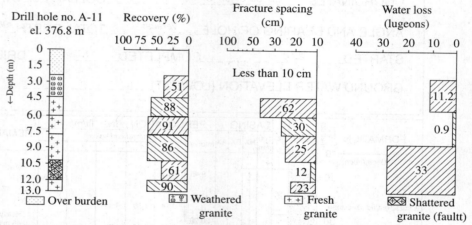

Fig. 11.22 Schematic presentation of drill hole logs showing core recovery, fracture spacing, and water loss indicated by pumping tests in rocks at different depths or test runs

11.7.8 Measurement of Water Table from Drill Holes

Drill holes may serve as piezometers in the study of a groundwater basin. Seepage from a reservoir or a dam foundation and sudden rush of water during tunnel or powerhouse constructions are some of the problems associated with groundwater level in the project sites. In hard rocks, the groundwater level is measured from uncased hole provided with an automatic recorder. It can also be measured by simply inserting a tape in a borehole. The top part of the overburden material is, however, kept cased and a cap is provided at the top. In soft rocks, the entire length of the borehole is kept cased, except the last 3 m or so covered by a perforated pipe allowing free flow of water. The measurement is taken from time to time throughout the year to measure the seasonal fluctuations. Several such measurements in boreholes of a project area covering the reservoir spread reveal the characteristics of the groundwater table and its gradient and flow direction.

Drill holes are used for tracer study using chemical and radioactive tracers related to the problem of reservoir competency (Section 15.6) and also to solve post-construction stage problems. Borehole camera installed in a hole furnishes important information on the fracture pattern and joint intensity of rocks at depth. In many of the tunnel and power projects, especially in the Himalayan region, extensometers are widely used through boreholes to measure the stress condition in rocks to fulfil the design requirement.

11.8 WATER PUMPING TESTS—APPROACH AND UTILITY

Water pumping test is taken up to measure the permeability of rocks at depth and to assess potential seepage through porous and jointed rocks of tunnels, underground powerhouses, and other subsurface structures. A perforated tube fitted with two rubber packers at the two ends is lowered to the depth through the drill hole in which the test is to be done. Tests are conducted in 1.5 m to 3 m sections. Water from a tank is pumped to the test section under varying pressures. A pressure gauge and a water meter record the applied pressure and water loss, respectively.

The most essential factor to be considered for this test is the use of the right type of packer, which can be tightly fixed to the rock wall of the drill hole. The quantum of water that permeates through the rock fractures in a test section per minute should be the same in the consecutive minute, provided there is uninterrupted flow. In practice, the test is conducted for 5 to 10 minutes, and the average value per minute of flow is taken so that in the case of clay filling in joints, it will be flushed off during the testing process. Recordings of water losses are done under increasing and decreasing order of pressures such as 2, 4, 6, 8, and 10 kg/cm². Loss of water is generally expressed in Lugeon (1 Lugeon = water loss of 1 litre per metre per minute under 10 kg/cm² pressure). Figure 11.23 shows the results (water loss in Lugeons) of water pumping test conducted in the hard rock of a dam site through different drill holes A-5, D-4, and K-8 applying pressures from 2, 4, 6, 8, and 10 kg/cm² in increasing (firm line) and decreasing (dotted line) orders.

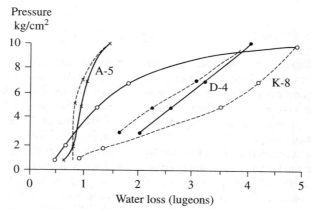

Fig. 11.23 Curves showing results (water loss in Lugeons) of water pumping test applying pressures in increasing (firm line) and decreasing (dotted line) orders

The most common weak features of hard rocks are the joints, faults, and shearing of rock, which create passage for permeation of water. Water pumping test result gives the permeability characteristic of rocks and helps to understand the adverse features in rocks due to jointing and shattering. In Fig. 11.23, the study of the water pumping test result through the drill hole K-8 provides evidence of the presence of a weak zone. The pumping test also suggests that the clay (or clay *gouge*) was flushed off during increasing order of pressure testing in hole K-8; however, during reverse order of pressure testing, high water loss resulted even at a low pressure indicating a weak zone in rock (fault or open joints) partially filled with clay. Figure 11.24 is an example where two drill holes (R–1 and D–6) accompanied by water pumping tests give clear evidence of a wide fault zone at the river section of the Umling dam site in Meghalaya. The presence of this fault was earlier inferred from the geological mapping of the dam site (Gangopadhyay 1981).

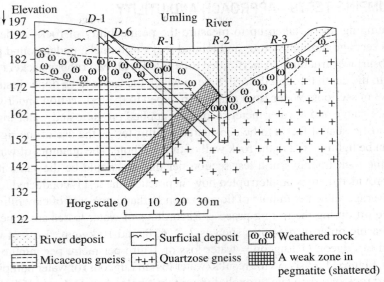

Fig. 11.24 Profile showing plan of drill holes taken up in a dam site accompanied by water pumping test to prove the presence of a fault

SUMMARY

- Engineering geological investigations in civil engineering projects such as dams, tunnels, and powerhouses are generally carried out in different stages/phases that include preliminary stage/planning phase, pre-construction stage/design phase, and construction stage.
- Preliminary stage investigation is taken up for selecting a suitable site for an engineering structure or finding the suitability of a site proposed by the engineers. The investigation begins with in-house study of aerial photographs and satellite imagery of the site followed by reconnaissance field visits to decide the suitability of the site.
- Pre-construction stage investigation includes detailed geomorphological and geological study by mapping and subsurface exploratory work. At the construction stage, large-scale geological maps are prepared for foundation areas, and in case of any weak zones, appropriate treatment is suggested.
- In some projects, even in post-construction stage, some geological defects may crop up endangering the stability of the structure. Geological investigation is undertaken to locate the defects and suggest corrective measures in such cases.

- Aerial photography and satellite imagery are important tools in geological investigations. When two aerial photographs of adjacent areas with around 60 per cent overlap are superimposed, a 3D view of the area can be observed under the stereoscope, which helps in the interpretation of the geomorphology and geology of a project site.
- Satellite imagery also facilitates general evaluation of geology including rock structures and even hydrological condition of the project sites.
- Geophysical survey, which is also a tool of geological investigation, is carried out by seismic, gravity, magnetic, resistivity, or radiometric methods, of which seismic method is mostly used to reveal the thickness of the overburden, bedrock depth, and subsurface rock structures such as faults, folds, and intrusive bodies.
- Subsoil investigation is generally conducted by pitting, trenching, and auger drilling at shallow depths and percussion drilling in deeper parts.
- Exploration by rotary drilling, followed by pumping test, is undertaken to obtain information on subsurface rock and nature of its permeability. The drill holes are also used to decipher the water

table and to prepare stratum contour map showing the nature and direction of groundwater flow.
- The soil and rock samples obtained from various exploratory works such as pitting, trenching, and drilling are tested in the laboratory for determining their engineering properties.

- After each stage of investigation, reports are submitted to the engineers with all information regarding field geology and the laboratory test results on soil and rock along with the geological maps and drill hole logs to facilitate engineering design and safe construction of the project.

EXERCISES

Multiple Choice Questions

Choose the correct answer from the choices given:

1. Preliminary engineering geological investigation is conducted by:
   (a) geophysical survey
   (b) study of aerial photographs and satellite imagery
   (c) exploratory drilling
2. The method of investigation to be followed for obtaining knowledge of geomorphic features of a project area is the study of:
   (a) contour plan of the project
   (b) geological survey geological map
   (c) the area by remote sensing method
3. In the planning stage, site investigation of an engineering project is taken up by:
   (a) in-house study of aerial photographs
   (b) visit to the site and plane table mapping of the site including surrounding areas
   (c) the study of the contour plan supplied by the project or Map of Survey of India
4. Construction stage geological mapping of a project site such as a dam is generally taken up to a scale of:
   (a) 1:10 to 1:100
   (b) 1:50 to 1:500
   (c) 1:500 to 1:1000
5. Exploratory drilling and core logging to obtain knowledge of subsurface features are taken up at the:
   (a) planning stage
   (b) design stage
   (c) construction stage
6. Seismic method of geophysical survey is important in engineering geological investigation as this can measure or detect:
   (a) overburden thickness
   (b) a fault zone
   (c) water resources

7. Magnetic survey is utilized to detect:
   (a) depth of ore deposit
   (b) unconformity between two rock formations
   (c) certain structural features like dome
8. Resistivity survey in engineering geological study is taken up to detect the:
   (a) presence of any ore deposit within the project area
   (b) depth of interface between two rock types
   (c) structure of folded rock
9. Rock types and geological structures can be studied using:
   (a) field mapping
   (b) rock core drilling
   (c) geophysical survey
10. The best suited method for studying subsurface weak zones such as a fault is:
    (a) field mapping
    (b) rock core drilling
    (c) geophysical survey
11. The structure of folded rocks can be studied by:
    (a) field mapping
    (b) rock core drilling
    (c) geophysical survey
12. Subsurface drill holes are taken up in different engineering projects following the general rules such as:
    (a) twice the height of the dam
    (b) at least 10 m in bedrock
    (c) both (a) and (b)
13. For testing of subsoil properties, undisturbed samples are collected:
    (a) by drilling with NX-size core bits
    (b) from drill holes with open drive sampler
    (c) from materials of wash boring
14. NX-size diamond bit core drilling is to be conducted in borehole to:
    (a) obtain good rock core recovery
    (b) utilize the borehole for water table measurement
    (c) utilize the borehole for permeability test

15. Study of changing landform due to landslides in a remote project area is generally taken up:
    (a) by helicopter flight
    (b) from remote sensing images of two periods (before and after the slides)
    (c) by survey of the site before and after the slide
16. Nuclear logging is taken up in engineering geology work:
    (a) to identify the presence of radioactive minerals
    (b) to identify zone of saturation and water bearing strata
    (c) for both (a) and (b)

Review Questions

1. What are dip and strike and what is the instrument used for measuring them? Draw the symbols of dip and strike for the following planes: inclined bed of sedimentary rocks, joint plane, foliation plane, horizontal dipping bed, and vertical dipping bed.
2. Describe the procedure of plane table mapping of a project site. Draw the symbols generally used in a geological map for the following:
   (a) *Rocks*: granite, basalt, sandstone, limestone, gneiss, and schist
   (b) *Structures*: vertical fault, plunging anticline, and plunging syncline
3. What are the different stages of engineering geological work? Discuss the work-activity of each stage.
4. Outline the method of investigation of a project area from the study of aerial photographs with respect to morphology, rock types, and rock structures.

5. Give an account of the remote sensing techniques of investigation of a project area. How does it help in the study of landform and geology of the area?
6. Describe the seismic method of geophysical study. How does it help in the evaluation of subsurface geology? Name the instrument used in seismic survey to investigate subsurface areas covering shallow depth.
7. What is radioactive logging? Explain the method of nuclear logging and its utility in engineering geological study.
8. What type of exploratory drilling is conducted in subsoil exploration? Discuss the method of subsoil study of shallow depth by pitting and trenching method and auger drilling method.
9. State the importance of rotary drilling in a project site. What types of bits are used in rock core drilling? What is the approach to obtain high core recovery from rotary drilling?
10. Explain the method of presentation of drill hole logs covering recovery percentage of rock cores, joints, and other structural features. What is the other utility of a borehole apart from obtaining cores of subsurface rocks?
11. How are pumping tests performed? Explain the method of determining permeability by pumping test. What is the unit for expressing water loss by pumping test?
12. Write short notes on magnetic survey, gravity survey, penetration test, wash boring, and percussion drilling.

Answers to Multiple Choice Questions

1. (b)	2. (c)	3. (b)	4. (a) and (b)	5. (a) and (b)	6. (a) and (b)	7. (b) and (c)
8. (b)	9. (a)	10. (c)	11. (b)	12. (c)	13. (b)	14. (a), (b), and (c)
15. (b)	16. (b)					

12 Construction Materials

LEARNING OBJECTIVES

After studying this chapter, the reader will be familiar with the following:

- Different types of construction materials including their characteristics and utilities
- Concrete aggregates and their testing procedure as specified by the Indian Standard
- Petrological study of alkali-reactive minerals and other deleterious materials
- Road metal and railway ballast and their properties, specification, and availability
- Source of different types of construction materials (rocks and soils) in India
- Exploration methods for stone quarry and their selection in rocky hill slopes

12.1 INTRODUCTION

This chapter deals with the different types of construction materials used in engineering structures. It provides an elaborate description of naturally occurring materials such as rocks, pebbles, gravels, sands, and clays, which are good construction materials. The chapter also furnishes details of use of rocks as blocks for masonry constructions and small fragments as road metals and railway ballasts and discusses the utilization of river pebbles, gravel, sand, and crushed stones for concrete aggregates. It gives an account of pozzolan, a naturally occurring material, and fly ash, an artificial product obtained from factory rejects, which are both used in concrete to reduce cement consumption. The chapter further analyses the importance of petrological testing of aggregates to find whether they contain alkali-reactive minerals and other deleterious materials and delineates the source of construction materials in India and their exploration methods.

12.2 PRINCIPAL TYPES OF CONSTRUCTION MATERIALS

The earth's surface contains vast reserves of varieties of materials such as rock and soil. These naturally occurring materials have been used for construction purposes from time immemorial. Civil constructions such as dams, bridges, embankments, highways, roads, and airports require large quantities of materials obtained from natural sources. The types and quantity of construction material to be used in a civil project depend on the engineering design of the specific structures. The following are the common types of naturally occurring materials used in civil engineering constructions:

(i) Rock (dimension stones and crushed stones)
(ii) Cobble, pebbles, and gravel (coarse aggregate)
(iii) Sand (fine aggregate)
(iv) Soil and clay
(v) Pozzolan

Rocks are the naturally occurring materials that are most abundantly used in civil constructions. Hard and durable rocks free from deleterious materials are selected for use as dimension stones or crushed stones. Boulder, cobble, shingle, gravel, sand, and silt are the products of rock erosion and are available in plenty in river valleys and other recent deposits for their use as aggregates (Fig. 12.1). Soil and clay used in the construction of an earth dam occur as capping of parent rock or as deposit of transported materials. Volcanic rock is the source of natural pozzolan used as partial substitute of cement. *Kankar*, which contains lime, has surface occurrence in many parts of North India and are used as mortar in house constructions. Artificial materials that are used for construction purposes include industrial waste (fly ash), burnt shale, and brick including *surki* (powdered brick).

Fig. 12.1 Source of construction materials from river terrace deposits: gravel, sand, and boulders

In addition, iron, steel, and cement are other important materials used in various engineering constructions. Iron ore that occurs in geological formations is required to extract iron and make steel. Cement is manufactured from limestone. Mixing of cement with aggregate and water in a fixed proportion yields concrete, which is as hard as rock and is used in a multitude of civil constructions such as dams, buildings, and roads. Instead of concrete, *colcrete* is used in some places in the construction of dams, weirs, and barrages. The process of making colcrete involves the preparation of a watery mixture of cement and sand, which acts like a colloidal paste immiscible with external water, and then mixing this colloidal paste with stone chips under pressure.

12.3 CHARACTERISTICS AND USAGE OF DIFFERENT TYPES OF CONSTRUCTION MATERIALS

Hard and durable materials are available abundantly from natural sources for use as construction materials for buildings or other engineering structures. Hard rocks are generally used as blocks for masonry works and by crushing them to meet the construction needs. Many of the construction materials are also received from artificial sources such as factory waste (fly ash) for different types of construction purposes as stated below.

12.3.1 Dimension Stones for Building Stone, Facing Stone, and Decoration

Dimension stones are large irregular blocks of rock that are obtained from rock quarry faces. Dense rocks with low porosity and high strength but amenable to dressing and polishing are good dimension stones suitable for use as facing stone of walls, building stone for houses, and the like. In general, dimension stones are hard and durable enough to withstand natural wear and tear. Ideally, these stones should be resistant to fire, heat, water, and wind action. The rocks are quarried from selected parts of hill slopes by mild blasting that produces large blocks of rock of uniform character. Special techniques of rock splitting are followed to procure even-surfaced large slabs. Once the dimension stones are obtained from the quarry face, these are cut and dressed to pieces of specific size and shape according to the construction need. For use in walls and floors of buildings, big blocks of rocks are nowadays sliced by diamond saw to various thickness and rectangular or square shapes and then given a mirror-like polish. In some buildings, leaf, creeper, animal, and human figures are carved in the stone blocks before using them to decorate the building walls.

Dimension stones are mainly used as *building stones* for construction of entire buildings or to cover parts of buildings. In historic times, many buildings, royal palaces, huge pillars, and religious edifices were built of large slabs or blocks of dimension stones. The irregular stone slabs obtained from the quarry face were first dressed into rectangular slabs or blocks and then used in the *masonry works* of the buildings. The mortar used for bonding the slabs includes plastic clay, lime (obtained from burnt shell or *kankar*), *surki* (powdered brick), and other indigenous materials such as cowdung and molasses. Use of stone blocks for constructing buildings is not popular nowadays, except in some religious and prestigious buildings.

Today, most buildings in cities use reinforced concrete or bricks for their constructions. However, dimension stones such as large dressed blocks of granite and marble with mirror-like polish are still widely used for surface covering of important buildings to provide an aesthetic appearance. Until the early parts of twentieth century, large quantities of rocks of different sizes obtained from quarrying hard durable rocks were used as *rubbles* for masonry dam construction. In general, the irregular rock blocks are properly sized by hand dressing before use for masonry work. The construction of masonry dams has been reduced in recent times with consequent reduction in quarrying dimension stones for use as masonry blocks. Large quantities of dimension stones in the form of even-surfaced slabs are used for *pavement* purposes in highways. Dimensional stones (Fig. 12.2) are also used in the *protective work* of river banks, sea coasts, and harbours against erosion by river or sea water.

Fig. 12.2 Dimension stones (gneissic granite) piled up for engineering construction

During historic times, there was a spur of activity in constructing numerous edifices and royal buildings in many parts of the world using excellent quality dimension stones (building stones) such as pink granite, white marble, and red sandstone possessing high strength. Monolithic blocks of huge dimensions were carved and placed in the gateways and walls of many historic monuments. In India, hundreds of monuments built through the ages by varied types of building stones can be seen even today withstanding the vagaries of nature. It can be safely presumed they will remain for several more centuries, bearing testimony to the profound knowledge of our ancient builders in the choice of rocks for building purposes (Gangopadhyay 1988).

Fig.12.3 Dressed blocks of Vindhyan sandstone widely used in constructing ancient buildings and other masonry works

The mauve-coloured Vindhyan sandstones are excellent building stones (Fig.12.3) that were used in many historic buildings and are still being used for decorating building walls and floors. The milk white Makrana marble of Rajasthan, which was used in the Taj Mahal, is extensively mined for prestigious building constructions and carving. Blocks of granite and gneiss of huge dimensions mined from peninsular India are today exported to foreign countries such as the US. These blocks are sliced into various sizes and polished to obtain a glittering surface (see Fig. 2.5) and are then marketed to be used in facing of building entrances and for other decorative purposes.

12.3.2 Crushed Stones for Road Metal, Railway Ballasts, and Rip-rap

Crushed stones are irregular rock fragments generally of smaller sizes compared to dimension stones. The rock chunks obtained from blasting quarry faces are further crushed and screened into proper sizes depending upon the nature of utilization. In fact, a rock quarry may produce large rock chunks as well as small fragments, with the large blocks being used as dimension stones and small fragments as crushed stones with or without further crushing. Good strength and high durability are the two primary characteristics that crushed stone pieces must possess for use in any engineering construction. The mineralogical character, texture, and packing of mineral grains in rocks are responsible for the strength and durability properties of crushing stones.

Crushed stones are mainly used as railway *ballasts, stone chips* for highways, roads, airports, and *coarse aggregates* of concrete. In roads and airports where stone chips are used in the bases and sub-bases, the stones withstand the stress due to fast-moving and heavy wheel loads in the structural design of the pavement. The action of repeated wheel loads due to wheel movement tends to cause impact and abrasion of the stone. It is therefore important that the crushed stones used for road construction have high strength and durability. Large quantities of hard and sound crushed stones or fragmented rocks of varying dimensions are required for constructing *rock-fill dams* where adequate supply of materials such as boulders, pebbles, and gravels are not available from local sources. Fragmented rocks are also used as *rip-rap* materials to cover the upstream faces of earth dams for protection from erosion due to water action. The 10 m layer of rip-rap material for 206 m-high Tehri rock-fill dam located in Uttaranchal (Lesser Himalayas) used blasted or crushed large dimension quartzite boulders for its upstream slope protection.

12.3.3 Suitability of Rocks for Engineering Construction

The mere presence of hard rock in a hill slope or surface outcrop does not indicate the suitability of the rock for construction purposes. Field study is necessary to find its structural behaviour and weathering condition. Rock outcrops with excessive planes of discontinuities and highly weathered rocks are not suitable for engineering construction. The behaviour of the rock under varied conditions is the foremost consideration in finding its suitability for construction purposes. In general, suitability is determined from laboratory tests of rock properties that include density, porosity, permeability, and strength. The mineralogical and textural properties determined from petrological studies are also considered in the selection of construction materials (Fig. 12.4). In general, the rock types that possess high strength and durability are suitable for use as construction materials. However, some rocks contain excessive deleterious materials including chemically reactive minerals and these are considered unsuitable for engineering constructions.

Strong and durable rocks for use as dimension stones or crushed stones for various engineering

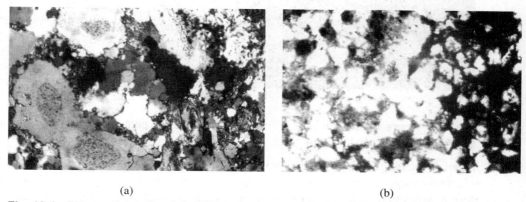

(a)   (b)

Fig. 12.4 Thin sections of rock (× 70) under microscope showing deleterious minerals: (a) schistose rock with chemically reactive minerals; and (b) sandstone possessing large percentage of weathered biotite and iron oxide

constructions are obtained in plenty from igneous and metamorphic rock suites. Igneous rocks such as granite, charnockite, basalt, and dolerite and metamorphic rocks such as gneiss, quartzite, quartzitic phyllite, and khondalite are sufficiently dense and have low absorption and high compressive strength. These rocks are widely used in engineering constructions of buildings, dams, roads, and airports. In the absence of very strong and durable rocks, moderately strong rocks such as schist, phyllite, slate, dolomite, sandstone, and limestone are also used for various construction purposes depending upon the design requirement.

In selecting quarry site for rocks, the locally available outcrops of hard and durable rocks are considered first. If suitable rocks are not available locally, they are to be brought from a distant place, provided it is economically viable for the project. In some cases, the design of the structure is changed when stones are not available locally. The absence of suitable rocks resulted in the change in the design of the Gumti dam in Tripura in favour of a brick–concrete structure using burnt bricks manufactured from locally available sandy soil.

12.3.4 Aggregates for Concrete

Aggregate is used to make concrete by mixing with cement and different types of naturally obtained materials such as gravels and sand to meet the requirements of different types of construction works. The materials used should be such that its strength achieved after making concrete will have sufficient bearing strength to withstand the structure. The material used should not be deleterious in nature such as chalcedony, chert, strained quartz, and so on. Generally, aggregate used may be coarse or fine depending upon the need as detailed below.

Coarse aggregates Aggregates constituted of cobble, pebble, gravel, and sand (Fig.12.5) are mixed with Portland cement in varied proportions by weight to make concrete. The size fractions of aggregates and their proportions are selected such that there remains no void after setting of the concrete with cement. The quality of the aggregate materials is the main deciding factor to achieve the desired strength of concrete. The aggregate should be free from deleterious

(a)   (b)

Fig.12.5 Aggregate for concrete: (a) deposit of cobbles and pebbles; and (b) sieved heap of gravel and coarse sand

materials including cement-reactive minerals. The principal engineering properties considered in selecting concrete aggregate include crushing strength, impact, abrasion, and soundness values. Recent deposits of cobbles and shingles derived from hard and durable rocks are the natural sources of coarse aggregate for concrete. These materials are obtained in profuse quantity from river valleys, terraces, and glacial deposits. In addition, crushed rocks are also used as coarse aggregates. Irrespective of the source of the materials, their properties are determined from the results of laboratory tests including petrographic study to decide their suitability for aggregate. The bulk materials for aggregate obtained from river deposits or crushed stones are screened to different size fractions according to design specification prior to their use in concrete.

The quality of the coarse aggregates varies from place to place depending upon the source rocks from which these have been derived. Terrains of igneous and metamorphic rocks comprising granite, basalt, dolerite, gneiss, quartzite, and marble are the main sources of good quality coarse aggregate materials obtained from river deposits as well as crushed rocks. Erosion and transportation are two prime factors associated with formations of the deposits of aggregate materials. Riverborne coarse aggregate materials having round and smooth surfaces are favoured for engineering use. This is because they consume less cement and have better bonding capacity compared to crushed rocks having angular and uneven surfaces.

Construction material survey in parts of Yamuna basin could find sufficient quantity of suitable quality aggregate for the Lakhwar Vyasi dam (Fig. 12.6). In general, deposits of coarse aggregate including pebbles originated from mica schists or of glacial origin carry a lot of silt and clay particles. However, if the available quantity of the pebble and gravel is not adequate enough, or if these materials are not available at all, crushed rocks are used to meet the demand of coarse aggregate after screening and washing of the clay coating. In the 90 m-high Rihand dam (gravity dam) in Uttar Pradesh, the requirement of 1.7 million cubic metres of coarse aggregate for concrete was met using crushed rock by quarrying a hill of gneissic granite at a distance of 5 km from the dam site (Rao 1975).

Fine aggregates Deposits of sand are generally abundant in river valleys. An igneous terrain gives rise to coarse-grained sands consisting mostly of quartz, feldspar, and some other hard minerals. Good quality sands are also produced from metamorphic rocks containing micas in addition to quartz and feldspars. However, sedimentary rocks produce deposits with more clayey materials that need to be removed before use as aggregate material. In general, sand and gravel deposits contain deleterious materials such as organic matter (coal, peat), clay lump,

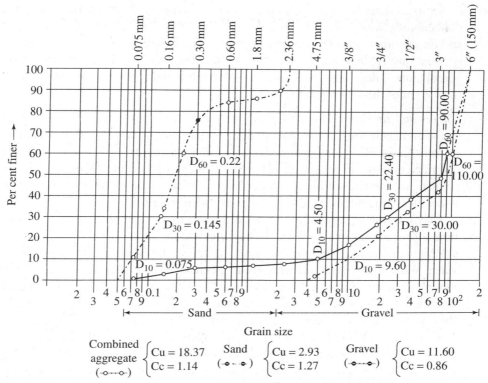

Fig. 12.6 Grading curves of aggregate drawn from results of construction material survey in parts of Yamuna valley (Shome and Kaistha 1992)

soft shale, and micas in varying quantities. According to Indian Standard (IS), these deleterious materials should be less than 5 per cent by weight of the total materials. Screening and washing of the sands are generally taken up to reduce the deleterious materials by removing the fine clay and mica particles.

Wherever required, quartz sands obtained from crushed quartzite are blended with river sand and gravel to improve their quality by lowering the percentage of very fine particles such as silt, clay, and micas. In the Umiam project of Meghalaya, the sand deposits were found to contain exceptionally high percentage of micas derived from the country rocks such as phyllite and schist. It was difficult to find a suitable quality of fine aggregate for concrete. Screening and washing could not bring down the mica content of the river sand to the required level. Quartzite available from a local hill was then quarried and crushed to the requisite size and blended with the river sand to get the right proportion. Similarly, in the Little Rangit Project of Sikkim, the presence of a high percentage of mica (nearly 14 per cent) and sufficient amount of clay and silt (about 3 per cent) in the sand of the river bed made it necessary to blend crushed quartzite with it to use as fine aggregate for concrete.

12.3.5 Soil and Clay

Weathering and decomposition of rocks produce soil and clay, which are in large demand in engineering construction of earth dams, embankments, and bases of roads and airports. Depending on the water-retaining capacity, soils on earth's surface may be of *pervious, semi-pervious,* or *impervious* type. Pervious soil is constituted of homogeneous material free from large voids, will not consolidate excessively under the superincumbent weight of the fill, and has high angle of internal friction, low cohesive strength, and permeability (K) greater than 50×10^{-4} cm/sec.

Semi-pervious soil has medium cohesion, medium angle of internal friction, and medium permeability, with K varying from 0.01×10^{-4} cm/sec to 50×10^{-4} cm/sec. The impervious soil on the other hand possess low angle of internal friction, high cohesive strength, and low permeability (K less than 0.01×10^{-4} cm/sec. Impervious soil is free from potential path of percolation. It does not allow water loss through it and will not consolidate excessively under the weight of the superincumbent fill.

Clayey soil or clay The clayey soil is used in the construction of embankments, retaining walls, and abutments. More than 50 per cent particles of clayey soil or clay are less than two microns in size. Clay is constituted of either non-clay minerals such as quartz, feldspar, calcite, and mica or clay minerals such as kaolinite, illite, and montmorillonite. Clay minerals require study by the methods of X-ray diffraction or differential thermal analysis (DTA) or examination under electronic microscope for their proper identification.

Clay minerals are crystalline hydrated silicates of aluminium. In some clay minerals, sodium, iron, and magnesium may be present. The clayey soil constituted of large percentage of clay minerals exhibit high plasticity, low shearing strength, and volume change (swelling) when saturated with water. When this type of soil with swelling behaviour is used in engineering constructions such as embankments, it may result in settlement in the body of the fill and slope failure in saturated condition. Hence, while selecting borrow areas of clay for an engineering structure, the quality of the clay especially with respect to its expansive nature needs to be thoroughly studied.

Clayey soil or clay has distinctive origin. Clayey soil such as the *black cotton soil* occupies surface areas of large parts of central and south India. Such clay deposits may extend to a dept of 3 m to 10 m. Clayey soil is found in deltaic deposits of the river Ganga and other rivers su as the Godavari and Cauvery. Clay also occurs in extensive areas of the eastern and wes coasts including lake (still water) deposits. Table 12.1 provides the typical analytical res clayey soil or clay of different origins.

Table 12.1 Properties of clayey soil or clay of different origins (Rao 1968)

Type of clayey soil	Clay fraction (%)	Liquid limit	Plastic limit
Black cotton soil	55	73	40
Clay of delta deposit	66	80	47
Clay of coastal deposit	80	90	63
Clay of still water basin	55	80	40

12.3.6 Pozzolan

Pozzolan is a naturally occurring substance consisting of siliceous or alumi has no cementitious value in itself, but when mixed with lime it becomes cem with lime at atmospheric temperature in the presence of water to produce a compound. This cheap substance is mixed with cement to reduce the cons in concrete and thus lower the cost of concrete. It also retards the alkali when mixed with Portland cement in a fixed proportion. The other benef include increased watertightness, lower shrinkage possibility, diminished improved resistance to attack from sulphate soil and to cracking.

Volcanic ash, tuff, diatomaceous earth, chert, and pumice stone pozzolans. It has been found that kaolinitic clay develops good calcination at the optimum temperature of 600–700°C, provided it is of certain geological formations depending upon their mineralogical characters show good pozzolanic characteristics on calcination at varied temperatures. Clays consisting predominantly of montmorillonite and illite need a temperature range of 700–800°C and 900–1000°C, respectively, to obtain optimum reactivity similar to pozzolan. All these natural materials are ground to powder before adding to cement in a predetermined proportion for making concrete. When used as a replacement of cement up to 20–25 per cent, it has been found that, similar to clays, natural pozzolans after calcination will retain all desirable qualities of concrete. In Bhakra concrete dam in Punjab, Siwalik clay stone that develops a pozzolan character after calcination at 871°C was used. This reduced the cement consumption in concrete up to 20 per cent without deterioration in the quality of concrete.

In India, Deccan traps of volcanic origin contain beds of ash and tuff that are used as pozzolans. Volcanic ash and pumice are also found in Barren and Nicobar Islands. Diatomaceous earth and radiolarian chert have been reported from Andaman Islands. Sedimentary formations such as Cuddapah, Vindhyan, Gondwana, and Tertiary formations including Siwalik possess clay and shale having pozzolanic properties. However, shale and clay beds of all these formations are not equally reactive. Fuller's earth and bentonitic clay of pozzolanic character occur in Malani Rhyolite of Rajasthan and in parts of Jammu and Srinagar. The kaolinitic clay formed from the decomposition of pegmatite and porphyritic granite occurs in Archaean rocks in different parts of India. In addition, lithomarge and low-grade bauxite of Gujarat, Maharashtra, and Jammu and Kashmir are also utilized as pozzolans.

12.3.7 Fly Ash (Artificial Pozzolan)

In addition to natural occurrence, pozzolans are available from artificial sources such as industrial waste, blast furnace slag, calcined clay or shale, and ground brickbats called *surki*. The leftover burnt coal of thermal plants is the *ash* that is known as *fly ash* in its industrial use. It is grey in colour resembling the colour of cement and the fineness of particles (< 63 microns) is also similar to that of cement. Fly ash is constituted of silica, alumina, and iron oxide in major quantities and calcium oxide and water-soluble salts in minor quantities. In fact, its chemical and mineralogical compositions are similar to natural pozzolan. Fly ash is well utilized when obtained from thermal plants close to the project area. This industrial waste, which is commonly heaped up at the backyard of the plants, can be procured very cheaply.

Analysis of six fly ash samples from Nellore thermal plant in Andhra Pradesh shows quartz, feldspar, and iron oxide as the main mineral contents, with average specific gravity 2.09 and compressive strength (lime reactivity) 36.4 kg/cm². The chemical analyses of three samples show that the oxides and salts present in fly ash are well within the specified limit of its use with Portland cement in a predetermined proportion for making concrete. The study result indicated that the fly ash if mixed in suitable proportions, replacing cement up to 20 per cent, does not change the strength or quality of concrete in any way but registered reduced sulphate attack, increased water tightness, and resistance to cracking. Moreover, the strength of the mortar using 20 per cent fly ash surpasses the mortar strength without fly ash after 90 days of curing (Lingam, Narasimhulu, and Ramakrishna 1968).

12.4 LABORATORY TESTS OF AGGREGATES WITH INDIAN STANDARD SPECIFICATION

Engineering properties such as crushing strength, impact, abrasion, and soundness values of aggregate materials are determined from the results of laboratory tests. The methods of tests are briefly stated in the following subsections.

12.4.1 Aggregate Crushing Test

Aggregate crushing test measures the resistance capacity of the aggregate under static load. Determination of crushing strength involves the crushing of the material by compression in the laboratory. The aggregate material of size range 12.5–10 mm is taken in a 150 mm-diameter steel mould up to a depth of 100 mm and subjected to a universal compressive load of 40 tonnes. The loading is continued for 10 minutes and the weight of the fines passing through the 2 mm-mesh sieve is measured. If the original weight of the test sample is A and the weight of the crushed materials (less than 2 mm in size) is B, the crushing value of the aggregate is obtained from the simple calculation, $(B/A) \times 100$. Thus, the percentage loss of the material due to crushing making fines or powdery material (less than 2 mm in size) is the measure of the resistance to crushing. A low value of aggregate crushing test implies that the material has good resistance capacity. If this value is less than 10 per cent, the aggregate material is very strong, but if the value is more than 35 per cent, it is considered to be weak.

12.4.2 Aggregate Impact Test

The aggregate impact test determines the aggregate impact value, which is a relative measure of resistance of an aggregate to sudden shock or impact. This value of resistance due to sudden impact generally differs from that obtained from slow compressive load applied in aggregate crushing test. The impact testing machine consists of a steel cup of 100 mm internal diameter and 50 mm depth and is provided with a steel hammer weighing 13.5–14 kg. The hammer slides freely between vertical guides so arranged that the lower (cylindrical) part of the hammer is above and concentric with the steel cup.

The aggregate impact value is obtained by testing the aggregate material that passes a 12.5 mm-mesh sieve and is retained over a 10 mm-mesh sieve. The sample is dried in an oven between 100°C and 110°C for four hours, cooled, and then weighed. The weighed sample of aggregate is taken in the cup of the impact testing machine and pounded by the steel hammer through 15 blows from a standard height of 380 mm. The weight of the powdery material passing through the 2.36 mm IS sieve is then measured accurately. This provides the weight loss of the test sample under impact. The percentage of this weight loss with respect to the original weight of the aggregate material gives the aggregate impact value.

12.4.3 Los Angeles Abrasion Test

Los Angeles abrasion test is intended to determine the abrasive resistance of aggregate materials such as gravel, sand, and crushed rock. It measures the percentage loss by weight of a standard sample after it is subjected to abrasion in the Los Angeles machine. This machine consists of a hollow steel cylinder of inside diameter 750 mm and inside height 500 mm and having several steel balls (48 mm diameter and weighing about 400 g) inside for abrading purpose. Bulk sample is first sieved into different size fractions. The quantity of the material to be charged and the number of steel balls to be placed inside the machine are decided according to the size range of the test sample.

Aggregate of 2.5 kg and 11 steel balls are required for testing 12.5–10 mm-size and 20–12.5 mm-size samples. The measured quantity of 2.5 kg aggregate (size 12.5–10.0 mm) is taken for test. The sample is washed thoroughly in water, dried in an oven between 105°C and 110°C, cooled, and then placed inside the cylinder and rotated at 30 to 33 revolutions per minute (rpm) until 5000 revolutions are completed. The percentage of abraded material (powdery material) by weight passing through 1.70 mm IS sieve in terms of total weight gives the abrasion value. The abrasion value is high in soft material and low in hard aggregate.

12.4.4 Deval Abrasion (or Attrition) Test

Deval abrasion test measures the resistance to wear of the aggregate material when subjected to abrasion under rotational condition. In this test, the sample weighing 5kg is divided into two equal parts of 12.5–10 mm and 20–12.5 mm size range. Then, 2.5kg sample of the coarser fraction is dried in an oven, cooled, and then placed in 30° inclined buckets. It is then rotated at 30 rpm for 3 minutes. The portion of the abraded sample remaining in the pan below the 2 mm IS sieve is collected and measured. The loss of the particles expressed as a percentage loss of the original sample gives the abrasion value as in the Los Angeles abrasion test. At least two tests are conducted and their average value is taken.

12.4.5 Soundness (or Sodium Sulphate) Test

Soundness test is intended to find the ability of the aggregate to restrict weathering. A saturated solution of sodium sulphate is prepared (dissolving 350g of anhydrous sodium sulphate in 1 l of water) and kept for at least 48 hours before use. The test is conducted on both coarse and fine aggregates (gravel and sand). The bulk sample of aggregate is first sieved to different fractions such as 10–4.75 mm, 4.74–2.00 mm, 2.00–0.425 mm, and 0.425–0.075 mm. About 115g of sample is necessary for one test of fine aggregate but heavier samples are needed for coarser fractions.

An aggregate sample of 10.0–4.75 mm fraction is taken for testing. The sample is washed in water, then dried at a temperature of 110°C, and cooled at room temperature. A measured quantity of 100g aggregate is immersed in sodium sulphate solution taken in a porcelain bowel. After saturation for 18 hours, the sulphate solution is decanted and the sample is thoroughly washed with water. A few drops of barium chloride ($BaCl_2$) is mixed with the sample and washed with water to clean the sodium sulphate coating from the sample. The sample is then dried, cooled, and the test repeated. In this way, after five cycles of the test, the sample is finally washed, dried, and sieved in a 4.75 mm IS sieve. The portion passing through the sieve indicates the weight loss of the sample. The percentage weight loss (five-cycle test) with respect to the weight of the original sample is a measure of soundness. Results of tests conducted in geotechnical laboratories of the Geological Survey of India (GSI) for crushing values, Los Angeles abrasion, and impact and soundness values of aggregate materials (crushed rocks or river gravels) of various rock types used in different projects in India are given in Table 12.2.

Table 12.2   Engineering properties of aggregate materials (crushed rocks) used in different projects in India (Ray, Mehta, and Ashraf 1972; Anand 1991)

Source *rock type	Specific gravity	Absorption (%)	Crushing value (%)	Los Angeles abrasion (%)	Soundness value (%)	Impact value (%)
(1) Quartzite	2.75	0.45	21.0	39.20	–	–
(2) Gneiss	2.71–2.81	0.17–0.24	15.1–15.7	17.12–17.33	–	–
(3) Quartzite	2.67	0.33	10.9	20.40	0.21	–
(4) Dolomite	2.83	0.19	9.0	17.50	nil	–

(Contd)

Table 12.2 *(Contd)*

Source *rock type	Specific gravity	Absorption (%)	Crushing value (%)	Los Angeles abrasion (%)	Soundness value (%)	Impact value (%)
(5) Sandstone	2.66–2.71	0.07–1.33	–	–	0.38–1.67	5.4–10.2
(6) Sandstone	2.23–2.38	6.9–9.0	–	–	100	–
(7) Gneiss	2.59–2.85	1.00–1.67	38.3–54.1	39.2–56.6	–	–
(8) Limestone	2.75–2.76	0.25–0.26	–	33.9–43.7	0.33–1.80	24.4–34.5
(9) Quartzite	2.64	0.72	18.1	39.8	–	17.1
(10) Sandstone	2.61	1.82	29.0	57.3	–	29.5

*(1) Hirakud dam and Orissa; (2) Hirakud dam; (3) Kosi project and Bihar; (4) Kosi project; (5) Calcareous sandstone, Gumti project, and Tripura; (6) Argillaceous sandstone and Gumti project; (7) Pancheswar project and Uttar Pradesh; (8) Mandira dam and Orissa; and (9) and (10) Dhaleswari project and Mizoram

12.5 MATERIALS FOR USE AS RAILWAY BALLAST AND ROAD METAL

Hard and durable rocks that can resist abrasion and wearing are generally suitable for railway ballast. There are several types of rocks such as basalt, granite, and charnockite that are used for railway ballasts as these possess properties specified by the Bureau of Indian Standards (BIS) and Indian Railway Standard (IRS). The details are further discussed below.

12.5.1 Suitable Rock Types and Desired Properties

Hard rocks with good strength are crushed to obtain suitable size pieces for use as *railway ballasts*. The process includes screening of the crushed rocks to obtain the desired size fraction. The remaining size fractions including fines and dusts and associated weeds are removed. The size and the thickness of the ballast layers to be placed for rail line depend on the nature of the railway tracks, that is, whether broad gauge (BG), medium gauge (MG), or narrow gauge (NG), which decides the load factor and thrust on the ballasts. The bearing capacity of the subgrade materials and drainage arrangement are given important consideration during placing the rail. The unit weight, dry density, compressive strength, impact, abrasion, and other properties of the ballasts are determined from laboratory tests (see Sections 12.4.1 and 12.4.4) before their placement. The most desired characteristics of the ballasts are the resistance to impact and the bearing capacity to sudden thrust, which are imposed by the moving engine and loaded wagons. The railway ballasts should also have resistance to wear and tear or abrasion. The desired impact and abrasive properties specified by Indian Railway Standard (IRS) are given in Section 12.5.2.

The rocks mostly used as broken pieces for *road metals* must be very hard, tough, and resistant to abrasion. Hardness is indicative of the power of resistance and toughness implies the capacity to withstand impact. Abrasion is related to the power of resisting forces from grinding and fracturing. Road metals of good quality will be such that with the movement of loaded vehicles or trains, there will be no grinding of the stone chips used and no crack will develop in the road. The material to be used as road aggregate should also have good binding capacity. Section 12.5.2 provides the required specification of abrasion, impact, and other properties of road metals fixed by the Bureau of Indian Standards (BIS). Basic igneous rocks such as basalt found in the Deccan traps, Rajmahal Hills, and many other parts of India are best for use as dimension stone and chips for road construction. The rock is very hard and dense for use for macadam and tarred roads (see Section 19.2).

Basalt and dolerite broken to size are also used as good quality railway ballast. Granite, gneiss, charnockite, basalt, dolerite, sandstone with siliceous matrix (such as Vindhyan sandstone), quartzose phyllite, and so on, which are available in most parts of India in abundant quantities, provide good quality railway ballasts and also serve as good road metals. The engineering properties of these rocks are shown in Table 10.1. High strength and low absorption as seen in the tested samples indicate that they are good quality construction materials for use as road metals and also very tough and hence are suitable for use as railway ballasts. Some siliceous varieties of schists, limestone, and dolomite can also be used when available locally, but if micaceous materials are present in high percentage and easily erodible and soluble matters occur in sufficiently high percentage in limestone and dolomite, they should not be used. In fact, the stone chips for use as road metals as well as rail ballasts are selected depending upon the local availability and their suitability as per standard specification.

12.5.2 Specification of Concrete Aggregate by Indian Standard for Road

Table 12.3 provides the standard for aggregate fixed by BIS (No. 6579–1981) and Indian Road Congress (IRC) (1988) for use in road construction and that fixed by Indian Railways for use as railway ballasts.

Table 12.3 Standard for use of aggregate as road metals and railway ballasts

Test type	BIS	Value (%)	IRC	Value (%)	IRS	Value (%)
Abrasion	Wearing surface	40	Water bound Macadam (WBM) sub-base	60	BG and MG	30
	Overlay	50	Black top base	50	NG	35
			Surfacing	40		
			Bituminous coarse	35		
Impact	Wearing surface	30	WBM sub-base	50	MG and NG	20
	Overlay	35	Black top base	40	NG	30
			Surfacing	30		
			Bituminous coarse	30		
Flakiness index	All coarse	15	Bituminous coarse	15		
Water absorption	All coarse	0.6	Bituminous coarse	2		

*BIS, Bureau of Indian Standards; IRC, Indian Road Congress; and IRS, Indian Railway Standard

Extensive gravel (shingle) deposits available from two sources, namely Murti river bed and Jaldhaka streams, were selected for use as aggregates for the Jaldhaka project of North Bengal. The study of the shingles indicated the weighed percentage of each rock type in Murti shingles to be quartzite 34.8 per cent, vein quartz 12.2 per cent, phyllite 21 per cent, gneiss 22.5 per cent, and mica schist 9.4 per cent and in Jaldhaka shingles to be granite gneiss 49.7 per cent, mica schist 38.7 per cent, and quartzite 11.3 per cent. Materials of both the sites were of suitable quality (Table 12.4) but Murti aggregates were used in maximum quantity without any crushing and grading as these were well graded and rounded.

Table 12.4 Characteristics of aggregate used

Type of test	Murti	Jaldhaka
Specific gravity	2.72	2.77
Absorption	0.35	0.60%
Aggregate crushing value	21.5	22.8
Soundness (5 cycles)	Negligible	Negligible

Of the various Indian rocks, basalt and dolerite are abundantly used as road metals and railway ballasts because of their wide availability and their characteristics such as high specific gravity, low water absorption, and high compressive strength (see Table 10.1). The crushing, abrasion, impact, and flakiness index values are also well within the BIS and IRC specifications. Lateritic rock of comparatively low strength that occurs in many places in India as capping of basic rocks are also used in highway and road constructions when available locally. Several samples of laterite tested for road construction purposes show that its abrasion value is 39.0–58.5 per cent, crushing value is 32.5–46.1 per cent and impact value is 26.8–45.1 per cent. These values are within the specified limit for use as road metal as shown in the Table 12.3.

12.6 DELETERIOUS MATERIALS AND ALKALI–AGGREGATE REACTION

Concrete is made by adding cement to aggregate materials (crushed rocks gravel, sand, etc.) in a fixed proportion to achieve the strength of a hard rock. The quality of the aggregate materials in concrete mix is the main factor that provides strength to concrete. *Deleterious materials* are those rock or mineral particles in aggregate that produce adverse effects on concrete due to their weak characteristics and chemical reaction with alkali in cement. Some aggregate materials react with cement in concrete. This cement–aggregate reaction, commonly known as *alkali–aggregate reaction*, reduces the strength and durability of the concrete. The alkali–aggregate reaction produces silica gel, which causes swelling pressure inside the concrete; as a result, concrete develops cracks necessitating costly repair and even replacement of the engineering structure.

Silica minerals such as tridymite, cristobalite, opal, chert, chalcedony, and strained quartz in aggregate materials are primarily responsible for the alkali–aggregate reaction. Crushed volcanic rocks such as glassy and cryptocrystalline rhyolite, dacite, and andesite and some varieties of phyllite also react with alkali in cement. Crushed basalt having good strength and high resistance to abrasion is suitable for use as coarse aggregate for concrete. However, some varieties, especially the amygdaloidal variety, may contain chloropheite, palagonite, volcanic ash, chert, and chalcedony that cause alkali–aggregate reaction and are therefore avoided from use as concrete aggregate. The Rajmahal basalt and the Deccan basalt were safely used as concrete aggregate in Farakka barrage in West Bengal and Koyna dam in Maharashtra, respectively, but Deccan basalt around Ukai project of Maharashtra could not be used due to the presence of excessive alkali-reactive minerals.

In addition to these chemically reactive minerals and rocks, deleterious materials also include micaceous rocks, shale and friable clayey rocks, sulphide and sulphate minerals, and organic matter. Gypsum and clay occurring as coating of aggregate material is detrimental to concrete. Good aggregates are those that possess adequate strength and durability, high density, low porosity and permeability, and good capability of fire protection and thermal insulation. Commonly used materials such as cobbles, gravels, sand, and crushed stones derived from hard and durable rocks (igneous and metamorphic) are good aggregates. However, gravels and sand deposits contain certain deleterious materials when derived from sedimentary rocks such as greywacke. While selecting a site for aggregate materials (crushed rock or river deposit), care should be taken to ensure that the site does not contain deleterious including cement-reactive minerals more than the IS-specified limit. In practice, the following actions are taken to solve the problem of deleterious material in aggregate:

(i) Crushed rock comprising excessive deleterious materials (beyond the permissible limit of 5 per cent) that promote alkali–aggregate reaction should be avoided from use as aggregate even if the rock is hard and possesses good strength.

(ii) Petrographic study helps to identify the type and content of deleterious materials present in aggregate including clay and mineral coating. Mica flakes and coatings of clay, calcium carbonate, gypsum, and so forth are removable by washing with water. When these materials are present in aggregate in excess, these are to be removed or reduced to sufficient extent by screening and washing with water.

(iii) If significant quantity of alkali-reactive deleterious material is present in the aggregate and if there is no alternative source to get better quality aggregate, then instead of Portland cement, low alkali cement, especially pozzolan, is to be used to retard the chemical action. Blending of some innocuous material brings down the content of the deleterious material within the specified limit.

12.7 PETROGRAPHIC STUDY OF AGGREGATE

The quality and durability of concrete depend on the type of rocks and minerals present in the aggregate materials. Petrographic study of aggregate is undertaken to identify the rock types and minerals present in it with the ultimate aim of identifying the percentage of minerals that are harmful for the concrete. Materials such as organic matter, calcium carbonate, and gypsum are considered deleterious. Minerals such as opal, chert, and chalcedony are chemically reactive with the alkali in cement. Coarse-size aggregate materials are examined visually and with the help of a hand lens of 10 to 25 times magnification. Fine aggregate (sand) is studied under the microscope taking the grains under oil immersion or fixing the grains on a glass slide.

The aggregate materials (crushed rocks or recent deposits) are first sieved into different size fractions. The weights of individual fractions are taken and the average weight is calculated. Petrographic study of each fraction is then undertaken. Whenever required, rock pieces, cobble, pebble, and coarse gravel are fragmented by geological hammer to examine the minerals and texture. The study includes the following aspects:

(i) Identifying the types of rocks and minerals and their physical characteristics, say, granite or gneiss—fresh to slightly weathered, strong, and durable—and phyllite or sandstone—moderately weathered, medium grained, soft. Weak visual and microscopic examination is conducted if any coating is present and the type of coating such as coal, clay, and silt or minerals such as calcium carbonate, gypsum, and opal is identified.

(ii) Ascertaining whether the coatings are easily washable by water or requires chemical treatment is also ascertained. For example, kaoline coating is easily washable by water but removal of iron stains requires use of chemicals.

(iii) Observing the particle shape, angularity, and smoothness. In engineering use, round and smooth surface grains are preferred to angular and uneven surface as the former consumes less cement.

(iv) Examining the fixed number of grains (about 200) generally by hand lens (for coarser particles) and under microscope (for fine grains). This is necessary for identification and percentage determination of alkali-reactive minerals such as chert, chalcedony, and opal. Special care is taken to ascertain the possible presence of strain quartz.

At the end of the study, it is necessary to make gradation of the materials, such as good, satisfactory, and poor. Good aggregate should have good strength, good durability, and high resistive capacity. The satisfactory type aggregate possesses moderate strength and durability and satisfactory physical character. The poor type contains particles of low strength and poor durability under climatic condition likely to cause cracking of concrete. The study result is reported in a tabular form containing factual data with respect to the different rock types.

In addition to natural aggregate materials, petrographic study by optical microscope and/or scanning electron microscope is undertaken for *concrete samples* to find if there is any sign of chemical reaction that may develop after some years, causing cracks in concrete. In general, the rim of the grains in concrete may show formation of gel or reactive scars or microcracks. Petrographic study of aggregate material of Sardar Sarovar (Narmada) dam in Gujarat showed a presence of 90–95 per cent basalt (Deccan) associated with secondary silica, zeolite, and calcite and rest (5–10%) of granite, quartzite, and sandstone. The aggregate containing deleterious material content (2.1–3.3%) is well within the limit of safe use. In addition, electron microscope study of concrete (10 samples) that are 5–10 years old showed no trace of gel or rim formation suggesting the absence of alkali reaction (Patel and Joshi 2004). Petrographic study of the concrete of Srisailam dam in Andhra Pradesh led to the conclusion that the aggregate used contains strained quartz (42.6–58.2%) and the concrete is prone to alkali–aggregate reaction (Raju 2004).

12.8 SOURCE OF CONSTRUCTION MATERIALS IN INDIA

India is endowed with rich reserves of good quality construction materials spread over different parts of the country. Each of the three geomorphic divisions of India, namely the peninsular plateau, the extra-peninsular region, and the Indo-Gangetic plain, has distinctive geological characteristics with respect to its rock formations and recent deposits (Chapter 24). The peninsular and extra-peninsular regions contain several hills and mountains incised by deep valleys. The rock formations of these two regions including the recent deposits of river valleys are sources of good quality rocks and aggregate used in various civil engineering works of India.

Large tracks of the peninsular plateau and parts of Bihar and Orissa comprise Archaeans, the oldest rock formations of India. Archaeans produce the best quality construction materials such as granite, charnockite, gneiss, and khondalite, which can be used as dimension stones or aggregate after crushing. The rivers of the peninsula containing very thick deposits of gravel and sand derived from the Archaean rocks provide good source of aggregate materials. More than 50 concrete and masonry dams constructed in the peninsular plateau have safely used Archaean rocks for engineering works. The crystalline limestone and marble of different hues available in Rajasthan, Madhya Pradesh, and parts of Mysore provide excellent ornamental building stones, especially for architectural works of temples and aristocrat buildings.

Next to Archaeans, the rocks in the Deccan traps cover the maximum areas of the peninsula. Massive basalt, amygdaloidal basalt, agglomerate, and tuff are the main rock types of the

Deccan traps. The Deccan basalt of massive variety can be safely used for aggregate and also for masonry works, but due to the presence of alkali-reactive minerals, amygdaloidal basalt is not found suitable for concrete aggregate. The volcanic ash and tuff bed of the Deccan traps are a good source of pozzolan, which is used as a substitute for cement. The basalts of both Deccan and Rajmahal Hills are also used as stone chips for road metals and railway ballasts.

Except the Archaeans and Deccan traps, other rock formations of the peninsula include Cuddapah, Vindhyan, and the Gondwana systems. The Cuddapah system is composed of metamorphosed sediments such as quartzite, sandstone, limestone, slate, and shale that occupy a large track of Andhra Pradesh and parts of Madhya Pradesh. The Cuddapah rocks, except shale, possess high-to-medium strength and hardness and are good for construction purposes. The Vindhyan system comprising sandstone, limestone, and shale covers most parts of central India. The Vindhyan sandstone was extensively used in building constructions during historic times and is also used at present for various engineering constructions including dams and road pavements. Vindhyan limestone is also utilized in making cement.

The Gondwana rocks consisting of sandstone and shale occur in extensive areas of Eastern and Central India covering Barakar, Son, and Narmada valleys. The argillaceous sandstone and shale of Gondwana are soft and are not suitable for construction purpose, but sandstones of siliceous matrix are sufficiently hard and durable and can be used in engineering works. The Tawa dam in Madhya Pradesh used the locally available medium hard Gondwana sandstone for its masonry construction. Tertiary rocks comprising soft sandstone, shale, and clay stone occur in eastern and western coasts. Shale and clay stone are used as pozzolans after calcinations. The silty and clayey soil derived from the tertiary sedimentaries is well suited for construction of earth dams.

The extra-peninsular region (the Himalayan terrain) is characterized by hundreds of ridges and intersected valleys that contribute vast reserve of rocks and aggregate for engineering constructions. Metamorphic rocks such as gneiss, schist, quartzite, slate, and phyllite with intrusive granite are the main rock types of the Lesser Himalayas with plenty of cobbles, gravels, and sands occurring in river and terrace deposits. Several projects have been completed and some are under construction using these rocks and the river deposits. Tertiary sedimentary rocks (Siwaliks) comprising sandstone, limestone, and shale (clay stone) and conglomerate occupy a large track of the southern fringe of the Himalayas. The Siwalik rocks have low strength and durability and are not good quality construction materials. Because of lack of availability of suitable aggregate materials in Siwalik rocks, the Bhakra concrete dam in Punjab used river terrace deposit of nearly five to eight kilometres downstream of the dam site as concrete aggregate and calcined clay stone as pozzolan for partial replacement of cement. The deposit of the river Chenab comprising dolomite and quartzite has been used as aggregate and fill material in the construction of Salal dam in Jammu and Kashmir. The Indo-Gangetic alluvium extends from Assam in the east to Punjab in the west covering parts of West Bengal, Bihar, Orissa, and Uttar Pradesh. The entire plain is devoid of rock and coarse aggregate, which would require transportation of these materials from long distance for construction work in this track. Clayey soil of this alluvial terrain is, however, suitably used as semi-pervious and impervious materials for earth dams. The sand, silt, and clay of the alluvium are utilized in making bricks for building constructions.

12.9 EXPLORATION FOR CONSTRUCTION MATERIALS AND SELECTING QUARRY SITES

The topographic feature is a major factor in the exploration for construction materials. Study of aerial photographs and remote sensing followed by field check help in the understanding of the geomorphic and geological features of the terrain and the identification of potential sites of construction materials within a short time. Once the potential sites are located from such studies, geological maps are prepared of these sites, delineating areas of rock outcrops, soil, or overburden cover including the nature of overburden material or soil. Subsurface exploration is then taken up following different techniques depending upon the depth and types of available construction materials. The exploration of soil, clay, and aggregate material is confined to shallow subsurface part, but quarry site for stones requires drilling to deeper parts.

Shallow subsurface exploration constitutes pitting and trenching within top 3–6m of soil or loose river or terrace deposits. Samples of the materials are taken from selected parts of the pit or trench for laboratory tests. The method of exploration for materials in loose deposits in deeper parts includes power auger drilling with samplers. The materials collected in the samplers from different depths are tested for their quality in the laboratory. The quantitative estimation of construction material is done from a simple calculation based on the available area and depth of the deposit above the groundwater level. Several holes may need to be drilled and depth of samples studied to find the suitability of materials and quantity available. The stability of the site where the construction material is located also needs to be investigated.

12.9.1 Selection of Quarry Sites

Selecting sites for quarrying of rocks requires a systematic approach. A hill face having exposed rock free from joins and fractures as far as practicable is first selected. If the hill slope is covered with overburden material or weathered rock, exploratory drilling is done to evaluate the depth of fresh and sound rock. Hill face exploration requires rock drilling with NX-size diamond bits that facilitates good core recovery allowing study of the quality and quantity of the rock available from the site for construction purpose. The drill holes should penetrate at least 6m in the fresh rock. A contoured geological map of the selected site is prepared, and several geological sections are drawn from drilling data that facilitate estimation of quantity of available fresh rock and the required depth of quarrying.

The core samples obtained from drill holes are tested in the laboratory for determination of engineering properties. Once the quality and quantity are found to fulfil the engineering need, the site is taken up for quarrying the required quantity of stones. The top overburden and weathered rock are removed first. In a small quarry, separation of rock chunks from rocky surface is done by placing tools such as crowbar along divisional planes (bedding, joints, and foliation) and then applying force using a sledge hammer. For large quantity of rocks, several shallow holes are first drilled in the rocky hill slope and then the face is blasted by controlled blasting. Pre-splitting method is followed to obtain big rock blocks along predetermined planes.

12.9.2 Adverse Effects of Quarrying

It may be mentioned that, similar to surface mining, the quarrying to meet the demand of construction materials such as dimension stones and crushed stones for engineering structures such as masonry dams and rip-rap of earth dams create environmental degradation. The exploratory work needs construction of roads to bring machineries and to carry the quarried materials to the construction sites. Excavation including blasting of rocks in hill slopes creates land instability and may initiate landslides. Quarrying for better rock also involves going down similar to subsurface mining. After completion of the surface and subsurface quarrying, the landform is left to nature, which hastens up the processes of weathering and erosion involving landslides causing further degradation of the environment.

In quarrying construction materials, when the overburden material from the surface is removed, it exposes the soil to weather. This further causes particles to become airborne leading to wind erosion and may spread particles of noxious materials, that may cause diseases of the respiratory tract in human beings. Quarrying causes physical destruction of the landscape by creating waste rock piles and open pits and causes disturbances to wildlife and plant species. There may be contamination of surface water passing through the waste piles of rock and processing plants resulting in water pollution. The sediment level in streams is increased by deposition of the transported soil from waste disposal area and by increased soil erosion from abandoned surface pits and underground excavation.

SUMMARY

- Rock, sand, gravel, clay, and pozzolan are the major naturally occurring materials used in engineering constructions. Geological search for construction materials for an engineering project is conducted by consulting existing geological maps and reconnaissance visit to the project area.
- Detailed investigation of potential quarry sites for construction materials includes finding the nature of topography, groundwater condition, stability of the site, and environmental impact of quarrying. Once a site for construction materials is selected, exploratory work is taken up by pitting and trenching for shallow deposits and by power drilling for subsoil deposits. Core drilling is taken up for rock quarry sites to study the rock condition at depths and collect specimens for laboratory tests.
- Hard and durable rocks are necessary for the construction of buildings and masonry dams, rip-rap of embankments, railways, and roads. Large-size stone blocks obtained from quarry sites are later sliced to requisite sizes for use as building stones and pavement slabs.
- The fill materials required for rock-fill dam are procured from unconsolidated recent deposits or by crushing of hard rocks. Granite, basalt, dolerite, gneiss, charnockite, quartzite, and sandstone, which are available in different parts of India, are good quality rocks used for construction purposes in various engineering projects.
- Construction materials should be free from deleterious materials such as organic matter,

excessive mica, pyrite, chert, and chalcedony. The material should also be free from alkali–aggregate reaction. The suitability of the construction materials for different purposes is decided after determining their properties by laboratory tests.
- Pervious and semi-pervious soils required for earth dams and embankments are found in plenty in the igneous and metamorphic terrains, whereas impervious type of soil required for core of earth structures occurs in areas with clay–shale beds.
- Pozzolanic material is found in ash and tuff beds of volcanic rocks such as basalt. Artificial pozzolan such as fly ash is procured from the thermal plant sites for use as a substitute of cement. Blast furnace slag, burnt shale, or clay can also be used, but laboratory tests are required to determine its properties and the requisite proportion to use as a replacement for cement.
- Samples of aggregates are tested in the laboratory and studied under microscope for determining the possible presence of deleterious, cement–alkali reactive materials. While selecting sites for construction materials, it is to be seen that the cost of quarrying and transportation of materials from quarry sites or borrow areas to the construction site is economically justifiable.
- The geological report on construction material will accompany a geological map showing the locations of selected sites of construction materials and contain factual data on assessed quality and quantity of the materials available in each quarry site.

EXERCISES

Multiple Choice Questions

Choose the correct answer from the choices given:

1. State which of the following is artificial construction material:
   (a) shingles

   (b) brick
   (c) clay

2. State which of the following is natural construction material:
   (a) fly ash
   (b) gravel
   (c) *kankar*

3. The type of construction material required for highway and runways construction is:
   (a) Deccan basalt
   (b) Cretaceous sandstone
   (c) mud stone

4. The Indian rock that was abundantly used in ancient India for constructing royal buildings and several monuments is:
   (a) Vindhyan sandstone
   (b) Gondwana sandstone
   (c) Cretaceous sandstone

5. The rock type that was used in the construction of Taj Mahal is:
   (a) polished pink granite
   (b) Makrana marble of Rajasthan
   (c) Eocene sandstone

6. The material used in preparing aggregate of concrete is:
   (a) gravel
   (b) clay
   (c) mica

7. The correct source of pozzolan is:
   (a) factory waste
   (b) Deccan trap bed
   (c) old crystalline rocks

8. One of the deleterious mineral that causes cement–alkali reaction if used as concrete aggregate is:
   (a) opal
   (b) quartz
   (c) muscovite

9. It is possible to recognize the presence of alkali-reactive minerals in construction materials by:
   (a) chemical analysis
   (b) looking with magnifying glass
   (c) petrological study

10. One of the following is a good dimensional rock:
    (a) phyllite
    (b) shale
    (c) granite

Review Questions

1. What are the common types of naturally occurring construction materials? Name the materials used in engineering constructions. Explain the process of manufacture and use of colcrete instead of concrete.

2. Give an account of the various uses of rocks as blocks or crushed stones. Give the names and occurrence of some Indian rocks that are preferred for building stones and facing stones.

3. What types of materials are used for coarse and fine aggregates for concrete? Where are these materials available?

4. What is pozzolan and what is the purpose of using it? Name the materials that can be used as pozzolan. Where are these materials available?

5. State briefly the laboratory methods of testing of aggregate materials for the determination of the following:
   (a) Aggregate crushing strength
   (b) Aggregate impact
   (c) Los Angeles abrasion values
   (d) Deval attrition values

6. What is the purpose of the soundness test of aggregate? Describe briefly the procedure of this test in the laboratory.

7. Give a detailed account of the materials used as road metals and railway ballasts.

8. State the criteria specified by the BIS for use of aggregate.

9. What is an alkali–aggregate reaction? Name the minerals responsible for such reaction when present in the aggregate. In addition to these minerals what are the other materials considered deleterious when present in the concrete aggregate?

10. State briefly the rock formation and places where construction materials are available in India.

11. Discuss the approaches and techniques of exploration of construction materials such as rocks, soil, clay, and aggregate materials for an engineering project.

Answers to Multiple Choice Questions

1. (b)   2. (b)   3. (a)   4. (a)   5. (b)   6. (a)   7. (b)   8. (a)   9. (c)
10. (c)

13 Treatment of Rocks and Soil by Grouting

●　●　●　●　●　●　**LEARNING OBJECTIVES**　●　●　●　●　●　●　●

After studying this chapter, the reader will be familiar with the following:

- Different aspects of grouting in rocks and other materials
- Types of grouting for foundation and nature of grout mix
- Equipment and packers used in grouting and grouting pressure
- Ingredients of grout mix and their ratio of use in rocks
- Approaches of strengthening foundations of engineering structures
- Pattern of grouting with reference to rock fractures and soil pores
- Factors considered for finding the efficacy of grouting

13.1 INTRODUCTION

Grouting is a process of strengthening weak rocks and unconsolidated soil of a project site. Grouting plays an important role in arresting the leakage of water through fractured rocks or porous materials of a dam foundation or reservoir basin. In a dam foundation, curtain grouting, consolidation grouting, and blanket grouting are generally taken up to strengthen weak rocks and to contain leakage. This chapter provides a detailed discussion on the different aspects of grouting including the ingredients of grout and the proportion of cement, water, and other admixtures and chemicals added for quick setting or retarding action. It describes with illustration the equipment used in grouting and explains the factors that control grout intake and pressure conditions. The chapter further elaborates the method of proving the efficacy of grouting in a site.

13.2 GEOTECHNICAL CONSIDERATIONS IN GROUTING

Grouting is a method of sealing fracture openings and pore spaces in rocks and unconsolidated deposits by injecting a mixture of water with cement and/or other suitable substances. In Indian Standard (IS) (1356–1987), the term *grouting* has been defined as 'a process of injecting under pressure a slurry of fluid grout (cement–water mixture), or other suitable materials into the mass of a defective rock formation through a borehole to fissures and cracks in the hope that all fissures, joints, and cavities will be sealed off against water in rock'.

The principal purposes of grouting a soil mass or rock are the enhancement of bearing capacity and the reduction of permeability of the mass. Grouting is widely applied in the treatment of defective features in foundations of dams, bridges, powerhouses, and other engineering structures that include weak rocks created by fault, shearing, jointing, and fracturing. It arrests seepage of water through porous and fractured rocks and imparts cohesion and strength to unconsolidated deposits to make them monolithic. Though grouting is an important tool used for the treatment of structural sites, other measures such as providing drainage, relief well, compaction of foundation materials, excavating slurry trenches, and concrete cut-off are also taken up in combination with grouting to obtain optimal results.

13.2.1 Relation of Grouting to Rock Types, Rock Structures, and Overburden

The rock types and rock structures of a site of engineering structure influence the grouting condition. Igneous and metamorphic rocks are commonly massive in nature, but they possess several sets of joints and other structural defects due to faults and folds. The volume of grout intake depends on the open spaces and interconnections of these joints and fractures. The foliation planes and cleavage in metamorphic rocks such as schist and slate provide passageways for the entry of sufficient quantity of grout. The contact planes of intrusive igneous bodies such as dykes and sills with country rocks are weak planes requiring grouting when exposed in a structural site. Basalts sometimes contain beds of pumice, scoria, agglomerate, and tuffs responsible for low strength and high permeability.

Sedimentary rocks such as conglomerate, sandstone, and siltstone are sufficiently porous and permeable and are amenable to grouting treatment. The soft sedimentary rocks such as shale and claystone are generally impermeable, but they develop permeability when highly jointed. If such soft and jointed rocks occur in the foundation, they need to be excavated or strengthened by grouting. Overburden materials formed of disintegrated bedrock and deposits of alluvial, colluvial, aeolian, and glacial origin are highly porous and permeable. Foundation grouting in combination with other treatments (e.g., compaction and drainage) is necessary to reduce permeability and improve stability of these deposits for the safety of the structures. Drainage under a dam represents a row of small diameter holes (spaced at 2 m to 5 m intervals) communicating with the tail water through a discharge device such as a pipe or drain filled with sandy materials.

13.2.2 Problems of Grouting Cavities

Solution of limestone, gypsum, and anhydride by surface and subsurface water creates cavities through which there may be copious flow of water. The solution cavities are generally found to develop along some structural planes. Soluble rocks affected by faults, shear zones, bedding planes, and joints are more prone to solution effect by movement of ground water along these structural planes. In some places, solution of limestone forms karst conditions in underground stratum. Boreholes are drilled, followed by pump testing, to locate and assess the permeability of the soluble rocks. In some holes, the test results may initially show low permeability due to infilling by inert materials such as sand and clay, but with increasing pressure the infillings are removed when large flow takes place. The treatment by grouting of cavities in a karst country is an expensive process, but it has to be taken up in all projects that have the problem of leakage through solution cavities.

Many engineering constructions such as dam foundations, reservoirs, tunnels, and underground structures face problems from groundwater. Broadly, these problems are associated with groundwater flow through the open spaces in jointed rocks, open bedding planes, karstic limestone, and faulted and sheared rocks. Observation of water table is very important to understand groundwater hazards and plan grouting in subsurface rocks and overburden materials.

Orientation of faults, intensity, and inclination of joints and other structural details that help planning of grouting are obtained from outcrop mapping and logging of drill cores. Water pumping tests are conducted to measure the permeability of rocks or soil mass of an engineering site and accordingly the grouting programme is planned. In case an artesian flow is detected in a borehole, the pressure condition of the zone is to be measured so that specific data can be obtained on the required high pressure for grouting this section.

13.3 DIFFERENT TYPES OF GROUTING INCLUDING THEIR PATTERNS

Strengthening of rock foundation of a dam is done generally by grout holes spaced at 3 m to 6 m intervals later staggered by intermediate holes using varied pressures and extended to different depths controlled by geological conditions. It is a common practice to use shallow holes (B-holes) for foundation grouting at low pressure before placing concrete cover. Supplementary holes (C-holes) inclined downstream for grouting from heel are used applying high pressure after adequate foundation cover. High pressure grouting through holes (A-holes) extended to sufficient depths to suit the geological condition are taken up from the gallery by vertical or inclined (70°–80°) holes after attaining sufficient concrete cover of foundation rock. The A-holes are drilled and grouted in a single line or multiple lines to create a curtain to seepage path. Drainage holes are also provided from the gallery in a row.

In addition to these three types of grout holes (A, B, and C), grouting is also undertaken to fulfil specific purposes. For example, faults or fractures are treated by high pressure grouting through vertical or inclined holes. The holes for these high pressure grouting are drilled before foundation cover, and pipes are kept in the holes through which grouting is taken up after placing foundation concrete. Figure 13.1 shows the strengthening of foundation rocks by grouting through different types of grout holes. The dotted lines indicate grout holes for treating the fault by high pressure grouting (F–F).

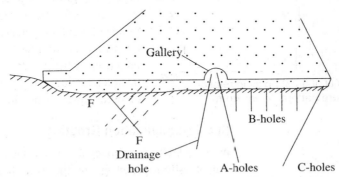

Fig. 13.1 Strengthening of foundation rocks by grouting through different types of grout holes

The grouting processes followed in the treatment of defects in rocks and porous overburden materials include the following:

  (i) Curtain grouting
 (ii) Consolidation grouting
(iii) Blanket or area grouting
 (iv) Contact grouting
  (v) Special-purpose grouting

13.3.1 Curtain Grouting

Curtain grouting has wide applications in the construction of dams and reservoirs. It is carried out to cut off seepage under dams or other structures as if by providing a curtain against seepage path. Controlling of seepage is accomplished by drilling and grouting through one or more lines of grout holes in the foundation to specified depths usually parallel to the dam axis or normal to the direction of water movement (Fig. 13.2). This creates a barrier against seepage of water by filling the voids or water passages with grout.

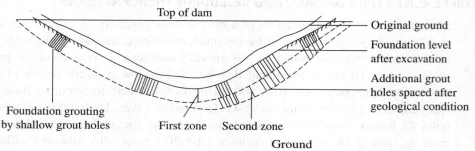

Fig. 13.2 Profile along dam axis prior to foundation cover showing foundation treatment by curtain grouting

As regards the depth of curtain grouting, a thumb rule is to grout to a depth of two-thirds of the hydrostatic head of the dam. Rock cores obtained from exploratory work and water percolation tests at the site provide data regarding the depth extent of the zone of intensely fissured and highly porous rocks. The percolation test data is a good guide to fix the depth of grout curtain. In the absence of sufficient subsurface data, Wahlstrom (1974) has mentioned about the *closure pattern* that includes grouting to a depth of one-third the dam height plus 15 m depth. Once grouting through the initial holes at fixed intervals is complete, more holes are drilled at intermediate positions with gradual reduction in heights to ensure that grout intake in the holes gradually reduces until it is insignificant in the holes towards the closure (Fig. 13.3).

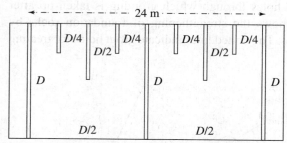

Fig. 13.3 Depths D of grout holes in relation to dam height in closure pattern (Wahlstrom)

13.3.2 Consolidation Grouting

When a loose rock mass is consolidated by grouting, it is called consolidation grouting. If the foundation rocks are intensely fractured and highly permeable, consolidation grouting is carried out covering the affected area. The bedrock at the top part is generally open fractured and as such consolidation grouting is commonly carried out to shallow depth (Fig. 13.4). The solids in the grout mixture settle in the void spaces to seal the fracture openings and consolidate the rock to increased strength.

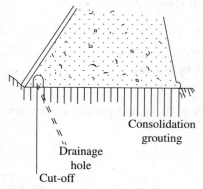

Fig. 13.4 Profile across dam axis consolidation grouting of foundation rocks

The grouting operation should be taken up after foundation excavation of a dam site (and not after dam construction) when location and orientation of the holes are fixed after geological mapping of the site. If the foundation rock contains clay seams, these are excavated before injecting grout to make the consolidation grouting effective. In case of extremely fractured and seamy rocks, it may be necessary to excavate the foundation to remove them and then take up grouting. Some interconnected open joints may cause wastage of large quantities of grout by leading the grout out of the foundation. In such cases, a single line grouting is to be undertaken at low pressure at the first instance in the periphery of foundation to seal the leakage path.

13.3.3 Blanket or Area Grouting

The process of grouting through a specified area of shallow depth for the purpose of sealing the voids of unconsolidated material or fracture openings of rocks of a site such as a dam foundation or reservoir surface is known as blanket or area grouting. The top few metres of the foundation or a part of the reservoir rocks may be found defective due to high porosity and/or fracturing when blanket grouting is undertaken only for that depth or area covered by the defective rocks. The main purpose of blanket grouting is reducing permeability and stopping water seepage. In general, the holes for blanket grouting are restricted to 6 m to 10 m of subsurface area from the foundation surface or ground surface of reservoir sites (Fig. 13.5). The grout holes are directed normal to the surface, but in special cases, where the rocks are steeply dipping, angle holes can be resorted to for intersecting the bedding and joint planes through which seepage is expected. Blanket grouting of a dam site is always to be taken up before the dam is constructed or the reservoir is filled.

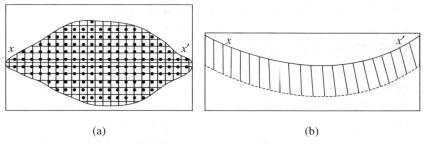

(a)   (b)

Fig. 13.5 Blanket grouting of dam foundation: (a) plan of the dam showing grout holes arranged in a grid pattern; and (b) section along dam axis (x–x') showing grout holes extended from foundation to subsurface strata restricted generally to 10 m

13.3.4 Contact Grouting

When grouting under pressure is undertaken to seal a gap and obtain a strong contact between an engineering structure and the foundation rock mass, it is known as contact grouting. In this process, a neat cement grout is injected at the contact of the concrete structure and the adjacent surface. Shrinkage of the concrete as it sets creates voids and seepage paths along the contact. The cement grout fills the voids and seals the seepage paths. It is a common practice to undertake contact grouting at the interface of a concrete dam with the abutment rock and also in the crown of a lined tunnel to make a watertight bond between the roof area and concrete. In contact grouting of some sites, header pipes are installed into the concrete at the place of its contact with rock during construction, and grouting is done through the pipes after the structure is raised. This treatment can also be done by drilling holes after the completion of construction.

13.3.5 Special-purpose Grouting

Special-purpose grouting is done to treat adverse geological conditions such as fault zones or highly fractured zones at depths or to stabilize subsurface materials at shafts or deep structures.

The execution of special purpose grouting is done after preparing a plan and section of the subsurface weak zone where such grouting is necessary. For example, to grout a vertical or steeply inclined fault zone, the grout holes should be inclined, close spaced, and directed to intersect the fault plane (Fig. 13.6). The geometry of the geological structure and the extent of the weak feature need to be evaluated by exploratory investigation for designing special purpose grouting. In order to seal very fine fissures in the concrete of a dam or tunnel lining, chemical grouts are used on account of their excessive penetration property.

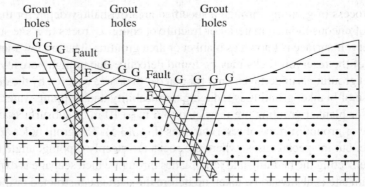

Fig. 13.6 Special-purpose grouting—treatment of fault zones (F–F) by injecting grout through inclined holes (G)

13.4 GROUTING EQUIPMENT AND PACKERS

Grouting equipment includes a drilling unit with a pump and air compressor to force grout into the drill holes, generally EX to NX size (38 mm to 76 mm diameter). A tank or drum mixes water with cement and/or other ingredients to make the grout. Two gauges, one for recording the applied pressure and another for measuring the quantity of grout flow, are attached with the grout hole. Lightweight and compact type of drill that can be operated by hand is suitable for shallow holes (Fig. 13.7). This compact drill can be used for making grout holes from adit, tunnel shaft, and gallery.

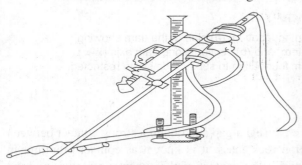

Fig. 13.7 Hand-operated lightweight and compact type of drill suitable for making shallow grout holes

Fig. 13.8 Drill rods for injecting grout

Hard rocks with close spacing joints are successfully grouted through EX and AX holes using small diameter drill rods (Fig. 13.8). For boring holes in dam foundations and tunnel roofs, wagon drills are utilized. The wagon drill is composed of a drill head mounted on leads that are supported on a track-, wheel-, or skid-mounted chassis. For underground work, generally electric- or air-powered drilling and grouting equipment are used. The most common type of drilling machine used for boring and grouting of deep holes is rotary drilling (Section 11.7). Percussion drilling is also taken up using various

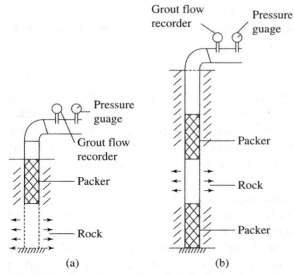

Fig. 13.9 Grouting packers: (a) single packer and (b) double packer

kinds of bits (Section 11.6). Percussion drilling is faster but its main disadvantage is that it causes caving of walls in soft rocks.

The packers used in grouting operation are made of stiff rubbers that have a watertight grip with the wall of the grout hole. There are two methods of using packers for carrying out grouting operation. In one method (single packer method), grouting operation is carried out in stages starting from the top and then going down with progress of drilling. The packer can be fixed from the beginning in rock and then inject grout when 3 m (or as desired) depth of drilling is completed (Fig. 13.9a). This method also gives an accurate result of the grout intake depending on the condition of rock in specific sections with respect to its permeability character.

In the other method (double packer method), the holes are drilled up to the full depth and then grouting is done starting from the bottom or from the top and going up or down accordingly at fixed intervals by attaching two packers. The two packers are so fixed that a stretch of 3 m to 6 m barren rock remains between them through which the grout is injected so as to penetrate the fractures of the surrounding areas (Fig. 13.9b). In the double packer method, since the full depth of the hole is drilled before beginning the grouting operation, there are chances for collapsing of wall rocks of the holes, especially if the rocks are soft or broken in nature.

13.5 BASIC INGREDIENTS OF GROUT AND ADMIXTURES

Grout ingredients are varied in nature. The most common grout is made up of Portland cement mixed with water in different proportions. In fact, the dimension of the cracks or fractures in rock determines the additional substances or admixture to be mixed with the cement–water grout.

13.5.1 Ingredients

Grout is an aqueous suspension of solids used for injecting under pressure into the bedrock or unconsolidated materials. Portland cement is the basic ingredient of cement grout, which is universally used for foundation treatment of dams and other engineering structures. The following are some specific types of Portland cement that are also available:

(i) Quick setting cement that has setting time one-third of normal cement
(ii) Cement that is resistant to sulphate attack
(iii) Cement of very fine particles that can penetrate fine cracks

Several *admixtures* in varying percentages are used with cement grout. They act as accelerators, retarders, or fillers. Calcium chloride is an *accelerator* that is mixed with cement grout when early setting is necessary. The use of calcium chloride also prevents sulphate attack and alkali–silica reaction. Some organic chemicals (acidic salts) are used as *retarders* that offset the undesirable quickening process and prolong the injection and placement time. Sand, silt, rock flour, fly ash, diatomite, pumice, barite, and other fine materials are used as *fillers* with cement grout to reduce cement consumption.

13.5.2 Grout Mixture

The ratio of water to cement in a grout mixture influences its viscosity, permeability, and setting time. Depending upon the ratio of cement to water content by volume, the grout is termed *thin* or *thick*. In thin grout, the water–cement ratio may be as low as 20:1, whereas in thick grout it may be 1:1 or more. With the use of accelerator or retarder materials or chemicals, the water–cement ratio will vary. Foundation grouting is mostly conducted with a grout mixture of Portland cement, bentonite, and water.

A *cement–bentonite–water grout* also contains ingredients such as sand, clay, and pozzolan. The presence of bentonite provides expansive (swelling) capacity in the grout when saturated with water. *Bentonite–water grout* without any cement is used in grouting a reservoir periphery to reduce the permeability of overburden and rocks with the ultimate aim of arresting leakage from the reservoir. *Silicate gel grout* that contains sodium silicate and calcium chloride solution and/or sodium carbonate solution is a quick setting grout suitable for use in packing of fine and medium sands.

13.5.3 Chemical Grout and Epoxy Resin with Instances of Their Use

Chemical grouts are solutions composed of two or more types of chemicals, which after entering into the fine fissures react to form a gel that fills the pores or open spaces. For example, sodium silicate solution mixed with another chemical solution (e.g., calcium chloride) is used as a chemical grout to convert the silicate into an insoluble gel by a catalyst dissolved in the second chemical. The gel acts as a sealant of the fine voids and fissures. While reducing the permeability, it helps to increase the strength rapidly. If the foundation contains silt or sand pockets, treatment by chemical grouting is necessary for gaining rapid bondage of the loose materials and improving the shear strength. *Epoxy resin* and *polyester resin* are used in repairing cracks in concrete and are also employed in rock bolting.

In a part of the Obra dam in Uttar Pradesh, a sandy layer of river bed between two concrete diaphragms was treated using chemical grout. In the sluice blocks of the Rana Pratap Sagar project in Rajasthan, a montmorillonite seam was treated by chemical grout. In the silty shale of the saddle dam foundation of this project, grouting was done by *silicate gel* mixed with bentonite and sawdust. In the Hirakud dam in Orissa, a horizontal crack of the vertical wall of the operational gallery had to be treated with a chemical grout to check leakage. In the Konar dam in Bihar, under Damodar Valley Corporation, a widening crack in the dam wall threatening leakage of reservoir water was successfully grouted by use of epoxy resin.

13.6 GROUTING APPROACHES FOR VARIOUS ENGINEERING STRUCTURES

Test grouting is necessary to design the plan of grouting in a project site including its cost estimation. The following points are given due consideration while planning to undertake test grouting:

(i) Where different types of rocks are present, test grouting should be done for each type of rock.

(ii) Different types of grout mixtures are to be used in test grouting so that the mixture giving the best result can be applied during the actual operation.

(iii) Permeability test and core drilling are to be conducted before and after the tests to ascertain the effectiveness of grouting.

13.6.1 Grouting of Concrete Dam Foundation

Both consolidation and curtain grouting are undertaken in the foundation of concrete dams comprising porous or fractured rocks (Fig. 13.10). Consolidation grouting is taken up when the foundation grade is attained after excavation and the foundation area is cleaned properly. A single line of curtain grouting at the upstream end of the dam is generally satisfactory in a concrete dam as it is founded on competent rocks. In case the foundation rock is of lower strength or contains erratic joints, multiple lines of grouting are undertaken. The distance between two adjacent lines will be 2 m or slightly less. Initially, the grout holes in each line are spaced at 3 m to 6 m intervals. The holes in the adjacent rows may then be staggered.

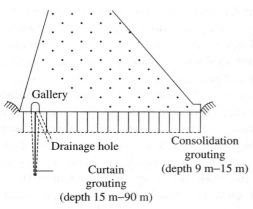

Fig. 13.10 Profile across axis of concrete dam showing consolidation grouting of foundation, curtain grouting, and drainage holes taken up from gallery

Curtain grouting is done after the dam is raised to a certain height when the foundation rock is already under the load of the concrete mass, which enables injection of grout under high pressure without disrupting the rock. The holes for grout curtain are drilled from the gallery when drainage holes are also drilled. Grouting of the dam should always be completed before filling the reservoir so as to avoid additional pressure and seepage possibility with the rise of water level.

13.6.2 Grouting in Earth and Rock-fill Dams

In earth or rock-fill dams, curtain grouting is taken up before beginning the dam construction. Low pressure (gravity pressure) is used for the upper zone avoiding any erosion of the dam foundation. If any sensitive material such as swelling clay is present in the fill material, it is better to remove this material and execute the grouting before the final foundation preparation. The grout holes are drilled from a concrete filling of a shallow trench (grout cap) in parts of the foundation. If the grouting is carried out from a grout cap at the upstream end (heel of the dam), this may be done after the completion of the dam.

13.6.3 Grout Curtain with Drainage Holes

The grout curtains in dams must be supplemented by drainage arrangement. In spite of thorough grouting, some small percentage of fissures may remain in the grouted rock that may act as seepage path under reservoir conditions. In gravity dams, grouting is possible from the inspection gallery by drilling new holes or through the drainage holes. However, in case of earth dams and rock-fill dams, it is difficult to undertake such grouting after the construction of the dam, and hence, preventive measures against possible seepage from these dam foundations are taken up by means of blanket or curtain grouting prior to dam construction. It is a common practice to

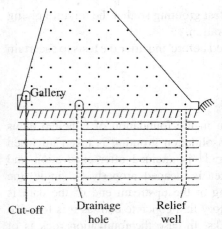

Cut-off   Drainage   Relief
           hole   well

Fig. 13.11 Grouting of foundation extending cut-off, drainage, and relief wells until the end of porous beds

excavate wells or construct drainage prisms downstream of the grout curtain to intercept the seepage water into the hole and thus reduce the built-up pressure (Fig. 13.11).

13.6.4 Reservoir Rim Grouting

Grouting of reservoir rim is taken up if there is a chance of seepage of stored water to another valley. This may happen in case a constricted land separates the reservoir from a lower valley. Presence of porous conglomerate bed, limestone or other soluble rocks, and structures such as faults in the reservoir area may cause substantial leakage. Investigations carried out to find the water tightness of a reservoir include geological mapping, exploratory drilling, and tracer study. In order to control leakage, generally a single line of curtain grouting is taken up along the reservoir rim. The exact extent and depth of curtain are decided from the geological condition of the reservoir rocks. Test grouting is first carried out for designing and estimating the grout pattern. Grouting is completed before filling the reservoir as the rise in water level may aggravate the leakage problem and create difficulty in grouting operation. Reservoir projects in karstic limestone in different parts of the world including the Obra dam of India have undergone very extensive and expensive grouting to arrest leakage of reservoir water (Section 22.7).

13.6.5 Grouting in Tunnels and Underground Chambers

Grouting in tunnels and underground chambers is done to consolidate the wall rocks and prevent seepage of water. In tunnels, pressure grouting is usually done by cement grout. The grout holes are directed so as to penetrate maximum number of joints and fracture planes in rocks of the underground structures. As the structural planes are visible in underground construction, it is easy to orient the grout hole for effective grouting of defective rocks. The grout holes in tunnels are drilled in a ring pattern starting from the invert level and proceeding upwards to cover the crown. The length and spacing of the holes depend on the rock condition and may be 6 m to 9 m deep and spaced at 3 m to 6 m intervals. After a lining is provided to the tunnel or underground chamber, contact grouting is necessary to seal the gaps between the grouted rocks and the concrete or steel linings.

13.7 GROUTING PLAN AND PATTERN WITH INDIAN EXAMPLES

The locations of the grout holes including depth of penetration are decided from the geological condition of the site. In hard rocks, 38 mm to 76 mm (EX to NX) size holes are good enough, but in soft rocks, larger diameter holes are chosen for effective grouting. Large diameter holes are drilled when there is a chance of caving of walls. The *spacing* is decided from the characteristics and frequency of the fractures in rocks. Primarily, the grout holes are spaced at 3 m intervals, but if grout intake is unabated, intermediate holes spaced at 1.5 m intervals are taken up. The close-spaced holes allow penetration of the grout covering more subsurface areas and thus help nearly complete sealing of the voids of porous materials or rock strata. The depth of the curtain should be sufficient to minimize seepage and reduce uplift pressure.

13.7.1 Geological Approach in Preparing Plan for Grouting

The geological approach of making a plan for foundation grouting will be to prepare large scale maps (scale 1:100 or larger) for foundation blocks, plotting details of structural elements including attitudes of the joints, fractures, shear zones, and so forth. With the help of this map and due consideration of water percolation test data, the plan for foundation grouting is prepared. Whether grouting should be undertaken before or after concreting depends on the nature of the rock types and may be decided after grouting a test hole. It is better to complete the grouting prior to concreting when all structural features are visible; however, if there is a chance of upheaval, grouting is to be carried out after one or two metres of concreting or masonry work to prevent the heaving effect.

13.7.2 Geological Considerations in Planning of Foundation Grouting

Based on the geological conditions, a broad programme on grouting operation is prepared during the planning or design stage of project investigation. Initially grout holes are spaced 3 m to 6 m intervals (Depending on the grout intake in these holes, intermediate holes at the mid-point of the two holes are then drilled and the entire foundation area grouted. The depths of grout holes for dam foundation are deduced from dam height and rock condition at depth, generally restricted within 15 m. If the rock is soft, consolidation grouting is to be planned up to the depth at which firm bedrock is available. The pressure applied for consolidation grouting in sound rocks starts from an initial low 2 kg/cm^2 and then it may be raised progressively up to 10 kg/cm^2. However, in very soft rock, the grout intake may be by gravity flow or by applying pressure limited to 1 kg/cm^2 as in the Siwalik sand-rocks of Beas dam site.

Fig. 13.12 Joints and other divisional planes in drill cores of subsurface rocks

The structural set-up of the rock is the prime factor that controls the grout intake. Unless the foundation rock is homogeneous in nature, the uniform pattern as in the grid may not produce the desired result. Study of rock cores obtained from exploratory drill holes with respect to spacing, dip, and depth of joints helps in planning of grouting pattern for dam foundations (Fig. 13.12). For example, in rocks with vertical joints, only oblique holes can intersect a good number of joints and ensure the flow of grout. A similar situation was found in the Umiam dam foundation of Meghalaya where the dip of joints in foundation rocks, quartzite and phyllite, is vertical. A grid pattern of grouting with vertical holes as per engineering design did not show any significant grout intake in foundation rocks. A plan was then made for grouting on geological consideration suggesting angle holes and fixing specific locations of the holes including directions and depths to penetrate for each grout hole. The plan produced high grout intake ensuring consolidation and enhanced strength of the foundation rocks [Appendix A.1 C(2)].

13.7.3 Statistics of Foundation Grouting of Major Indian Dams

The results of both consolidation and curtain grouting of some major Indian dams with respect to their depths of grouting, pressures applied, the total grout consumption, and various such aspects have been recorded in Table 13.1. In case of curtain grouting, the holes

Table 13.1 Foundation grouting of some Indian dams (Hukku 1975; Sanganeria 1975; Chowdhury 1975; C. Rao 1975)

Project name	Structure (height)	Foundation rock	Grouting (grout)	Depth and spacing	Pressure (kg/cm²)	Other observations
Bhakra, Punjab	Concrete (226 m)	Sandstone, siltstone	Consolid Curtain (cement)	15 m 90 m	Twice H	Cement consumption one million bags
Rihand, Uttar Pradesh	Concrete (90 m)	Gneissic granite	Curtain (cement)	30 m (R) 22 m (A)	7 to 14	No consoled grouting because of sound rock
Rana Pratap Sagar, Rajasthan	Masonry (54 m)	Sandstone, shale	Consolid Curtain (cement)	9 m (6 m) 67% H	2.1 1.5 H	Hexagonal pattern; average grout intake 24 kg/m
Beas, Punjab	Rock-fill (76 m)	Siltstone, sand-rock	Cut-off (cement)	5 rows (3 m)	Gravity to 0.7	Pressure applied only in cracked sand-rock
Lower Sileru, Andhra Pradesh	Masonry (72 m)	Khondalite	Consolid (cement)	9 m (6 m)	0.2–0.5	Hole inclined 30° to vertical; joint 40°–70°
Tenughat, Bihar	Composite (51 m)	Gneiss, sandstone	Consolid Curtain (cement–bentonite)	9 m 12–24 m	2–14 10	Vertical and inclined grout holes; efficacy proved from post-construction permeability
Obra, Uttar Pradesh	Earth and rock-fill (62 m)	Limestone, shale	Curtain Foundation (cement–bentonite chemical)	15–17 m (inclined 30 to vertical hole)	Nil to 24 (consumption in limestone 90 kg/m and in shale 24 kg/m)	Grouting by packer method bottom to top; grouting in stages; chemical grouting in river bed sands between two diaphragms; cavity and fault grouting
Ukai, Gujarat	Earth and masonry (69 m)	Basalt, dolerite (Deccan)	Earth Cut-off Masonry Consolid (cement)	15 m (3 m) 9 m (7.6 m)	3.5	Average grout intake 83 kg/m Grout holes grid pattern; curtain grouting up to 24 m depth
Nagarjuna Sagar, Andhra Pradesh	Masonry gravity (125 m)	Gneiss, quartzite	Curtain (cement)	9 m (6 m)	Grout take *	C:W 1:8 and 1:12; * negligible in gneiss; heavy in quartzite
Kadana, Gujarat	Masonry and earth (66 m)	Quartzite	Consolid Curtain Earth (Cement)	9 m (3 m) 0.6 H 8–10 m	3.5–7.0 1.5–2.3	C:W 1:20, 1:10, 1:5, and 1:1 increased gradually; efficacy: water loss down from 20 to 4 Lugeons

H, Hydraulic head; R, River bed; A, Abutment; C, Cement; W, Water; and Consolid, consolidation are drilled in single, double, or triple rows and spaced 3 m to 6 m apart. The depths of the holes are kept at 30 per cent to 70 per cent of the head under full reservoir level (FRL) and maximum pressure used is generally 1.5 to 2 times the hydraulic head. For example, in the 226 m high Bhakra dam of Punjab (Section 21.6) the depth of the grout curtain provided and the maximum pressure used are 40 per cent (90 m deep) and twice the hydraulic head, respectively, whereas the depth of the curtain was kept at 67 per cent and grouting pressure adopted was 1.5 times the hydraulic pressure in the Rana Pratap Sagar dam of Rajasthan [see Appendix A.1 B(1)].

13.8 EFFECTIVE PRESSURE AND ROCK MASS PERMEABILITY

The application of correct pressure for grouting operation requires knowledge of strength properties of rocks or materials to be grouted and good judgement about the effect of pressure on them. In consolidation grouting to shallow depth, pressures are generally applied depending upon the nature of rock condition and structural set-up so that the foundation rock does not get fractured. Application of high pressure is necessary in curtain grouting extended to deep parts in hard rock so that more grout can enter into the rock fissures.

Under full storage condition, the pressure due to the head of reservoir may cause seepage of water by removing the clay filling. As such, the suggested grouting pressure should not be arbitrary. In general, it should be 1.5 to 2 times the designed hydraulic head but not less than the head. In the foundation of the Aliyar dam of Tamil Nadu consisting of hard but highly jointed rock (granite gneiss), the applied pressure was less than the hydraulic head. It created seepage trouble in the post-construction stage.

At the entry point of grout into fissures, it maintains the same pressure as applied, indicated by a pressure gauge. As the grout tends to penetrate the fissure openings, it exerts pressure to cause upheaval of the rock above flat plane of fissure. If the rock is soft in nature, there is a chance of further opening of the fissure plane. The friction along the open fissure reduces the fluid pressure. The reduction in pressure is also a function of the distance of travel of the grout. The more the distance covered, the less will be the effective pressure. At every point of the grout flow, the overburden pressure counteracts to stop upheaval due to the applied pressure. It is important to note that if long and flat fissure plane is cut across by the grout hole, application of very low pressure should be considered. A common feature of cutting an open plane of fissure is the fall of gauge pressure and large quantity of flow of grout. In such a case, the pressure is to be reduced to such an extent that the flow rate remains the same.

It is emphasized that the success of grouting regarding intake of sufficient quantity of grout depends upon the effective grouting programme with respect to rock attitudes. Some rocks with very low permeability may not take grout at all. Thus, prior estimation of permeability is an indicator of possible grout intake of the rock. In soft rocks, grouting is always done under low pressure. Shale is porous but not permeable and it is a natural barrier against leakage of water. Use of high pressure produces fissure in soft impermeable rocks such as shale. In Kopili project, Meghalaya, there was a high intake of grout in impervious shale and not the underlying limestone bed which was cavernous and fractured. This unusual situation is caused due to grouting under high pressure that caused fracturing of shale and thus more grout intake. In hard and fractured rocks, the frequency and apertures of fractures are the principal avenues for grout intake. The permeability of such rocks is dependent on fracture apertures. The cement–water grout (having cement particles of size about 100 microns) cannot penetrate if the pore or fissure spaces are less than 100 microns. Observation has revealed that a fractured rock starts with thin cement grout, but takes thicker grouts with increasing permeability imposed by the use of progressively higher pressures. In other words, there is more cement penetration with increasing permeability. The Lugeon values of pre-grouting permeability tests bear the relation shown in Table 13.2 with the proportion of cement in water–cement grout as advocated by Houlsby (Attewell and Farmer 1976) from a *rule of thumb* decision.

Table 13.2 Relation between permeability and water–cement ratio

Permeability (Lugeons)	Water–cement ratio
10	3:1
10–30	2:1
30–60	1:1
60	0.8:1

There is, however, no strict correlation as to the grout intake and the effective pressure. During pressure grouting in hard rock by application of increasing pressures, the apertures are enlarged to some extent allowing more intake of grout.

13.9 TREATMENT OF CAVITY BY GROUTING

In an engineering project, cavities may occur below the dam foundation or in the reservoir area. These are generally formed in limestone by solution effect or in conglomerate created by removal of loose bonding of cementing materials. Some old quarries or mines also create cavities by collapsing. It is very difficult to grout properly under such situation as it is covered by thick overburden (collapsed) materials. Hidden below alluvium or located in soluble rocks, identification of cavities and their filling is a challenging job. An open joint or voids in rock can be successfully grouted using cement grout, but if there is a cavity filled with clay, grouting may not be effective. In addition, a single hole may be unable to trace the cavity. A detailed exploratory work is necessary to locate the cavity and prove its dimension before grouting is undertaken. Usually, geophysical search, tracer study, and drilling with water permeability tests provide clues to the presence of subsurface cavities and solution channels.

The treatment of cavities depends on the dimensions. If the cavities are of considerable size, they are first packed in by ordinary fill material. A sand mixture is usually used with cement grout to seal the cavity. *Intermittent grouting*, which is a process of grouting at intervals of several hours, is used in grouting of cavities. Before the cavity filling operation is taken up, the maximum amount of grout to be injected should be predetermined based on the knowledge of the cavity dimensions indicated from exploratory data. The cavities are first grouted or packed by sand mixture (the ratio of sand to cement being 2:1 or thicker), the sand being the filler. They are then grouted by cement–bentonite mixture. Bentonite in the grout mix reduces the permeability of the rock and thereby contains or reduces seepage. The grout should be rather thick so that it can seal cavities and provide bondage to varied size grains in the rock. If the cavity is big, the thick grout can fill it to stabilize and strengthen the rock. Refusal of grout intake suggests that grout filling has been effective.

13.10 EFFICACY OF GROUTING

The following four factors are considered to examine the efficacy of grouting:

(i) Location and direction of the grout holes
(ii) Depth of penetration
(iii) Applied pressure
(iv) Control of grout consistency

The geological set-up of the area determines the locations where grouting operation is to be taken up. The exploratory work done at the planning stage helps to reveal the subsurface geology, and accordingly, the nature of grouting required and the depth to which grouting is to be extended are planned. Vertical grout holes are always preferred for ease of drilling and grout injection. These holes are, however, effective only where the rock strata are of uniform nature and structural planes including bedding are nearly horizontal. If the area for grouting consists of vertical or steeply dipping bedded rock or weak zones, the grout holes are to be inclined and so oriented that they intersect nearly normal to the bedding or foliation plane or the weak zones. The pressure to be applied should be such that it does not damage the rock condition. In practice, initially low pressure is applied and thin grout is used but gradually the pressure is increased and thicker grout used. However, thick grout should be used from the beginning if the subsurface region consists of large openings in structurally weak rocks or those created by solution effect.

The dimension of open spaces in rocks determines the grout consumption or efficacy of grouting. In the case of use of cement slurry as grout in finely fissured rock, the size of the cement particles should be less than the openings of the rocks for their settlement into the gaps or openings. Openings greater than the capillary size permit free flow of grout water, but obstruct the entrance of solid suspensions of the grout. It is also necessary to see that the openings are interconnected to allow flow of the grout under pressure. In the case of the presence of sufficiently open spaces, the cement slurry can travel a fair distance through them and becomes effective in sealing the open spaces. It may also be remembered that excessive grout intake through a hole may sometimes create a false impression. The grout may pass on to a different place travelling a long distance through crevices in the rock.

After the grouting work is completed at a site, the efficacy of grouting is determined through some test holes drilled for rock coring and later conducting permeability test through them. Appearance of stringers or threads of cement in the cores is indicative of the penetration of grout. A comparison of the pre-grouting and post-grouting permeability test results help to ascertain the efficacy of grouting. The relatively low water loss in post-grouting permeability test indicates that grouting has been done effectively. In the Kadana dam in Gujarat consisting of quartzite, curtain grouting was undertaken starting with very thin grout and gradually going to thicker grout with cement–water ratio from 1:20 to 1:1 using pressures progressively increasing from 3.5 kg/cm^2 to 7.0 kg/cm^2 (Table 13.1). The water permeability tests in the pre-grouting and post-grouting operations showed reduction in average permeability from 10 Lugeons to 3.5 Lugeons, confirming the effectiveness of grouting.

- Grouting is an effective method of treatment of soil and rock by injecting a mixture of cement and water (called grout) under pressure. The grout after penetrating the soil pores and fracture openings of rock gets hardened and, in the process, seals the voids in the soil or rock arresting leakage and enhancing its strength.
- Several types of grouting such as curtain grouting, consolidation grouting, blanket grouting, and special-purpose grouting are adopted for the treatment of the weak zones in the soil or rocks in project sites.
- Curtain grouting is taken up at the extreme upstream part of a dam foundation to contain seepage and avoid uplift pressure. Consolidation grouting is undertaken at shallow depths to increase the bearing capacity of the unconsolidated material and fissured rocks.
- Blanket grouting is done in specific stretches of a project site such as parts of a reservoir to reduce permeability and seal the leakage path. Special purpose grouting is adopted in structurally disturbed rocks such as a fault zone.
- The ingredients of a grout are basically water and cement in varied proportions such as 1:1 (thick grout) and 5:1 (thin grout). In the preparation of a grout, sometimes some specific admixture is used with cement to obtain early setting or to retard the setting time of grout. Inert substances such as sand, rock powder, fly ash, and diatoms are at times used

with the grout mix to reduce cement consumption and economize the treatment process.
- Bentonite, a swelling type of clay, is mixed with cement grout to treat caving of boreholes and cavernous rocks along reservoir rims. In special cases, a chemical grout such as epoxy resin or polyester resin is used for grouting foundations containing pockets of loose sand and silt.
- The equipment used in grouting operations consists of a drilling unit with pump and compressor that can drill grout holes. A perforated pipe fitted with rubber packers is inserted at the two ends of the grout hole for injecting grout to the test section. Two gauges are fixed with the grout hole, one for the measurement of the applied pressure and the other for monitoring grout consumption.
- The nature of rock and geological structures are the guiding factors in deciding the grouting patterns that include fixing rows, spacing, depths, inclination, and direction of grout holes.
- The pressure to be applied in a grouting operation is dependent on the permeability, hardness, and strength of the rock or soil mass being treated by grouting. In general, pressures used in grouting foundation rocks of dam and reservoir projects are restricted to 1.5 to 2 times the hydraulic head.
- The efficacy of grout is estimated from the comparison of permeability value measured prior to the start of grouting and that measured after the completion of the grouting operation.

Multiple Choice Questions

Choose the correct answer from the choices given:

1. Grout is:
   (a) a mixture of water and rock powder
   (b) a clayey admixture of soil and water
   (c) cement and water mixed in different proportions
2. The formation amenable to grouting is:
   (a) consolidated strata
   (b) unconsolidated strata
   (c) hard and massive rock
3. The rocky foundation that needs pressure grouting is:
   (a) shale and clay stone
   (b) gneiss with several sets of open joints
   (c) sandstone with siliceous matrix

4. The A-holes in a dam foundation grouting are:
   (a) shallow holes of low pressure
   (b) inclined holes from heel
   (c) high-pressure deep holes
5. The B-holes in a dam foundation grouting are:
   (a) shallow holes of low pressure
   (b) inclined holes from heel
   (c) high-pressure deep holes
6. The C-holes in a dam foundation grating are:
   (a) shallow holes of low pressure
   (b) inclined holes from heel
   (c) high-pressure deep holes
7. To treat a vertical fault trending east–west, the orientation of the grout holes should be:
   (a) a deep vertical hole
   (b) an angle hole (60°) directed towards east or west

(c) an angle hole (60°) directed towards south or north

8. In a 60 m-high concrete dam, the depth of the first row of grout holes in the closure pattern of grouting having no test data on permeability of rocks will be:
   (a) 25 m
   (b) 35 m
   (c) 45 m

9. An example of chemical grout is:
   (a) cement–water mix
   (b) bentonite water
   (c) silicate gel

10. In a blanket grouting of a dam site, the common practice is to restrict the depth of grouting to:
    (a) 6–10 m
    (b) 10–20 m
    (c) 15 m

11. In a permeability test of foundation rocks, the pressure could be raised up to 25 Lugeons. The water: cement ratio of grout for use in foundation grouting to start with should be:
    (a) 1:1
    (b) 1:2
    (c) 2:1

12. To find the efficacy of grouting, it was observed that permeability before grouting was 10 Lugeons. After completion of grouting under different pressures, it was found that permeability was 2 Lugeons but no further pressure could be raised; hence the grouting was:
    (a) effective
    (b) non-effective
    (c) inconclusive

Review Questions

1. What is grouting? Discuss the influence of rock types, rock structures, and nature of materials in grouting. State how solution cavities and groundwater may cause problems in grouting.

2. Give a short account of the different types of grouting with respect to pattern, spacing, and depth of grout holes.

3. What are the different types of equipment used in grouting? Explain the approach of their operation. State the method of using packers during grouting operation.

4. State the basic ingredients of grout and admixture. What is chemical grout and under what site condition is it used?

5. Discuss the methods of grouting of foundations of different types of dams and tunnel rocks.

6. What type of pattern is followed in foundation grouting? How does the geological condition of a site guide in taking decisions on grouting pattern? What will be the pattern of grout holes for rocks with vertical open joints?

7. What is the general rule of applying pressure while undertaking grouting operation in a project site? Elaborate your answer with different geological and structural conditions of the grouting site. What is the general rule of water–cement ratio in a grout mix for using in fractured rocks having permeability varying between 10 and 30 Lugeons?

8. Explain the method of cavity grouting. At what stage of grouting operation in a cavity is the success of the grouting operation confirmed?

9. Give a short account of the different factors considered to find the efficacy of grouting. After the grouting operation is completed in a site, what is the approach of confirming that the site has taken sufficient grout?

10. Write short notes on accelerator, retarder, filler, epoxy resin, silicate gel, and intermittent grouting.

Answers to Multiple Choice Questions

1. (c)	2. (b)	3. (b)	4. (c)	5. (a)	6. (b)	7. (b)	8. (b)	9. (c)
10. (a)	11. (c)	12. (a)						

14 Dams and Spillways

●　●　●　●　●　●　●　●　　**LEARNING OBJECTIVES**　●　●　●　●　●　●　●　●

After studying this chapter, the reader will be familiar with the following:

- Different types of dams and their utilities and functions
- Different types of spillways and their workings
- Role of topography and geology in the design of a dam
- Adverse effect of fault on dam foundation and its treatment
- Causative factors of dam disaster with examples
- Selection of a dam site and the investigation methods involved
- Search for construction materials for different types of dam
- River diversion and construction approaches for dams

14.1 INTRODUCTION

This chapter deals with the different aspects of dam construction, their utilities and functions, and the investigation approach during site selection and construction stages as well as during post-construction stage in some cases. It provides an elaborate description on the sources of construction materials and selection process of a dam site, which involves the study of the project area with aerial photography, remote sensing, and geological mapping. The chapter also explains the methods of geophysical and drilling exploration of the site for evaluation of subsurface data, especially to meet design. It further highlights the adverse effects of faults, joints, and other weak geological features in a dam foundation and suggests their remedial measures. In addition, it elaborates the causative factors of dam failures with examples.

14.2 TERMINOLOGY AND BASIC ASPECTS OF DAM CONSTRUCTION

A dam is a water-retention structure suitably built across a river valley to block the flow of river water and form an artificial lake in its upstream part called *reservoir*. Construction of a dam covers the river valley or *channel section* and parts of the two sides of the valley called *abutments*. To the engineers, abutments are the parts of the dam that rest on side slopes. The top of the dam is known as its *crest* and the difference in levels between the top reservoir surface and the crest measures the *freeboard* of the dam. The upstream end of the dam where it meets the foundation is the *heel* and the downstream part of its contact with base is the *toe* of the dam.

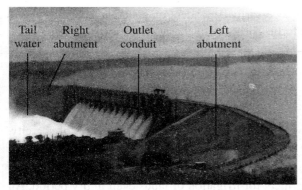

Fig. 14.1 Concrete gravity dam showing rocky abutments and tailwater discharge

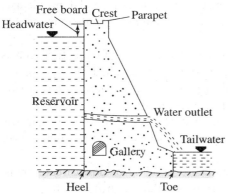

Fig. 14.2 Cross section of a concrete gravity dam illustrating different parts

Figure 14.1 shows a concrete gravity dam built across a narrow river valley with rocky abutments and tailwater discharge through the conduit outlets

Blocking of the river flow causes difference in levels of water surfaces near the dam structure. The high reach of water flow at the upstream part of the dam is called the *headwater* and water at the downstream base of the dam that results from backwater effect and discharge through a spillway or conduit outlet is called *tailwater*. A long open space called *gallery* is kept inside the dam for movement of men and machineries for taking up grouting and drainage works and also installation of instruments to check the function of the dam (Fig. 14.2).

14.2.1 Acting Forces and Design Principles of Dam Safety

When a dam is constructed, the impounded water exerts pressure on the upstream wall of the dam tending to push it downstream whereas the gravity tends to resist it. The larger the dam with respect to its height, the greater will be the pressure from reservoir water. The gravitational force of a dam depends on the density of the dam body. In addition, the force generated by an earthquake acts adversely on a dam. Seeping of water through the foundation causes up-thrust or *uplift* at the base or from the dam body itself. In fact, many types of forces act on the dam such as water thrust, uplift, load of sediment deposit, dead weight of dam, wind pressure, and seismic force. These forces concentrate at the dam foundation and affect the stability of the structure. The overall result is the development of deformation (strain) in dam foundation and possible displacement (shear) of the structure. The dam is so designed that it possesses sufficient strength to withstand all the forces acting on it.

The forces to be resisted by a gravity dam fall into two categories. First, forces such as weight of the dam and water pressure are calculated from the specific gravities or densities of the materials and properties of pore pressure. Second, forces such as earthquakes, silt, and load can be predicted on the basis of assumptions of varying reliabilities. For consideration of stability, it is assumed that the dam is composed of individual transverse vertical elements each of which carries its load to the foundation without transfer of load from or to its adjacent elements and the vertical stress varies linearly. The *uplift pressure* distribution in the body of the dam is assumed to be equal to water head at the upstream face and zero at the downstream face. It is also assumed that earthquakes have no effect on uplift forces.

For ensuring the stability of the structure, the principal design approach is such that the foundation must have enough strength or load-bearing capacity to resist the hydrodynamic and seepage forces. The vertical force (the load of the dam) must be high enough to avoid the adverse effects of the horizontal (shear) force. The denser the material of the dam, the more will be the capacity to shear resistance. The limiting factor for a concrete dam is that large cracks should not appear in concrete (which is not necessarily dependent on concrete technology) due to overloading, irregular settlement, and so forth. These are the basic principles of design for the safety of the structure. In practice, however, several factors are considered in the design of a dam.

14.2.2 Utilities of Dams

Harnessing the water resources of a country is the main aim of constructing a dam. A dam is the most essential infrastructure for water resource management. Dams ensure adequate water supply by storing it during the time of surplus and supplying it in the days of scarcity. Construction of a dam ensures availability of water in the reservoir throughout the year, which serves the following purposes:

(i) The stored water is supplied to the farmers for irrigation of lands even in dry period. An unfertile land or arid region is irrigated to produce crops by diverting the reservoir water by a system of canals.

(ii) City and town people are supplied water from the reservoir for drinking and domestic use. The supply remains uninterrupted even during the summer months when there is scarcity of water.

(iii) The artificial lake created by damming of the river is used as a centre of recreation such as for boating and fishing activities. It facilitates growth of tourism and helps navigation.

(iv) Water is needed in industrial units associated with various production works. The demand for water in industries is mostly met by water from the reservoir.

(vi) Many of the dams with large reservoir capacity release water for generation of hydroelectric power. People in both rural and urban areas and many industries benefit from the power generated.

(v) Heavy rains in the upstream parts of many rivers cause flooding of downstream areas. Building of dam stores the excess water in the reservoir and thus controls floods.

In spite of plenty of water in the innumerable river systems of India, the country suffers from draught or floods in some place or the other. About 80 per cent of the annual rainfall of India occurs during the monsoon months when the water can be utilized from the run of the river. However, the rainfall is not evenly distributed, resulting in floods in some parts and drought in some other parts of the country. Hence, if the water is stored in reservoirs by building dams, this will reduce flood havoc and serve the need of water for drinking, irrigation, power generation, and industry throughout the year. Several dams in India are built to serve such multiple purposes, though flood control is the primary purpose of many of them. The Damodar Valley Scheme is the first multi-purpose river valley project in India that mitigates the floods of the river Damodar, irrigates vast areas of Jharkhand and West Bengal, and generates electricity for use by these two states.

14.3 TYPES OF DAMS AND THEIR FUNCTIONS

Dams are broadly classified into the following four groups on the basis of the types of material used in their construction:

(i) Concrete dam
(ii) Masonry dam
(iii) Rock-fill dam
(iv) Earth dam

Table 14.1 provides a list of the different types of dams in India that are more than 100 m in height.

Table 14.1 Different types of dams in India that are more than 100 m high

State	Name of dam	Year of completion	Type of dam*	Height (m)	Length (m)	Design purpose**
Andhra Pradesh	Srisailam	1984	PGM	145	512	H
	Nagarjuna Sagar	1960	PGM	125	4865	I & H
Himachal Pradesh	Pong	1974	TE	133	1956	I & H
Punjab	Bhakra	1962	PG	226	518	I & H
Himachal Pradesh	Chamera	1994	TE	141	240	H
Jammu and Kashmir	Salal	1986	ER	116	630	I & H
Karnataka	Supa	1987	TE	101	322	–
Kerala	Idukki	1974	PG (A)	169	366	H
	Cheruthoni	1976	PG	138	650	H
	Kakki	1966	PG	114	336	H
Maharashtra	Koyna	1964	TE & PG	103	805	H
Tamil Nadu	Sholayar	1971	PG	105	1244	–
Uttar Pradesh	Lakhwar	NA	PG	204	452	I
	Ramganga	1974	TE	128	715	I & H

*TE, earth dam; ER, rock-fill dam; PG, concrete; PG (A), concrete arch dam; and PGM, masonry cum concrete
**H, hydroelectric power generation and I, irrigation

14.3.1 Concrete Dams

Depending on their bearing strength and stability, *concrete dams* are subdivided into three categories, namely gravity, arch, and buttress dams.

Gravity dams have straight or slightly curved axes and are generally triangular or trapezoidal in cross section with bed widths about two-thirds their heights. The downstream face of these dams can be straight or curved. The thick concrete body of a gravity dam provides the necessary weight to withstand the forces due to water thrust and uplift pressure. A narrow valley with sound rock is the ideal place for constructing a concrete gravity dam (Figs 14.1 and 14.3), but it can be successfully built in soft sedimentary strata also after strengthening the foundation.

The 225.5 m-high Bhakra dam in Punjab, which is one of the highest concrete gravity dams in the world, has been built on shale–sandstone rocks of Siwalik formation (Fig. 14.4). Figure 14.3 shows a concrete gravity dam under construction across the river Umiam in Meghalaya having hard quartzite and arenaceous phyllite in the abutments and base.

Arch dams are arched or curved in their configuration and hence the name. The 169m-high Idukki dam in Kerala constructed on massive granite gneiss of a canyon site is an example of large arch gravity dam (Fig. 14.5). In fact, arch dams are curvilinear in plans with the convex

Fig. 14.3 Concrete gravity dam under construction across the river Umiam, Meghalaya

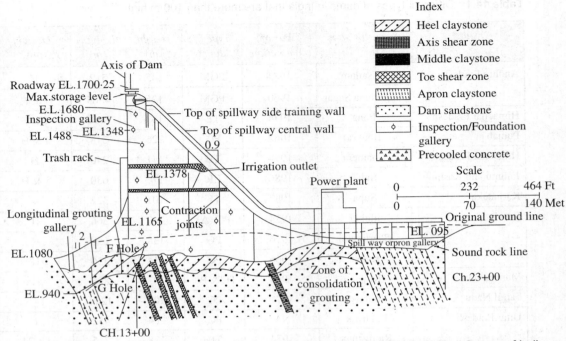

Fig. 14.4 Different parts and foundation geology of Bhakra dam in Punjab (Geological Survey of India, Misc. Pub. No. 29, SL. PL. 2)

Fig. 14.5 Idukki arch gravity dam in Kerala

side facing headwater, see Fig. 14.6 (a). In cross section (vertical section), these dams are relatively thin and curved in shape Fig. 14.6 (b). In an arch dam, a part of the thrust due to reservoir water is distributed to the abutment rock by the arch action. The thickness and curvature of the body of an arch dam are so designed that more than half of the acting load is transferred to the abutment. As such, the abutment and foundation rocks should have sufficient strength to bear the load. In case more than half the dam load is kept on the foundation, it is to be called an *arched gravity dam.* Arch dams are generally constructed in narrow valleys or canyons such that the heights are generally more than the axial lengths, see Fig. 14.6 (c).

Buttress dams are massive, thin, or arched slabs supported by vertical walls or buttresses. The slabs having upstream slopes bear the water thrust while the buttresses acting normal to the planes of the slabs take the load of the headwater and transfer it to the foundation. Steep beams or girders spaced between the buttresses prevent their bending. The structural parts of the buttress dams are made of concrete or reinforced concrete. Depending upon the shape, the buttress dams may be of flat slab type as shown in Figs 14.7 (a) and (c) or arch type shown in Figs 14.7 (b) and (d). Buttress dams require firm rock in the river bed where they buttress. The interior part being hollow, these dams consume less concrete than other types of dams; hence, their construction cost is comparatively less.

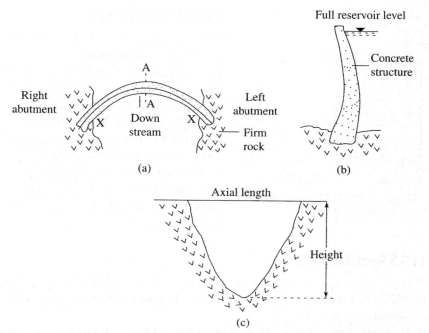

Fig. 14.6 Arch gravity dam: (a) plan view; (b) cross section across the arch dam (A–A); and (c) cross section along the axis (X–X) of the arch dam

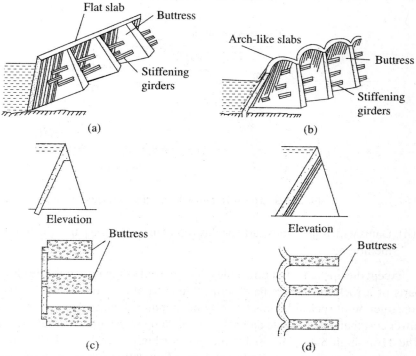

Fig. 14.7 Buttress dam: (a) three-dimensional view of flat slab type dam; (b) three-dimensional view of arch type dam; (c) elevation and section of flat slab type dam; and (d) elevation and section of arch type dam

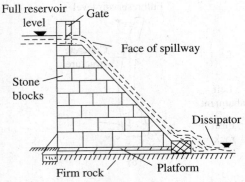

Fig. 14.8 Masonry dam made of cut blocks of sound rock having gated spillway

14.3.2 Masonry Dams

Masonry dams (Fig. 14.8) are made of big undressed blocks of rocks including river boulders bound together by concrete. The rocks selected for construction of masonry dams are non-porous and the binding among the rock fragments or boulders creates a watertight body. As such, the entire body of the masonry dam acts as an impermeable barrier against leakage of reservoir water. Only the cemented joint portions among the blocks must be watertight to avoid seepage. Many old dams in India are stone masonry structures. The 139 m-high Srisailam dam in Andhra Pradesh is an example of masonry construction.

14.3.3 Rock-fill Dams

Rock-fill dams are made up of an admixture of ground spoils, river deposits (e.g., gravels, pebbles, and boulders), and crushed rocks. They are trapezoidal in cross section with side slope commonly in excess of 1:1.

Depending upon the nature of impermeable membrane provided in the rock-fill dams, they are grouped into the following two types:

  (i) Dams with concrete wall in the upstream face, which creates the impermeable barrier against leakage (Fig. 14.9)

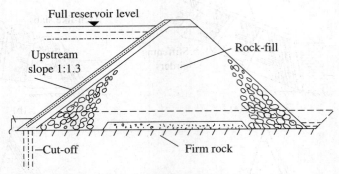

Fig. 14.9 Rock-fill dam with concreted upstream slope

  (ii) Dams with impervious earth or clay (core) in the central part, which acts as the impermeable membrane

Except the upstream impermeable concrete wall or the impermeable clay core, the remaining parts of a rock-fill dam remain permeable even after compression. In the past two decades, construction of rock-fill dams has become popular as large dams of this type can be built at a lower cost than the concrete dams. Rock-fill dam is specially constructed in high seismic zones. The 116m-high Salal dam in Jammu and Kashmir—see Fig. 14.10, showing the removal of loose river deposits by power shovel during foundation preparation—is an example of a rock-fill dam in a high seismic zone.

Fig. 14.10 Foundation preparation of rock-fill dam of the Salal project of Jammu and Kashmir

14.3.4 Earth Dams

Earth dams are trapezoidal in cross section with gentle side slopes (Fig. 14.11). An earth dam is generally constructed in a broad valley where it is very expensive to build a concrete or masonry structure. The earth dam also does not require rocky foundation and it can be founded on firm soil. However, availability of adequate quantity of homogeneous earth of both pervious and semi-pervious types is to be assured for constructing an earth dam.

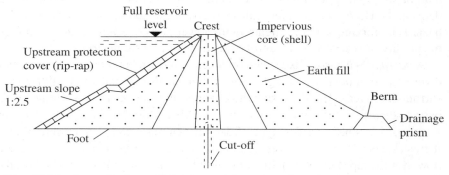

Fig. 14.11 An earth dam with impermeable core and gentle slopes

Fig. 14.12 An earth dam immediately after compaction of earth and stone pitching

Depending upon the material and competence of the foundation, the slopes provided in earth dams commonly vary from 1:1.5 to 1:3 (vertical: horizontal). The material excavated from the *borrow pits* is *hauled* to the dam site by earth movers and compressed into graded layers as per design (Fig. 14.12). To attain complete water tightness, a clay core that acts as the impermeable membrane is provided. Most dams in India are earth dams. Earth of suitable quality is available in plenty in most parts of India. The cost of construction of an earth dam including quarrying, placement, and compression is the least. A dam made of earth and fill material is known as an *embankment dam.*

14.3.5 Different Types of Dams, Barrages, and Weirs

A dam is called a *composite dam* if a part of it is made of earth or rock-fill and the other parts are built of concrete or masonry works. In a composite dam, the spillway is constructed in the concrete or masonry dam portion. The 4600 m-long Hirakud dam of Orissa is a composite type dam composed of earth for a stretch of 3650 m and the rest is of concrete and masonry construction.

A dam constructed of excavated natural materials or industrial wastes is an *embankment dam*. If more than 50 per cent of the total volume of an embankment dam is made of compacted fine-grained materials obtained from borrow areas, it is known as earth-filled or earth dam. In fact, an embankment dam is a massive water barrier having a semi-permanent water proof natural covering for its surface and a dense waterproof core.

A *tailing dam* is a special type of dam (concrete or earth filled) that retains the tailing slurry of mine waste in the pond created by the dam. *Tailings* are the materials leftover after the process of separating the valuable fraction from the worthless fraction after mining of an ore. Water is spilled through the spillway of a tailing dam in a conventional way.

Burning of coal in coal-based power plants produces a huge stock pile of fly ash, which is disposed in a pond that forms a huge deposit of ash pond. The wet disposal of fly ash involves constructing a large pond and filling it with the ash slurry. Around the ash pond, an artificial barrier is created in the form of an embankment dam and the structure is known as *fly ash pond dam* or simply *ash pond dam*. It is a zoned construction with clay core. There are many such dams in the US for which safety rules have been enacted so that the material does not cause any hazards by mixing with the stream in the lower parts. Design of such dam specially takes care of stability of structure and control of seepage of fly ash materials.

A *barrage* is a barrier like a dam built across a river, but it has no fixed reservoir. The barrage diverts a part of the river through a network of canals especially to facilitate irrigation of land of surrounding areas. The Farakka barrage across the Ganges in West Bengal is an example. This barrage diverts water from the Ganges to the river Hooghly through feeder canals.

A *weir* is a small overflow type of dam that causes a complete closure of the stream channel. It creates a barrier across a river creating a pool of water behind the structure but allows water to flow over the top (Fig. 14.13). In its simplest form, a weir consists of a bulkhead of timber, metal, or concrete with an opening of fixed dimension generally cut like a V-notch at the top edge. A weir is commonly constructed to alter the flow regime of the river with a view to prevent flood and help navigation. It also allows measurement of water discharge of small- and medium-size rivers.

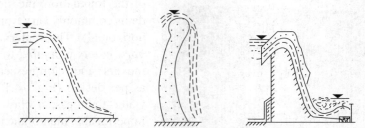

Fig. 14.13   Different types of weirs allowing free flow of water from top

14.4 SPILLWAYS AND GATES

S*pillways* that discharge water from upstream to downstream part are the most essential structures required for proper functioning of dams (Fig. 14.14). It is a means of releasing floodwater without causing any damage to the dam. The spillway may be an integral part of the dam constructed at the centre or at the side of the dam axis. In some dam projects, especially in earth dams, the spillway is located in a separate place outside the body of the dam.

A spillway has three parts. The *head structure* functions as the water inlet being provided with guide walls and bulkhead, the *discharge structure* carries the flow from its upper reaches to lower parts, and the *terminal structure* acts as an outlet or discharge to the tailwater. The spillway can be free flowing or it may be under controlled flow by a gated structure. The five major types of spillways are described in the following subsections.

14.4.1 Normal Spillway

Figure 14.14 shows a normal spillway. It is a concrete structure in the form of a channel at the upper part followed by a steep *chute*. The chute is rectangular or trapezoidal in shape (cross section) with training walls through which the reservoir water flows to the downstream valley (Fig. 14.15). In a normal spillway, the flow of water is generally regulated by a gate at the crest of the dam.

Fig. 14.14 Spillway with chute and discharge of water to the downstream part of North Koel concrete dam (see full view of dam in Fig. 14.24)

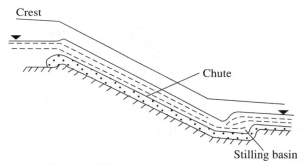

Fig. 14.15 Sectional view of normal spillway showing chute and stilling basin

14.4.2 Pipe Spillway

In pipe spillway (Fig. 14.16), the discharge structure is made of a pipe passing through the dam body. The pipe is laid for some lengths of the dam in a dig and cover manner for the low headwater projects.

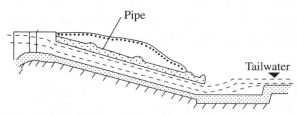

Fig. 14.16 Sectional view of pipe spillway showing flow of tailwater

14.4.3 Tunnel Spillway

In a tunnel spillway (Fig. 14.17), a tunnel punctures the upper end working as intake of discharge and the lower end (almost horizontal) discharges the water. It is provided with a gate chamber for free flow or flow under pressure.

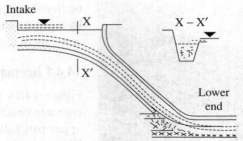

Fig. 14.17 Tunnel spillway from upper end into lower end with a sectional view of intake (X–X')

14.3.4 Glory Hole (or Shaft) Spillway

The glory spillway (Fig. 14.18) is in the form of a vertical or steeply inclined shaft for water to flow freely or under pressure. Its head structure may be gated or ungated.

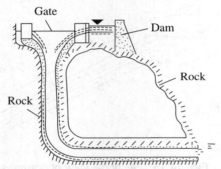

Fig. 14.18 Glory hole spillway with gate structure

14.4.5 Side Channel Spillway

The side channel spillway structure (Fig. 14.19) is located towards the left or right abutment of the dam. The upper and lower portions of the channel are joined by a chute for regular flow of water from the reservoir to meet the main river at the downstream part. The portion of the chute consists of an access channel, a head structure in the form of a trough. The head structure is a weir generally broad crested. In an ungated chute, the weir structure is designed with a curvilinear crest to ensure specified discharge.

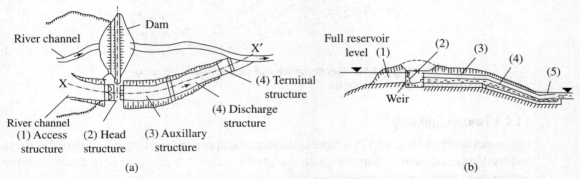

Fig. 14.19 Channel spillway: (a) plan of a side channel spillway; and (b) cross section along x–x

A spillway structure terminates in a *stilling basin* where the intensity of the discharge is reduced by means of a design device. The end of the spillway may be in a ledge to provide overflow condition or a cantilever extension. In a concrete dam, the end device may be a *ski jump* that projects the jet at some distance away from the toe of the dam. The spillway of a dam is constructed in a separate place when the dam body does not possess sufficient space for the discharge or if the downstream bedrock is of soft nature threatening deep scouring at the toe of the dam.

In an earth dam of high headwater flow, it is not possible to construct a concrete structure such as a spillway on the body of the dam for the safety of the structure. A separate spillway is then constructed for the earth dam at a place having a favourable topography and sound rock suitable for founding a concrete structure. For a small earth dam with low water head, however, a conduit or pipe type spillway can be made through the body of the dam. The prime criterion in the design of a spillway is that it should safely discharge the floodwater of the river without any damage to the dam. Inadequate spillway capacity may cause abnormal rise in the upstream floodwater level leading to overtopping of the non-overflow section and damage to the structure.

14.4.6 Outlet Works

A portion of the reservoir water may be diverted through a tunnel or conduit for the purpose of supply to irrigation canals, water supply schemes, or hydroelectric plants for power generation. This water diversion structure is termed the *outlet*. The outlet may be in the form of a tunnel through the abutments to carry water for the distribution system or it may be a conduit made of pipes through the body of the dam (Fig. 14.1). The water enters through the conduit inlets located at the submerged top or bottom orifices at the reservoir end installed with service gates and discharged to the downstream parts (Fig. 14.20). The concrete dam of Bhakra has such a conduit type of outlet for distribution of water to irrigation canals to various users in the downstream areas (Fig. 14.4).

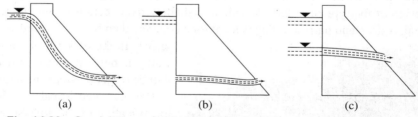

(a)   (b)   (c)

Fig. 14.20   Conduit type of outlets in the body of the dam with entry of water through a gated structure: (a) near full reservoir level; (b) near dead storage level; and (c) at the other part of the upstream end

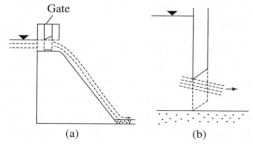

Gate

(a)   (b)

Fig. 14.21   Gates of a dam: (a) controlled type spillway with gate (plank) at the crest; and (b) plain gate

14.4.7 Gates of Different Types and Their Functions

Gates are mobile structures that facilitate closing or opening passes for water to flow down. They can be opened completely or partially to allow controlled discharge. A *controlled* type spillway is provided with gates in the dam crest, which carries pressures to the piers and abutments, see Fig. 14.21(a). The gates attached to the spillways are mainly of two types, namely plain gate and radial gate. They may be automatic or manual. In the automatic type, the gates open automatically by a float device when water rises to a certain level.

Plain gates   These are commonly known as planks and work by translation movement along slots in piers across the spillway crest. It may be of lift or slide type, as shown in Fig. 14.21(b).

Radial gates   *These* operate by rotary turning on hinges, see Fig. 14.22.

In addition to these service types of gates, an emergency gate is also provided for use if the service gate fails. A spillway may also be of *uncontrolled* type without any gate. In such cases, high discharge may harm the dam crest or scour the downstream parts. (Fig. 14.22)

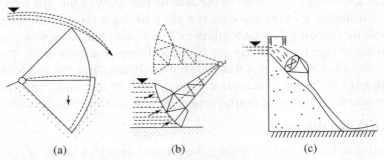

(a)   (b)   (c)

Fig. 14.22   Radial gate: (a) radial gate in normal position;
(b) operation of a radial gate; and (c) spillway with radial gate

14.5 INFLUENCE OF GEOMORPHOLOGY AND GEOLOGY IN THE DESIGN OF A DAM

A dam is designed following the accepted safety norms and keeping the costs as minimal as possible. Configuration of the river valley at the dam site, nature of foundation rocks or materials, depth to bedrock, and availability of construction materials are essential considerations in the design of the type of a dam. A wide and shallow river section is suitable for an embankment dam. A deep and narrow valley is ideal for a concrete dam having a chord-to-height ratio of 2:1 or 3:1. In the rock-fill or earth dams, the valley walls in both upstream and downstream sides will be of gentle slope, generally less than 1:1.

Precise geological evaluation including identification of the adverse geological features of the dam site is necessary to help the engineer design suitable type of the dam after adopting measures to rectify the defects of the dam foundation. The following points are important when designing a dam:

Fig. 14.23   A dam site on the North Koel river comprising fresh and massive granite gneiss before construction of a high concrete dam

(i) The foundation must be very stable so that it does not cause any settlement. Massive rocks free from faults, shear zones, joints, weathering, and decomposition available at the river bed or at a shallow depth are favourable for the foundation of a concrete dam. Figure 14.23 shows a dam site on the river North Koel in Bihar comprising fresh and massive granite gneiss before construction of a high concrete dam and Fig. 14.24 shows the constructed concrete dam.

Fig. 14.24 Constructed concrete dam at the site on the river

(ii) If the site consists of a thick mantle of overburden at the river bed or at the abutments, the site is not considered suitable for a concrete or masonry dam as the firm bedrock is available at a very deep level. If the overburden material of a dam site is the in situ disintegrated product of igneous or metamorphic rocks and the deposit is impervious in nature, an earth dam may be constructed over the deposit after excavating the loose river deposit.

(iii) It is, however, necessary to conduct in-situ strength and permeability test of the deposit for designing the suitable structure. Similarly, if the dam site rock is weathered or decomposed to a great extent, the site is generally not considered for construction of concrete or masonry dam. It is necessary to carry out exploratory work at the site to find the depth of fresh rock for the purpose of economic design of the dam. Whether a dam site has any buried channel below the proposed foundation is also to be explored as this may result in leakage under reservoir condition.

(iv) The nature and disposition of the layered rock of a dam site should be known for planning and designing a structure. Soft rock is not suitable at the foundation of a high dam, especially for concrete and masonry types. Shale and sandstone occurring as alternate beds with low downstream dip may bring distress or failure of a dam. Such site requires appropriate protective measures in the design of the dam against slide of the structure.

(v) Unfavourable geological structures such as joints and shear zones are commonly found in all types of foundation rocks. Fault or large zones of crushing present in the foundation will create problems in construction of a dam. If weak structures such as faults, thrusts, shear zones, and joints are present in a dam site rock, they need to be precisely evaluated for adopting corrective measures in the design of the dam.

(vi) When a sound bedrock occurs in the dam site, either exposed on the river bed or below the river deposit at a shallow depth, any type of dam can be constructed. However, in the case of a concrete dam, sound rock foundation is most essential. Besides, the disposition of bedding, foliation, joints, or like structures with respect to their strike and dip need to be thoroughly studied as low angle down–dip direction of planar rock structure along downstream direction of a dam is considered unfavourable.

(vii) When a dam site consists of a very thick zone of weathered and decomposed rock including thick unconsolidated deposit, it is suitable for earth or rock-fill dam in which the load of the dam is spread over a wide foundation; however, a cut-off (impermeable membrane) is provided by impervious earth or clay that extends down to impermeable bedrock or sufficiently below in natural earth ensuring water tightness.

(ix) Availability of suitable quality stone chips and sand for concrete aggregate is necessary for a concrete dam. The foundation rock must possess sufficient bearing strength, roughly four times the calculated design load due to the dam. Similar to the concrete dam, the foundation of a masonry dam also requires firm rock to withstand the acting forces. However, masonry dams also need large quantities of dimension stones from quarry sites of nearby locality or boulders in river bed.

(xii) Rock-fill dams require large quantities of rock fragments and also earth. Quarry sites for construction of dams should be available within a short distance so that the haulage charge is not prohibitive. Borrow areas for homogeneous and impervious type of earth or clay are necessary for construction of an earth dam. In fact, availability of desired quality and quantity of construction materials and the cost involved in their quarrying and haulage are the primary concerns for selection of type of the dam.

14.6 ADVERSE EFFECTS OF FAULTS IN DAM FOUNDATION AND ITS TREATMENT

The harmful effects of a fault, when present in a dam foundation, are well known to the engineers and geologists. When a rock mass is subjected to fault movement, it creates a zone of shattered and pulverized rock, very often containing certain thickness of clayey product called *gouge* that has very low modulus of elasticity. If a heavy engineering structure such as a concrete dam is constructed on this weak zone created by the fault, a stress concentration will take place in the dam concrete and the weak foundation rock of the fault zone will be deformed. As a safety measure against such an adverse feature, a plug is provided in such a weak zone as detailed in the following subsections.

14.6.1 Treatment of a Fault by Plug and Its Depth Calculation

The plug used to treat the fault zone is a seal generally of concrete mass introduced to the soft zone after removing the soft materials from the fault zone by excavation. On the basis of the study by the US Bureau of Reclamation, the following formula has been framed to compute the safe depth of plug that produces elastic qualities and bearing strength equal to that of a sound rock:

$D = 0.00066BH + 1.5$ for dam height $H > 46$ m

$D = 0.3B + 1.5$ for dam height $H <$ or $= 46$ m

Here H is the height (in metres) of the dam above the general foundation, B is the width (in metres) of the weak zone, and D is the depth (in metres) of excavation of weak zone. In the case of gouge seam, D will be greater than $0.1H$.

This method of fault treatment was devised after studies of faults in the foundations of Shasta dam and Friant dam in the US. In these two dams, the modulus of elasticity of the weak zones was 1/2 to 1/5 times of that of the surrounding sound rocks of the foundation. In the case of a high-gravity dam, the shear friction factor governs the stability of the dam. The foundation area occupied by a fault in a dam will have low shearing resistance, which may lead to deformation or settlement. In order to increase the shearing resistance, the following measures are necessary:

(a) Provide a plug having equal shearing resistance as the sound rock
(b) Widen the base of the dam

In fact, the treatment of a fault zone is carried out considering the effect of deformation due to fault. The effectiveness of the plug and the optimum depth of the plug can be determined only after proper investigation of the stress distribution in foundation rock, which is affected by width, deformability, location, and strike of the faults and the depth of the strengthening plug.

14.6.2 Stresses Along Faults at Different Dispositions

A fault zone lying under a dam is composed of weak gouge materials with low modulus of elasticity as compared to sound rocks and hence more deformable. The pressure of a fault zone induces more stress in a foundation as compared to a normal foundation. A fault zone in a foundation is thus a source of weakness to the foundation and careful studies are warranted to tackle the problem. The following observations are based on a computer study on faults under gravity dams (Biswas 1970).

(a) **Upstream vertical fault** When a fault is located on the upstream of the dam outside the dam base, it does not pose a serious foundation problem. Vertical stresses induced are of low orders and the fault zone does not warrant elaborate treatment measures. However, care should be taken to ensure that the horizontal tensile stress is within the safe limit that is pronounced up to a depth equal to only half the dam height. Therefore, in an upstream vertical fault, only the zone extended downstream of the fault needs treatment.

(b) **Central fault inclined upstream at an angle of 45°** Such an orientation of a fault can be considered as harmless from structural point of view. When the orientation of a fault is such that only a small part of the fault stretch is within the dam base, the stresses induced are similar to a normal foundation. Of course, this hypothesis is true when the fault is inclined to upstream. In such cases, a dam can be built with minor foundation treatment. More treatment may be necessary to control seepage, which is dependent on the permeability of the rock mass.

(c) **Centrally located vertical fault** In this case, foundation stresses induced are more than the stress due to the fault sloping upstream. The fault is entirely inside the dam base indicating compression on the downstream and tension on the upstream sides. Under this circumstance, probe is necessary for the whole foundation mass and proper treatment measures should be adopted.

(d) **Central fault inclined downstream at an angle of 45°** This orientation of fault deserves full attention and may pose serious problems to design engineers. Presence of this type of fault induces larger stresses in the foundation. Increase in stresses is 75 per cent to 100 per cent as compared to a normal foundation. Though horizontal normal stress does not change much, vertical normal stress and principal stresses tend to reach alarming proportions.

(e) **Fault at downstream toe** Contrary to the three cases described in cases (a), (b), and (c), stresses on the upstream side of a fault are more pronounced than those on the downstream side. Therefore, careful foundation treatment is necessary for the foundation under the dam. Tensile stress is also induced on the upstream of the dam, which is smaller than that due to central, vertical, and sloping downstream faults as in cases (c) and (d).

(f) **Effect of depth of plug** In all these cases stated, stress along the fault and also maximum stress in the foundation reduce with increases in depth of plug. When a fault is located upstream of the dam, the stress along the fault increases with increase in depth of plug. However, maximum stress induced in the foundation gets reduced with increase in the plug. The reduction in stress for every 5 m increase in depth of plug is not very appreciable. The current study is for depths of plugs of 10 m, 15 m, 20 m, and 25 m. In case the depth of plug is increased further, appreciable reduction in stresses will result.

From the study, it is concluded that a fault on the upstream of the dam or that sloping upstream from the foundation (cases (a) and (b) poses less problem than a central fault, a fault sloping downstream, or a downstream fault (cases (c), (d), and (e). In these three cases, the fault needs elaborate foundation treatment. As the fault shifts upstream to downstream of the dam, foundation hazard increases, the most acute case being the fault at the downstream toe. In the extreme case, it is desirable to provide a plug of depth 20 per cent to 30 per cent of dam height. This has been testified by the photo-elastic studies of Fumio Ishi, Ragguichi Ho Iida, and Isao Shibata (1967).

14.7 CAUSATIVE FACTORS OF DAM DISASTERS

A majority of the dam failures throughout the world is due to adverse geological conditions of the dam foundation. An analysis of the records of dam disasters suggests that apart from geological causes, the safety of a dam is also endangered by other causes such as faulty spillway design, natural calamities, and use of wrong construction materials.

14.7.1 Geological Causes

Poorly compacted sedimentary rocks or other soft rocks when present in the toe or abutment of a dam may lead to scouring. The bedrock, even with high strength, may be highly fractured. Some foundation rocks may be porous and permeable allowing seepage or leakage of water from the reservoir. If the foundation is not watertight, passage of water or *piping* through it will develop *uplift pressure* and bring about failure of the dam. The 25 m-high Dudhawa dam in Madhya Pradesh developed leakage through a porous sand zone below the foundation rock. After a period of three years, when it almost led to the failure of the dam as a result of uplift pressure, the reservoir was emptied and elaborate treatment was undertaken for its safety. The 61 m-high Malpasset concrete arch dam in France is a classic example of failure due to uplift pressure. The foundation comprising sheared and jointed mica schist was ruptured resulting in a catastrophic flood in 1959 that killed 344 people.

Significant foundation problems arise from the presence of thin beds or laminations in sedimentary rocks such as clayey siltstone, shale, claystone, and silty shale in the foundation. These *soft layered rocks* have low compressive strength and are incapable of withstanding the load of heavy structures and may lead to plastic deformation or shear failure. Smooth surfaces of the shale, siltstone, or sandstone dipping at a low angle accentuate sliding tendencies and the dam may *slide along the dipping plane*. The 26 m-high and 1.6 km-long Tigra concrete dam in Madhya Pradesh, consisting of alternating beds of sandstone and shale with low downstream dip, failed due to sliding after filling the reservoir in 1919.

Porous conglomerate and breccia containing clay or calcareous matrix if present in the foundation undergo disintegration as a result of saturation with reservoir water and in the process lose their strength. This endangers the stability of the dam. With absorption of water, the clay, silt, and calcareous matter in the matrix of these rocks are washed away providing passageways for extensive *leakage*. Similarly, if the foundation consists of faulted rock causing crushing and fracturing of the rock, the storage water may cause large-scale *seepage* through the crushed rocks, washing away the soft materials. This creates voids in the foundation rock and the dam may fail by *settlement*. St. Francis Dam in the US built on soft schistose rock traversed by a fault failed within two years of its construction.

When incompetent rocks such as shale or claystone occur with competent sandstone beds, they may bring about failure of the dam by differential *settlement*. The presence of soft shale bed alternating with hard sandstone beds in the foundation of the 21 m-high Kedar Nala dam in Madhya Pradesh caused breach of the dam after a heavy rainfall within four months of its construction in 1964.

There are several instances of dam failures where cavernous limestone was present in the foundation. Caverns in reservoirs lead to the problem of competency of the reservoir. When water is impounded in the reservoir, the river detritus that generally fills the caverns is washed away giving way to large-scale leakage and even making the reservoir completely dry.

A landslide is associated with the adverse geology of affected areas such as a fault zone. Strong vibrations due to the impact of *slide* create waves in the reservoir water that may pass over the top of the dam and sometimes cause breach of an embankment dam and bring about its failure. A concrete dam may survive such breach but the downstream inhabitants will have to face the devastating effect of water that passes over the dam. In 1963, the Vaiont dam in Italy failed causing the death of 1900 people. This 266 m-high thin arch dam sustained no damage in the main shell or abutments. The catastrophe occurred due to a slide caused by the defective geological features of the reservoir rocks.

Creep and slope failures are regular features in the steep abutment slopes constituted of beds of shale and sandstone. Sensitive clay (containing montmorillonite) in foundation rocks causes swelling and slaking under saturated condition and may cause *settlement* and slide. Large *landslide* or failure of hill slope in the storage basin immediate upstream of a dam may have disastrous consequences. The right abutment of Jaldhaka barrage in West Bengal was affected by a landslide during the 1964 monsoon, and a huge quantity of the slide debris chocked the intake structure of the weir at the toe of the dam. In 1962, after heavy rains, the debris caused by a huge landslide blocked the stream causing floods and total damage of the Leimakhong weir, powerhouse, and other engineering structures in Manipur.

14.7.2 Other Causes

Faulty design of spillway is the common cause of failure of dam by overtopping. If the volume of floodwater exceeds the calculated discharge, the spillway cannot cater to the need and overtopping or overflow takes place. In case of an earth or a rock-fill dam, the water overflowing the crest of the dam will erode the dam threatening damage or failure of the dam. If the spillway is part of a concrete dam, the excess flow will inundate downstream areas. In general, hydrological data of 50 to 100 years is required to predict peak flow in the design of a spillway. However, in practice, due to insufficient data, the engineers may underestimate the peak flow and design the spillway with assumed value. Such design may lead to overtopping of dam. The 6 m-high Kali Sind masonry weir in Rajasthan breached in 1960 due to floodwater overtopping the structure. It caused movement of masonry blocks about 30 m downstream.

Failure of the dam may also take place due to lack of proper drainage arrangement in the *design*. Absence of drainage results in the saturation of the foundation material. The 18 m-high Asti dam in Maharashtra constructed in 1883 and again in 1923 failed by saturation of the foundation and consequent settlement of the structure that slipped in the downstream direction.

Natural calamities such as earthquakes may develop cracks in dams creating avenues for leakage of large volumes of storage water. It may also bring about complete damage of the dam if affected by an earthquake of high intensity. At every dam project of India, proper earthquake factors are incorporated in the design of the dam depending upon the seismic history of the area. The Koyna dam provided with an earthquake factor of 0.05g (vertical acceleration) escaped failure during the earthquake of 1967, but some cracks developed in several structures such as the underground powerhouse, spillway bridge road, and a century-old bridge over the

Koyna River. The earthquake in 1971 caused liquefaction of the fill material in the upstream wall of the San Fernando dam in the US (California) and the dam failed.

If *substandard materials* are used in the construction of dams or if the construction material of an embankment dam is not properly compacted, this may result in internal erosion of dam material and failure of the dam. The failure of the Teton earth dam in the US (Idaho) in 1976 is an example. Permeable loess material was used as the core of this dam, which was founded on fissured rhyolite that caused internal erosion or piping and eventual failure of the dam.

A short history of failures of several Indian dams has been furnished in Section 21.7. Table 14.2 shows an assessment of the failure of various dams of the world and that of India in relation to major causative factors (Chowdhury 1999).

Table 14.2 Percentage of dam failures of total world and India (in brackets) and their causes

Major causative factors related to dam failures	Percentage of failure	
	Earth dam	Masonry/concrete dam
Uplift pressure and sliding	29 (11)	18 (20)
Insufficient spillway	6 (11)	–
Piping, etc.	23 (33)	40 (20)
Spilling/overtopping	–	21 (30)
Saturation/settlement	6 (11)	–
Seepage along structural planes	6 (11)	–
Faulty design/poor construction	29 (23)	15 (20)
Miscellaneous	–	6 (10)

14.8 PRELIMINARY INVESTIGATION AND SELECTION OF A DAM SITE

Engineering geological investigation of a dam site is carried out to reveal the geological conditions of the site with the ultimate aim of assessing the suitability of the site for the construction of the proposed dam. Adverse geological conditions may affect the feasibility of the project or enhance the construction cost. Engineering geological investigation is necessary to identify the geological weakness of a site so that suitable measures can be adopted to rectify the defects of the dam foundation. In fact, most of the engineering problems of the site are geological or geology-dependent. A detailed geological study is very important to recognize these problems prior to selection of the site for dam construction. In general, engineering study at a dam site is undertaken in different stages.

The geological set-up of a dam site is variable from place to place. Every dam site has its own characteristics. The purpose of investigation is to unravel these characteristics for their consideration in the design of the dam. In the preceding sections, the geological factors associated with dam disasters have been highlighted. While selecting a dam site, it is to be found whether the site under investigation has any such geological weakness. In fact, many dam sites have adverse geological conditions that create problems in building the dam in the proposed site. Thick mantle of weathering or decomposition of rock, presence of soft rock in the foundation, structural defects such as faults, thrusts, shear zones, and intensive joints are the common defects of a dam site. Karstic condition and deep zone of kaolinization at dam foundation have caused problems in some projects. A highly permeable boulder deposit or a buried channel below the dam foundation threatens excessive leakage of reservoir water. Thick deposit of moraine and avalanche are special problems of high-altitude dam sites, especially in the Himalayan terrain.

In practice, engineers propose one or more alignments, and accordingly geological investigation is carried out on the site or sites. At the preliminary stage of engineering geological investigation, geological maps of the different alternative sites are prepared to a scale of 1:1000 or so, plotting the major geological set-up and main structural features. The merits and demerits of the different sites are then evaluated and the best site from geological as well as engineering considerations is selected for further study. Photo-geological study including application of remote sensing techniques is very helpful in the preliminary stage investigation to prepare a map of regional geology as well as local geology, covering large areas in and around the dam site within a short time. The study of aerial or Landsat images will provide a quick appraisal of the broad features of site conditions with respect to the landform, rock types, and geological structures.

Geological mapping (systematic geological survey maps) covering the project area is carried out at this initial stage of investigation to understand the regional geology and broad structural elements. The slope characteristics of the valley walls and the nature of rock exposed or the approximate thickness of the overburden at the dam site are recorded. Special attention is provided for the identification of weak features such as a fold, fault, shear zone, clay seam, and slide scar in the site. The field study helps to understand the merits and demerits of alternative alignments and choose the most favourable one. The engineering geologist in the course of investigation may also find a geologically better site than the one proposed by the engineers. The final selection of the dam site will, however, depend upon the consideration of the engineering aspects in relation to the geological conditions of the site including the cost factor.

14.9 DETAILED INVESTIGATION OF A DAM SITE FOR DESIGN PURPOSES

After selection of the dam site, a detailed study is undertaken to reveal the surface and subsurface geology of the site that facilitates design of the structure and estimation of project cost. The first step in this investigation is the preparation of a large-scale map (scale generally, 1:100 or 1:500), plotting all the lithological units and structural features of the dam site covering at least 100 m areas around the layout of the dam. In case a fault traverses the dam site, its disposition with reference to dam axis and other details such as thickness of crushed zone are to be recorded in the map. If the river bed at the dam site is covered by a river deposit, the character of the deposit including its thickness is recorded.

The next step is the exploratory work such as pitting, trenching, and drilling to assess the depth to bedrock and the rock types of the dam site. The most important of the exploratory works is the drilling to reveal the subsurface geology of the dam site including geological structures from the logging of the drill cores. A plan is prepared plotting in a map the location of the drill holes including their inclination and direction and the approximate depth to penetrate is suggested. In general, the drill holes are made vertical; however, due to water cover in the river bed, inclined holes are drilled from the two sides so as to pass through the central part of the river.

The drill core logging helps to understand the subsurface geological condition of the site including bedrock and fresh rock depths and draw profiles for foundation grade of the proposed dam. In addition to drilling, drifts are made at the abutments in case of concrete dams to estimate the weathering profile and examine abutment condition. Geophysical method is also applied in some sites to reveal the subsurface geology, especially adverse structural features such as faults and folds and the presence of a large cavity and buried channel.

Laboratory tests on rocks and soil of the site are also conducted for quantitative assessment of their engineering properties to meet the design requirement. Undisturbed samples of unconsolidated deposit and rock cores are used in the testing. In some sites for concrete dams with weak foundation condition or complicated geological set-up, in situ test is carried out to understand the rock behaviour under stress condition. In general, plate load test is conducted to find the load-bearing capacity of soil mass and soft band or zones of weathered rock of dam foundation.

NX-drill holes with diamond bits are necessary to obtain maximum recovery of rock cores in hard rock. The logging of rock cores is done with diagrammatic representation and recording the following information: (a) rock type or types, (b) depth to bedrock or thickness of overburden material, (c) the thickness of weathered rock or the depth to fresh or foundation grade rock, (d) the percentage of core recovery and rock quality designation, (e) fracture frequency of cores, and (f) shear zone, clay seam, and any other weak feature, if any. Field permeability tests are also conducted through the bore holes to assess the permeability characteristics of the porous and fractured rocks. The logging of rock cores and results of permeability tests are presented schematically (Figs 11.21 and 11.11), which help the engineers in designing a structure. Some portions of the cores obtained from drilling are sent to the rock testing laboratory for conducting tests on various engineering properties such as the density, porosity, and compressive strength required for design purposes.

Geological structures observed in the surface in and around the dam site and that revealed from the drill holes are studied in detail and also plotted in geological maps. Many of the structural features such as faults and shear zones are weak features and their presence may lead to stability and leakage problems for the proposed dam. The dam site may be covered by river deposit but study of the surrounding area might suggest the presence of a fold or fault at the dam site, which can be substantiated from the drill holes. Morphological expression also helps in the interpretation such as a straight river valley being indicative of a probable fault passing along the valley, which is proved from drill holes. The details of the structures such as faults and folds that may affect the rocks of the dam site are studied in detail with respect to the gouge zone, clay seams, tension cracks, and so on and probable treatment for the affected zones are proposed.

14.10 SOURCE OF BUILDING MATERIALS FOR DIFFERENT TYPES OF DAMS

Different types of building materials are needed according to the engineering structures. For example, a concrete dam needs concrete which is made of cement and aggregates generally available from natural sources. Similarly, a masonry dam requires durable rock blocks for building masonry structures. The following describes the details.

14.10.1 Building Materials for Concrete Dams

The *aggregate materials* such as cobbles, pebbles (shingles), gravel, sand, or crushed rock are mixed with cement in different proportions to make concrete of required strength as in the design of a concrete dam. River bed and terrace deposits are the natural sources of the concrete aggregate materials. The materials for concrete aggregate are also obtained by excavation of the exposed rocks and crushing them and screening into proper sizes. The river-borne shingles and gravels due to their long transportation by the river water develop smooth and round surfaces that help better bonding and low consumption of cement. Pozzolan and fly ash, where available, are used as a substitute of cement in certain proportions. In practice, concrete used in the construction of a dam may have more than one type of mix. Table 14.3 provides a list of different mixes (weight of aggregates and cement) used in the Umiam concrete dam in Meghalaya for one cubic metre of concrete.

Table 14.3 Concrete mix per cubic metre of concrete in Umiam dam, Meghalaya

Size (mm)	Aggregate type*	Mix A-1 (kg)	Mix AP-1 (kg)	Mix B-1 (kg)	Mix B-2 (kg)
150–75	Cobble	385	385	–	–
75–40	Pebble	355	355	485	–
40–20	Pebble	335	335	405	545
20–4.76	Pebble	250	250	340	480
< 4.76	Sand	355	355	405	480
	Cement	140	120	175	195
	Fly ash	–	25	–	–

*Coarse aggregate constitutes quartzite, meta-dolerite, and arenaceous phyllite having strength above 1760 kg/cm² and the silt and clay (particles size < 0.15 mm) content is 2.7%.

The aggregate materials are first subjected to various tests to find their suitability for use in the dam construction. The tests on coarse aggregate and rocks include density, absorption, strength (compression, shear, and tension), abrasion, impact, soundness, modulus of elasticity, and chemical reactivity. Petrological tests are also conducted to find the mineral content in relation to deleterious materials. The materials for earth dams are tested for properties such as liquid and plastic limits, consolidation, cohesion, shear, angle of internal friction, permeability, and slacking. Indian Standard Institution (ISI) has provided the specification for coarse aggregate materials, which is followed in making concrete for dam construction (Section 12.4). In practice, the specification may be changed to some extent depending upon the availability of the aggregate materials and the design requirement of the concrete strength.

14.10.2 Boulders and Rock Fragments for Masonry Dams

Masonry structures are constructed using large boulders of river deposits. When boulders of the requisite size and quality are not available in river beds, these are obtained in large fragments or slabs by quarrying rocky hill faces and are used in masonry works after chiselling into blocks. The main considerations in masonry dam construction are availability and suitability of the rocks conforming to the specification of the ISI. All common igneous and metamorphic rocks such as granite, basalt, charnockite, granite gneiss, quartzite, and quartzitic phyllite (Tables 10.1 and 10.3 of Section 10.4) are hard, sound, and durable and possess high strength fulfilling the required specifications such as $G > 2.6$ and compressive strength > 1760 kg/cm². Sedimentary rocks such as Vindhyan sandstone (Table 10.2) and quartzitic sandstone (Cuddapah) and some varieties of dolomite are also suitably used as masonry blocks for dam construction.

On account of the absence of requisite quantity and quality of construction materials, the design of the proposed 31 m-high Gumti dam at Tripura had to be changed to a brick masonry structure with a thick concrete lining all around. The bricks for the dam were made out of local clay. Even for concrete aggregate, crushed bricks (8000 m³) had to be mixed with crushed calcareous sandstone pebbles (12,000 m³) collected from the stream beds nearly 30 km away from the dam site.

14.10.3 Fill Materials for Rock-fill Dams

Compared to masonry dams, rock-fill dams are more favoured in modern time as they are more economical. A rock-fill dam uses the entire deposit in a river or terrace as fill materials constituted of boulders, pebbles, gravels, sand, and clay. The rock fragments quarried from nearby rocky hill slopes are also used. In the mountainous terrains, the river valleys, hill slopes, and terraces contain plenty of construction material for rock-fill dams, and if the dam is proposed in the valley, the haulage charge will also be less. However, the overall quality of the bulk material will conform to the design requirement.

The Salal rock-fill dam in Jammu and Kashmir used dolomite boulders as the fill materials taken from the river deposits upstream and downstream of the dam site and also from the terrace deposits of the locality. In addition, stripping of the abutments also gave large quantities of rocks for the rock-fill. However, as dolomite beds were highly fractured, massive Murre sandstones were used for rip-rap purpose. The demand for nearly 1.7 million cubic metres of impervious material required for the core of the dam was met by using the clayey soil from three borrow areas located at a distance of 10 km to 11 km from the dam site.

14.10.4 Sandy and Clayey Material for Earth Dams

Sand, silt, and clay, the decomposed products of rocks, are available in plenty in most parts of the projects, especially in the reservoir basins. The decayed rock of hill slopes especially in areas composed of sedimentary rocks are generally devoid of pebbles and boulders and are found to have good quality material for use in earth dams. Both pervious and semi-pervious types of materials are necessary for the earth dams. After removal of overburden layers, the rock surfaces are generally semi-pervious in nature. Earth samples are collected from borrow pits for testing their properties. Depending upon the quantity required for the dam, the borrow areas are selected in the reservoirs and in the foothill areas in and around the project.

Fig. 14.25 Rip-rap constituted of slabs of hard and durable rocks covers upstream slope of an earth dam in Bihar

14.10.5 Rip-rap for Protecting Dam Slopes

Rip-rap constitutes slabs of rock used to cover the upstream slope of an earth or a rock-fill dam to protect the dam body from the eroding process of nature and wave action (Fig. 14.25). The rocks for the rip-rap should be hard and durable. They are collected from the river bed or terrace deposits after screening out unwanted fractions. If river boulders are not available in sufficient quantity, they are obtained from quarrying the hard rocks available locally. The cost of transportation is an important factor to be considered while selecting quarry sites for rip-rap.

Loosely compact boulder beds in sedimentary formation are also mined and boulders separated by screening to use as rip-rap materials. The desirable compressive strength of the rip-rap is 350 kg/cm^2, which is attained by many rock types including limestone, sandstone, phyllite, and schist (Tables 10.1, 10.2, and 10. 3 of Section 10.4). The Vindhyan sandstones having density 2.65 kg/cm^3, absorption 0.85–1.94 per cent, and compressive strength 550–1975 kg/cm^2 are used as masonry blocks as well as rip-raps. In the absence of any durable rock, concrete blocks are also used as rip-rap material.

14.10.6 Impervious Core Materials for a Dam

The availability of materials to be used for the core or shell of rock-fill and earth dams varies from project to project depending upon the country rocks and their decomposition or the nature of overburden material. Ideally, this should be a well-graded mixture of sand gravel and fines with highly plastic tough clay, having plasticity index more than 20. However, if the earth contains grains in different proportions, say, sand and silt in large quantities, then it may be blended with clayey earth to achieve the required compactness and impervious character. In some projects, sufficient quantity of suitable earth for the shell may not be available. In such cases, crushed soft rocks are mixed as per design requirement for use as core materials of rock-fill and earth dams. The Beas project in Himachal Pradesh successfully utilized for the core material 50 per cent crushed sandy-rock mixed with 50 per cent clay–shale exhibiting the properties shown in Table 14.4 (Ray, Mehta, and Ashraf 1972).

Table 14.4 Properties of sand-rock and clay–shale used in the Beas project

Sand-rock		*Clay–shale*	
Unit dry weight	1.88 gm/cm^3	Liquid limit	19.2–34.6
Unit saturated weight	2.17 gm/cm^3	Plastic limit	14.8–21.6
Cohesion	nil	Cohesion	14,408–23,826 kg/m^2
Angle of internal friction	30°	Angle of internal friction	18°–35°

14.10.7 Guidelines for Selecting Sites for Building Materials

Selection of sites for available construction materials is an important part of planning stage investigation. The availability of the required quality and quantity of the construction materials is the deciding factor for the proposed type of dam. It may not be possible to construct a masonry or rock-fill dam if suitable quality and sufficient quantity of rocks are not available in the nearby areas. Similarly, paucity of earth and impervious clay may restrict the possibility of constructing an earth dam. In general, borrow areas for earth may be searched in reservoir basins and rocks in the hill slopes. As a construction material, the word *stone* is generally used by the engineers, which is rock to geologists. Quarry sites for rocks requiring blasting to obtain rock fragments are selected at least one kilometre away from the dam site so that the blasting does not affect the dam.

While selecting quarry sites for aggregate materials for a concrete dam from river beds or hill slopes, care should be taken to ensure that they do not contain minerals reactive with cement used in making the concrete. It is necessary that samples of concrete made with the stone chips and sand of the selected quarry sites are tested in the laboratory to ensure that it does not show any alkali–aggregate reaction. It involves the petrological study of a thin section of the concrete sample under microscope to find alkali reactivity.

As stated earlier, borrow areas for both pervious and semi-pervious earth may be searched in the reservoir basin. Impervious earth is especially necessary for the core of rock-fill and earth dams. It is available in the areas represented by sedimentary rocks such as shale or clay stone, the decay of which produces a thick mantle of impervious earth overlying the bedrock. Partly decayed products of rocks found below the topsoil in the periphery of the reservoir spread or in the hill slope of the surrounding areas provide suitable semi-pervious earth required for earth dams.

Earth containing high percentage of organic material should be avoided. Similarly, borrow area for clay should not contain harmful clay minerals such as montmorillonite and bentonite, which generally occur in basaltic terrain. In some areas such as ancient glaciated regions, overconsolidated clay may be present, which develops high pressure under saturated conditions and may cause harmful effects on dam body if it is made of this clay. The earth or clay is to be tested in the laboratory for the types of clay minerals present in it. If the earth possesses a considerable quantity of harmful clay minerals such as montmorillonite, it will cause excessive swelling under saturated conditions, which may develop sufficient expansion pressure to breach the dam body.

For rock-fill dams, fill materials such as broken rocks or boulders and earth of huge quantity are obtained from debris of hill slopes or from boulder deposits of river beds. Huge quantities of spoils such as broken rocks and earth are available from the excavation of diversion channels or tunnels in a dam project. In earth and rock-fill dams, spillway excavation also provides large quantities of broken rocks and earth. Cutting of hill slopes for new roads by diversion of old roads from reservoir areas also supplies spoils such as earth, clay, and rock fragments of varying dimensions. The broken rocks including the earth and clay obtained from these sources may meet the demand of fill materials to a great extent and thus reduce the construction cost of a rock-fill dam.

Fig. 14.26   Removal of river detritus after diversion of river

14.11 RIVER DIVERSION AND CONSTRUCTION WORK

Construction work of a dam can start only after diversion of the river water. Diversion work is generally done during the dry season when the quantity of flow in river water is less than in the rainy season. Figure 14.26 shows earth movers and trucks being used in the removal of river detritus after diversion of the river at the beginning of the construction of the Umiam concrete dam. High-power pumps with long pipes are in operation for flushing away accumulated water from the site. See Fig. 14.3 for the Umiam dam under construction. The method of diversion depends upon the morphology and geology of the dam site and its immediate surrounding areas as detailed in the following subsections.

14.11.1 Methods of Diverting Rivers

The first step is the construction of a *cofferdam*, which is a small dam used for the temporary purpose of diverting river water from its original path for a certain distance. The diversion is done by two ways. In the first method, a cofferdam is constructed at the upstream part of the dam foundation and the river water is diverted by means of a tunnel or channel through the bank. The second method is the in-river diversion where the cofferdam restricts the flow to a side by contraction of the river channel to 30 per cent to 60 per cent. The cofferdam, though a small structure, is designed with due consideration of the river discharge, stability factor, and scouring effect.

The choice of diversion method depends on the discharge capacity of the river and the morphology as well as geology of the site. If the discharge is large and valley abutment consists of hard rock, a tunnel alignment is always taken up. The tunnel diversion is more justified when the tunnel is planned for use as a service spillway. The cofferdam constructed at the upstream part for diverting the water should be located at least 10 m upstream. Sometimes, there may be a need to build another cofferdam at the downstream part to arrest the back flow of the water from the tunnel or channel outlet. The channel used for diverting the water may be one or more in number and excavated through the abutment slope. A channel diversion is more economical but the presence of hard and stable rock is necessary to avoid the chance of scouring and sliding.

In the in-river diversion method, the foundation work of the dam is executed in the river channel while the cofferdam remains partly or fully in river water. This method is normally taken up if the foundation is made of cohesive material and the proposed dam is of medium height. The cofferdam may be of overtopped or non-overtopped type with respect to floodwater. In the fixed crested type, the construction can be continued round the year. However, in the overflow type, the work needs to be stopped for a few months when due to heavy rain, flood discharge passes over the cofferdam to inundate the foundation. After a certain lapse of time, the construction work begins with dewatering of the foundation. The diversion method may also be a combination of both the methods where the stream water flows partly through the side and partly through the channel over the abutment.

14.11.2 Foundation Preparation

The most important part in dam construction is the *foundation preparation*, which starts when the dam site becomes dry after diversion of the river water. The engineering geologist remains constantly present at the site to decide on the *foundation grade* rock on which the dam can be safely founded. The first step in the foundation preparation is removal of the river detritus or overburden material from the river bed. The bedrock necessary to found a concrete dam is obtained below this loose deposit, but the rock may be in a state of decomposition, which is to be removed by excavation. If blasting is used to remove the decomposed rocks from the foundation, this must be of mild nature so that the vibration does not develop cracks in the fresh and sound rock of dam site including abutments.

This is followed by *striping* of the weathered rock surface including clay seams, fault gouge, and pockets of soft materials from the foundation. The resident geologist is to identify weak zones and get them removed until fresh rock is available. The cavities or openings formed from striping are subjected to *dental treatment*, which involves filling them by concrete. If a fault passes through the foundation, its disposition and attitudes are studied carefully, and it is treated by removing the fractured and gouge material from the fault zone and then providing concrete plug.

The foundation rocks may be porous or jointed to different extents when treatment by grouting is adopted. Consolidation grouting is undertaken in the foundation that seals the void spaces or openings in porous and jointed rocks and enhances the load-bearing capacity of the foundation. In addition, curtain grouting is taken up and drainage holes are provided at the upstream end of a concrete dam to guard against seepage and prevent uplift pressure. The first type of grouting is done by comparatively shallow holes, but deep holes are taken up for curtain grouting from the gallery of the dam (see Sections 13.6 and 13.7).

14.11.3 Construction Approaches for Different Types of Dams

After treating the weak zones and strengthening the rocks by grouting, the foundation becomes ready for laying concrete for construction of concrete dams. A batching plant located within a short distance from the dam site mixes aggregate materials including stone chips or pebbles, sand, and cement in fixed proportions (as per design) to attain the strength as required by the design of the dam (Fig. 14.27). The concrete from the batching plant is directly placed on the foundation rocks in layers until the design height is attained. Cylinders or cubes of the concrete are tested

Fig. 14.27 A batching plant where aggregates and cement are mixed in fixed proportions to make concrete of design strength

in the laboratory to ensure the desired strength. To facilitate concreting work, the foundation is marked into different segments and concrete is laid in these blocks as per plan. The concreting work of the spillway and other structures within the dam is also taken up side by side so that the dam attains its full height along with its appurtenant structures as per the design plan.

The foundation preparation for the construction of a rock-fill dam is more or less the same as that of the concrete dam. After removal of the overburden material, the dam is founded on bedrock having adequate strength to withstand the forces acting on it. The rock fragments of different sizes hauled from the quarry sites and dumped on the foundation are compressed with the help of huge machineries. The work of placement and compression of the impervious earth in the central part or concreting of the upstream face (as per design) is continued side by side with the construction of the rock-fill dam.

The earth dam is built on natural earth and does not require rock foundation. The loose materials from the top surface are, however, removed by excavation to obtain the underlying layer of earth and spread over and rolled by further layers suitable for earth dam foundation. Figure 14.28 shows an earth dam with layers of rolled earth and the upper central trench is to place impervious clay in the core. Earth from borrow areas (with predetermined suitability from the field and laboratory tests as outlined in Chapter 6) for construction of the dam are then transported to the dam site and dumped on the foundation and compressed into layers. The most important consideration is the design aspect of the compression including the central clay core. Figure 14.29 shows an earth dam in Bihar under construction with excavated hill slope material and earth partially compressed for foundation grade in partly weathered rock. A trench has been dug at the central part (core) for laying impervious clay. The compression reduces the permeability and increases the density and strength of the earth so that there will be no settlement during the lifespan of the dam. It is to be ensured that the dam body becomes impermeable to the extent that there is no passage for water to cause the problem of piping.

Fig. 14.28 Earth dam with layers of rolled earth

Fig. 14.29 An earth dam in Bihar showing partially compressed excavated hill slope material and earth

Once the foundation area or a part of it is ready to lay the concrete or dump earth or rock, it is the primary duty of the resident geologist to undertake geological mapping to a sufficiently large scale plotting all the geological details of the foundation. This map is an important record of the dam foundation. In case of any post-construction problems related to weakness of foundation or leakage through shear zones and so on, this map acts as a guide to identify the probable zone or geological feature responsible for causing the problem and thus helps to take appropriate remedial measures to solve the problem.

When the construction of the proposed dam reaches a height that is more than that of the cofferdam, the latter becomes defunct. The diversion tunnel or channel excavated for temporary diversion of river is then of no use. Hence, the tunnel or channel used for diversion of river is sealed at its two ends to stop further flow. The river water is then stored in the reservoir behind the dam for various utilities.

Several instruments are installed in a dam to monitor its behaviour in the post-construction period. The following are the common instruments:

(i) Piezometric cells placed in observation wells and drainage bore holes to measure the uplift and pore water pressure
(ii) Extensometer, clinometer, deformation meter, and tiltmeter to record the movement of the foundation
(iii) Stress and strain meters for measuring the stress and strain in foundation or dam body
(iv) Seismographs for measuring micro-seismic movement
(v) Discharge measuring devices to monitor if any seepage takes place.

The instruments monitor abrupt behaviour of the dam, if any, during the post-construction period and accordingly safety measures are undertaken.

14.12 POST-CONSTRUCTION WORK

Engineering geological investigation is necessary for many of the post-construction problems of a dam. If the problems are geological in nature or created by natural calamities such as floods, landslides, and earthquakes, then the engineering geologists attend to such problems to find a rational solution. Seepage is a common problem in the post-construction stage. Tracer study is conducted to delineate the zone and grouting is recommended to seal the path of seepage. Blockage of drainage holes or malfunctioning of grout curtain may develop uplift pressure of serious consequence in a dam after its completion. In a 49 m-high masonry dam in Madhya Pradesh, the drainage holes were choked and the grout curtain did not function properly. Excessive leakage was observed in the dam after construction and it required the lowering of the reservoir level and revitalization of drain holes.

Percolation of water through the body of a masonry or concrete dam along the construction joint, contraction joint, mortar joint, foundation contacts, and cracks is a common phenomenon. This causes seepage of water and consequent scouring of dam body and development of uplift pressure. Gradually, the condition of the dam deteriorates leading to large leakage and distress of the structures. Safe design of the dam for a stable structure is required to eliminate such possibilities that result in the failure of the structure.

The filling of reservoirs may trigger landslides or even earthquakes that affect the dam or its appurtenant structures. These aspects require in-depth geological study and immediate advice for suitable treatment. Scouring or cavity formation at the downstream part of the dam due to spilled water is a common problem in many dams detected after some years of dam construction. The Sikasar project in Madhya Pradesh experienced scouring due to the presence of soft shale with quartzite in the spill channel foundation. It required providing flexible sausage as training wall to stop the scouring for saving an earth dam close to the spill channel.

In general, with time, the construction material of the dam gets degraded due to improper maintenance. Common causes of material deterioration are leaching of lime from cement, alkali–aggregate reaction, and sulphate in rock causing surface spalling which affect the safety of the structure. The Hirakud dam after its completion developed cracks due to alkali–aggregate reaction. In the Dudhganga dam in Maharashtra, the pyrite in quartzite caused swelling and spalling in the dam body. The Tong barrage in Madhya Pradesh with gypsiferous shale in the construction material caused cavitations in the post-construction period requiring geological investigation and immediate treatment.

During the construction of a dam, several instruments are installed in a dam to understand its functioning. If any geology-related problem is indicated in the post-construction stage by the behaviour of the instrument, the attention of the engineering geologist is immediately drawn for finding a solution to the problem in cooperation with the engineers.

SUMMARY

- A dam is constructed across a river to create an artificial lake for storing water, which can be used for irrigation, water supply, generation of electricity, and industrial purposes. Dams are mainly of four types, namely concrete, masonry, rock-fill, and earth dams.
- A dam is provided with a spillway structure generally having a gate to facilitate controlled discharge of water from the reservoir to the downstream side whenever needed, especially during floods.
- A concrete dam is constructed in a narrow river valley with firm rocks.
- A masonry dam is built of blocks of hard and durable rocks bound by concrete. It needs firm bedrocks for its foundation.
- A rock-fill dam is constructed by compacting unconsolidated materials of river beds, terraces, and hill slopes and providing an impervious barrier through the dam.
- An earth dam is generally located in a wide valley and founded on decomposed rocks or in-situ soil, but availability of large quantities of pervious, semi-pervious, and impervious earth is required in the neighbouring areas for its construction.
- The following work-approaches are necessary for a dam site investigation.
- After reconnaissance of several possible sites, select the most favourable one from consideration of geology and cost.
- Prepare geological map (pre-construction stage) covering the dam site and the immediate areas around it, and depending upon the site condition give a programme for subsurface exploration of the dam site by diamond drilling to facilitate maximum core recovery.
- From the exploratory work, record the overburden thickness, depth of foundation grade rock, joint frequency, and other defective features of the site and represent them in drill core logs.
- Search quarry sites (sufficient quality and quantity) for construction materials needed for the proposed type of dam and locate their positions in a map.
- Conduct field permeability tests for earth and clay. In addition, arrange laboratory testing of samples of rock and other construction materials for their engineering properties including alkali–aggregate reaction for concrete aggregate.
- Submit detailed feasibility report with all geological information along with geological maps, geological cross sections, drill hole logs, and laboratory and field testing data on rocks and soils.
- Construction work starts after diversion of river water from the dam site. Suggest suitable alignment for diversion tunnel or channel to remove river water. Day-to-day inspection is to be done and rectification measures are to be suggested for geological defects while working in the construction site.
- To found a concrete or masonry dam, the loose river deposit is to be excavated until sound rock is available. The embankment dam may be laid on in situ or natural earth, which should be homogeneous and fairly impervious nature. However, after compression, the foundation material must attain the strength as per design requirement. Ensure that the cut-off for the impermeable clay core for embankment dam is extended down to impervious bedrock.
- In the construction stage, supervise the foundation preparation work for building dam, which involves excavation of overburden and

decomposed rock and striping of clay seams and other weak zones followed by dental treatment and grouting.

- With the progress of construction work, prepare large-scale (say, 1:100) contoured geological map and geological cross sections along and across the dam axis showing in detail the rock types, structures, joint pattern, and weak zones.
- While attending to any post-construction problems of a dam, consult the geological map and recorded data to evaluate the probable reason of the problem and accordingly suggest corrective measures.

EXERCISES

Multiple Choice Questions

Choose the correct answer from the choices given:

1. The type of dam that requires an impermeable membrane is:
   (a) rock-fill-dam
   (b) masonry dam
   (c) arch gravity dam

2. The type of dam in which the cost of construction is the least is:
   (a) concrete dam
   (b) rock-fill dam
   (c) earth dam

3. The type of dam constructed in a narrow valley with hard rock having good strength is:
   (a) masonry dam
   (b) rock-fill dam
   (c) arch gravity dam

4. The dam built in a broad valley covered with alluvial deposit is:
   (a) masonry dam
   (b) earth dam
   (c) rock-fill dam

5. The type of dam designed for construction in an area of high seismicity is:
   (a) concrete dam
   (b) rock-fill dam
   (c) earth dam

6. The dam constructed in a mining area to retain the mine waste separated from valuable fractions is:
   (a) buttress dam
   (b) tailing dam
   (c) earth dam

Review Questions

1. What is a dam and what is its utility? Draw the sectional view of a dam and show the positions of the toe, heel, crest, parapet, abutment, gallery, and gate.

2. Give an account of the major types of dams and their functions. What is the distinction between a straight gravity dam and an arch gravity dam?

3. What are the construction materials necessary for different types of dams? What will be your approach to search for these materials?

4. What is a spillway and what is a chute? Name and describe briefly the major types of spillways.

5. What are the utilities of gates? Describe the functions of two major types of gates.

6. Explain how morphology, geology, and availability of construction materials influence the design of a dam.

7. What are the main causative factors of dam disasters? Give some examples of dam disasters due to geological causes.

8. What are the different stages of dam construction and what types of work are done at these stages? How can an investigation stage work help to solve post-construction problems in a dam?

9. Explain with details the following: barrage, weir, uplift pressure, rip-rap, cave materials, fill materials, and cofferdam.

10. How is the river diverted prior to a dam construction? What precautions are to be taken in foundation preparation for a concrete dam?

11. Describe in detail the different stages of investigation that are required to be taken up for building a dam.

Answers to Multiple Choice Questions

1. (a)   2. (c)   3. (c)   4. (b)   5. (b)   6. (b)

15 Reservoirs

● ● ● ● ● ● ● LEARNING OBJECTIVES ● ● ● ● ● ● ●

After studying this chapter, the reader will be familiar with the following:

- Selection of a reservoir site with respect to its topography and geology
- Effect of reservoir creation on the geohydrology of the neighbouring areas
- Sedimentation and degradation processes of reservoir capacity
- Basic aspects of pumped storage scheme and its construction with examples
- Environmental effects due to the creation of a reservoir

15.1 INTRODUCTION

A reservoir is created by the construction of a dam for storage and supply of water for irrigation, drinking, hydroelectricity generation, and various other purposes. Construction of a reservoir brings about changes in the geo-hydrological condition of the neighbouring areas as well as social and ecological changes of the reservoir area. Over a period of time, the storage capacity of a reservoir dwindles due to siltation. Investigation of a reservoir site lays emphasis on ensuring that no leakage takes place and no national assets such as minerals of economic importance and historical monuments are submerged under storage condition. In recent times, the pumped storage scheme, which uses the same water in two reservoirs to generate electricity, is found economical when suitable topography is available. This chapter deals with all these aspects associated with reservoir construction.

15.2 CREATION AND FUNCTIONS OF RESERVOIRS

A reservoir is a man-made water body created by constructing a dam across a river (Fig. 15.1). The reservoir water is used to meet the various needs of the people and industry. A reservoir stores water during excess flow in the river and releases it in times of scarcity when the natural flow of the river is inadequate.

The flow pattern of the river and its tributaries decides the shape of the water body of the lake. The reservoirs of upland hilly terrains are generally spoon shaped with tail ends extending for several kilometres along the upstream valley. In plains, the reservoir spread is erratic due to its spread following the river channel and its tributaries.

Fig. 15.1 Reservoir of the arch dam of Idukki hydroelectric project of Kerala

The storage capacity of a reservoir depends on the rainfall and run-off in the surrounding catchment areas. The rainfall of the catchment area must be adequate enough to fill the reservoir up to the maximum level.

The height of a dam decides the *maximum (full) reservoir level* (FRL) up to which water can be stored. In practice, however, water level is kept below this level in normal time. If the quantity of water (*storage* up to FRL) tends to exceed this level due to heavy rains in the catchment area, the excess water is released downstream of the dam through the spillway. The reservoir level below which water does not go down even during dry season is called the drawdown level, *dead capacity level* (DCL), or *dead storage level* (DSL). The volume of water between the FRL and DCL is the *active capacity* or *live storage* of a reservoir. The quantity of water that remains below the DCL is the *dead storage capacity*. The reservoir spread at FRL is called the *water surface area* (Fig. 15.2).

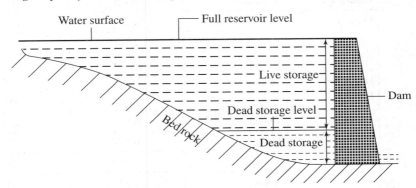

Fig. 15.2 Reservoir created by a dam showing nomenclature of different parts

The portion of rain that flows as run-off over the catchment area ultimately drains into the river causing increase in the reservoir water. Thus, the volume of storage water depends to a great extent on the rainfall and run-off of the catchment area. Table 15.1 shows the annual run-off and rainfall in the catchment area and the flow characteristics of the river Barakar, which feeds the reservoirs of Maithon and Tilaiya dams in the Damodar Valley Corporation (DVC) (Gangopadhyay and Majumdar 1989).

Table 15.1 Drainage area and flow characteristics of two reservoirs fed by river Barakar*

Reservoir	Drainage area (km²)	Maximum flow (m³/sec)	Average flow (m³/sec)	Storage length (km)	Submerged area (ha)	Total storage (m³ × 10⁶)
Maithon	6293	7080	101.94	26	10,720	1361.8
Tilaiya	984	1370	15.85	23	5908	394.7

*Annual rainfall in the catchment area is 1245 mm and annual run-off 3181 million m³

In general, a dam project for storage of water is beneficial to the human community. Impounded water in a reservoir provides hydroelectric power, which is rather cheap compared to other sources of power. The beneficial functions of reservoirs are many. Irrigation, water supply, flood control, transportation, fish farming, tourism development, and several other benefits can be obtained from the reservoir. India has constructed several high dams in different states with large reservoir capacity (Table 15.2). Nearly 92 per cent of Indian dams are built to utilize the water of the reservoirs for irrigation purposes, 2 per cent for hydroelectric power, and the rest for multiple purposes including water supply, power generation, irrigation, and flood control.

Table 15.2 Reservoir capacities of major dams in some states of India

State	Reservoir	Catchment area (km^2)	Live reservoir capacity $(10^6 \times m^3)$
Andhra Pradesh	Srisailam	206,242	8287
	Nagarjuna Sagar	215,192	6841
	Sriram Sagar	91,750	2300
Chhattisgarh	Mahanadi	3670	767
Gujarat	Ukai	62,225	6615
	Sabarmati	5540	778
	Kadana	25,520	1472
Himachal Pradesh	Gobind Sagar	56,980	6655
	Pong dam	12,560	7119
Jharkhand	Tenughat	4481	821
	Maithon	6294	571
	Panchet	10,966	223
	Konar	997	275
	Tilaiya	984	319
Rajasthan	Mahi Bajaj Sagar	6149	1883
	Jhakar	1010	132
	Rana Pratap Sagar	24,864	1573
Tamil Nadu	Lower Bhavani	4200	929
	Mettur	42,215	2647
Uttar Pradesh	Matatila	21	707
Uttaranchal	Ramganga	3134	2196
	Rihand	13,333	8967
West Bengal	Mayurakshi	1847	547
	Kangsabati	3584	914

*Source: Central Water Commission

Distribution of reservoir water to the users involves laying pipelines or excavation of tunnels or channels with intake and outlet structures at the two ends that regulate the quantum of flow. In a hydroelectric power station, huge quantity of water from the reservoir is allowed to pass through penstock pipes and fall on turbines under high head for generation of electricity. Sometimes, reservoir water is diverted to the downstream part of the dam to discharge into the normal course of the river from where the water is taken for irrigation by channel. When a dam is built for flood control, the storage level of water is kept far below the FRL so that large parts of the reservoir remain empty. This helps the excess flow to accumulate in the empty space during floods and to release it to the downstream valley to save the localities around the reservoir from flood havoc.

Creation of a reservoir has several adverse effects too. Villages, forests, cultivated lands, and even roads, rails, mineral deposits, and archaeological and historical monuments may be submerged with the construction of reservoirs. The major impact is the environmental degradation. The equilibrium of the wildlife and environment is disturbed and the landscape changed with degrading shore. The filling of reservoirs influences the biological regime of the surrounding terrains. Submergence of forest and farmlands produces biological wastes stored in the reservoir. These biological products supply food to the fish and help their growth. Fish farming is possible if such organic matter occurs in the reservoir in sufficient quantity. However, excessive growth of weeds and algae and their decay cause pollution of the lake water with deteriorating effect on its fish population.

15.3 DEGRADATION OF CATCHMENT AND RESERVOIR RIM AREAS

Construction of a large storage structure degrades the catchment area. Degradation negatively affects the dam and the reservoir. Once a reservoir is filled with water, both the surface and the groundwater regimes are disturbed. The water table of the areas around the reservoir rises. The alluvium of the reservoir valley gets saturated and the water body exerts increased pressure on the basin bottom. Under the new situation, the previously dry lands adjacent to the reservoir rim get saturated with seepage of water. The saturation of the ground by backwater along with wave action on water surface accentuates land subsidence and slides along the periphery of the reservoir.

The reservoir may be in hilly areas or in lowlands. The lowland areas undergo maximum degradation with the first filling of the reservoir when abrading action of saturated ground brings about rapid change in the morphology of peripheral areas. The reservoir floor also gets rapid filling from the denuded shore materials. After a lapse of certain time, the rate of siltation and erosional effect is reduced and a stable condition is obtained. Lowlands of alluvial type influence more deposition of sediments compared to hilly regions.

In mountainous regions, the water body of a reservoir covering the narrow valleys is generally of low capacity. The filling of reservoir in this terrain causes infiltration of water through the debris and joints in rocks at the hill slope, initiating rock fall and debris slide. The fallen debris will be easily abraded and transported to the reservoir floor. Due to selective siltation, the tail end of the reservoir of a hilly terrain may be choked along the rim areas. The size of the depositing materials may be big and thus the volume of deposition may be sufficient, and the capacity of the reservoir will be reduced within a short period.

The creation of a reservoir by constructing a dam in the upstream part of a river controls its downstream flow pattern to a great extent. Daily fluctuation of the run-off below the dam may be associated with the peak load operation of the power plant resulting in irregular stream flow. The river bed of the lower reaches changes its hydrological regime. An intensive scouring of the river valley takes place by the water with silts spilled from the reservoir during floods. If the flow with load of silt is small, it will cause only limited scouring. However, excessive flood discharge results in deep scouring of the downstream land. The flood water inundates the housing areas of the downstream valley causing distress to the people. Another effect is the drop in water level, which affects agriculture and water supply to the adjacent areas.

15.4 EROSION OF CATCHMENT AREA AND RESERVOIR SEDIMENTATION

Eroion of catchment area is caused bv various natural aqgencies of which rainfall is the main agent. After a shower, the eroded materials from the catchment area get deposited in the reservoir basin resulting is a decrease in reservoir capacity. The following gives a detailed description of this aspect.

15.4.1 Relation of Erosion with Rainfall

The varied natural agencies and geological processes such as extensive weathering of rock and erosion of slope materials affected by landslide and mass movement are responsible for change in landform of catchment areas. This is especially true in high rainfall and mountainous areas where steep slopes and considerable velocities of the stream water facilitate rapid transportation of the eroded materials as bed load and partly in suspension that ultimately accumulate in the reservoirs. In an attempt to quantify the erosion of a terrain caused by various processes, Murthy (1977) gave an equation that can be expressed as

$$E = IRS^{1.35} L^{0.35} P^{1.75}$$

The equation shows that E (soil loss) will be of increasing order with increasing values of I, R, S, L, and P. Here, I is the inherent erodibility of the soil, R is a factor describing cover condition, S is the degree of slope in per cent, L is the length of the slope in feet, and P is the maximum 30 minute-intensity rainfall (in inches), covering the last 2 years. In Eastern Himalayas, the annual rainfall is as high as 500 cm. There are records of rainfall up to 78 cm per hour continued for eight hours and of more than 1 cm rain per hour prolonged continuously for days together in the past 70 years. All these records indicate very high P value for the terrain. The steep slope S with great length L and unstable hill cover materials (to provide a high value of R) has been established from a study of mountainous terrain (Gangopadhyay 1978). The study indicated that hill slopes with values higher than 19° are liable to slide during monsoon. This unstable mountain slope material accounts for a high value of R. Thus, all values of P, L, S, R, and L being high, it is expected that E, the soil loss of the catchment area due to erosion, will also be high and consequently it will result in high rate of sedimentation of the reservoir (Fig. 15.3).

Fig. 15.3 Umiam reservoir in Meghalaya (see Fig. 14.3 for dam site and Appendix A.3 for description)

Studies estimated that 40,000 m³ of silt gets deposited in the Umiam Lake in Meghalaya every year. Such high rate of siltation decreases the storage capacity and increases water loss by evaporation.

15.4.2 Sedimentation (Siltation) Rate

All rivers carry suspended sediments consisting of sand, silt, and clay-sized particles. The area below the DCL is the place of accumulation of sediments transported from the surrounding areas. The rate of sediment deposition in the reservoirs varies from river to river. The sedimentation or siltation becomes intensive during rainy season due to increased run-off capacity. Silt deposited in the reservoir from its surrounding areas reduces its capacity and as such shortens the life of the dam. If soft rock or loose sandy and silty soil is present in the catchments, rains wash the materials easily and streams carry the large load of silt and deposit it into the reservoir. The huge quantity of debris deposited in the reservoir basin from slides in the peripheral regions decreases the life of the reservoir.

Deforestation of lands around the reservoir areas enhances soil erosion and increases the silt flow in the reservoir basins. In such reservoirs of anticipated high siltation, an auxiliary dam is constructed upstream of the main dam to reduce the siltation. The quantity of silt carried by a stream that will be deposited in the reservoir after construction of dam is calculated in the design of the dam based on some assumptions. The statistics of siltation rate of some Indian dams show that the actual rate of siltation is many times more than the assumed rate, and as a result the life expectancy is reduced to a great extent (Table 15.3). Life of reservoir refers to the physical life based on the rates of siltation assumed at the designed stage and later observed.

Table 15.3   Siltation rate and life expectancy of important reservoirs of India*

Name of reservoir	Annual rate of siltation (ha m/1000 km²)		Percentage of assumed reservoir life actually available
	Assumed	Observed	
Bhakra	4.29	5.95	72.20
Tungabhadra	4.29	5.98	78.77
Matatila	1.33	4.33	30.25
Panchet	6.67	10.48	63.88
Maithon	9.05	12.39	72.85
Mayurakshi	3.75	16.48	22.70
Shivaji Sagar	6.67	15.24	44.00
Hirakud	2.52	6.60	38.09
Gandhi Sagar	3.61	9.64	37.01

*Data from World Commission on Dams adapted from Public Account Committee India
(PAC 1982–83:103)

15.4.3 Measurement of Siltation

The annual rate of silt deposition in a reservoir under storage condition is measured by survey method. First, a contour map of the reservoir area is prepared by ground survey prior to storing water (empty condition). Permanent reference points (benchmarks) are kept on the ground on both sides of the reservoir. Second, contour map of the reservoir area under storage condition is then prepared at the given time interval after reservoir filling following the *range method*. The directions (ranges) are first marked on the map with reference to the permanent reference points coinciding with certain fixed feature of the previous survey map.

The depth of the river with respect to the water surface of the time is measured by means of a sounding pole or the *echo-sounding* instrument while moving in a country boat from one bank to the other. The boat carries a long wire (with distance marked) rolled in a wheel. As the boat moves, the wheel unrolls the wire to facilitate measurement of the distance covered. A man with walkie-talkie guides the direction of movements of the boat. Once the contour map for the storage condition is prepared, the volume of silt deposit in the given time is known from the planimetric measurement of the areas between two contours. The silt deposit of a reservoir basin is generally expressed as 'annual rate of siltation' in ha/1000 km². From the given time in years, the annual rate of silt deposit in the total reservoir area (in km²) can be obtained from dividing the total volume (ha m) by the number of years.

15.4.4 Siltation Study by Remote Sensing and Aerial Photographs

Satellite imageries and aerial photographs are important tools in the investigation of reservoirs including siltation. The false colour composite images of Landsat 2, 3, and 4 depict the land use, land cover, geomorphology and drainage pattern, and the reservoir water directly affecting the inflow sediments in the reservoirs in the case of DVC projects. High siltation zone was observed at the meeting of the river Barakar with the river Damodar because the sediment-carrying capacity is suddenly slowed down and silt spread gradually into the reservoir mouth (Srivastava and Rao 1991)

Study of aerial photograph is carried out in the reservoir basin covering the catchment area for a rapid evaluation of the geological and hydrological conditions. The stereoscopic study of overlapping photographs enables assessment of the water storage capacity of a reservoir basin for a particular dam height. The study examines the catchment area and reveals the characteristics that influence the reservoir under storage conditions. These include morphological features, whether hilly or lowland; stream gradient and its effect on run-off; drainage pattern; ground slope and ground use, viz. whether terrace cultivation; erosional intensity along reservoir periphery, whether protective measures are necessary; forest cover, which retard the run-off; and river features such as meandering river, oxbow lake, and silt bar. Geological characteristics such as overburden cover, rock outcrops, structural pattern, rock and soil types, slide prone areas, and peaty and marshy ground can be easily identified from the study of aerial photographs. Other information such as whether the reservoir areas contain construction material for the dam and appurtenant structures can also be obtained from aerial photographic interpretation followed by field observations.

15.5 RESERVOIR CAPACITY AND RESERVOIR LIFE

The volume of water that a reservoir can hold is known as the *reservoir capacity*. It is calculated from the reservoir cover and depth and expressed in hectare metre; 1 hectare metre (ha m) = 10,000 m³ (base 100 m length × 100 m width × water column 1 m high). The reservoir capacity varies with deposition of silt from the catchment area and the rate of siltation is measured following the method as outlined in Section 15.4.3. The *reservoir life* is the expected number of years for which the reservoir will be able to hold substantial quantity of water.

The period up to which the reservoir can be used should be decided by factors controlling the catchment area such as the topography of the catchment area and the nature of soil and rock.

The drainage area covered by the soil and each type of rock is to be considered separately as their erodibility values will be different. The reservoir life 'assumed' in the design very often differs from the 'observed' life of reservoir computed from the rate of siltation after few years. In some projects, the assumed and observed values differ widely.

Though siltation in a reservoir depends on the morphology, geology, and hydrology of the drainage area, sediment outflow also has a role in it. A portion of the incoming sediments flows down the dam. This reduces the actual deposit in the reservoir with respect to total intake in the storage area. As seen from particle study, the silt and clay size particles mostly (95 per cent) remain in suspension and the outflow is very small compared to sediment deposits. Hence, desiltation measures are necessary if the inflow is excessive. It is difficult and will be a costly proposition to remove the silt from the reservoir floor by dredging. Sluicing is generally done as a desilting measure. In this procedure, the outlet gates are opened from time to time and sediments are forced out at a high speed of water. Desilting basins are constructed adjacent to the reservoirs to allow settlement of sediments, which reduces the quantity of sediment flow to the reservoir.

15.6 SALIENT ASPECTS OF RESERVOIR INVESTIGATION

'Reservoirs must hold water. Not all of them do, and there are many ways by which water may be lost. Most of them are geologic or of geology dependence', observed the veteran engineering geologist Dr Charles P. Berkey (1950). It can be easily understood that if the reservoir does not hold water, the very purpose of constructing a dam is defeated. In other words, the reservoir area must be watertight to prevent escape of water to the downstream valley. Ground saturation and infiltration of water through fractures in rock at the initial stage and evaporation during the summer months account for loss of reservoir water to some extent. Adverse geological conditions such as presence of faults, cavernous limestone, conglomerate, or highly permeable stratified rocks may cause enormous leakage from the reservoirs. Geological investigation for reservoirs is, therefore, directed towards finding the prevalence of such conditions and the possibility of leakage after water is impounded in the reservoir.

15.6.1 Different Possibilities of Reservoir Leakage

There are several instances of leakage from reservoirs related to adverse geological conditions such as the Dudhawa and Kedar Nala dams in Madhya Pradesh, Nagarjuna Sagar dam in Andhra Pradesh, and Hirakud dam in Orissa (Appendix A). The following geological attributes are mainly responsible for leakage of reservoir water:

(i) The reservoir located in *karstic or cavernous limestone* terrain may become incompetent to hold water. In karstic region, subterranean channels or tunnels created by solution of limestone may extend from the storage area down to the adjacent lower valleys threatening escape of the entire reservoir water (Fig. 15.4). A thorough investigation is, therefore, necessary to prove the water tightness of a reservoir in karstic limestone.

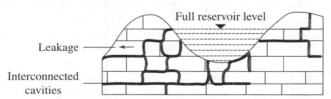

Fig. 15.4 Leakage of reservoir water to lower valley through interconnected caverns in reservoir limestone

(ii) *Buried channel* contains loose sand and silt of very old channel or paleo-river deposits through which large-scale leakage may take place under reservoir condition. Such channel deposits occurring in the foundation rock or reservoir periphery and connected to lower valley may cause excessive leakage under reservoir condition, see Fig. 15.5 (a). The exact location of the zone is traced by exploratory work and accordingly grouting is done to seal the leakage path. Walters (1971) has given an account of the presence of epigenetic gorges (buried valleys) in dam projects from different parts of the world and the leakage problems faced.

The Moosakhand dam in Uttar Pradesh is an example of facing leakage and uplift problem below the dam due to the presence of a buried channel filled with loose till material and peat. The affected zone was treated by grouting and providing relief wells. In three river valley projects of Arunachal Pradesh, geotechnical investigation with the aid of subsurface exploration revealed the presence of thick deposits of buried channels, see Fig. 15.5 (b) posing problems of foundation stability of dam and powerhouse and leakage of water from the reservoirs. After considering different aspects of the geotechnical problems created by the buried channel, it was planned to consolidate the deposits by grouting to render these suitable for foundation and arrest leakage (Krishnamurthy, Sinha, and Bhatia 1992).

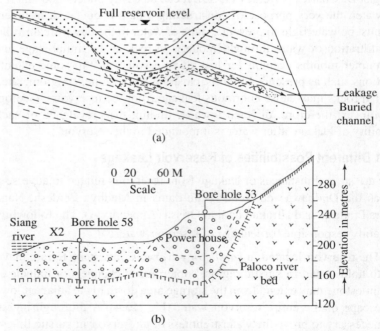

Fig. 15.5 Leakage due to buried channel: (a) buried channel in reservoir rock showing passage of leakages to lower valley; and (b) thick deposit of buried channel in a project site of Himalayan terrain (Krishnamurthy, Sinha, and Bhatia 1992)

(iii) Sedimentary rocks such as conglomerate, breccia, and porous sandstone or unconsolidated deposits such as boulder and gravel present in the reservoir rims or at the dam site due to their high permeability may cause large-scale leakage of the reservoir water (Fig. 15.6). The soft clay materials from conglomerate matrix are washed away after filling the reservoir and the rock disintegrates creating openings for escape of water.

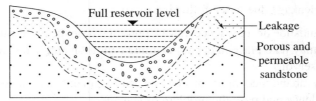

Fig. 15.6 Leakage of reservoir water through porous and permeable rocks

(iv) Even very hard rocks of a reservoir area affected by fault, shear zones, and extensive joints may pose problems of leakage. If highly jointed rocks occur in the reservoir rims and extend to adjacent valleys, they may cause substantial leakage of reservoir water. The presence of a fault or wide shear zone passing through a dam site or reservoir to lower valleys leads to a potential danger of leakage. The shattered materials of the shear zone or the clay gouge of the fault plane are washed away under reservoir condition creating avenues for water to bypass the fault zones (Fig. 15.7). In Umiam project of Meghalaya, under storage condition, substantial leakage took place through a narrow zone of fissured quartzite dividing the reservoir from a downstream valley. Grouting in the divide could not stop the leakage, but after nearly a year, it became negligible because of silt deposition into the fissures of the rock.

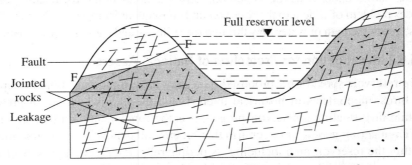

Fig. 15.7 Fractured rock including fault (F–F) creating passage for leakage of water from the reservoir

(v) The hydraulic gradient or the direction of groundwater movement is also the direction of possible leakage from the reservoir to the adjacent lower valleys. However, if the water table in the lower valley is above the FRL, generally there is no danger of reservoir leakage. The Tenughat reservoir in Bihar is separated from the Bokaro valley by a narrow ridge, but since the water table of this valley was higher than that of the reservoir, no leakage took place after reservoir filling (Chowdhury 1975).

(vi) The presence of groundwater influences the capacity of the reservoir. Caution is to be taken to differentiate the actual water table from the perched water table that occurs in jointed rocks overlain by impervious shale. When punctured by a bore hole, the water will rise above the shale showing the actual water table. Attewell and Farmer (1976) observed that if the phreatic water at the reservoir periphery after storage is found to be at a higher level than the FRL, the groundwater will flow towards the reservoir increasing its potential capacity.

15.6.2 Problems from Slide and Resultant Sedimentation

In addition to the reasons given in Section 15.6.1, a reservoir may be vulnerable to sedimentation along with leakage due to the following reasons:

(i) Old landslide or subsidence may be present within the reservoir spread along which further slide may take place under water-filled condition and deposition of the slid material into the reservoir.

(ii) If any abandoned mine or coal measure remains within the reservoir, it may initiate subsidence and pose the problem of leakage.

(iii) If residual soil or overburden material of great thickness lies in the reservoir rim, it may create an avenue for leakage through the periphery.

(iv) In some cases, selection of borrow areas and removal of impervious layer of earth from the reservoir may expose fractured rocks through which leakage may take place.

(v) Vibration due to blasting in the quarry sites close to reservoir rims may cause the problem of landslide. The debris thus accumulated in the reservoir rims may also cause the problem of leakage and sedimentation in the reservoir.

Under the new regime of the reservoir, the water table rises and the soil or detritus covering the peripheral lands of the reservoir gets saturated, causing instability. This may lead to frequent creep failures or debris slide from the periphery. If the rim area is rocky but faulted or folded and possesses old slide scars, these may cause large-scale rock slides. Layered rocks in the peripheral regions of the reservoir with bedding planes dipping towards the adjacent valleys are liable to create slides under reservoir condition. Unstable lands in the peripheral region of the reservoir poses problem of slide into the reservoir and consequent increase in siltation. The slide debris may also block the path of water flow through the spillway and cause failure of dam due to overtopping.

15.6.3 Investigation for Protection of National Assets

The reservoir spread of a proposed dam may have important mineral deposits, rare animals or birds, monuments of historical interest, and the like, which are of great national interest. While investigating a reservoir project, it is the fundamental duty of the investigator such as an engineering geologist to confirm that no mineral deposit of economic importance is present within the reservoir area including its peripheral parts. Once the reservoir is full, the mineral deposit, if any, will remain buried under the reservoir water causing a great loss of national resource. During reconnaissance or feasibility stage investigation, it is essential to find whether there is any mineral deposit in the reservoir area so that the planner can take a decision on the project. The Chandil dam reservoir in Bihar across the river Subarnarekha was allowed to submerge an abandoned gold mine after providing dykes.

Within the reservoir area of a dam project, there may be sites with buildings of historical importance. A forest with rare animals, birds, and trees such as a tiger project, a bird sanctuary, or reserved forest with rare species of plants may exist in the storage area. Sites of archaeological interest may be also present in the reservoir area. These aspects are to be investigated at the preliminary stage by reconnaissance so that these sites are either avoided or plans are made to re-store them, if possible, in a suitable place. The Silent Valley project of Kerala was abandoned due to the existence of rare birds and plants in the hills within the proposed project area. The Aswan dam required removal of archaeological treasures to a different site. Several historical temples near the Nagarjuna Sagar dam in Andhra Pradesh were shifted, piece by piece, to another place to avoid their submergence by reservoir water.

15.7 PUMPED STORAGE SCHEME WITH CASE STUDY

Development of hydroelectric power by pumped storage scheme is an energy storing system. The basic concept of the scheme lies in storage of base load energy for later release as premium peaking power. Where the load demand between day and night fluctuates to a great extent, this scheme is found to be attractive. The motor or generator used in this method is such that it can synchronize rapidly. The main purpose of pumped storage scheme is to meet peak electricity demand.

15.7.1 Basic Aspects

The scheme consists of an upper reservoir and a lower reservoir connected by a water conductor system and a powerhouse. During the period of low demand, the power generated by power plants is utilized to pump water of the lower reservoir to the higher one. During peak demand time, water from the upper reservoir is conveyed through the turbines to generate electricity to meet the power demand.

The scheme needs 3 kWh pumping energy to generate 2 kWh energy, which may appear to be uneconomical. However, the economic advantage comes from converting low value, low cost off-peak energy to high value, high cost peak energy. The following are the main advantages of the scheme:

(i) It can be operated to full output.
(ii) It can be used for storing surplus power in the upper reservoir.
(iii) It can save fuel by drawing pumping energy from large thermal power plants.
(iv) It imparts flexibility to the power system.

15.7.2 Investigation for Pumped Storage Scheme

Engineering geological investigation to select sites for different structures of the pumped storage scheme should satisfy the following engineering, topographical, and geological conditions:

(i) The fundamental need for the successful operation of the pumped storage scheme is locating the project sites near the existing power plant where peak load demand is to be met by generating additional power. It is also to be seen that 'no demand' or 'off-peak' power is available for a long continuous period without failure. Selection of structural sites of the pumped storage project close to a thermal plant will ensure smooth operation of the scheme without loss of energy in transmission.
(ii) In the pumped storage scheme, the morphological expression gets prime consideration in drawing the project scheme. The topographical positions of the upper and lower reservoirs should be such that sufficient head is obtained from the level difference of their locations. Normally, the head will be nearly one-fifth of the horizontal distance between them. If the topography provides higher head due to decreasing horizontal distance, the required storage of water will be less and the size of the tunnel as well as the power plant building will be reduced. This will reduce the construction cost of the project.
(iii) The other important considerations are the hydrology and geology. There should be sufficient water in the reservoirs for 10 hours of peak operation. The data on discharge of the streams contributing water to the reservoirs and the annual rainfall of the area is considered in the design of available storage. Geotechnical study should ensure that the reservoir is watertight so that there is no water loss. Leakage of reservoir water may bring about failure of the scheme for want of sufficient water. The sites should preferably

be located in igneous and metamorphic terrain with hard and compact rock showing no significant permeability and absence of any weak zone such as fault and shearing. Quarry sites for sufficient quantity and suitable quality of construction materials for the structures should be available in the vicinity.

15.7.3 Case Studies on Pumped Storage Schemes

The pumped storage scheme has been successfully implemented in many places. The following subsections analyse the implementation of the scheme in two such areas in India.

Kadampari Pumped Storage Scheme in Tamil Nadu

This scheme with an underground powerhouse is under operation since 1987. Prior to implementing this scheme, the Tamil Nadu grid was to cater to a heavy share of agricultural load for five hours a day but demand during the rest of the time covering the night was very low. The pumped storage scheme was then taken up in the Kadampari area with suitable topography and geology. The reservoirs and other structural sites consist of charnockite and migmatite gneiss, which are very hard and sound rocks traversed by some steep dipping joints. The underground powerhouse of the scheme located 200 m below the ground has generation capacity of 4×100 MW power. The power generation was trouble free for three years but then the project faced an accidental trouble due to fire in the powerhouse. The power generation was stopped for some time when abnormal seepage and buckling of power shaft were observed. The powerhouse was dewatered and suitable rectification measures adopted.

Purulia Pumped Storage Scheme of West Bengal

In the West Bengal power system, the peak demand is very high but lasts for a short duration during daytime. The full demand cannot be met during peak time due to shortage of power whereas there is surplus during the night. The Purulia pumped storage scheme was taken up to meet this peak demand by utilizing surplus generation from the thermal stations. The scheme consists of an underground powerhouse of 800 MW capacity with 4×200 MW reversible turbines. The heights of the upper and lower dams are 77 m and 85 m, respectively, and the live storage capacities of the upper and lower reservoirs are 12.48 million m^3 and 11.60 million m^3, respectively. The total length of the water conductor system that includes lined tunnel and steel penstocks is 850 m (Fig. 15.8).

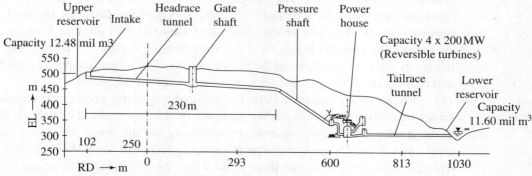

Fig. 15.8 A typical section of the Purulia pumped storage scheme (Bagchi 1991)

The project area under the Purulia pumped storage scheme comes within the confines of Ayodhya hills of Chota Nagpur plateau. The scheme utilizes water of the tributaries of River Subarnarekha. All the structural sites such as dams, reservoirs, tunnels, and powerhouse have been located in areas with favourable topography and geology, the entire project site being composed of hard and sound gneissic rocks. The joints in the rocks are tight and in many places filled with quartz–felspathic materials. Rocks occur at the surface or at the subsurface under shallow cover of overburden as proved from exploratory drill holes. The reservoirs cover forest areas but without significant inundation of the flora. There were no displacements of the inhabitants and as such environmental degradation as a result of the implementation of the scheme is avoided (Bagchi 1991).

15.8 PETROLOGICAL STUDY OF SUSPENDED SILTS IN RIVER WATER

The high content of sediments carried through rivers causes problem of rapid filling of reservoir. The rivers of mountainous terrains such as the Himalayas carry high content of sediments during times of snow melt and rainfall. While the coarse-size materials are taken by traction load, sufficient quantity of finer sediments such as sand, silt, and clay are taken by suspension. An estimation of silt flow through the river Teesta in Sikkim shows the height of the suspended particles to be as high as 5148 cubic metres per square kilometre per year. If such high quantity of suspended sediments passes through the tunnel, it may choke parts of the tunnel obstructing water movement.

If silts containing abundant minerals of high hardness (> 5 in Mohs scale of hardness) and flowing at a high speed are allowed to enter the powerhouse of a hydroelectric project, they may abrade the penstock pipes and blades of hydro turbines. In the Jaldhaka project of Darjeeling, Himalayas, excessive silt flow of reservoir water not only choked a part of the pressure tunnel, but also caused abrasion of the turbine to such an extent that it required repairing of the machine, hampering power generation for a considerable time.

This calls for analyses of the suspended sediments, especially of mountainous areas with respect to their quantity, particle size, and shape but the main aim is the determination of hardness of the different sized particles. This can be done from the identification of the minerals in the sediments by microscopic study and then computing their statistical distribution. Microscopic analysis and statistical distribution of suspended particles with respect to hardness were carried out in many Himalayan projects for design purposes (Gangopadhyay 1980). The data was helpful for adopting measures against heavy inflow of silt and deciding on material hardness of turbines.

15.8.1 Procedure for Determining Hardness of Particles

Daily samples of suspended silt are collected from the central part of the river section close to the proposed dam site while routine measurements of river discharge are taken by the engineers. In the laboratory, the quantity of silt measured each day is recorded in grams per cubic metre and thus the monthly and annual quantities are obtained. The total quantity of sediments of the four wet months (June–September) is subjected to mechanical analyses (using standard sieves) in four size grades (> 0.5 mm, 0.5–0.25 mm, 0.25–0.10 mm, < 0.10 mm). The representative samples containing 400 to 500 grains of each size grade are taken. The grains are studied under the microscope to identify the minerals and then computed for hardness H (Mohs scale of hardness) and classified into four groups:

Group A ($H < 5$)—muscovite, biotite, chlorite, calcite, and magnetite

Group B (H 5–6.5)—apatite, ambhibolite, pyroxene, ilmenite, hematite, pyrite, and rutile

Group C (*H* 7)—quartz
Group D (*H* 7–7.5)—epidote, garnet, tourmaline, zircon, sillimanite, kyanite, and staurolite

15.8.2 Example for Hardness Measurement of River Silt

In the Sindh hydroelectric project in the Western Himalayas, the average quantity of suspended particles carried by the river per year was measured to be 1320 g/m^3, the monthly maximum being 481 g/m^3 (August) and the minimum being 29 g/m^3 (December). The percentage weights for samples of four months (May to October) in different size grades were computed. The variation found is shown in Table 15.4.

Table 15.4 Percentage weights for various size grades of suspended particles of Sindh River

Size grade (mm)	Percentage by weight
> 0.50	0.1–3
0.5–0.25	1–7
0.25–0.10	28–77
< 0.10	35–70

The statistical count of percentage by number with respect to hardness of silt particles under different size groups is shown in Table 15.5.

Table 15.5 Statistical count by percentage for various hardness of suspended particles of Sindh River (average values of analyses of ten samples)

Particle size (mm)	Hardness H in Mohs scale			
	1–5	5.5–6.5	7	7–7.5
> 0.50 mm	8	16	74	2
0.5–0.25 mm	7	14	75	4
0.25–0.10 mm	6	17	72	5
< 0.10 mm	8	16	73	3

The study reveals that the maximum number of grains (average 77 per cent) possesses hardness *H* in the range 7–7.5 (including quartz grains), having very high abrasive capacity. On an average, 15 per cent grains have medium hardness (5.5–6.5), which can abrade iron easily, and only 8 per cent of the particles with low hardness of *H* less than 5 have no deleterious effect. The grains studied under microscope showed that about 95 per cent of them were of irregular shape with sharp edges. The overall analyses of the data indicated requirement of adopting measures to reduce the suspended silt content and save probable damage of the hydro turbines from abrasion.

Modal tests and other studies have revealed that water containing more than 0.2 kg/m^3, particle size of more than 0.25 mm, and mineral hardness of more than 5 will be very harmful. Smaller particles of less than 0.1 mm may also be harmful depending upon their shape and hardness. Silt quantity of 0.1 kg/m^3 or more may damage the turbine parts depending upon the total head. If a turbine with a high head and high rpm is allowed to run under low head (less than 60 per cent of the rated capacity), considerable damage by cavitations and erosion may take place in its runner guide vane and draft tube under the thrust of water-borne silts.

It is of practical importance for the engineering geologist to undertake the following studies of the suspended sediments to understand the magnitude of the problem and make proper planning to avoid such problems:

(i) Measure the quantity of silt carried by the stream at its catchment area.
(ii) Study the size of the suspended particles, including its angularity.
(iii) Find the hardness of the particles and interpret the results of their effect on the hydro turbines.

15.9 RESERVOIR-RELATED EARTHQUAKES

Many reservoirs are responsible for causing earthquakes. When a large dam is constructed, the impounded water exerts pressure on the reservoir floor, which may trigger earthquakes. Such reservoir-induced seismicity (RIS) was first recorded in Quedd Fodda dam in Algeria in 1932. The 221 m-high Hoover dam in the US experienced an earthquake of maximum intensity *M* 5 on Richter scale after four years of reservoir filling in 1935. The investigation on the Hoover earthquake indicated that it was of RIS type, the pressure of reservoir water on the floor having a distinct relation with the earthquake activity. Since then, investigation has proved that earthquakes of different magnitudes have occurred after impoundment of large reservoirs in different parts of the world (see Section 20.8).

In India, the 104 m-high Koyna dam in Maharashtra experienced RIS of varying magnitudes from the very start of reservoir filling in 1962 and the maximum intensity of 6.3 was felt in 1967. The RIS of Koyna was so destructive that it damaged nearly 80 per cent of buildings in Koyna town, caused the death of about 200 people, and injured nearly 1500 persons. The tremor was felt even in Mumbai about 230 km from Koyna.

Presence of faults and segmented rocks in the reservoir area increases the chance of earthquake (RIS) under water-filled condition. In the Koyna reservoir area, igneous rocks have intruded the parent rocks of older metamorphics. The boundaries of the intrusive body and the older rocks are affected by a thrust that occurs within a shallow depth (less than 3 km). The change of pore pressure induced by reservoir water reduced the strength of the rocks leading to failure along the thrust plane and resulted in earthquake. A network of digital and analogue seismographs installed at Koyna recorded distinct pattern in the earthquakes induced by the reservoir. The foreshock and aftershock pattern and the decaying rate of the aftershocks differ from those in normal earthquakes. It was observed that the intensity of RIS is dependent on the fluctuation of water depth in the reservoirs. It increases during the rainy season with increased rate of filling water and decreases during the summer season.

A dam with very deep reservoir basin that can hold large volume of water may also induce earthquake. Gupta Harsh (1992) after studying the RIS in different large reservoirs of the world came to the conclusion that any dam with a reservoir depth of more than 85 m has a chance of causing high-intensity earthquakes and as such the dams (higher than 85 m) should be designed so that they can withstand earthquake shock up to 6.5 in Richter scale.

The optimum structural set-up is the gravity faulting environment (and not thrust faulting) when even a relatively meagre water load may act as a triggering effect for RIS. There are 11 large reservoirs in the Himalayan terrain. However, there is no evidence of RIS from any of these impounded reservoirs. Non-occurrence of RIS in the vicinity of Himalayan reservoirs is comprehended primarily due to thrust fault environment, which is non-conductive for RIS (Gupta 1992).

15.10 ENVIRONMENTAL IMPACT OF CREATION OF A RESERVOIR

In the construction of a project, the most desirable aspect is to improve the human lives and at the same time minimize environmental degradation. While dams with reservoir are needed for multi-purpose benefits including power generation, it is very important that environmental study is undertaken such that the problems related to displacement of the people, loss of forest and farm land, and ecological imbalance are properly understood at the early stage of project planning.

15.10.1 Salient Aspects of Environmental Changes

The environmental changes caused by the creation of a reservoir are mainly of the following types:

Morphological change The reservoir water causes downstream degradation by erosional effect and brings about deterioration of the catchment area enhancing silting of the reservoir and consequent shortening of its life.

Social change It causes displacement of the population and their old habitats by the submergence of their lands.

Ecological change It disturbs the harmony of the biological process and imposes biogenic degradation.

The water of the reservoir due to construction of a dam submerges upstream valley and dries up the downstream parts to a great extent, obliterating the prevailing conditions. If the reservoir is located in a narrow and deep valley, it may hold a large quantity of water with less submergence of lands and less environmental problems. However, it depends on the overall morphology of the storage area. In general, the larger the dam, the more will be the area covered by the reservoir involving submergence of agricultural lands and populated villages.

Construction of the reservoir causes deterioration of the soil and rock of its surrounding areas by saturation of water. It influences rapid erosion of the surface soil and porous rock of catchment area and deposition of the sediments in the reservoir water. The storage of water in a reservoir upsets the equilibrium of the river by changing discharge characteristics and transport capacity of sediments including the bed load. Under reservoir condition, the water pressure of the submerged land and forest gets transmitted partly or wholly to the adjacent catchment area causing degradation of land, which becomes prone to erosion.

The hydrological regime is changed with creation of the reservoir causing a rise in groundwater table. This increases seepage of water into the soil and alteration of soil chemistry. The seeped water by capillary action tends to go up into the soil pores and after evaporation leaves some incrustations of harmful effect on crop growth. The salt incrustation also prevents growth of plant roots.

If quarry sites and borrow areas are located in the reservoir periphery instead of inside the storage area, this may cause easy erosion of the barren surface of the abandoned quarry sites. It increases sedimentation in the reservoir water. Slide-affected areas around the reservoir are vulnerable to more erosion. Most of the storage dams are constructed for power generation and flood control, but rapid siltation of the storage area may affect power generation and enhance the chance of flooding of downstream valley. Above all, it affects the people of the area due to submergence resulting from construction of dam. Hence, in the planning stage, it needs to be decided whether, after spending huge funds, the construction of the storage dam will be overall beneficial or will be detrimental to the people and the environment.

The fauna and flora of a land are greatly affected by the spread of reservoir water. There is a change in the aquatic life too due to turbidity, shortage of organic matter, and dissolved mineral content. The deep water becomes depleted in oxygen. The downstream part of the dam is deprived of fertilizing silt, affecting crop production.

The impact is apparent on the people, animals, and plants of the terrain that will be covered by the dam and reservoir project. Building of a dam requires cutting forest trees and excavation of hill slopes for construction materials causing ecological disturbances. Construction of the Idukki dam across the river Periyar in Kerala reduced the forest land and degraded the vegetation cover with submergence by reservoir water. The Silent Valley project of Kerala was stalled due to the harmful effect of biosphere reserve.

15.10.2 Measures Taken to Minimize Adverse Environmental Effects

It is necessary to consider whether these environmental effects are adverse to overall benefits of the dam and reservoir project. The following measures are generally adopted or required to be taken up to minimize the negative environmental effects:

(i) Large dams having chances of causing environmental problems are nowadays given serious consideration. The environmental effect may be reduced by construction of small dams and reservoirs. It is agreed that small dams are more cost-effective than large dams.

(ii) In the catchment area, planting of trees is taken up to arrest soil erosion and prevent environmental degradation. Afforestation and pasturing reduce ecological imbalance. In the Dulhasti project of Jammu and Kashmir, Chamera project of Himachal Pradesh, and Rangit project of Sikkim, a huge number of trees (100 to 500 times the original) were planted in the catchment area. Such restoration work of the forest resources is a part of environmental management needed in reservoir projects.

(iii) Reduction of siltation to the reservoir should be aimed so that it does not reduce the generation capacity of power. Measures towards reducing silting of the reservoir include plantation, construction of check dams, making contour bunds, benching of the hill face, and wire-netting of the slide-prone ground slopes. The quarry sites or borrow areas may be located within the reservoir areas; it will remain under water minimizing the erosion of the adjacent areas of the reservoir.

(iv) People displaced from the reservoir area are to be rehabilitated by providing them house and land for irrigation or facilities of job in accordance with their professional need. In the question of settlement, 'land for land' is generally accepted. Lands given to cultivators should be similar to pre-settlement conditions or at least not of poorer quality.

The engineering geologist while investigating a reservoir project records the factors of environmental impact by preparation of a land use map at the preliminary stage of investigation. The engineers or planners are to consider these factors revealed from the environmental study before taking decision on the execution of the project. Many environmental measures may bring about more benefit to the people by improving the quality of their lives. To achieve this end, a harmonious development of the hydropower and other benefits of the reservoir projects, maintaining essential ecological processes is necessary.

SUMMARY

- A reservoir is a water body created by constructing a dam across a river. The water of the reservoir can be used for irrigation, power generation, pisciculture, tourism, and even drinking purposes.
- Engineering geological investigation of a reservoir area includes geological mapping with special emphasis on leakage problem. If the reservoir area consists of porous and permeable rocks, detailed exploration of subsurface rocks by drilling and permeability tests are carried out to find the possibility of leakage under water-filled condition.
- Geophysical survey and tracer study are also undertaken in karstic limestone terrain to assess cavernous characters and flow characteristics of reservoir rocks. While investigating a reservoir area, it is to be seen that mineral deposits, historical buildings, and forest areas having rare wildlife do not fall under the submergence area.
- In an area where topography is suitable for two reservoir sites with sufficient head between them, and if demand of power in the area during the day varies sufficiently from that at night, it is economical to generate hydroelectric power by the pumped storage scheme.
- In the pumped storage scheme, the water stored in the upper reservoir is led to the powerhouse below to generate electricity during daytime and then released to the lower reservoir from where water is pumped back to the upper reservoir by night. Engineering geological investigation under this scheme involves the study of two reservoir sites with the ultimate aim of finding the competency of the reservoirs to hold water.
- Creation of a reservoir has environmental impacts bringing about morphological, social, and ecological changes. The most obvious environmental impact is the displacement of people of the submergence area from their abodes and depriving them of their land for irrigation. Animals and birds of the reservoir area covered by forest get disturbed and may even become extinct if they belong to rare species.
- With filling of the reservoir, the groundwater table of the adjacent lands rises. This causes saturation of the lands that were dry earlier, resulting in rapid erosion and even landslides of the surrounding landform. The deposition of eroded and slid materials into the reservoir reduces the storage capacity and the reservoir life.
- In a hydroelectric project, the reservoir water containing minerals of high hardness may enter into the hydro turbine and may cause damage to turbine blades by abrasion. This problem is solved by petrological analysis of the silt particles with respect to their hardness and accordingly designing the turbine to withstand abrasion.
- Large reservoirs induce earthquakes because of the pressure of the stored water on the reservoir floor or slide along faults in the reservoir rocks. Suitable earthquake factor, therefore, needs to be incorporated in the design to save the storage structure from earthquakes.

EXERCISES

Multiple Choice Questions

Choose the correct answer from the choices given:

1. The kind of study of silt particles of reservoir water that helps to understand abrasion of hydro turbine of a powerhouse made of steel involves:
   (a) determination of index properties of the silt particles
   (b) petrological study for identification of the mineral and knowledge of its position in Mohs scale of hardness
   (c) Using a magnifying glass (x10) to observe the silt

2. The range of hardness of minerals in Mohs scale of hardness that causes abrasion of the steel blades of turbines is:
   (a) 9 to 10
   (b) 3 to 5
   (c) more than 6

3. The mineral that may damage a hydro turbine because of its abrasion is:
   (a) topaz
   (b) fluorspar
   (c) appatite

Review Questions

1. Discuss the creation of a reservoir. Draw a sketch to show a dam with reservoir and label the following: live storage, DSL, and FRL. What are the factors on which the volume of reservoir water is dependent?
2. What are the utilities of reservoir water? Explain the adverse effects of creating a reservoir.
3. 'Once a reservoir is filled with water, both the surface and groundwater regimes are disturbed.' Give your comment in favour of this remark. Describe the effect of reservoir water in the upstream mountainous region and the change in flow pattern and other consequences in the river regime in the downstream part.
4. Describe the effect of siltation on reservoir life. State the method of measurement of annual siltation rate of a reservoir. Explain how aerial photograph study and remote sensing may serve as important tools in reservoir investigation.
5. What do you understand by reservoir capacity and reservoir life? Explain the method of calculation of reservoir capacity. State the factors on which reservoir life is generally dependent.
6. Give a short account of the ways of possible leakage of water from a reservoir to lower valley. State the main aims and importance of investigation of reservoir area prior to its filling.
7. What is the basic concept of the pumped storage scheme? State the fundamental need of the pumped storage scheme for its successful implementation. Describe the geological and hydrological considerations of taking up such a scheme.
8. What is RIS? How is it developed after filling of a large reservoir with water? What is the minimum depth of water that makes a reservoir vulnerable to RIS?
9. What kinds of environmental degradation are associated with the creation of a reservoir? Describe the deteriorating effects of construction of a reservoir on the people living in the reservoir area and how their plight can be minimized.

Answers to Multiple Choice Questions

1. (b)   2. (c)   3. (a)

16

Tunnels

After studying this chapter, the reader will be familiar with the following:

- Different types of tunnels and terminology of tunnel parts
- Geotechnical aspects of tunnel construction through rocks
- Methods of tunnelling through soft rock or unconsolidated deposits
- Problems in tunnelling due to adverse geological conditions and their solution
- Investigation method for selecting tunnel alignment and subsequent works
- Methods and machinery of tunnel construction
- Determination of rock mass quality and design of support for tunnel

16.1 INTRODUCTION

Tunnels are constructed to serve various purposes such as transportation of road or rail traffic, carrying irrigation water, and sewage conveyance. The main aspects of discussion in this chapter are the methods of tunnelling in rock and soft ground and procedure of engineering geological investigation from the stage of selection of tunnel until the completion of tunnel construction. It highlights the problems in tunnelling due to adverse geological and structural features and suggests remedial measures for strengthening the tunnel rocks. The chapter also provides an account of the tunnelling machineries and their use in relation to rock types. It further delineates the excavation methods and support systems of tunnels from the estimation of rock mass quality.

16.2 COMPONENTS AND TYPES OF TUNNELS

A tunnel is a horizontal or slightly inclined passageway excavated through the subsurface rock or unconsolidated materials and open to ground surface at the two ends. The shape of a tunnel in cross section varies widely but most tunnels in India are circular, semicircular, or horseshoe shaped. The roof and floor of a tunnel are termed the *crown* and *invert*, respectively, and the two sides are the tunnel *walls*. The *spring line of a tunnel* is the meeting point of the roof arch and the sides (Fig. 16.1).

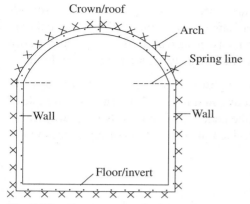

Fig. 16.1 Common terminology of a tunnel

In a tunnel, the *vault head* is the topmost part of the tunnel and the *support* is the structure used to take up the load. At the entry point of a horseshoe-shaped tunnel, steel ribs are erected as tunnel support and a masonry wall is provided above the vault head as protective measure against the overburden (Fig. 16.2). To a geologist, overburden is the unconsolidated soil or debris overlying the bedrock above a tunnel or other engineering structures. An engineer, however, considers the entire material (bedrock and unconsolidated material) overlying a tunnel as overburden.

Excavation work of a tunnel is generally executed from its two ends, known as inlet bulkhead or *inlet portal* and outlet bulkhead or *outlet portal*. *Shafts*, which are vertical passage from ground to tunnel, are also used to facilitate entry and excavation work of the tunnel. In some underwater tunnels, *cassions* are sunk to facilitate movement of men and equipment and boring of the tunnel in small segments. A *drift* is a small length tunnel that is bored nearly at the same level or close to the main tunnel line for the purpose of visual observation and instrumental tests of the rock quality with respect to tunnelling condition. Sometimes an *adit*, which is, in fact, a drift, provides passage for entry of men and machineries to conduct tunnelling work in additional faces (Fig. 16.3).

Depending upon the purpose of utility, tunnels are mainly of two types, namely hydraulic tunnels and traffic tunnels. Tunnels are also constructed for other purposes such as storage of fuel, nuclear installation, and disposal of hazardous materials.

Hydraulic tunnel In a hydraulic tunnel, the works of the tunnels are associated only with the passage of water. Hydraulic tunnels may be of different types. *Power tunnels* carry water to hydroelectric projects plants. *Irrigation tunnels*

Fig. 16.2 Inlet portal of a horseshoe-shaped tunnel with steel rib support and a masonry wall to arrest debris slide

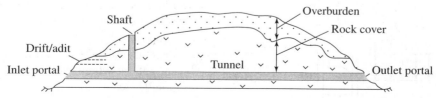

Fig. 16.3 Ground profile along a tunnel showing geological and common features

convey water from the storage behind a barrage or dam for irrigation purpose. *Diversion tunnels* serve the temporary purpose of river diversion for dam construction. Spillway tunnels are constructed for releasing water from the reservoir. Tunnels are also used for sewage conveyance or water supply. Numerous water conductor tunnels of all these types are prevalent in India.

Traffic tunnels These include tunnels used as highways, subways, and railways for transportation. The 3 km-long Banihal tunnel connecting Jammu and Srinagar is an example of highway traffic tunnel constructed in rock. Some of the major cities in India have railway traffic tunnels (e.g., Kolkata Metro Railway tunnel) constructed in subsoil to ease the congestion of traffic movement by surface transports.

16.3 TUNNELLING THROUGH ROCK

Tunnelling is done in rock or in soil mass. All long tunnels and mining tunnels are driven through the rocks generally following the normal method of blasting by detonating the charges through drill holes. Construction of tunnels in rocks may encounter several difficulties due to geological hazards related to rock structures such as faults, folds, joints, and shear zones. Engineering geological investigation is taken up through drilling exploration for evaluating the subsurface condition to plan and design the tunnel and for checking the rock conditions from the excavation stage until its successful completion. All these aspects are discussed in this section.

16.3.1 Rock Pressure and Arching Action in a Tunnel

Rock pressure is a combination of field of forces arising because of the weight of rock. Tunnelling disturbs the equilibrium of these forces and the entire mass of rock or some portion thereof above the tunnel heading tends to set in motion imposing pressure on the lining or support. The distributed pattern of the pressure is shown in Fig. 16.4. A low pressure zone is formed in the close vicinity 'A' of the tunnel heading where the rock is disturbed. This is followed by a higher stress zone at 'B' and then the pressure gradually decreases towards 'C' where it becomes equal to the vertical stress that depends on the density and thickness of the rock above the tunnel. This kind of rock behaviour forming an arch in pressure distribution is known as *arching* or the *arch action*. The magnitude of arching depends on the shape and size of the tunnel at the heading and various other factors related to properties of rock such as its strength and stratigraphy, density, and structures, especially the extent of fissuring. If the tunnel is driven below the water table, the rock pressure is increased due to hydrostatic pressure of water. Though the actual value of the rock pressure is measured by instrumentation, it can also be measured through theoretical approach.

As already mentioned, the rock pressure in a tunnel at depth is the combination of the field of forces arising due to the weight of the rock. When a tunnel is excavated, the equilibrium condition created by these forces is disturbed and the rock mass or some portions above the heading sets in motion or deforms exerting pressure on the support (Fig. 16.5). The amount of deformation depends upon the magnitude of confined stress and the rate at which the rock mass moves (generally small). After relieving, the confined condition is known *as pressure relief*, which may be slow or instantaneous. If the tunnel excavation is done at a

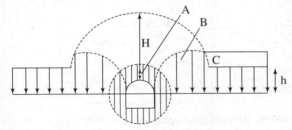

Fig. 16.4 Rock pressure and arching action

Fig. 16.5 Inside view of a tunnel showing deformed steel rib support due to overburden or rock pressure

great depth, the pressure release is instantaneous, causing bursting of rock called *popping*.

The nature of stratification and structures such as folds and faults has an important bearing on the rock pressure at tunnel grade. It has been observed that arching action is well developed in horizontally bedded stratified rock and also in intact igneous and metamorphic rocks. However, if the rocks are highly fractured or affected by faults and folds, the rock pressure on tunnel lining varies with changing rock conditions. While pre-construction stage engineering geological investigation is carried out by means of large-scale geological mapping along the tunnel alignment, care should be taken to prepare geological section evaluating the structural condition at tunnel grade.

16.3.2 Effect of Bedded Rocks on Tunnel Lining

Geotechnical evaluation of rock condition at depth and predicted rock pressure on the tunnel heading are very important at the planning stage for consideration of the data in the design of structure. The following are the schematic descriptions of the behaviour of rock mass and rock pressures on the tunnel lining or support because of the varying conditions of stratification and rock structures:

(i) In case of a tunnel passing through horizontal or low dipping beds, there will be uniform vertical pressure on tunnel lining, see Fig. 16.6(a).

(ii) If the bed is vertical dipping and tunnel line perpendicular to strike, the tunnelling will also experience uniform vertical pressure, see Fig. 16.6(b).

(iii) When the tunnel alignment is parallel to the strike of the vertical dipping strata, there will be heavy pressure on arch, see Fig. 16.6(c).

(iv) If the tunnel axis is perpendicular to the strike of the dipping strata, the lining will experience uniform vertical pressure with longitudinal thrust, shown in Fig. 16.6(d).

(v) In case the tunnel is aligned parallel to the strike of dipping beds, there will be pressure concentration on the sides of tunnel lining, shown in Fig. 16.6(e).

(vi) When the tunnel axis is oblique to the strike and dip of inclined beds, the pressure will be concentrated on the sides, see Fig. 16.6(f).

From the study of the relation of rock structures and attitudes of the sedimentary beds, it may be seen that a tunnel aligned perpendicular to strike of the sedimentary beds is comparatively in a better situation than that aligned parallel or oblique to it. Stratified rocks and *foliated metamorphic rocks* (e.g., mica schists) may also act adversely on tunnel lining depending upon the relation of foliation dip and strike with the tunnel axis. If the tunnel is aligned oblique to the strike of the foliation, a situation may arise where the main thrust will be on the side resulting in overbreak or partial collapse of side wall. Tunnelling through schist in Jaldhaka project in Darjeeling Himalayan region experienced this type of overbreaks because of foliation strike of schist being oblique to tunnel alignment and dip sidewise.

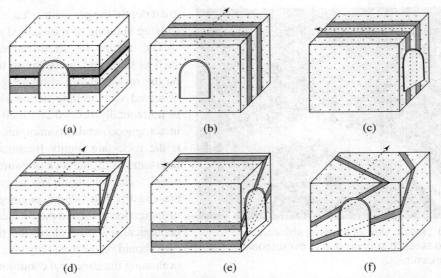

Fig. 16.6 Relation of attitude of stratified rock and tunnel lining: (a) tunnel passing through horizontal beds; (b) tunnel aligned perpendicular to the strike of vertical beds; (c) tunnel aligned parallel to the strike of vertical beds; (d) tunnel aligned perpendicular to the strike of dipping beds; (e) tunnel aligned parallel to the strike of dipping beds; and (f) tunnel aligned oblique to strike and dip of inclined beds

16.3.3 Effect of a Fault Traversing a Tunnel

A fault traversing a tunnel as shown in Fig. 16.7(a) has the following effects on a tunnel depending upon its location:

(i) If a tunnel is in the *hanging wall of a fault*, as shown in Fig. 16.7(b), it should be located away from the hanging wall side of the fault as far as practicable to avoid the offshoots of fault. If it is a reverse fault, the hanging wall block containing the tunnel moves, which is an unfavourable condition.

(ii) When a tunnel is in the *foot wall of a fault*, as shown in Fig. 16.7(c), in a normal fault, the foot wall block is subjected to movement. This position at the foot wall block is the worst scenario for tunnel driving.

(iii) When the tunnel alignment *traverses perpendicular to the trend of the fault* as shown Fig. 16.7(d), the zone of intersection of the tunnel and fault will cause problems, but this situation is better than the tunnel aligned oblique to the fault.

(iv) When the tunnel *traverses oblique to the fault*, it will meet a thicker zone of crushed rock, as shown in Fig. 16.7(e). It is necessary to take the tunnel through the position where the width of the fault zone is of minimum thickness, even by giving a kink to the tunnel alignment, if needed.

In all the cases of a tunnel traversing a fault, it is to be seen whether it is an active fault, which is always an unfavourable feature for tunnelling. An alignment for tunnel should be so chosen that it does not run along the fault zone. If a tunnel cuts the fault, the crushed rock and the gouge of fault zone will cause serious damage to the lining. A low angle fault (thrust) is the most damaging feature for the tunnel. The pressure of water flowing through the fault zone might cause overbreak and even collapse of the tunnel structure. The clayey gouge may swell with soaking of water and exert pressure on tunnel heading to cause deformation of rocks.

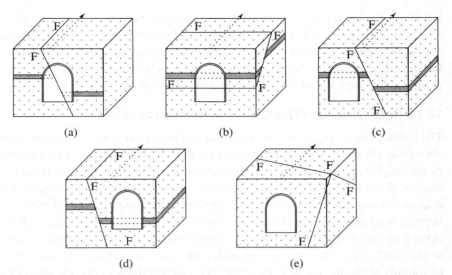

Fig. 16.7 Influence of fault on tunnel rocks: (a) tunnel traversed by fault plane; (b) tunnel located in hanging wall block; (c) tunnel located in foot wall block; (d) tunnel aligned normal to the trend of the fault plane; and (e) tunnel aligned oblique to the trend of fault

In case of an active fault traversing a tunnel, its reactivation causes immense destruction. The crushed material of a fault zone may flow out and deposit the sandy material on the tunnel floor hindering tunnelling progress. Shifting the tunnel alignment or selecting an alternative alignment away from a fault saves potential hazards that may have to be encountered during construction of a tunnel traversing a fault.

16.3.4 Effect of Folds on Tunnel Lining

The following are the effects of folds on tunnel lining:

(i) A tunnel will experience intense lateral pressure if it passes through an anticline at the portal portion with fold axis aligned nearly parallel to tunnel line (Fig. 16.8(a)). There will be more bending of the anticlines in the upper strata and more joints due to tension stresses compared to lower strata. As such, it is advisable to locate the tunnel in a deeper part where the joints will not create problem of tunnelling.

(ii) A tunnel passing through syncline, see Fig. 16.8(b) will experience more lateral pressure in the middle part away from the portals. If the rocks of a syncline passing through a tunnel is permeable in nature, surface water may permeate to the tunnel creating problem of dewatering (draining out water using pumps). The presence of such folded structure and its demerits need to be identified during geological investigation at the planning stage of a tunnel for necessary drainage arrangement in the design, as it may cause adverse problems during tunnel excavation.

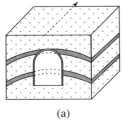 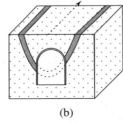

Fig. 16.8 Tunnel passing through folded rocks: (a) through an anticline; and (b) through a syncline

Bedded rocks with unfavourable strike and dips and structures such as folds or faults may lead to instability of tunnel rocks if located close to the hill slope. For example, a dipping bed towards the lower valley if traversed by a tunnel at right angles is likely to be affected by slope failure. In case of horizontal or vertical bed, there may not be any serious problem if the tunnel line is aligned parallel to it. However, if the strata or fault planes dip towards the slope or the folded rock with axial dip is inclined towards the slope, it will lead to the problem of slide, especially after saturation with water.

16.3.5 Rock Cover and Overbreaks in Relation to Joints

The rocks through which tunnels are excavated may be of soft sedimentary types such as shale, claystone, siltstone, and sandstone, or they may be hard igneous or metamorphic types such as granite, basalt, gneiss, schist, and charnockite or their weathering products, resulting in soft unconsolidated material. The *rock cover* of a tunnel should be differentiated from the zone of unconsolidated material or *overburden* present in the topmost parts of most of the tunnels. The vertical depth from the top of the tunnel up to the tunnel roof minus the thickness of overburden is the rock cover of a tunnel. The engineers generally consider the entire zone above the tunnel as the overburden. The thickness of rock cover is an important criterion in geological analysis of stability of the tunnel. When a tunnel is excavated through insufficient rock cover, the entire thickness of the overlying material may tend to cave in. However, if the tunnel has adequate cover of intact rock (viz. thickness of cover is more than twice of the tunnel's diameter or height), the arch action plays an important role in minimizing the chances of caving.

Tunnelling is to be done as per the size and shape in the design of the tunnel. The contractors are supposed to restrict excavation up to the *design cross section* of the tunnel bound by a line so that the finished tunnel has minimum thickness of concrete lining (Fig. 16.9). They may exceed little and excavate up to only a few centimetres from this line. This line up to which they are expected to excavate and for which they get paid is known as the *pay line*. Any break above this line is not payable, but if the breakage is less than the design line, it is to be excavated by the contractor without demanding any extra payment. Any extra breakage above the line is known as overbreak. In general, in hard and massive rocks, the overbreaks are small. The inside part of the tunnel is provided a smooth surface called *lining*. The lining reduces the wall surface roughness and allows smooth flow of water.

The attitudes of rock strata and intensity of divisional planes such as joints and their spacing are considered to be the major reasons for the possible overbreak of tunnels. Study of joints in hard rock (charnockite) in 2042 m-long Balimela head race tunnel in Orissa has indicated that where the rocks are traversed by two intersected sets of inclined joints, the spacing is close (5–30 cm) to very close (< 5 cm) or moderately close (30–100 cm), and tunnelling in these

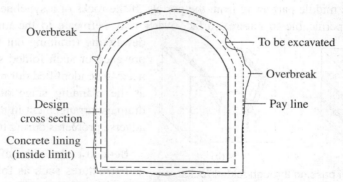

Fig. 16.9 Designed shape, size, and pay line of a tunnel

rocks has experienced the maximum overbreaks. However, where the rocks are traversed by only one set of joints or where the strike of joints is parallel to tunnel alignment, the spacing is wide (1–3 m) to very wide (>3 m) causing less overbreaks. Thus, the study on relationship between overbreaks and spacing led to the conclusion that closer the spacing, more are the overbreaks, and wider the spacing, less are the overbreaks (Fig. 16.10).

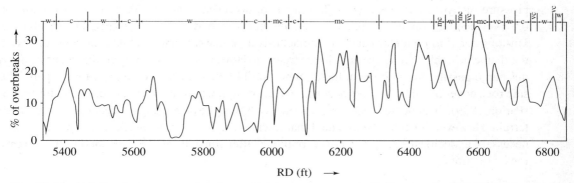

Fig. 16.10 Relation between overbreaks and joint spacing recorded in hard rocks of Balimela head race tunnel in Orissa: vc, very close; c, close; mc, moderately close; w, wide; and vw, very wide (Sinha, Pradhan, and Singh 1971)

An extensive study on the influence of joints in rocks of overbreak in hard rock (granite) of Idukki power tunnel (Seshagiri 1975) indicated that overbreaks are more associated with blast holes and not with joints in rock. The overbreaks in rock were classified into three zones in relation to the intensity of joints. In zones I and II, the joint spacing is more than 1.5 m and the joints had no contribution in overbreak. The reason for overbreak was attributed to large numbers of perimeter blast holes beyond the designed limit of the tunnel. Only in zone III, close-spaced and blocky joints contributed to the overbreak.

16.3.6 Relation of Overbreak with Tunnel Dimensions

Some authors such as Wahlstrom (1948) and Proctor and White (1946) have found the following relationship between tunnel diameter and extent of overbreak of tunnel rock guided by the inclination of joint and stratified rock:

(i) In horizontal strata traversed by inclined joints, the overbreak may be up to half the tunnel diameter.
(ii) In vertical strata with horizontal joints, the overbreak is restricted to one-fourth of the tunnel diameter.
(iii) In inclined strata traversed by joints at any angle, the caving in of rock is restricted to a maximum thickness of half the tunnel diameter.
(iv) In highly jointed or crushed rock, the overbreak may extend up to three times the diameter or three times the sum of height and width of the tunnel.
(v) It is also noticed that pulses due to blasting are not transmitted beyond the distance of twice the tunnel diameter.

Observation of nearly 100 tunnels by Judd has indicated that 'the rock types and tunnel sizes (cross sectional area) have but minor influence on the percentage of overbreaks; the construction procedures apparently are more important in this respect' (Krynine and Judd 1957). In some of the Indian tunnels too, it was observed that maximum overbreak results from excessive charge in blasting (as in Idukki tunnel rock) and not from geological causes.

According to the thumb rule, the rock cover above the tunnel (excluding overburden) should be at least twice the diameter in intact rock to prevent collapse of roof. However, it strictly depends upon the pressure condition at the tunnel. On the rock cover of water conductor tunnels, the conventional practice is to provide a vertical cover equal to the internal water pressure or hydrostatic head (H). Recent trend, however, is to provide less cover (as low as $0.5H$) depending on the nature of rock. In Indravati hydroelectric project, Orissa, the 4.3 km-long and 7 m-diameter head race tunnel driven in massive charnockite falls short of 'twice the diameter' rock cover in several long stretches. However, it was observed that a heavy flood discharge passing through the unlined tunnel for 24 hours caused no damage to any part. The safe rock cover of $0.4H$ provided according to the design was found to be effective for this tunnel (Parida 1999). It indicates that a comparatively lower pressure can be accepted in hard rock terrain. However, it cannot be assumed that it will be so for all projects. A cautious approach is necessary to decide the safe rock cover for tunnel after considering the rock condition including rock structures.

16.4 TUNNELLING THROUGH SOFT GROUND

The term soft ground tunnelling is used when the media of tunnelling is unconsolidated deposit or soil mass. The materials excavated are loose materials termed *muck* and the tunnelling is generally done by cutting and providing immediate support, commonly known as *cut and cover* method. Unlike rock tunnelling, the problem encountered is related to soil moisture and swelling pressure, and the main concern is to save the roof from collapse. All these aspects are discussed in detail in the following subsections.

16.4.1 Type of Material, Imposed Load, and Stability

Soft ground tunnelling is done in unconsolidated overburden materials, either alluvial deposits or products of decomposed rocks. The deposits include clay, silt, sand, gravel, and pebbles and are generally an admixture of them. The nature of the ground is the main consideration in soft ground tunnelling, as excavation through the soft zone may damage the surroundings and lead to the problem of subsidence in the build-up area. In soft ground tunnelling, the ground should possess sufficient stand-up time to provide support to the tunnel. In stiff clay, the stand-up time is few hours to a day when supports such as timber beams or steel plates are provided. Tunnelling through unconsolidated deposits such as clayey sand or sandy clay experiences a stand-up time of few hours, but in loose deposit, it is almost zero.

In tunnelling through the soft ground, the unconsolidated materials above the heading act like a fluid. The vertical pressure is equal to the total weight of the material overlying the tunnel. The horizontal pressure is assumed to be varying between one-third and two-thirds of the vertical load. The materials of the soft rock tunnelling have very low shearing and tensile stresses. The pressure on tunnel lining varies with time in increasing orders. The problem of soft rock tunnelling is more acute if the tunnel passes below the water table. Soft rock tunnelling below water table experiences constant flow of water and frequent overbreaks and makes it difficult to provide support before collapse, that is, within the very short stand-up time. The constant water flow is stopped by use of high power pump or compressed air. In addition, drainage arrangements are also made to keep the water table below the tunnel floor.

The stability ratio of soil (Nc) at a depth of Z in soft ground tunnelling is the ratio of effective soil pressure (Pz) at the depth Z to undrained shear strength of the soil (Su).

Thus, $Nc = Pz/Su$ (16.1)

If Nc is below five, the tunnel is considered safe, but if it is in more than five, immediate support is necessary. Use of compressed air is most effective in this sort of support (Sengupta 1991).

16.4.2 Method of Soft Ground Tunnelling

In the soft ground composed of loose overburden materials, tunnelling is nowadays done by conventional machines after strengthening the material by grouting. Heavy machines are now available for excavation in soft ground as well as in soft rock. If such machines are used for rock tunnelling, the tunnel section, after each sequence of advance, will require immediate support. Sometimes, caissons or *shafts* are sunk from the surface to enter the tunnel level to carry out the tunnelling work in different segments. The most commonly used methods of soft ground tunnelling are *cut and cover* method and *shield* method.

Cut and cover method is best suited for any kind of soft ground with limited overburden cover. This method of tunnelling involves the following steps:

- First vertical retaining walls are constructed at the two sides to protect the side walls from collapse.
- Second, the soil from the intermediate portions of the sided walls is excavated.
- Third, all sides are lined by reinforced concreting to appear like a reinforced cement concrete (RCC) box.
- Finally, the pit or cut portion remaining above the RCC box is back filled by earth and the road and other pre-existing structures are restored or reconstructed.

At the beginning of the work, retaining walls are embedded in the soil at the two sides. It is a difficult task and is performed following the 'diaphragm wall technique'. The technique is the same as that used in a dam site by cutting a slit and backfilling by suitable material such as bentonite-mixed concrete or RCC. The other problems are related to the type of soil and include earth pressure, heaving of floor, hydrostatic pressure or piping, and settlement of the soil at the bottom. These problems are solved by appropriate design for drainage arrangement and thickness of concrete walls.

The use of *shield method* is very successful in tunnelling through soft ground at any depth but the ground should be free from squeezing or swelling type of clay. It facilitates advancement by full-face excavation. In the shield method of tunnelling, the *shield*, (a jack powered metallic body), is used to shape the tunnel profile. It is driven into the soil in advance to permanent tunnel lining. The shield looks like a circular box with a cutting edge in the frontal part, which is driven in the soft ground for a short distance by a hydraulic jack. The rear part of the shield is equipped with an arrangement to provide iron wire support as the shield advances. The annular spaces of the supports are immediately filled by thick suspension type cement grout (water:cement 1:1). The entire section is then heavily reinforced, starting from bottom and moving up towards the roof cover. The shield method of tunnelling is suitable in sandy soil having stand-up time of about zero, as use of shield facilitates cutting the ground and providing immediate support by breasting. Operating a shield involves the following approaches:

(i) A shaft is excavated up to tunnel level and then the shield is introduced and operated from the tunnel level.

(ii) The stability ratio of soil is kept at $Nc = (Pz - Pa)/Su$, where Pa is the net air pressure and Su and Pz stand for undrained shear strength and effective pressure of soil, respectively, as in Eq. (16.1).

(iii) A tunnel lining is designed to support the load due to overburden pressure. In addition, it takes care of 'jacking thrust' arising from erection of tunnel by pushing of the shield by jack.

The 16.5 km-long Metro Railway tunnel of Calcutta, the first of its kind in India, is an example of tunnel construction in soft ground consisting of sandy and silty alluvial deposits. The cut and cover method was mainly followed, but in some sections *shield* method was also used in the excavation (Appendix B.5).

16.5 GEOLOGICAL HAZARDS IN TUNNELLING

In India different types of tunnels were constructed in various parts of the country. Geological problems faced during tunnelling through several rock types or unconsolidated materials were many. A description of the main problems is discussed below.

16.5.1 Geological Problems Due to Tunnelling with Indian Examples

The hazards encountered in the construction of a tunnel are not related to the type of the ground but depends on the geological conditions of the rocks or unconsolidated material of the ground through which the tunnel passes. The following are the problems generally encountered during tunnel construction:

(i) Overbreak including chimney formation and wall collapse

(ii) Spalling of tunnel rock

(iii) Flowing ground

(iv) Squeezing and heaving ground

(v) Temperature rise with presence of thermal spring

(vi) Gas flow

(vii) Seismic effect

Overbreak It is by far the major type of hazard in tunnel construction. It results when the tunnel pierces through soft rock or extensively jointed rocks. The chances of overbreak are more in fractured rock affected by tectonic movement. In the sedimentary rocks where the rocks are layered and alternated with hard and soft bands, overbreak is more prolific due to differential strength of the rocks and their attitude or configuration of the beds. In homogeneous rock free from fractures, the problem is less, but highly micaceous metamorphic rocks due to their less strength and more fragility (with high foliation character) are responsible for more overbreaks. Folding and faulting result in deformation and shattering of rocks. If a tunnel cuts across a fault or shear zone, it may cause serious overbreak deep inside the crown appearing like a *chimney* (Fig. 16.11). *Wall collapse* is a very common feature in many of the tunnels. This is caused mostly by unfavourable structure of tunnelling media. If the tunnel is aligned parallel to the strike of beds or foliation planes, the problem of collapse of wall rock including a part of the roof becomes more acute. This problem was frequently experienced when tunnelling through the tectonically disturbed rocks of the Himalayan terrain (Appendix B.3).

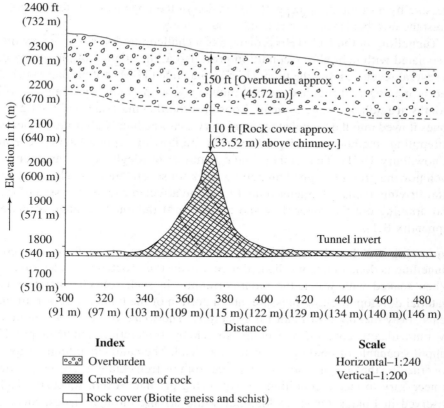

Fig. 16.11 A large chimney formation while tunnelling through the shear zone in the schistose rock of the Jaldhaka project (Chowdhury 1971)

Spalling It is the process of splitting of layers of rock from the tunnel wall and roof under saturation with water for a prolonged period and undergoing some chemical reaction. The sedimentary tuff is made of argillaceous as well as arenaceous (sandy) materials, the shaley matter being 55 per cent in the tuff. The slaking characteristics of these matters may cause spalling and squeezing of the incompetent members after excavation of tunnel. The slaking characters can be known by collecting samples from surface and drill cores and testing in the laboratory. Heavy supporting is required to arrest tunnel rocks affected by spalling. Spalling conditions in strong quartzite and squeezing ground condition in softer phyllite rock were experienced in the Dul Hasti tunnel in Jammu and Kashmir.

Flowing ground Flowing ground condition inside a tunnel is a very adverse geological situation. Such a situation arises when entrapped water under hydraulic head gushes out immediately after tunnelling. During rainy season, this problem may become serious with large flow of water through porous rock with water table above the tunnel level. The problem is aggravated when large quantities of silt and clay flow out from tunnel rocks with the water and deposit the load to choke a part of the tunnel. It creates a situation of repeated mucking, aggravating and delaying the tunnelling work. Since natural arching cannot be formed in loose material under flowing condition, the removal of muck will lead to further problems. In sedimentary rocks, some thin beds or pockets of loose porous sandstone may cause such problem of flowing ground. The approach to tackle the situation would be to allow the muck to flow out of cavity, if required by a small diameter subsidiary tunnel, and to reduce the water pressure by maximum drainage. Refrigeration of the impounded water is a good technique to arrest the flow but is yet to be applied in our country.

Tunnelling in the Dul Hasti project was a challenging job where heavy ingress of water associated with debris flow condition prevailed at the contact of the quartzite and phyllite in several reaches. Even the tunnel boring machine (TBM) got partially buried under heavy debris making it difficult to work. Flowing ground condition stopped work for several months at Jaldhaka Stage I tunnel. Loose overburden materials and rock gouge from shear zones flowed into the tunnel at regular intervals immediately after muck-clearing operation. Forepoling method was adopted to cross the flowing ground having several shear zones (Chowdhury 1971). The Umiam Stage II tunnel in Meghalaya is another instance where such flowing ground stopped tunnelling work for several months. The work could start only after driving a small diameter tunnel from the affected part by the side of the main tunnel and draining out the water flow and clearance of the muck through it in wet season (see Appendix B.1).

Squeezing and heaving Squeezing and heaving ground condition is encountered when tunnelling is done in unconsolidated rock or claystone containing deleterious clay minerals. When soaked with water, the montmorillonite in such soft clay band or unconsolidated material develop swelling pressure, and a portion of the tunnel wall or roof may squeeze or heave away making it difficult for tunnelling. The *squeezing ground* condition is prevented by immediately covering the entire stretch by shotcreting and then providing steel rib supports capable of holding the distributed load. The cumulative tunnel support required is governed by the designed stress computed from instrumentation. The use of yielding support is necessary to tackle conditions of squeezing ground. Severe squeezing and heaving was observed in Loktak tunnel in Manipur while driving through shales. Steel supports were twisted and tunnel diameter reduced due to squeezing. Heaving of floor was also equally heavy. Rock bolts and shotcreting were adopted to tackle such adverse geological conditions (Chowdhury 1996).

Thermal springs Thermal springs may be met in some hilly terrain during rock tunnelling through very deep parts. This may cause rushing of hot water inside the tunnel leading to problems in tunnelling. There will be a temperature increase in the deeper parts of underground, and the heat may be prohibitive for excavation work. In the Kolar gold field, when underground mining is continued below a depth of 4 km, the *popping action* (shooting of rock fragments under pressure) is frequently encountered. The temperature is so high at that depth that excavation could proceed only after providing air-conditioning facility.

Gas flow Gas flow is encountered in many tunnels of the world, especially in mining tunnels. Suffocating gas generated from blasting in the absence of proper ventilation arrangement may hamper the working in the tunnel. At times, during tunnel boring, there is an accumulation of poisonous gas from some organic sources such as the peaty bogs or swamps close to tunnel line. One such gas is methane (marsh gas), which is inflammable and causes danger to workers by igniting into flame as observed in coal mines but is generally exhausted quickly. Such hazards can be avoided by providing proper ventilation arrangements to divert the gas to surface by pipes and blowers. Loktak project of north-eastern India during tunnel construction experienced natural gas flow through the pores of shale from a part of the tunnel, which was, however, stopped within a short time.

Seismic effect   Seismic effect causes serous hazards to tunnels. The intensity of earthquakes that occurred in an area in the past needs to be known to suggest future possibility of ground movement affecting the tunnel. Seismic zoning map (see the figure and table given in Section 20.7.1) gives the status of earthquakes experienced in different parts of India. The most vulnerable area is the Himalayan terrain where several faults and thrusts are present. Movement of ground along these planes of tectonic activity due to stress relief may trigger earthquakes. Most of the major earthquakes in India such as that of Bhuj, Gujarat, in 2001 are generated by the movement of Indian plate. Even peninsular India, which was so long considered to be a stable land, has been affected by earthquakes, for example, the Koyna earthquake.

During the investigation on Yamuna hydroelectric project in the Himalayan terrain of Uttar Pradesh, J.B. Auden (1942) observed that the effect of earthquakes is generally restricted within the top 30 m of the ground. Hence, if the tunnel grade is constructed sufficiently below this depth, the earthquake may not cause serious damage to the tunnel. A tunnel may pass a thrust or major fault in the extra-peninsular region; the release of accumulated stress brings about movement along these planes. The intensity of probable earthquake in an area is known from seismic zoning map and accordingly suitable design is prepared for tunnelling through the terrain.

16.6 DIFFERENT STAGES OF GEOTECHNICAL WORKS FOR TUNNEL

As discussed in the foregoing paragraphs, tunnelling through rocks or unconsolidated materials undergoes different types of problems. An engineering geologist is required to give guidance as to how to select the best tunnel alignment and the nature of exploratory works. The engineering geologist is also required to know about the subsurface condition of the tunnel and several other geotechnical works for successful tunnelling.

16.6.1 Selection of Tunnel Alignment

The first step in selecting a tunnel alignment is the evaluation of the geomorphology and geology of the terrain. A straight tunnel is always desirable. If a bend is inevitable, the angle at the turning should not be less than 120°. However, there should not be any bend near the two portals. The exit and inlet ends should be in hard rock that is free from fault or subsidence. In most cases, the alignment for water conductor tunnel in hydroelectric projects is fixed by the engineers on the basis of engineering analysis and topography. However, an alternative can be suggested on the considerations of geology and hydrology. For example, a tunnel alignment selected through rocks having very low rock pressure and are free from major tectonic and structural disturbances and devoid of karst, slump, rock slide, and other adverse effects are well accepted by the designer.

If the tunnel alignment passes through a valley or by the side of a hill with insufficient rock cover, the tunnel grade in the valley portion should be depressed to avoid problem of instability. If the physiography permits, the tunnel should be aligned across the strikes of the bedding or foliation planes of the rock, as it attributes more stability in tunnelling due to arching action. Aligning the tunnel parallel to the bedding or foliation brings about the problem of side wall collapse. A knowledge of the structure and tectonic history of the region helps to anticipate the likely geological problems to be encountered during tunnel construction. In fact, a correct appraisal of the weak structural features that may affect safe tunnelling is of utmost importance for design consideration.

After the preliminary selection of the tunnel line, a large scale (say, 1:1000 to 1:2000) geological map of the terrain covering the tunnel line and its surrounding areas is prepared. This map should have contours at close intervals to facilitate drawing of ground profile along the tunnel line. It is, however, desirable that the cross section of the ground along the tunnel alignment is prepared by actual topographical survey. In general, the engineers prepare this section plotting the height and length of the section in different scales, but to facilitate geological plotting, the scale in both the directions should be the same. With the help of a geological map along and around the tunnel alignment, a geological section is drawn by projecting the various strata and other features to the tunnel grade in the ground profile (Fig. 16.12).

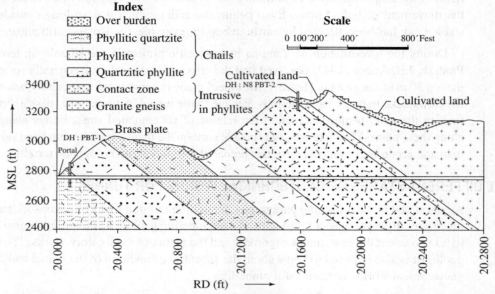

Fig. 16.12 Tentative geological section along the Pandoh Baggi tunnel in Himachal Pradesh showing different rock formations including overburden at the tunnel grade (Tiku 1971)

For presenting the correct geology of igneous and metamorphic terrains in the tunnel grade, knowledge of the rock types and structural features at depth is necessary. In the geological section, the tunnel grade will be clearly marked by a single or double line, and the rock type and structural features to be obtained at depth up to the tunnel grade and below will be shown. Further demarcation is made for fresh rock, weathered rocks, and overburden consisting of soil or loose sandy materials of the topmost part, if present. This clearly brings out the width of the fresh rock above the tunnel crown, which is a very important criterion in the interpretation of tunnelling condition.

At the early stage investigation for selecting tunnel alignment, care should be taken to ensure that no mineral deposit of economic value is affected by tunnel construction. If any monument of archaeological or historical importance exists over the tunnel alignment, the safety of the structure is to be ensured. While investigating an earthquake prone area, thorough analysis is required on the 'seismic zone' depending upon earlier shocks, especially if it is in the Himalayan terrain. Accordingly, the designs of the tunnel would include the necessary safety factor to safeguard the tunnel from any damage due earthquake tremors.

16.6.2 Subsurface Exploration

Subsurface data of rock conditions derived from exploratory work at the tunnel level of tunnel alignment is necessary to meet the design requirement instead of depending on geological interpretation. Once the study of surface geological study is completed, exploratory work by means of drill holes and drifts along the tunnel alignment is taken up to obtain knowledge of the subsurface geology. The number and locations of drill holes are so chosen that these provide data on the rock types, presence of fault and fold, extent of fracturing at depth, and other adverse features at depth anticipated from surface mapping.

Angle drill holes are necessary to prove the extent of weak zone at depth due to vertical dipping fault traced from surface geology. The locations of such drill holes including their inclinations and directions of boring are precisely fixed so as to cut across the anticipated adverse structures such as fault at tunnel grade. The drill holes will be extended up to tunnel grade rock and depths further than 6 m. In case the tunnel grade occurs at a very deep part, only a few holes will be drilled up to 10 m in fresh rock to understand the overburden thickness and rock types at depth. Horizontal holes are also drilled from valley slopes if the tunnel passes at a short distance.

Water percolation test (pumping tests) is conducted through each drill hole to understand the permeability of rocks due to porous characters, fracture opening, shearing, and shattering. Many of the drill holes are maintained to use as piezometers for measuring groundwater levels for knowledge of fluctuations in the water table in different seasons. A graphic presentation of the percolation test data is given with drill core logs (Section 11.7, Fig. 11.22), and the groundwater table is marked by a line in the geological section to help design the tunnel and construction approach.

In some projects, geophysical survey, especially by seismic refraction method, is carried out to understand the subsurface geology. A hole needs to be drilled to prove the nature of anomaly detected from geophysical study. Geophysical survey is especially important to detect the anticipated presence of fault or other weak zones along the tunnel alignment, which may be hidden under overburden.

The exploratory programme of a tunnel project also includes excavation of few drifts close to the inlet and outlet portals of the tunnel or any other place as necessary to obtain geological details. The test drifts facilitate a view of the rock condition with respect to its nature and structural attitudes and degree of weathering in the subsurface. If the tunnel passes through great depth limiting the drilling to only few holes, drift is taken up from valley floor and extended up to tunnel grade, which facilitates visual observation of the rock condition at tunnel grade. A drift may also act as an adit for movement of machineries in the construction time. A log of the test drift presents schematically the rock types, their attitudes, and all other geological features observed.

The core samples of rock obtained from drilling or those collected from the drifts are tested in the laboratory to determine their density, strength, porosity, and other engineering properties. In case of unconsolidated material or clay, it is necessary to test for swelling properties and clay minerals (e.g., montmorillonite) that causes overbreak under saturated condition. The frequency of joints, the attitude of rocks, the nature of weathering, and similar geological data obtained from the study of drill holes and drifts provide important information on the expected overbreaks and other possible hazards. The subsurface information obtained from drill core logs and the laboratory test data is necessary for design of the tunnel.

Use of borehole camera for borehole photography helps to understand the nature of jointing and other adverse geological features at depth and to design the tunnel support. In all large tunnels, especially those passing through faults and thrusts, behaviour of stress is measured by inserting extensometers or other stress-measuring instruments in the tunnel grade rock through the drill holes at the planning stage investigation. The quantitative evaluation of the rock load in tunnel is measured by installing load cells in appropriate place. Flat jack is a simple instrument used for measuring the in situ elastic modulus of tunnel rock. The deformation of rock that takes place by applying pressure through jack can be directly recorded for use in the design of tunnel.

16.6.3 Construction Stage Work: Three-dimensional Tunnel Logging

In the construction stage, the most important work is the mapping or plotting of the geological features inside the tunnel covering the two walls and arch section and is known as *tunnel logging* or *three-dimensional (3D) log of tunnel*. The work includes plotting of the rock types and structural features such as dip and strike of beds, foliation planes, joints, faults, clay gouge, fold, shear zones, and seepage observed after tunnel boring. In general, the rock faces in the two side walls and the arch sections inside the tunnel is mapped by tape and compass and the data plotted in a plan.

The plan for tunnel logging is prepared as shown in Fig. 16.13(a) where the portion a–a′ represents the floor, a–b and a′–b′ the two walls, b–c and b′–c the half arch sections, and o–o′ the central line showing distances at intervals of 3 m. In the plot, a dipping bed (DB) in the tunnel arch (portion c–b and c–b′) will appear curved in the plan, but a vertical bed (VB) will appear straight. A folded cut-out of this plan will provide a 3D appearance of a tunnel when the two ends c and c are joined together, keeping a–b and a′–b′ standing in vertical position on the floor a–a′. Figure 16.13(b) shows the 3D tunnel log prepared from plotting geological and structural features of tunnel rocks on such a plan. Figure 16.14 shows the 3D geological log of a part of the head race tunnel in Maneri Bhali hydroelectric project of Uttar Pradesh.

During tunnel logging, while using any magnetic instrument such as a compass for plotting the rock attitudes, it is to be ensured that there are no iron materials that will distort the data. The presentation of the data may be done in different ways. The tunnel log (scale 1:100 to 1:200) shows the geological features of the roof and the two walls of the tunnel. Tunnel logging should be done with the progress of tunnelling, preferably immediately after excavation and mucking of certain parts, and should not be done after the completion of the entire tunnel. In some tunnels, shotcreting and bolting are necessary immediately after excavation before muck removal. In such cases, 'face logging' is done after each blast and a composite map is prepared on the basis of these face logs later on. The tunnel log (or the map of tunnel interior) thus prepared is helpful to decide on the required tunnel support, lining, and treatment of weak zones. The geological report along with the tunnel log and associated diagrams helps to take up these remedial measures.

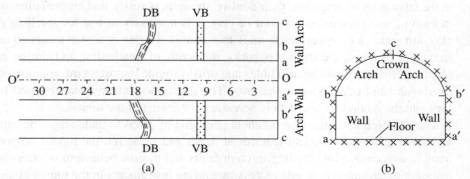

Fig. 16.13 Tunnel logging: (a) plan for tunnel logging that includes plotting of geological features of the walls and arch section of a tunnel (VB, vertical bed; DB, dipping bed); and (b) tunnel section along c–o–c

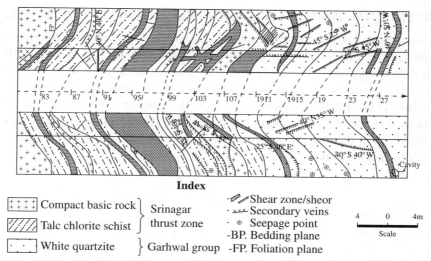

Index

+ + + + Compact basic rock	Srinagar
//// Talc chlorite schist	thrust zone
· · · White quartzite	Garhwal group

Shear zone/sheor
Secondary veins
⊕ Seepage point
-BP. Bedding plane
-FP. Foliation plane

4   0   4m

Scale

Fig. 16.14   3D geological log of a part of the headrace tunnel in Maneri Bhali hydroelectric project, Uttar Pradesh (Negi 1992)

16.6.4 Other Geological Activities of Tunnel Work

In addition to tunnel logging, the other engineering geological activities at the construction stage are related to the following aspects:

(a) A clear appraisal is given to the engineers on the exact position and dimension of the weak zones (due to fault, shearing, fracturing, etc.) that present a threat of overbreak and require remedial measures. In a folded rock, joints and fractures may be intense where strengthening of rock by grouting or shotcreting is to be taken up as remedial measure. At places, rock bolting may also be required, especially where hard rocks form detached blocks due to joints and faults. RCC lining is provided in extensively jointed rock.

(b) The construction stage work also ascertains whether support is to be provided, whether it is of immediate requirement, and the portion of tunnel section (with respect to distance from inlet portal) where support is needed. The geologist should also suggest the type of support to be provided. Immediate support is necessary to prevent danger of heavy overbreaks or roof collapse where soft seams, clay gouge, and squeezing ground are detected during tunnel logging. Load test in such section decides the actual weight the supports can withstand.

(c) In some cases, sudden inflow of water with silt and mud water may halt tunnel excavation. In such cases, the geologist should suggest ways to tackle the situation. In general, use of high-power pump may decrease the water flow if the water (perched water) is trapped in some porous rock zone, but in case of groundwater flow it is difficult to deplete the water table. A small diameter pilot tunnel may divert the water. In the dry season, however, the flow diminishes or completely stops when the silt can be removed and tunnel boring work continued.

(d) In many tunnel projects, the conditions of the portals at the two ends of the tunnel require special study because of their vulnerability to complete collapse due to insufficient rock cover. The method of cut and cover by reinforced concrete is adopted for weak portal rocks.

(e) Tunnelling in rock suggests that a circular tunnel should have a minimum rock cover of twice the tunnel diameter. In a horseshoe tunnel, the minimum rock cover should be three times the sum of the width and height of the tunnel to avoid or minimize the chance of complete roof collapse.

16.7 CONVENTIONAL METHODS AND MACHINERIES USED IN TUNNELLING

In the seventeenth century, gun powder was used for tunnel construction in India. The nineteenth century saw the use of dynamite for rock blasting by mechanical drills with tungsten carbide bits. Normally, jacks or *wagon drills* are used for boring holes for blasting. A self-propelled jumbo provided with ladder for drilling and safety cover for the driller is used for accurate drilling of blast holes. The jumbo moves to the heading of the tunnel and drills several holes in the rock as per designed pattern. The pattern and depths of the holes may vary depending on the dimension of the tunnel and type of rock, but a typical hole is about 3 m deep and only a few centimetres in diameter (Fig. 16.15).

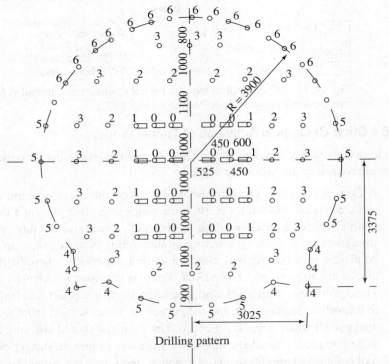

Drilling pattern

Delay no.	Description	Depth (m)	No.of holes	Bottom charge	Column charge	Total stricks per hole	Total charge in Kg
0	Baby cut	1.0	8	5	2	7	7
0	Baby cut	2.5	8	10	7	14	17
1	Cut holes	4.0	8	13	16	25	29
2	Side,bottom	3.0	15	18	18	16	36
3	Lifters	3.0	15	18	18	16	36
4	Trimmers	3.0	8	–	6	4½	6
5	Trimmers	3.0	12	–	9	4½	9
6	Trimmers	3.0	15	–	11	4½	11
			93	64	87		151

Fig. 16.15 Drill hole pattern for blasting of 7 m-diameter tunnel of Yamuna hydroelectric project (Kodkade 1971)

After completing drill holes, the next step is packing explosives into the holes and detonating the charges. The mucking operation is then undertaken by laying trolley lines or by dumpers. Buckets mounted on wheels and hauled on trolley line by locomotives operated on battery are also used in the mucking. The process of drilling, blasting, and mucking is repeated and thus the tunnel advances. With advances of tunnel, permanent concrete lining and supports are provided wherever necessary (e.g., zones of jointed and fractured rocks).

16.7.1 Conventional Method of Tunnelling by Tunnel Boring Machine

The conventional drilling and blasting method is commonly followed in small diameter tunnel driving in rocks but a TBM is deployed to excavate large diameter tunnels with circular cross sections. Considerable speed and economy are experienced in tunnelling nowadays by use of TBMs. A typical TBM is circular in cross section and can be used in boring both hard and soft rocks in tunnels of varying dimensions, even as large as 15 m diameter. This machine can cut smooth tunnel walls with limited disturbances to the surrounding rocks and thus helps in reducing the cost of lining. TBM is an expensive machine but for a large diameter and long tunnel, the overall cost of using this machine is less than tunnelling by conventional method of drilling and blasting. The Dul Hasti hydroelectric project of Jammu and Kashmir deployed this machine for full-face tunnel boring in the excavation of 6.8 km-long tunnel some decades back. The large diameter circular-shaped head race tunnel of Parbati hydroelectric project of Himachal Pradesh was also bored by a TBM.

16.7.2 Tunnelling by Road Header Machine

Another machine used in tunnelling is the road header, which, in contrast to TBM, can cut variable-shaped tunnels and not only circular tunnel section. This machine has been found advantageous and convenient to operate in tunnelling through moderately hard rocks of compression strength less than 140 MPs to soft rocks for a limited length of less than 2 km. The basic cutting tool for a road header is a large milling head mounted on a boom, which in turn is mounted on tracks or within a shield (Fig. 16.16). Unlike TBM, the operation of road header does not require drill and blast for excavation. This has been used in Subansiri hydroelectric project of Assam and Arunachal Pradesh. In view of the proximity of the four tunnels to each other, blasting was not possible in this project as it might have developed cracks in the rock (Siwalik sandstone). Instead, the road header was used for cutting in the entire operation. Deployment of heavy-duty road header improved the production though it generated lot of dust, which necessitated the use of foam for dust depression.

(a)   (b)

Fig. 16.16 Road header machine: (a) view before commissioning; and (b) in operation inside a tunnel (Chowdhury and Das 2009)

16.7.3 Shield Method of Tunnel Excavation

Shield, which is used for tunnelling in unconsolidated material, is also suitable to operate in very soft rock. Tunnel with soft rock (up to 4 in Mohs scale of hardness) can be excavated by an electrically operated *mole*. In a mole, a powered hydraulic jack holds the frame of the boring machine in position when a second set of powerful jacks push a rotating disc (cutting tool) against the rock face. The rotating disc cuts the end and crushes the soft rocks into bits. The crushed material including clayey muck is carried by a conveyor belt and taken out by trains of mining cars. Sometimes, buckets carrying the material are lifted and taken out through shafts operated under pulley system. In large-diameter tunnels, mucking is done by means of dumpers with rubber tyres and overhead loaders.

Many mechanized excavation methods have been developed by specialized manufacturers in the US and Europe. 'Jet piercing plasma joint, laser beam and electron beams method used in western countries have good prospects in shattering tunnel rocks Under mechanical method, fluid erosion, ultrasonic and abrasion were expected to show promise. Nuclear energy for mass excavation with no muck to handle, due to crater of ditch formation, is also heard of but its use in open excavation is yet to be demonstrated,' observed Murti (1971). The position has not changed yet, even after so many decades of his observation.

16.8 EXCAVATION METHODS OF ROCK TUNNELLING AND SUPPORT SYSTEM

The excavation for tunnelling in rock is aimed at bringing about the exact shape of the tunnel as in the design. The tunnelling is carried out by forward increment in excavation consisting of drilling, blasting, ventilating, mucking, and installing the support. The time taken (in hours) to complete a *round* represents a *cycle*. The rock types including the structural features decide the extent of time or cycle necessary to install the support before the next round begins.

16.8.1 Full-Face, Top Heading and Bench, Side Drift and Multiple Drift Methods

The methods followed in tunnel excavation include *full-face*, *top heading and bench*, *side drift*, and *multiple drift* methods. All these methods involve blasting, which is preceded by drilling a set of holes (called blast holes) used for detonating purposes and filling them with inflammable substance. In full-face method, the entire face is blasted in one operation. It is most suitable for strong ground or smaller tunnels. In general, this method is followed in hard and in small-diameter tunnels. After completion of excavation, reinforcement by bolting and shotcreting is provided to the crown. Rock tunnels, especially large-diameter tunnels comprising stratified rock, or blocky and jointed rocks are constructed by top heading and bench method.

The top heading and bench method can be operated by excavating side drift or multiple drifts. The sequence of work involved in the top heading and bench method, shown in Fig. 16.17(a) includes drilling in the top two-thirds of the tunnel, loading of explosives, and properly detonating the rock to achieve sufficient breakage of excavated materials with minimum damage to surrounding rock. The tunnel is then ventilated to remove the gases generated from the explosion. Thereafter, the crown area is reinforced with shotcreting or rock bolting. The procedure for *bench* is taken up with the same sequence of work as in top heading. The centre portion of the lower one-third of the tunnel is drilled, blasted, mucked out, and then bolted or shotcreted for supporting the tunnel rocks. In the Malaprabha tunnel in Mysore, 41 holes were drilled in the top heading and 27 holes in the bench operation. Using gelatine (60–80% strength), blasting was done electrically with the help of electric delay detonators. Normal pull achieved was 1.5 m to 3.0 m. For 1.5 m pull, 60 kg gelatine blasted 71 m^3 of rock.

Large tunnels with poor-quality rock use the drift methods. In the side drift method, drifts are driven first in the two sides of the tunnel section providing supports, see Fig. 16.17(b).

Then, the middle section is excavated keeping the end supports intact. In the multiple drift method, tunnelling is done by driving several drifts, see Fig. 16.17(c). The side drift or multiple drift method of tunnelling is adopted in stiff or hard cohesive soil and well-consolidated soil. Initial shotcrete support is needed immediately after each round of drift excavation. Systematic reinforcement (welded wire fabric or fibres) is provided by a ring closure behind the excavation for stabilization of weak tunnelling media.

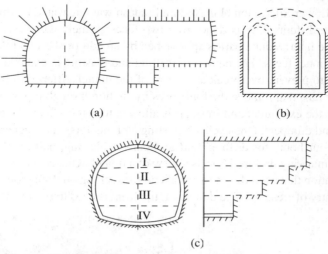

Fig. 16.17 Sketch showing various methods of tunnelling:
(a) top heading and bench method; (b) side drift method;
and (c) multiple drift method

The type of excavation and nature of reinforcement required for various types of tunnelling ground are shown in Table 16.1.

Table 16.1 Excavation approach and nature of reinforcement for varied ground conditions

Tunnelling ground	Excavation sequence	Reinforcement
Intact rock	Full-face or large top heading and bench	Spot bolting (fully grouted dowels)
Stratified rock	Top heading and bench	Systematic doweling or bolting in crown
Blocky and seamy ground	Top heading and bench	Bolting in top heading considering joint spacing
Squeezing or swelling ground	Top heading, bench, and invert	Doweling or bolting in top heading and bench considering joint spacing
Stiff or hard cohesive soil	Top heading, bench, and invert; subdivision into drifts (side drift)	Welded wire fabric or fibres; shell with full ring closure in invert
Well-consolidated non-cohesive soil	Top heading, bench, and invert; subdivision into drifts (multiple drifts)	Same as above

A typical sequence of tunnelling in rock includes drilling and blasting, mucking of the detached rocks, supporting the rock, and lining with concrete. The drilling and blasting pattern of rock is designed depending upon the rock condition and their structure. The number of drill holes and the distribution of charge density are adjusted to suit the geological condition of tunnel rock. In some large diameter tunnels, initially a small cavity (pilot tunnel) is made up to the crown to facilitate observation of the rock condition. According to the rock condition, drilling and blasting are continued to widen the cavity until the design dimension is achieved. The 9.75 m-diameter tunnel in Lower Sileru tunnel in Andhra Pradesh was excavated by this method.

In general, tunnelling is done from two faces, namely the intake and exit ends, but in long tunnels, the intermediate part is approached by an adit or shaft and the tunnel work is continued from additional faces. The pattern of drilling and blasting is planned to keep the vibration at the minimum to prevent overbreak at the rim of the tunnel. After each blast, ventilation is provided by blower fans that remove the foul gases by suction through pipes leading to the upper end of a shaft or to the exit outlet and fresh air is allowed to enter. This is followed by mucking operation by carts and dumpers. One cycle of blasting and mucking takes generally about ten hours, which includes four hours for drilling, half an hour for blasting, half an hour for ventilation, and five hours for mucking. Figure 16.18 is a view of a tunnel at the construction stage showing pipe used for ventilation that extends out to portal end for sucking out foul gases from inside after each blast and for entry of fresh air. The dumper at the front is used to remove the muck from the tunnel.

Fig. 16.18 A view of a tunnel at the construction stage showing pipe used for ventilation

16.8.2 Types of Tunnel Supports Including Rock Bolting

A support is necessary in a newly excavated tunnel portion to keep the opening stable and avert any collapse of roof or tunnel wall. This installed initial support system is important for safe tunnelling until lining is provided. The conventional supports provided in a tunnel can be one or more combination of the following types:

  (i) Timber planking
 (ii) Steel ribs
(iii) Steel ribs with pre-cast or cast-in-site concrete slabs
(iv) Shotcreting with or without wire mesh
 (v) Perfobolt with shotcreting

(vi) Forepoling

(vii) Rock bolting

The type of support provided depends on the size of the tunnel cavity, geological condition of rock, measured or anticipated rock load, and water flow conditions. In adverse condition of tunnel rock, *timber planks* or *steel rib* supports are provided immediately after tunnel driving to avoid any overbreak, see Fig. 16.19 (a). In general, the supports are kept permanently. If the tunnel rock is not very weak, the supports by timber logs or steel ribs may be of temporary nature until the rock becomes stable when the supports are removed and concrete lining is provided. The *steel rib support* is very common and may be taken up alone or in combination *with concrete lining* of tunnel interior and filling of the caved space between rock and the support by rubble or concrete, see Fig. 16.19 (b).

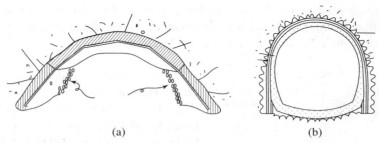

(a)   (b)

Fig. 16.19   Tunnel supports: (a) sequences (I, II, and III) of erecting steel ribs in jointed rocks and; (b) complete face of a tunnel with steel rib support and concrete lining

Supporting the rock by *shotcreting* is done in many tunnels for arresting the overbreak when the rocks are extensively jointed, having a tendency to collapse immediately. Shotcreting is the process of concrete spraying into the place by a nozzle. It is generally of 5–8 cm thickness and contains some additive with capacity of immediate setting. It prevents cavity formation in addition to collapse of rock. In case of soft and weak rocks, shotcreting of more thickness and with wire meshing is provided. Once shotcreting stops the tendency of caving, permanent support by steel rib is provided to contain the total rock load. Shotcreting can be made without or with wire mesh or reinforcing fibres.

Perfobolts are perforated hollow tubes that are filled with grout inserted into the drill holes. The grout is extruded to fill the annular space around the tube when a piece of reinforcing rod is pushed into the grout filling the tube. Perfobolt is provided after shotcreting has made it possible to keep the rock intact. In case of hard rock with blocky joints, perfobolts are successfully used for retaining the rock by extending the rods several metres inside sound rock.

Forepoling technique consists of providing an umbrella of suitable material (pipes or pointed rocks) over the reach of the tunnel where flowing ground condition is noticed or expected. Typically, these are devices in overlapping arrangement as shown in Fig. 16.20 (a). The aim of forepoling is to ensure that there is no movement of material above the umbrella during the removal of muck. The two ends of the forepole rods are supported, one end on steel ribs and the other on undisturbed muck, ahead of the heading, shown in Fig. 16.20 (b). Steel channels are also used when the materials of the flowing ground are clay gouge and sheared materials without any big rock fragments. In case rock chunks are present, multiple rods or forepoles are used. Forepoles such as rods or channels provide a complete umbrella over the entire zone

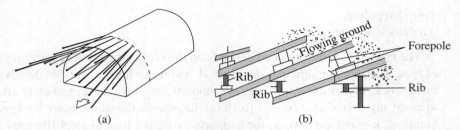

Fig. 16.20 Forepoling method of supporting flowing ground: (a) forepoles driven in tunnel crown to make an umbrella to arrest material movement; and (b) enlarged view of forepole arrangement

from where flow takes place. In multiple forepoling , there are added rows of forepoles to bear the pressure acting on the section. Forepoling techniques are used in extremely weak rock condition as noticed in many tunnels constructed in the Himalayan region (e.g., Jaldhaka tunnel in West Bengal, Beas tunnel in Himachal Pradesh).

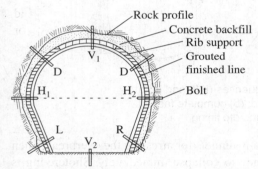

Fig. 16.21 A tunnel section showing rib support, grouted interior finish, and reinforcement of jointed rocks by rock bolting

Rock bolting is undertaken when the rock loads are excessive, especially in a large-diameter tunnel (Fig. 16.21). Rock bolting reinforces and supports partially detached, laminated, or otherwise incompetent rocks that would be subjected to failure by gravity and loss of frictional effect. The effectiveness of rock bolting can be improved by providing a wire net over the rock surface followed by concrete spray. Rock bolting of wedge type is undertaken along with shotcreting or grout ring to hold the rock in excavated profile before any breach. Rock bolting is applicable in hard but jointed rock and blocky and seamy rock, but not in crushed rock or in swelling rocks.

16.9 PRESSURE TUNNEL AND LINING

Pressure tunnels are subjected to high hydraulic pressure from the interior parts. The pressure of the rock and that of underground water work in compression before water is allowed to flow in the tunnel. However, when under operation, the hydrostatic pressure inside exceeds that from the outside. If the rock is hard, it can withstand the hydrostatic pressure developed inside. However, the tunnel needs watertight lining for smooth flow and for resisting the outer pressure. For tight contact of the lining with rock, cement grouting is to be carried out to seal the fracture openings and voids of porous rocks. A tighter contact is achieved by injecting cement grout under high pressure into the interface of lining and rock. Cement grouting behind the lining ensures its elastic interaction with the rock and uniform distribution of rock pressure on the lining. In the design of the lining for pressure tunnel, more consideration is given to the interior pressure due to the water flow under pressure.

The size and shape of a pressure tunnel are decided depending upon the hydraulic requirement and the geological condition along the tunnel alignment. The thickness of lining provided is based on the static load acting on it permanently or temporarily. The permanent load includes the rock pressure and the mass of the lining itself. The other prolonged temporary pressure indicates underground water pressure through the rock and inside water pressure when tunnel is in operation. Several other temporary pressures such as the blasting effects, rock creep, and grouting pressure act for a short time. The tunnel lining is designed in such a way that it has the capacity to resist the combination of all these loads. The linings are generally provided by plain cement concrete (PCC) or RCC and combined double layers having concrete outer and inner steel encase. The pressure tunnels are lined for free and smooth flowing and for preventing seepage.

If the rock in the water conductor tunnels is soft and erodes easily, RCC or PCC lining is provided. In general, all pressure tunnels are lined to ensure smooth flow of water under pressure. The concrete lining is provided after removing the timber or steel rib supports after a certain time gap. The lining work is completed in three stages; first, kerb concreting in two corners; second, lining the arch and sides; and finally the invert is lined. The steel kerb carries the load of concrete casting until concrete sets against the rock. In case no timber or steel support is provided, shotcreting will be done immediately after tunnelling and concrete lining will be undertaken shortly afterwards or simultaneously. The main consideration is that the developed load may be active if the tunnel passes through a fault, fold, or thrust zones. The actual rock load due to varying rock formations and structural set-up inside the tunnel is measured by instrumentation to help design of tunnel lining.

In the tunnel lining, the coarse aggregate material to be used for concrete must be of sound and durable nature and free from deleterious materials. The maximum size of the aggregate to use for concrete should be smaller than 37 mm. If coarse aggregate for tunnel is not available from the nearby rivers, this can be obtained by crushing the tunnel spoils of hard rock to sizes of 25–16 mm and 16–10 mm sizes. The thickness of lining provided in many of the tunnels varies between 20 cm and 76 cm. For example, it is 20 cm in the Yamuna tunnel in Uttar Pradesh, 25 cm in Beas–Sutlej link tunnel, 30 cm in Parambikulam tunnel in Tamil Nadu, and 76 cm in Koyna head race tunnel. The strength of the concrete used for lining is variable depending upon design requirement. The concrete used in the Beas tunnel lining has varying strengths of 140, 210, and 280 kg/cm².

16.10 ROCK MASS QUALITY AND SUPPORT REQUIREMENT

In designing a rock tunnel, the nature of rock mass with respect to its structural features in addition to its strength is considered. Fault, fold, joints, shear zones, etc. are harmful features of tunnel rocks. In order to bring stability of the tunnels and save from overbreaks, various types of supports are designed as detailed in the following paragraphs.

16.10.1 Design Aspects

The design aspect of a tunnel including excavation, lining, and support system depends upon the quality of the rock mass at the tunnel. Among the various geological factors that control the quality of rock mass, the following are considered while designing tunnel support:

  (i) Open joints, clay filling, porous and permeable layers, or foliation planes
 (ii) Unfavourable orientation of rock attitudes
(iii) Fault or shear zone or karstic condition
 (iv) Swelling type of clay
  (v) Build-up of water pressure
 (vi) Presence of in-situ stress or deformed rock

The support requirements suggested by Proctor (1971) for hard rock and firm rock with jointing are as follows:

  (i) Hard and intact rock—unsupported
 (ii) Hard stratified and schistose rock—unsupported

(iii) Massive, moderately jointed—unsupported in firm ground and very firm ground and steel rib support at 3 m interval or shotcrete in arch only in troublesome jointed zone

16.10.2 Rock Load System of Terzaghi for Tunnel Support

In most of the Indian tunnels, rocks are of varied nature and numerous problems are encountered during tunnelling. The rocks of the Himalayan terrain possess high stress and tunnelling is always difficult. The type of support required in a tunnel is dependent on the overall quality (or rock mass quality) that includes all weaknesses of rock. Several classifications of the rock mass quality were devised and modified by a number of authors. Terzaghi (1950b) was probably the first to suggest rock load based simply on geological observation of rock condition (Table 16.2). The rock load is expressed as a function of tunnel size.

Table 16.2 Terzaghi's classification of rock load at a depth of more than $1.5(B + H)$ (B is the width and H the height of tunnel)

Rock condition	Rock load in feet	Remarks
Hard and intact	Zero	Light lining required if spalling or popping occurs
Hard rock	$0-0.5B$	Light support; load may change erratically from point to point
Moderately jointed	$0.5-0.25B$	
Moderately blocky and seamy	$0.25B-0.35(B + H)$	No side pressure
Very blocky and seamy	$0.35-1.10(B + H)$	Little or no side pressure
Completely crushed	$1.10(B + H)$	Considerable side pressure; softening effect of seepage towards bottom of tunnel requires either complete support for lower ends of sides or circular ribs
Squeezing at moderate depth	$1.10-2.10(B + H)$	Heavy side pressure, invert strut required; circular ribs recommended
Squeezing at great depth	$2.10-4.50(B + H)$	
Swelling rock	Up to 250 feet	Concrete ribs required; in extreme cases, yielding support is used

The rock loads proposed for support and lining in Terazaghi's method is found to be very conservative, but it gives a good approximation of the rock and is easy to estimate. Moreover, it does not consider dip, strike, alteration, and other factors related to the rock conditions. Among the other classifications, the Norwegian Geotechnical Institute (NGI) classification of Barton, Lien, and Lunde (1974) and the geomechanics classification of Bieniawski (1974) are used in many countries including India from the 1980s to determine the rock mass quality and decide the nature of support requirement for tunnels. Both these methods (see Sections 10.8 and 10.9) are found to be satisfactory in forecasting the rock load and designing the support system.

16.10.3 Methods of Evaluating Tunnel Support by Tunnel Quality Index and Rock Mass Ratio Systems

Evaluation of support requirement for a tunnel as suggested in the NGI classification requires first the estimation of tunnel quality index Q (Table 10.5 of Section 10.8) and then finding its relation with *equivalent dimension De* of the tunnel from the chart. The value of *De* is obtained from the following relation:

De = Excavation span, diameter, or height (m)/Excavation support ratio (ESR)

The ESR values of different excavation types proposed by Barton, Lien, and Lunde (1975) fall into five categories (A to E) as shown in Table 16.3.

Table 16.3 ESR values of different types of openings

Excavation type	ESR value
A. Temporary mine openings	3–5
B. Permanent mine openings, water tunnels for hydro power (excluding high pressure penstocks) pilot tunnels, drifts, and headings for large excavations, V	1.6
C. Storage rooms, water treatment plants, minor road and railway tunnels, surge chambers, access tunnels	1.3
D. Power stations, major road and railway tunnels, civil defence chambers, and portals intersections	1.0
E. Underground nuclear power stations, sports and public facilities, and factories	0.8

Publication of Barton et al. (1980) provided additional information on the relation of unsupported spans, roof support pressure, and rock bolt length with tunnel quality index Q and other parameters as follows:

Maximum unsupported span = $2ESR\ Q^{0.4}$

The length L of rock bolts can be estimated from the relation:

$L = 2 + (0.15B/ESR)$

where B is the excavation width.

On the basis of the study on case records, the following relation was proposed for rock pressure on tunnel roof by NGI support system (Grimstad and Barton 1993).

$$P_{roof} = \frac{2Jn^{1/2} \times Q^{1/2}}{3Jr} s$$

Table 16.4 shows the tabulated data on the predicted support pressure as in NGI support system and actual measured data of rock pressure by Central Soil and Materials Research Station (CSMRS), India, for some of the Indian tunnels (Majumdar 1991).

Table 16.4 Comparison of measured pressure (CSMRS, India) with predicted pressure

Project name	Tunnel diameter (m)	Rock type	Predicted pressure (kg/cm²)	Measured pressure (kg/cm²)
Loktak Tunnel	4	Crushed shale (fair)	2.0	1.20
		Poor	3.3	1.58
		Exceptionally poor	7.5	4.08

(Contd)

Table 16.4 *(Contd)*

Maneri Bhali Tunnel	7	Sheared schist	2.3	2.0
		Foliated schist	1.2	0.8
		Fractured quartzite	1.4	0.6
Tehri Tunnel				
(a) Grade 1	13	Phyllite	0.8	0.25
(b) Grade 2	13	Phyllite	2.1	0.25
(c) Grade 3	13	Phyllite	3.85	1.24
Salal Tunnel	11	Dolomite	1.9	1.1

It can be seen from the table that compared to the predicted NGI support pressure, the measured pressure of CMRS is found to be slightly low in the conservative side.

The chart showing the relation between *De* and *Q* originally given by Barton, Lien, and Lunde (1974) was later updated by Grimstad and Barton (1993). The authors have also given the estimation of support requirement for the varied rock categories (1 to 9) shown as 'Reinforcement categories' in the updated chart given in Fig. 16.22.

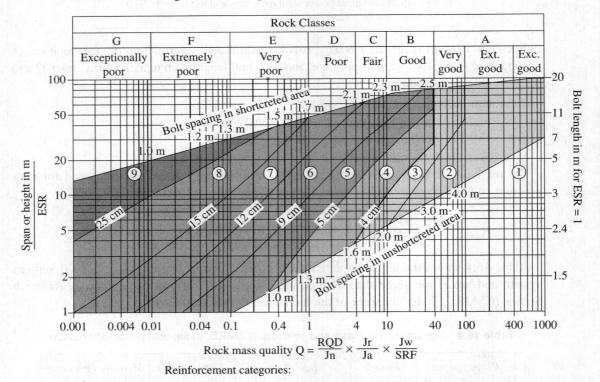

$$\text{Rock mass quality } Q = \frac{RQD}{Jn} \times \frac{Jr}{Ja} \times \frac{Jw}{SRF}$$

Reinforcement categories:

1) Unsupported
2) Spot bolting
3) Systematic bolting
4) Systematic bolting, (and unreinforced shotcrete, 4–10 cm)
5) Fibre reinforced shortcrete and bolting, 5–9 cm

6) Fibre reinforced shotcrete and bolting, 9–12 cm
7) Fibre reinforced shotcrete and bolting, 12–15 cm
8) Fibre reinforced shotcrete, > 15 cm, reinforced ribs of shotcrete and bolting
9) Cast concrete lining

Fig. 16.22 Estimated support categories based on the tunnelling quality index *Q* (Grimstad and Barton 1993)

According to Barton, tunnelling revolution has taken place with the development of stainless steel fibre reinforced shotcrete (*Sfr*) to spray in dense, low permeability concrete of 35 to 45 MPa in-situ quality and epoxy-coated and PVC-sleeved triple erosion protection rock bolts. The Norwegian method of tunnelling (NMT) support now follows the stainless steel fibre reinforced shotcrete instead of wire mesh as the use of latter in an irregular excavated tunnel surface may not achieve regular contact with rock. The support system proposed in the geomechanics classification of Bieniawski (1974) for horseshoe-shaped tunnel of 10 m width, vertical stress < 25 Mpa, and depth < 100 m has been presented in Table 16.5.

Table 16.5   Geomechanics classification for excavation and support requirement

Rock mass class (RMR)	Excavation	Support: rock bolts 20 mm diameter, fully bonded	Support: shotcrete	Support: steel sets
Very good rock I RMR 81–100	Full face; 3 m advance	None	None	None
Good rock II RMR 61–80	Full face; 1.0–1.5 m advance; complete support 20 m from face	Locally bolts in crown, 3 m-long, spaced 2.5 m with occasional mesh	50 mm in crown where required	None
Fair rock II RMR 41–60	Top heading and bench; 1.5–3.0 m advance in heading; commence support after each blast; complete support 10 m from face	Systematic bolts 4 m long, spaced 1.5–2.0 m in crown and walls with mesh in crown	50–100 mm in crown and 30 mm in side walls	None
Poor rock IV RMR 21–40	Top heading and bench, 1.0–1.5 m advance in heading; install support con-currently with exca vation, 10 m from face	Systematic bolts, 4–5 m long, spaced 1–1.5 m in crown and walls with wire mesh	100–150 mm in crown and 100 mm in sides	Light ribs spaced 1.5 m where required
Very poor rock V RMR < 20	Multiple drifts; 0.5–1.5 m advance in top heading; install support concurrently with excavation; shotcrete as soon as possible after blasting	Systematic bolts 5–6 m long, spaced 1–1.5 m in crown and walls with wire mesh; bolt invert	150–200 mm in crown, 150 mm on sides, and 50 mm on face	Medium to heavy ribs spaced 0.75 m with steel lagging and forepoling if required; close invert

Barton has suggested the use of the data of *Q* system for initial support system and subsequently modified based on observation. Plotting of *Q* values versus rock mass ratings (RMR) indicates that both are in conformity, thereby confirming the classification assessment of Majumdar (1991). Tilak (1999) after a review of several completed Indian tunnels since 1980 came to the conclusion that the recommendation made on the basis of the two rock mass classifications with respect to support requirement wherever adopted by the project was found to be beneficial in terms of cost and time. Consensus of opinion is that it is always better to take up at least some instrumentation or testing of rock load in the pre-investigation stage and compare the values computed from the two methods for better confidence of the methods.

SUMMARY

- Tunnels are basically of two types, namely hydraulic tunnels and traffic tunnels. Hydraulic tunnels are water conducting tunnels, whereas traffic tunnels serve as railways, highways, and subways for transportation.
- Construction of tunnel in rock is done in the sequence of boring, blasting, and mucking. In small-diameter tunnels, full-face tunnelling method is applied in which rocks from the entire face of the tunnel is removed by a single operation of blasting.
- Large-diameter tunnels need heading and benching method that includes removal of the top part and then the bottom parts by blasting operations.
- In soft and weak rocks, single drift or multiple drift method is followed in which tunnelling advances by providing supports on the two sides and then excavating the intermediate part by driving single or multiple drifts.
- The geological hazards encountered in rock tunnelling are mainly overbreaks, roof collapse, spalling of wall rock, flowing ground, squeezing, and heaving of tunnel rock.
- Tunnelling in soft ground is done either by cut and cover method or by shield method. The shield having a sharp cutting edge in front facilitates full-face excavation followed by immediate support and concrete lining.
- In the cut and cover method, the two sides of the tunnel are protected by retaining walls and the material (earth) of the intermediate portion is excavated using conventional machine. The pit thus formed is provided with RCC lining.
- Both morphology and rock structures are considered in geological investigations for selecting tunnel alignment. Large-scale geological mapping is carried out covering the area along two sides of tunnel alignment followed by exploration by core drilling to reveal subsurface rock condition.
- Geophysical survey by seismic method is conducted in some tunnel projects to determine bedrock profile and structural features such as folds, faults, and shear zones. With the help of data of subsurface exploration, a geological section is prepared along the tunnel alignment to help engineering design.
- An important part of geological investigation for tunnel is 3D mapping (tunnel log) of the tunnel rocks, especially recording of all adverse geological features and accordingly suggesting remedial measures for the weak stretches.
- The geological factors that are mainly responsible for the hazards in tunnel construction in rocks include rock stress and deformation, faults, open joints, clay filling, swelling clay, and divisional planes such as beddings and foliations.
- Several authors have classified the tunnel rocks based on some of these factors and suggested methods of safe tunnelling and suitable support systems. An overall analysis of all these classifications suggests that the recommendations given based on rock quality designation of NGI and RMR of CSMRS if adopted for the purpose of providing lining and support system will be most suitable and cost-effective in preventing tunnel hazards.

EXERCISES

Multiple Choice Questions

Choose the correct answer from the choices given:

1. Visual observation made in a tunnel is in:
   (a) observation adit
   (b) vertical shaft
   (c) Inlet portal

2. In a horizontal strata traversed by joints:
   (a) the overburden may be upto the same diameter of the tunnel.

   (b) the cavity in a rock is restricted to twice the thickness of the tunnel diameter.
   (c) it is observed that placidity due to the blasting does not transmit beyond the distances of the twice the diameter of the tunnel.

3. The correct statement among the following is:
   (a) In horizontal strata traversed by inclined joints, the overbreak may be upto the same tunnel diameter.

(b) In inclined strata traversed by joints at any angle, the caving in of rock is restricted to a minimum thickness of half the tunnel diameter.

(c) In highly jointed or crushed rock, the overbreak may extend up to three times the diameter or three times the sum of height and width of the tunnel.

4. Soft rock tunnelling is done in:
   (a) a consolidated overburden material
   (b) alluvial materials present as tunnelling material
   (c) tunnelling through soft ground – the unconsolidated material above the heading flows like a fluid

5. Most common method of soft tunnelling is
   (a) cut and cover method
   (b) Shield method
   (c) hydraulic pressure or piping, settlement at the bottom are no problem

6. The natural hazard that may occur during tunnelling is:
   (a) overbreaks
   (b) flowing ground
   (c) squeezing ground

Review Questions

1. Draw a section along a tunnel and show the following: inlet portal, outlet portals, rock cover, overburden, shaft, and drift. Also draw a across section of a tunnel to show positions of tunnel invert, wall, crown, spring line, and arch section.

2. Enumerate different types of tunnel and their functions. What is a pressure tunnel?

3. Discuss how the pressure on the heading of a tunnel is distributed after its excavation. Explain the arching action and the factor on which such action is dependent.

4. Describe the effect of bedded rocks on a tunnel lining for the following positions:
   (i) Tunnel traversing along horizontal beds
   (ii) Tunnel passing normal to strike of vertical dipping beds

   (iii) Tunnel passing normal to strike of beds dipping at an angle of 45°

5. Explain the effect of a fault on a tunnel passing through rocks for the following situations:
   (i) Tunnel located in the hanging wall of the fault
   (ii) Tunnel located in foot wall of the fault
   (iii) Tunnel aligned perpendicular to the trend of the fault
   (iv) Tunnel passing through an active fault

6. Describe the effect on the lining of a tunnel passing through rocks which are folded (i) as an anticline and (ii) as a syncline.

7. Explain the following terms with the help of a cross section of a tunnel: design cross section, concrete lining, pay line, and overbreaks.

8. State the relation of overbreaks in tunnel rocks with the attitudes of the joints in the rocks. What should be the minimum rock cover above a tunnel roof to avoid roof collapse?

9. Give a short account of tunnelling through soft ground composed of unconsolidated materials. What type of machinery is used in soft ground tunnelling? Describe the approach of operation of the machinery.

10. What types of problems are generally encountered in tunnel construction? Narrate these problems in relation to different geological conditions of the tunnelling media.

11. What will be your approach in selecting a tunnel alignment? Describe the methods followed in subsurface exploration along the tunnel alignment.

12. Explain the method of tunnel logging. Describe the geological activities in the construction stage of a tunnel.

13. Give an account of the conventional method of tunnel boring. Describe shield method and wagon drill method of tunnel excavation.

14. Write short notes on the following:
   (i) Full-face method of tunnelling
   (ii) Side drift method of tunnelling
   (iii) Multiple drift method of tunnelling

NUMERICAL EXERCISES

1. Study of rocks of a 3 m-diameter road tunnel provides the estimated Q value of 12 and RMR value of 72 (see the result for questions 2 and 3 in Exercises of Chapter 7). Give your opinion on (i) rock mass class, (ii) support system, and (iii) excavation type suitable for rocks of such a tunnel.

[Answer: (i) Good, (ii) spot bolt or local bolt with 2.5 to 3 m-length rods at 2.5 m spacing, and (iii) full-face type]

[*Hint:* Use the following steps:
1. Consult Table 16.3 for *ESR* value of road tunnel. It is 1.0. Then, Span/*ESR* is 3/1.0 = 3.

2. Now, see the chart in Fig. 16.22. For ordinate 3 (Span/*ESR*) and abscissa $Q = 12$, the rock type of tunnel is found to be 'good' to 'very good'.

3. So, the tunnel will only require bolting and the bolt size will be nearly 2.5 m-long rod and spacing 2.5 m. This is read from the chart (Fig. 16.22).

4. For the RMR value of 72, consult Table 16.5. It states clearly that tunnel rock is 'good' and will not need any steel support but 'local bolting by 3 m-long rod spaced at 2.5 m intervals' will do.

5. The RMR value suggests that it should follow 'full-face' tunnelling method (see Section 16.8.1)].

17

Powerhouses

● ● ● ● ● ● ● **LEARNING OBJECTIVES** ● ● ● ● ● ● ●

After studying this chapter, the reader will be familiar with the following:

- Different types of powerhouses and their functions
- Geotechnical method of investigation of surface powerhouse
- Details of investigation procedure for underground powerhouse
- Geological problems encountered during construction of powerhouses and their solution
- Investigation methods for thermal and nuclear powerhouses

17.1 INTRODUCTION

A powerhouse may be a surface or an underground construction. In the investigation of a surface powerhouse of a hydroelectric project, in the first instance, aerial photographs are studied to find the general topographical and geological features of the site. Detailed investigation is then followed by visit to the site and large-scale geological mapping. Investigation of an underground powerhouse is conducted by exploratory drilling. Geological logging of drill cores and their laboratory testing for strength properties are the most essential steps to understand the site conditions of an underground powerhouse. This chapter presents case studies on some Indian powerhouses and the geological problems encountered during their construction and the methods suggested to rectify the problems. It also provides guidelines for selecting a nuclear powerhouse site. In addition, the chapter highlights the problems of setting up a powerhouse site in the Himalayan terrain.

17.2 DIFFERENT TYPES OF POWERHOUSES AND GENERATION OF HYDROPOWER

Depending upon the sources of energy, there are three main categories of powerhouse—hydroelectric, thermal, and nuclear. In hydroelectric powerhouses, it is the water pressure that drives the turbines. In thermal power plants, fossil fuel such as coal, natural gas, or oil is used to generate steam for driving the turbines. The turbines are also driven directly by diesel-run motors for generation of power. Fissionable materials are used for generation of nuclear power.

17.2.1 Harnessing Hydropower from River Water

Hydropower converts the energy in flowing river water into electricity. The quantity of electricity generated is determined by the volume of water flow and the amount of head created by the dam. Head is the height from the turbine in the powerhouse to the water surface of the reservoir. Some of the advantages of a hydroelectric power scheme are that it is cheap to run, the running water (fuel) is not exhausted, it does not cause pollution, and the reservoir water that produces electricity can also be used for fishing, boating, and tourism.

High dam with large storage capacity reservoir is built to develop the hydropower potential of a river. The essential structure required for harnessing power from reservoir water is a powerhouse, which can be located within the body of the dam (Fig. 17.1) or near its toe (Fig. 17.2). In many hydroelectric power schemes, the powerhouse is sited in the downstream valley very close to or away from the dam. Depending upon the topographical features, when the powerhouse is located at a long distance from the dam site, the water from the reservoir is diverted to the power station through a power channel or a power tunnel. The source of energy is the work done by the water moving from the reservoir and falling to the powerhouse. The generation of power depends on the volume of available water and the head (or height) of falling water from the reservoir level to the power plant.

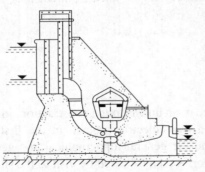

Fig. 17.1 Powerhouse within the dam body

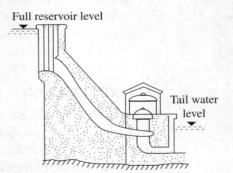

Full reservoir level

Tail water level

Fig. 17.2 Powerhouse at the toe of dam

The main device in the powerhouse is the turbine having huge blades that rotate under the thrust of water. The reservoir water is led through pipes (called *penstocks*) and then allowed to fall down to the turbine forcing movement of its blades. A generator attached to the turbine produces electricity, which is then distributed through transmission lines for domestic and industrial use. The power potential of a hydroelectric project is calculated as follows (Lewit 1982):

$$\text{Available power } (P) \text{ in kilowatt} = Q \times H \times g \times efficiency$$

where, Q is the discharge in m³/s, H is the head in m, and g is 9.81 m/s² and *efficiency* depends on the type of the turbine or the total project assessment.

The volume of water available from the reservoir remaining constant, the greater the head the more will be the power generation. The capacity of system in a powerhouse is expressed by the terms peak capacity and installed capacity. *Peak capacity* is the maximum kilowatt capacity with all power sources available at the time of occurrence of annual peak load in the system less power requirement at the generating station. *Installed capacity* is the sum of the name plate ratings of all generating equipment installed in the system. It is the maximum continuous kilowatt capacity of the system with all power sources available including thermal, nuclear, and hydro installations. When no restriction is imposed on the use of electric power, the maximum load in a system in a year is its *peak load* measured by actual deliveries at the generating station bus bar. While calculating the peak load, the losses due to distribution and transmission are not considered.

17.3 SURFACE POWERHOUSE FOR HYDROELECTRIC PROJECT

A typical hydropower plant includes a dam, a reservoir, penstocks (pipes), a powerhouse, and an electrical power substation. The dam stores water and creates the head; penstocks carry water from the reservoir to turbines inside the powerhouse; and the water rotates the turbines, which drive the generators that produce electricity. The electricity is then transmitted to a substation where transformers increase the voltage to allow transmission to homes, businesses, and factories (Fig. 17.3).

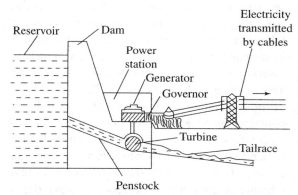

Fig. 17.3 Different components of a hydroelectric power plant

Hydroelectric powerhouse is an integral part of the overall features of a hydroelectric project consisting of a dam and reservoir or lake, with water conductor system (tunnel or channel), penstock, and tail race channel or tunnel (Fig. 17.4). Selection of the powerhouse site is done at the planning stage along with locating the other structural components. The components of a powerhouse are machine hall, service bay, entrance to service bay, unit bays, drainage gallery, tail race bed, cable trench, and cable shaft. During detailed mapping of powerhouse foundation, the geological features including the weaknesses of area covered by each of these components are to be recorded separately for design purposes.

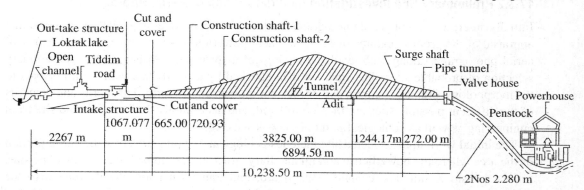

Fig. 17.4 Ground profile showing powerhouse, penstock, and long tunnel to carry lake water for generation of hydroelectric power by Loktak hydroelectric project of eastern India (Manipur)

17.3.1 Landform Characteristics of a Surface Powerhouse

A hydroelectric powerhouse may be a surface construction or an underground structure. The surface powerhouse is located in the immediate downstream part or away from the dam, depending upon suitable topography where high head is available. In hilly terrain, many hydroelectric projects involve construction of long tunnels to carry reservoir water to the powerhouse located in the lower valley to obtain sufficient head (Fig. 17.5). A surface powerhouse should preferably be located in a flat terrain or gentle slope to avoid instability of the site. The groundwater table should be low. An area with hard and sound rock is favourable for a powerhouse site where heavy machineries are installed. If the site contains normal soil, the powerhouse is commonly founded on raft foundation for a medium head hydroelectric project or on pile foundation for a very large powerhouse. The foundation should be homogeneous and of monolithic nature.

Fig. 17.5 Rammam hydel project of Eastern Himalayas with a long penstock leading water to an underground powerhouse to obtain available head 'H'

In the powerhouse site investigation, the main geological aspect that needs careful consideration is that no part of the foundation settles due to the load imposed by the machineries such as turbine and generator. Settlement of the foundation causes wear of the turbines. It may also damage the penstock connection with the power plant. Another essential point for consideration is the vibration of the foundation when machines such as turbines are in operation. In fact, vibration is responsible for the settlement of soil foundation and damage of powerhouse walls. The design of the foundation needs to consider methods for resisting the vibration. For powerhouse on soil foundation, the walls may be tied up with firm rock by means of steel rods to reduce the vibration to zero. The engineering geologist needs to find whether suitable rock is present at shallow depth close to the powerhouse for tying the rods.

17.3.2 Preliminary Site Investigation Including Study of Aerial Photographs

Initially, interpretation of aerial photographs under stereoscope gives a quick three-dimensional appraisal of the ground features of the area where powerhouse site is to be selected. Study of aerial photographs along with available geological map of the area helps in the preliminary selection of probable sites. The sites are then checked by ground observation with respect to the morphology, drainage pattern, soil or rock conditions, and so forth. The merits and demerits of the different possible sites are then considered, and accordingly the best possible site from engineering as well as geological point of view is selected.

The initial work in the selected site includes geological mapping covering the entire portion of the area where the powerhouse and all ancillary structures are to be located. Since the sites for the penstock, tail race channel, switchyard, and transformer are closely related to the powerhouse location, these sites are also covered by preliminary investigation with respect to their stability while selecting site for the powerhouse. The available rock types, their attitudes, major joints, and fault and shear zones if present are plotted during mapping. If the area has hill slope very close to the powerhouse site, the site should be studied with respect to slide possibility.

A hill slope that involves cutting for founding the powerhouse may lead to problems of rock fall and debris slide. In case any fault is detected, special care is to be taken to find the thickness or width of zone affected by crushing and whether any movement has taken place along the fault plane in recent times. Because of the soft nature of rock in a fault, a change in the design of the powerhouse layout is desired so that heavy equipment such as turbines does not directly rest on it. A site with recent activity should be rejected in favour of another site free from such faults.

17.3.3 Detailed Investigation by Subsurface Exploration

A drilling programme is planned after selection of the site based on geological investigation and engineering consideration. The plan of drilling generally includes one drill hole in each of the four corners and one hole each in the locations of each turbine unit. Additional holes will be required for weak zones due to fault, thrust, intense jointing, and shearing. As a rule, depths of the drill holes with respect to the foundation grade will be 1.5 times the width of the powerhouse. If the foundation grade is at a depth of 10 m from ground surface and the width of the powerhouse is 20 m, the depth of each drill hole should be 45 m. However, in case rock of uniformly same nature continues at depth (as in igneous rocks such as granite and charnockite), the depth of the drill holes may be limited to only 10 m in fresh rock. Exploratory drilling will be done using NX-size diamond core bits to obtain good recovery of rock cores and facilitate study of joints.

17.3.4 Laboratory Testing of Rocks for Strength Properties

Rock cores obtained from drilling at the powerhouse site are tested for determination of various engineering properties in the laboratory including triaxial shear strength, compressive strength, density, absorption, and elastic properties. In soil-covered areas, samplers are attached with the drilling rod to collect undisturbed soil samples. The soils obtained from the sampler are used for laboratory determination of shear, cohesion, angle of internal friction, and other soil properties. Tests for swelling and clay minerals are also to be carried out.

17.3.5 Large-scale Foundation Mapping and Study of Seismicity

A powerhouse foundation should be above the water table. The drill holes may be utilized for measuring the depth of water table. It is also necessary to find if there is any artesian condition. In folded strata, the synclinal part creates problem of artesian flow. In that case, the position of the powerhouse may be shifted or treatment is adopted. In the Pong powerhouse in Punjab, the sand-rock bands showed sub-artesian conditions in the foundation. To obtain relief of the artesian pressure, several 30 cm-diameter artesian wells at intervals of 15 m were provided around the powerhouse.

Geological features of the powerhouse are visible after excavation of the site for foundation-grade rock or soil. At this stage, large-scale (1:100) plane table geological mapping is undertaken covering each component of the power plant. The rock types, attitudes, and frequency of the joints, the openness or infilling material of the joints and fractures, and other weak zones such as shearing or iron staining, alteration, and kaolinization if present should be clearly brought out for suitable treatment. The load-bearing capacity of the foundation rock needs to be determined by conducting tests on certain parts of foundation grade where rock is apparently weaker (e.g., weathered or partly weathered rock) than the other parts. Ideally, a powerhouse foundation should be homogeneous and monolithic. Such a condition is rarely attained. Faults, joints, fractures, and other planes of discontinuities create heterogeneity in rocks. Consolidation grouting is generally taken up to strengthen and create a monolithic foundation.

Whether the powerhouse site and its adjoining areas have been affected by any earthquake in the past is to be studied from the available records and mentioned in the engineering geology report. The earthquake zone within which the powerhouse site falls is to be mentioned. The appropriate seismic factor wherever necessary is accordingly incorporated in the design of the structure.

17.4 INSTANCES OF GEOTECHNICAL PROBLEMS OF SURFACE POWERHOUSE

Geological defects due to weak rocks and adverse structural features of rock formation affect the stability of powerhouse foundation. In many cases, these defective features are visible only after exposing the foundation grade. The following presents some instances of such geological problems actually encountered in some powerhouse projects after foundation excavation and their solutions.

Powerhouse of Ramganga hydroelectric project of Uttar Pradesh

The powerhouse site consists of sand-rock and clay shale of Siwalik formation. The soft and fractured rock of the foundation was incompetent to bear the heavy load of the equipment. The variation in modulus of elasticity of foundation rocks threatened differential settlement. Moreover, the access of water through the open joints in the rocks posed problem of uplift pressure. To overcome these problems, the powerhouse was laid on 7 m-to 9 m-thick raft foundation (Chaturvedi and Mandwal 1978).

Bassi powerhouse of Himachal Pradesh

After excavating to the depth as per the design of powerhouse foundation of Bassi hydroelectric project, it was found that the rocks (sandstone and claystone) were partially weathered. Going further down was not practical from the design point of view and it was also uneconomical. In-situ bearing strength was then determined and it was found that the weathered rock had sufficient strength to bear the load. However, the rocks at the foundation grade were segmented by two cross faults. There was no evidence, however, of any recent movement along the faults. The crushed zones of the fault were removed up to a depth twice the width and backfilled with concrete. Later, the fault zone was grouted to a depth of 6 m by inclined drill holes (45° and 60°). At places where the turbine was directly on the fault zone, the turbine position was shifted by minor adjustment in the powerhouse layout. The soft claystone band was excavated from the foundation and backfilled with concrete. After all these treatments, the powerhouse was laid on reinforced concrete raft (Srivastava and Sondhi 1978) to provide required stability.

Balimela powerhouse of Orissa

Excavation of this powerhouse site encountered very hard and fresh gneissic rock at the foundation. The rock was initially considered to be suitable for founding the powerhouse without any treatment. Detailed geological study, however, indicated that the rocks were fragmented by several sets of joints that were either open or filled by clay material. These clay-filled joints and some shear zones provided heterogeneity to the foundation. Water percolation test in these jointed rocks showed sufficient water loss, indicating that the joints were open at depth and also interconnected. In view of these open joints in the rock, the foundation was grouted to a depth of 6 m using pressure between 0.5 kg/cm^2 and 2.10 kg/cm^2 with grout mix of cement-to-water ratio 1:2 and 1:5. In addition, the 10–30 cm-wide shear zones were treated by scooping out the soft material up to 50 cm depth and backfilled by concrete (Sinha, Pradhan, and Singh 1978).

17.5 UNDERGROUND POWERHOUSE OF A HYDROELECTRIC PROJECT

Underground powerhouse for hydroelectric project in India was first constructed in Maithon project of Damodar Valley Corporation in 1953. Other instances of underground powerhouses include the Koyna project of Maharashtra, Idukki project of Kerala, and Yamuna (Stage II) project of Uttar Pradesh constructed in the 1960s. Since then, several underground powerhouses were constructed for hydroelectric projects in different parts of India.

17.5.1 General Aspects

A stable landmass with outcrops of hard and sound rock mass free from fault, fold, and excessive joints and without any groundwater problem is ideal for locating underground powerhouse. Many of the problems encountered are the same as those in the construction of large tunnels in rock. However, unlike a tunnel, any settlement of the foundation or roof collapse may jeopardize the very functioning of the powerhouse including generation of power supply and also damage the machineries. Hence, even a minor geological weakness such as a shear zone in the roof arch or walls needs to be studied carefully. In the feasibility stage of engineering geological investigation for underground powerhouse site, as in the case of other engineering structural sites, study of aerial photographs is taken up in the first instance, followed by surface mapping. After selecting the site from ground survey, exploratory drilling is taken up.

The underground powerhouse site is explored by several NX-size drill holes. A programme of exploratory drill holes may be formulated in the same way as explained in connection with subsurface exploration for surface powerhouse but at least one exploratory drift will be driven close to the proposed underground chamber. This is necessary for visually examining the rock condition and conducting instrumental tests, especially for stress measurement. The drift may also be used to take up drill holes from its rear end so as to penetrate the proposed foundation grade of the powerhouse.

17.5.2 Special Considerations

The following aspects need special attention in engineering geological investigation of an underground powerhouse:

(i) Orientation of powerhouse
(ii) Rock stress condition at depth
(iii) Influence of faults and joints
(iv) Groundwater condition and water seepage
(v) Excavation method and support system

The dimension and shape of the subsurface cavity affect the stress condition of the rock. The orientation of the cavity with respect to the structural attitudes of rock influences the stability. The most stable configuration is achieved when the long axis of the cavity is kept parallel to the maximum principal stress and perpendicular to the direction of the main structural feature such as the bedding or foliation strike. The direction and magnitude of stress are obtained by over-coring methods. In rocks devoid of any bedding or foliation structures, the powerhouse cavity may be oriented with the long axis perpendicular or oblique to the trends of the maximum number of joints. Intact igneous rocks such as granite and charnockite, volcanic rocks such as massive basalt, or sparsely foliated metamorphic rocks such as gneiss or horizontally banded massive quartzite are suitable for underground powerhouse construction.

The stress condition in underground powerhouse rock influences overbreaks and collapse of rock from roof or slide of the wall rocks. The rock pressure at depth on the roof arch of underground cavity can be estimated following the empirical method of Norwegian Geotechnical Institute (NGI) or rock mass rating (RMR) system. However, due to the heterogeneity of the rock mass and irregularity in the shape of the cavities, determination of in-situ stress in powerhouse rock is highly essential. The test for in-situ stress may be carried out in rock in a drift close to the proposed foundation grade. The junctions of the powerhouse walls with pressure shafts, draft tube, and access tunnel are the zones of high concentration of stresses.

Quantitative evaluation of the in-situ stresses is necessary for design consideration. Extensometers are used to measure the stress in rock at depth through boreholes. The direction and magnitude of stress are obtained by over-coring methods. Flat jack test or uniaxial jack test method has been used to measure rock stress in many of the projects (Section 10.6). The method of photoelastic modelling is applied in assessing the stress condition in subsurface rock with the help of computer software. Tables 17.1 and 17.2 show the results of in-situ stress tests in rocks of Srisailam and Sardar Sarovar underground powerhouse sites of the Central Water and Power Research Station (CWPRS) applying flat jack techniques (Tilak 1999).

A fault may cause problems if it traverses the underground powerhouse rocks. The fault plane may initiate rock fall from the roof or wall. Any movement along the fault is detrimental to the functioning of the powerhouse equipment. In the case of presence of faults, some change of the layout of the powerhouse cavity is desired. To prevent the possibility of slide, heavy support is to be provided. If the rock exhibits several joints dipping towards the wall, there may be slide of parts of roof towards the walls. The close-spaced joints require supports after assessing the stress condition.

Table 17.1 In-situ stress test of rocks in Srisailam underground powerhouse site

Stress	Rock type	Flat jack test (by CWPRS)
Vertical	Quartzite	5.5 MPa
Horizontal	Quartzite	2.7 MPa

Table 17.2 In-situ stress test of rocks in Sardar Sarovar underground powerhouse

Stress	Rock type	Flat jack test (at a drift)
Vertical	Basalt	1.379 MPa
Horizontal	Basalt	1.171 MPa
Vertical	Dolerite	1.287 MPa
Horizontal	Dolerite	1.947 MPa

The orientation and spacing of the predominant joints can be studied from borehole camera through the drill hole used to get subsurface conditions of rocks of the powerhouse site. The infilling materials in joints or the fault gouge are the main harmful factors initiating slide due to their low cohesion and internal friction. Once the slide takes place, it accentuates further roof collapse. Presence of permeable beds, close-spaced joints, and fault zones in the foundation rocks provides passages for seepage of water. If the water table is high, there will be a constant problem of groundwater flowing into the cavity. Water pressure in the fractures and fault zone due to seepage of water may cause hydraulic failure of the rock mass of cavity. Pumping test is conducted through drill holes to determine the nature of permeability of jointed rock. The drill holes also serve as piezometers for water table.

The conventional approach of tunnel excavation in rocks (Section 16.8), especially by multiple drift method shown in Fig. 16.17(c), is followed in the excavation of underground powerhouse. The sequence of excavation generally starts from top heading and then proceeds downwards to bench and invert (Fig. 17.6). Prior to excavation, the blasting pattern is designed depending upon the rock condition by making specific changes in the distribution of charges in the periphery. Conventional supports include steel sets and rock bolts followed by thick concrete lining. In recent times, more preference is given to rock bolting and shotcreting. If the rock is very hard and sound, it is kept unlined. As in a tunnel, tunnelling quality index Q and RMR values of rock (Section 16.10) play an important role in deciding the support system for underground power plants in India with provision of cable anchor wherever required. *Pre-stressed anchoring* is done in large underground openings by first drilling holes as per design angle and pre-grouting them. The cables are then installed and stressed against the anchorage.

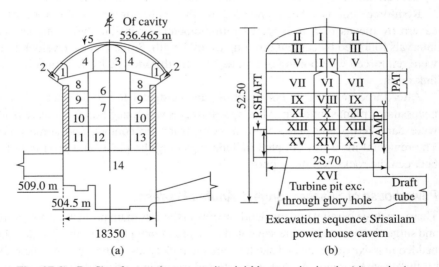

Fig. 17.6 Profile of powerhouse cavity: (a) Yamuna hydroelectric project, Uttar Pradesh; and (b) Srisailam hydroelectric project, Andhra Pradesh, with the numbers indicating the sequence of excavation followed from top downwards

17.6 INSTANCES OF GEOLOGICAL PROBLEMS IN UNDERGROUND POWERHOUSES

The geotechnical problems faced by some underground hydroelectric powerhouses during construction and post-construction stages and the measures adopted to overcome the difficulties of adverse geology are outlined in the following paragraphs based on Tilak (1999).

Powerhouse of Koyna project, Maharashtra

The powerhouse of Koyna hydroelectric project is located underground in the Deccan trap. Due to horizontal lava flow, any orientation of the powerhouse cavity was possible. The powerhouse consists of three chambers, namely the machine hall, the valve house, and the transformer hall, and the span measures 15 m, 6 m, and 12 m, respectively. The machine hall arch encountered a bole layer that was treated by 3 m to 5 m-deep bolts, introducing each at intervals of 1 m, and providing 91 cm-thick concrete lining of the wall rock.

Excavation started by a drift towards the roof, which was then widened until the level of the haunch. The arch portion was then lined by concrete. The remaining portion was then excavated for benching following pre-split blasting technique. The condition was dry during the entire excavation. The walls of the chambers were left unlined because of very hard basalt and no dilation or rock fall was observed.

Powerhouse of Kadamparai project, Tamil Nadu

Kadamparai hydroelectric project with underground powerhouse, the first pumped storage scheme of India, was completed in 1987 but faced post-construction problems. It has four 400 MW units 200 m below the ground level. The dimension of the cavity is 129 m × 21 m × 34 m. Transformer bays are located on both sides of the long axis of the powerhouse. Charnockite and gneiss are the rock types of the powerhouse site. The three adits, power cable shafts, and access tunnels are unlined, whereas surge shafts and pressure shafts are lined.

Reinforcement has been provided to the arch portion of the roof of the powerhouse cavern by means of 5 m-long, 20 mm-diameter expansion shell bolts spaced at 2 m × 2 m intervals followed by cement grouting to full length. Some cracks developed in the cavern walls were reinforced with bolting of 5 m to 7 m length followed by 75 mm-thick guniting and chain link mesh.

After three years of its construction, an accidental fire took place in the powerhouse that necessitated plant shutdown. After dewatering, it was found that there were places where linings were damaged, some plates were missing from the contact, and some pipes were buckled. The remedial measures adopted include replacing the plates and grouting of the contact zone between the rocks and the supports.

Powerhouse of Srisailam project, Andhra Pradesh

There are three parallel underground caverns in the Srisailam project for powerhouse, transformer, and surge chamber, which are separated from one another by 30 m-thick rock pillars. Horizontally-bedded massive quartzite and siltstone are the rock types of the caverns. The rocks are intersected by three sets of joints. A fault was detected traversing a part of the powerhouse site but it was avoided by a little shift of the cavity layout towards downstream. Estimation of the rock quality indicated an *RMR* value of 61 and *Q* value of 25, indicating good condition of rock. Due to the presence of a shear zone cutting the flat-bedded rock, the powerhouse cavern experienced some rock falls. The excavation was done by multiple drifts from roof towards the floor.

The supports provided include the following (Tilak 1999):

In the roof arch   9 m-long, 22 mm-diameter rock bolting by staggered holes spaced 3 m × 1.5 m and 4.8 m-long 22 mm-diameter rock bolt, staggered holes spaced 1.5 × 1.5 m
In the walls   6 m to10 m-long, 25 mm-diameter grouted rock bolts with staggered holes spaced 2 m × 1.5 m
At the bottom parts   2 m-long, 25 mm-diameter rock anchors drilled 1.5 m in rock spaced at 2 m intervals
Inside lining of cavity   70 mm thick steel fibre reinforced shotcrete

17.7 THERMAL POWERHOUSE

Hydroelectric power is not sufficient to meet the country's demand of electricity. Hence coal in addition to diesel and gas based power is also required to meet the need. Of these, low grade coals available from various coal-fields are mainly utilised that have been discussed below.

17.7.1 Basic Needs

A substantial portion of the power demand in India is met by the thermal power produced mostly by use of low-grade coal as fuel. There are several power stations that use diesel or natural gas. A thermal powerhouse requires water for cooling the turbines. In the absence of natural water source such as a lake near the powerhouse site, a dam is constructed to create a reservoir to supply water to the power plants.

Majority of thermal powerhouses of India are coal based. When the powerhouse is operated by coal, a regular supply of large quantities of coal should be ensured. Railway facility is needed for supply of coal from the coal mining areas. In general, low-grade coal with high ash content is used in the thermal plants. A dumping place or big pit is required for deposition of large quantities of ash. The ash produced is mixed with water to form slurry, which is poured into the pit. The place where the ash is dumped should be periodically investigated with respect to the effect of chemicals (sulphates and sulphurous acid) that may leach out from the ground to contaminate groundwater, especially during the rainy season. To avoid contamination, large silos made of concrete are used instead of pits. The ash is picked up from the boilers by vacuum suction and then the compressed air system will eject the ash into the silos.

When gas is used for the turbines, it requires laying of pipelines for receiving the supply from the source. The terrain through which the pipelines pass will need geological investigation to avoid problems of slide and subsidence and this causes interruption in the supply of the fuel to run the turbine.

17.7.2 Site Investigation by Mapping and Subsurface Drilling

Engineering geology helps in the selection of site for the powerhouse and ancillary structures. In general, engineers provide some probable sites and the most suitable one from geological consideration is to be selected. The geological study includes mapping of the area with special care to reveal the characteristics of the rock with respect to its homogeneity and structures such as fault and whether there is any recent movement along the fault. Drill holes are taken up to study the subsurface condition of rock. If the area is covered by soil, details regarding the thickness and nature of soil are to be determined by drill holes followed by laboratory testing of soil. Samples are collected by soil exploration for laboratory determination of clay minerals and swelling property. A swelling type of clay may develop heaving of the foundation, resulting in crack and settlement of foundation. Drill holes are used for measurement of groundwater table. An area with high water table is likely to encounter problem of constant dewatering and needs waterproofing treatment at depth. The site should be avoided if the groundwater always remains higher than the design base of the powerhouse where equipment will be installed.

A thermal powerhouse requires installation of heavy machineries covering a large area. An ideal condition for thermal powerhouse is uniform presence of firm rock at a depth of 3 m to 4 m under the floor at the main powerhouse and also in the areas outside the power plant for locating machineries such as fuel oil handling, water treatment plant, and crusher house, where the maximum strength of the foundation needed is of the order of 10–15 tonnes/m^2. Such an occurrence of rock may rarely be found. Extensive excavation is required for founding the structure for feed pump and cable gallery and locating cooling water pump at the subsurface region. If rocks are exposed throughout the area, it will involve blasting of the rock, which is not only expensive but also a time-consuming process. Engineers, therefore, favour a soil-covered area rather than a rocky area. According to Nag, Manier, and Reddy (1978), 'Geological exploration in that area can help in localizing a site with most favourable foundation conditions for the power plant itself. Pile foundation is quicker and cheaper than rock foundation except in unlikely possibility of rock being situated at an uniform depth of about 3 to 4 m below floor level in the plant area.'

Location of many of the ancillary structures such as the ash pump house, hoppers, and wagon trippers will involve cut and cover to an extent of 12 m to 15 m depth. Excavation in rocks of such magnitude is observed to be uneconomical, and hence, soil is preferred. The presence of rock is desired in limited areas near the surface where boiler, precipitator, and chimney are located. The required bearing strength for the thermal plant foundation is 25–30 tonnes/m^2. Such bearing capacity is generally not expected in soil. Therefore, heavy machineries are installed in soil-covered areas on pile or raft foundations. Soil of bearing capacity upto 60 tonnes/m^2 can be achieved by driving piles. Testing of soil is needed for the actual loading and for designing the length of the piles. After pile driving, in-situ bearing strength of the soil is carried out to determine whether the piling has been successful to achieve the design load-bearing capacity. The raft foundation is taken up in areas where ancillary structures will be installed.

Foundation for the high-rise chimney is to be explored by drill holes. A rocky base is desirable for founding the chimney to withstand the high pressure of its own load. Turbodynamos and motor generator are different from conventional types. Using high-speed turbodynamos involves foundation problem due to vibratory stress. For the high-speed machine, in-situ dynamic properties of soil need to be determined in addition to other parameters such as density, cohesion, angle of internal friction, and Poisson's ratio for consideration in the foundation design. To arrest vibration of foundation during the operation of machineries, the powerhouse structure is to be tied with firm bedrock as done in hydropower plants.

17.8 NUCLEAR POWERHOUSE

Nuclear power is necessary to cope with the increasing demand for power. Because of the danger of harmful radiation from nuclear powerhouse, it is highly essential to locate the plant structure on stable foundation as well as avoid radiation hazard. Since nuclear powerhouse uses fissionable materials, it requires construction of heavy and massive structure for radioactive protection.

17.8.1 Guidelines of Atomic Energy Commission on Site Selection

There are several guidelines laid down by the Atomic Energy Commission (AEC) of India. Three of the major criteria are as follows:

(i) The powerhouse should be located about 5 km away from areas of major habitation.
(ii) Source of water capable of continuous supply of about 70–85 m^3/s should be available near the plant site for cooling and flashing the waste from fissionable material.

(iii) The foundation of the structure should be very rigid without any chance of differential settlement in the future. It should behave uniformly throughout the long period of the power plant's life.

17.8.2 Method of Engineering Geological Investigations

Photogeological studies are taken up in the first instance for selecting suitable site for founding the structures for nuclear power station keeping in view the necessary guidelines provided by the AEC. Alternative possibilities in a locality are also recorded from the photogeological study for further consideration by ground checking. In practice, the area for locating the atomic power plant is proposed by the AEC team in collaboration with power plant engineers. The engineering geologist needs to study the area and select the structural site for founding the machineries. The study includes reconnaissance traverse, recording of topographical features, geological mapping, logging of drill holes, and identifying geohydrological conditions.

An important aspect of the study is to ascertain whether the area is affected by any earthquake. In fact, seismicity is the most important factor in searching the location for a nuclear power plant as any seismic movement in future will be highly detrimental to the structures of the power station. Study of aerial photographs helps to determine whether there is any evidence of recent earthquake movement in the area such as the presence of a fault of recent or sub-recent age. Evaluation of local and regional stability is another important aspect of geotechnical study. Regional stability is decided from the tectonic activity, seismic activity, and hydrothermal activity within a radius of 300 km from the proposed site (Demin and Xun 1982). Based on the stability of the proposed site, the base of the foundation is located on firm ground, or the structure is designed with suitable seismic coefficient.

Once the area is covered by reconnaissance visit, large-scale geological mapping is undertaken at the possible site for the powerhouse including ancillary structures. Once the site is finally selected or accepted from engineering point of view, exploratory drill holes are taken up for finding the rock condition or nature of soil at depths. The drill holes are used for the study of groundwater table and also the fluctuations in groundwater levels. Laboratory tests are conducted to find the engineering properties of the rock and soil samples. In addition, field load testing is carried out for the rock after removing the outer weathered zone. This in-situ bearing strength of rock is necessary for engineering design.

The following conditions should be satisfied for a nuclear power station:

(i) The rock should be hard and sound and of homogeneous character without any major discontinuities.

(ii) The powerhouse site, including its surrounding areas, should be a stable landmass and should not be affected by any creep, slide, or subsidence.

(iii) There should be no effect of fault movement in the geologically recent or sub-recent times.

(iv) In case any fault is traced, its effect and whether it is still active should be determined by installing a seismograph along the fault.

(v) The groundwater table should be low. It must be lower than the powerhouse foundation during the rainy season.

An ideal area satisfying all these conditions may not be available for locating a nuclear powerhouse. The geological report brings out all the available topographical, geological, and hydrological features and wherever needed suggests remedial measures. A homogeneous nature of rock foundation is desired for nuclear power station. If soil or soft rock is present in the powerhouse site, a raft foundation is to be provided for installing nuclear reactors and other equipment. The isolation or disposal of radioactive waste needs to be done in a very deep subsurface sound rock such as granite, charnockite, granulite, and basalt, which are available in many parts of India.

17.8.3 Problems of Site Selection in Himalayan Area

Unlike peninsular India, the geological condition of the extra-peninsular (Himalayan) region is different, where rocks are affected by thrusts and faults, and earthquakes are frequent. There are several records of earthquakes of different magnitudes in the Himalayan terrain, causing displacement along fault planes, sliding, and subsidence of grounds and damage of structures. In general, the Himalayan areas are prone to seismic tremors.

Study of the foothill areas (Siwaliks) of the Himalayas in search of suitable locations for nuclear power stations indicated a nature of the same geological and structural set-up in all the possible sites. The area comprises soft rocks such as sandstone, sand-rock, clay shale, and boulder conglomerate of Siwalik formation. Sand-rock and clay shale are especially of very low strength. The bedrocks (Siwaliks) are covered by thick terrace deposits consisting of river-borne boulders, gravels, and fines. The rocks were subjected to movement in geological past by thrust and fault along which seismic activity might have taken place during recent times. If any fault and thrust are detected close to the proposed sites, instrumental tests are necessary to determine their activity with respect to movement in recent to sub-recent (late Pleistocene) times. There are some tear faults (e.g., Yamuna tear) concealed under the overburden materials where problem of groundwater will be intense (Hukku et al. 1975).

17.8.4 Problems of Locating an Atomic Power Plant on Alluvium—Case Study

Geotechnical investigation was carried out in parts of the Hissar district of Haryana for locating an atomic power plant. The investigation focussed on the physiography, geology, geohydrology, and seismicity of the area. Topographically, the area is nearly a flat country with some sand dunes. The average annual rainfall in the area is about 50 cm. The climate is sub-tropical, semi-arid, and dry. Geologically, the terrain is represented by Indo-Gangetic alluvium and most parts of the area are covered by thick aeolian deposits. The aeolian deposits are underlain by Archaean granite and gneiss but the area is devoid of any rock outcrop. Bedrock could be traced from a well about 250 m below the alluvium. Six holes drilled to a depth of about 60 m in the area met only alluvium comprising sandy silt, silty clay, stiff (plastic) clay, and *kankar*. Standard penetration test through the boreholes indicated presence of stiff clay bed at a depth of 20 m and extended down to 26 m. The water table was recorded at a depth of 1.6 m to 6 m from the surface. Laboratory tests indicated sulphate content in groundwater.

From overall consideration of the geology, geohydrology, and stability of the area, it was recommended that the power plant and ancillary structures should be constructed at a depth of 20 m providing raft foundation on concrete piles. It was also suggested that a special type of cement be used for construction purpose as sulphate content in groundwater is detrimental to ordinary cement concrete. The area falls under IS seismic zone III for which a horizontal seismic coefficient of $0.34g$ was found to be necessary and was proposed for inclusion in the design of the structure. Providing an inclination of 1.5 H:1 V to the cut slopes with 3 m-wide berms at every 8 m-height intervals and arranging proper drainage system were recommended as remedial measures to arrest potential danger of slope failures. Use of high-power pumps was suggested for lowering the groundwater table to avoid development of pore water pressure and heaving below the structural foundation (Mandwal 2005).

SUMMARY

- A powerhouse is constructed to install turbine and other machineries for generating electricity. The source of energy that drives the machineries for production of electricity is water pressure in hydroelectric power plants, fossil fuel such as coal, diesel, or natural gas in thermal power plants, and fissionable material in nuclear power plants.
- In general, flat and stable landform containing homogeneous soil free from swelling clay or a rocky area free from planes of discontinuities is suitable for construction of a powerhouse.
- Engineering geological investigation in a powerhouse site is carried out by large-scale (1:100) plane table mapping of the area, plotting detailed geological and morphological features. The site is then explored by diamond drilling using NX-size core bits. The drill holes are utilized for measuring groundwater table and the rock cores are tested for strength and other properties.
- The underground powerhouse site in addition needs determination of in-situ stress of rocks at powerhouse cavity level. If the rocks possess high stress and unfavourable rock structures indicating possibility of slides after excavation, the engineers are suggested to design the structure making provisions for proper support including rock bolting.
- A thermal powerhouse is located in an area that has assured water supply from a lake or an artificial reservoir. A site comprising firm rock at a shallow depth (3 m to 4 m) below the floor of the thermal power plant is the most desired condition. However, if it involves excavation deeper than 10 m, it is preferred to construct the powerhouse in soil, providing pile and raft foundations to avoid the expensive process of rock blasting.
- Investigation for the thermal plant site is taken up by site visit followed by geological mapping and exploratory drilling. If the foundation is to be laid on soil, the thickness of soil deposit is recorded from drill holes, and subsoil samples are tested in the laboratory for their engineering properties to facilitate design of the structures.
- Suitability of a nuclear powerhouse site is decided after detailed investigation on the physiography, geology, geohydrology, and seismicity of the area following the guidelines given by the AEC.
- The location of the nuclear plant must be 5 km away from areas of habitation. Continuous supply of large quantities of water is to be ensured near the plant area for cooling and flashing of waste. The foundation should be on homogeneous rock free from fissures. The water table should be below the foundation level of the structure. A stable landmass where there has been no seismic event in the recent past is needed for setting up a nuclear powerhouse.

EXERCISES

Multiple Choice Questions

Choose the correct answer from the choices given:

1. Selection of a surface or underground powerhouse site of a project is to be done at the:
   (a) planning stage
   (b) design stage
   (c) after study of aerial phots of the project area

2. The surface powerhouse in a hydroelectric project is built at:
   (a) the toe of the dam
   (b) any place at the toe or away from the dam
   (c) where hard rock is available near the dam

3. Geological mapping should preferably be done by:
   (a) surface observation to a scale of 1:10
   (b) plane table survey to a scale of 1:100
   (c) from the study of aerial photo

4. For thermal powerhouse, the economic foundation will be:
   (a) pile foundation in soil deposit
   (b) rock foundation
   (c) soil deposit of low bearing strength
5. A nuclear powerhouse is to be located:
   (a) close to an urban area
   (b) at least 5 km away from areas of habitation
   (c) near populated area to reduce transmission loss
6. The main problem in locating nuclear powerhouse in the Himalayan terrain is the presence of:
   (a) soft rocks of Siwalik formation
   (b) thrust, fault, and seismicity
   (c) transportation of power from mountainous area to plains

Review Questions

1. Name the three main categories of powerhouse. State the type of source materials required to generate power in different categories of powerhouse.
2. Describe how river water is used to generate electricity. Where should a powerhouse for harnessing power in a hydroelectric project be located? What are the machines used to generate power? Write the equation used to calculate power production in kilowatts in a hydroelectric project.
3. What are the basic aspects of a surface power-house of a hydroelectric project? Describe the topographical, geological, and hydrogeological conditions favourable for a surface powerhouse site.
4. Give a short account of engineering geological approaches in surface and subsurface investigation including laboratory testing of rock samples necessary in locating a powerhouse site.
5. What are the adverse geological features generally observed in a powerhouse site? How are these geological defects rectified?
6. State the geological conditions that need to be considered in selecting an underground powerhouse site. Describe the excavation method including the sequence of excavation of an underground powerhouse.
7. Give an account of the geological problems generally faced in the construction of an underground powerhouse. How can these problems be solved?
8. What is the basic need of a thermal powerhouse for generating power? Describe the approach of geological investigation in selecting a thermal powerhouse site.
9. Enumerate the basic aspects of locating nuclear powerhouse as specified by the AEC. Describe the essential geological conditions and the method of geological investigation required in selecting a nuclear powerhouse site.
10. Explain the geological and structural set-up and the natural phenomenon that cause problems in locating nuclear power plants in the Himalayan terrain.

Answers to Multiple Choice Questions

1. (a)   2. (b)   3. (b)   4. (a)   5. (b)   6. (b)

18 Bridges

● ● ● ● ● ● **LEARNING OBJECTIVES** ● ● ● ● ● ● ●

After studying this chapter, the reader will be familiar with the following:

- Functions of superstructure and sub-structure of a bridge
- Different types of bridges and the forces acting on them
- Bridge components and support systems
- Investigation methods and building materials of a bridge
- Construction approach of a bridge in different reaches of a river
- Case studies on bridge foundation problems and their remedial measures

18.1 INTRODUCTION

This chapter deals with the geotechnical approach of site selection for a bridge that depends upon factors such as geomorphic features, foundation conditions, and available construction materials. In the design of a bridge, it needs to be ensured that the foundation of the bridge piers can support the static load of the bridge superstructure and the live load of the moving vehicles or trains. The site investigation should be made by exploratory drilling to evaluate the subsurface geological conditions and take decisions on the foundation capability of supporting materials to bear the design load. Case studies on several bridges have indicated that the main cause of subsidence of piers and even failure of a bridge is related to weak foundation. The chapter suggests the measures to be taken to strengthen the foundation.

18.2 BASICS OF A BRIDGE

The engineering structure of a bridge constructed across a river is characterized by a superstructure and substructure. The superstructure serves as a link between the two sides of a valley, canyon, or body of water such asa river. It connects roads and railways at the two terminals and facilitates traffic movement. The substructure transmits the superimposed load to the underlying foundation materials through supports. The design of a bridge depends on its function, the terrain condition, the materials to be used in the construction, and the available funds for building the bridge.

Fig. 18.1 A single-span steel beam bridge constructed over a high-altitude stream

To cross a small valley, a bridge may be made of only a single *girder* or a *beam* placed on two supports on the two sides called *abutments*. If the river valley is very wide, intermediate supports called *piers* are required. Thus, between the two abutments, a bridge may have a single pier or multiple piers and accordingly it will be designated as a double-span or multi-span bridge. In a multi-span bridge, the length of the span between two piers may or may not be equal. A long, continuous single girder may cover the entire length of the bridge, or there may be several girders, each resting on two piers. Fig. 18.1 shows a single-span steel beam bridge constructed over a high-altitude stream, resting on two stepped abutments that support the bridge and also provide slope protection.

The forces acting on a bridge are of two types, namely static and live loads. The static load is provided by the weight of the material of the bridge including the supports. The live load is generated by the vehicular and train movements over the bridge. The total load is ultimately transferred to the foundation. Hence, the foundation should be strong enough to bear both the static and dynamic loads. Past records of earthquakes of the area where the bridge is to be constructed are studied and suitable seismic factor is incorporated accordingly in the design of the structure to save it from damage from earthquakes.

The materials generally used in the construction of bridges include timber, steel beam, steel cable, concrete, and reinforced concrete. A long bridge constructed over a valley with no apparent stream or with a small flow of water below is called a *viaduct*. An *aqueduct* generally means a water supply or navigable channel or canal constructed to convey water (see Section 19.6), but the term also applies to any bridge or viaduct that transports water across a gap. A description of the aqueduct crossing the river Narmada is included in the case study in Section 18.11.

18.3 MAJOR TYPES OF BRIDGES AND ACTING FORCES

There are five major types of bridges, namely girder or beam bridge, cantilever bridge, arch bridge, suspension bridge, and cable-stayed bridge. The magnitude of load of the bridge and the forces acting on the support system considered in the design depend on the type of bridge as explained in the following subsections.

18.3.1 Girder or Beam Bridge

The simplest type of bridge has two supports to hold a beam or girder. This is known as a single-span girder bridge or a beam bridge, as shown in Fig. 18.1. A beam bridge is made of

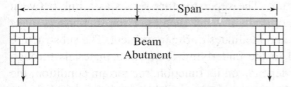

Fig. 18.2 A single-span girder or beam bridge with abutment piers

wooden plank or materials such as steel or reinforced concrete. The two abutments that support the bridge are concrete or masonry structures. In a single-span girder bridge, the two abutments take the load acting vertically and transmit it to the foundation (Fig. 18.2). In the multi-span girder bridge, in addition to the abutments, the load is distributed

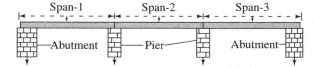

Fig. 18.3 A multi-span girder bridge with abutment and river bed piers

Fig. 18.4 A multi-span bridge over a large river with piers founded in river bed

Fig. 18.5 An arch bridge

to the piers vertically (Fig. 18.3). A multi-span bridge over a large river requires construction of several piers in the river bed below the water level (Fig. 18.4).

18.3.2 Arch Bridge

In an arch bridge, the two ends of the bridge are carried outwards curving to meet the ground on the two sides of the valley on which the bridge is built (Fig. 18.5). In this type of bridge, the weight is carried outwards along two paths curving towards the ground. As a result, the ground around the abutments is squeezed and pushes back on the abutment. Thus, in addition to vertical force, an arch bridge works by transferring its load partially into the horizontal thrust restrained by the abutments at either side, as shown in Fig. 18.6(a). Fig. 18.6(b) shows a double-arch bridge with *spandrel* deck above the arch. The area between the deck and the arch is known as *spandrel*. Fig. 18.7 shows a spandrel deck arch bridge with piers founded in river bed below water.

18.3.3 Cantilever Bridge

In the cantilever bridge, the end girders or beams have extended arms to hold a relatively small beam. As in Fig. 18.8, the two beams of the bridge, firmly anchored with the piers, support another beam that is the middle deck of the traffic-way. In a cantilever bridge, the total weight is transferred vertically to the two piers and nearly no load is imposed on the abutments.

18.3.4 Suspension Bridge

The deck or traffic-way of a suspension bridge is hung by cables that hang from towers. The cable transfers the weight to the towers, and the towers transfer the weight to the ground. In a suspension bridge, the stress is transferred vertically to the foundation but the bridge is anchored with the bedrock or a large concrete block on two sides to provide stability (Fig. 18.9).

18.3.5 Cable-stayed Bridge

Cable-stayed bridge is in some respects similar to suspension bridge, both having a suspended deck structure and towers called pylons. In cable-stayed bridge, the cables from the towers go directly to the road deck instead of spanning from tower to tower (Fig. 18.10). The towers form the prime load-bearing structure and do not require firm anchorage to withstand the horizontal pull of the cables as in the suspension bridge in which the deck merely hangs from the suspenders. In the cable-stayed bridge, the deck is in compression, under pull towards

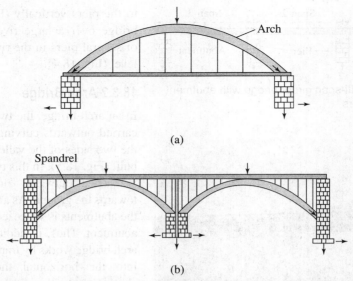

Fig. 18.6 Arch bridges: (a) solid-ribbed single-arch bridge; and
(b) spandrel deck Double-arch bridge

Fig. 18.7 Spandrel deck arch bridge with piers
founded in river bed below water

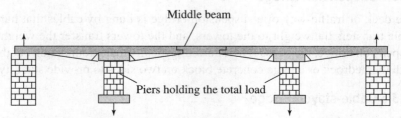

Fig. 18.8 Cantilever bridge showing piers bearing the load of super
structure

the towers, and has to be stiff at all stages of construction and use. A cable-stayed bridge is
essentially similar to a cantilever bridge, but it has shorter span and requires less cable.

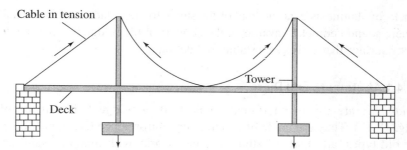

Fig. 18.9 Suspension bridge

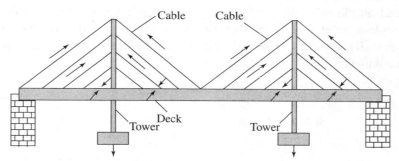

Fig. 18.10 Cable-stayed bridge

The *Vidyasagar setu* (Fig. 18.11) is a golden gate bridge constructed across the river Ganga connecting the twin cities of Kolkata and Howrah. It is a cable-stayed bridge 457 m long and 115 m wide with nine lanes and is supported by 121 tension cables. It is the world's third largest cable-stayed bridge that stands on four pylons on a 100 m-deep foundation in alluvial deposit.

Fig. 18.11 Vidyasagar *setu* showing the long suspended deck standing on pylons

18.4 SUPPORTS AND FOUNDATIONS OF BRIDGES

A bridge has a superstructure and substructure. The superstructure imposes vertical load on the foundation, and therefore, the foundation must be very rigid to withstand the load. The concrete blocks of abutments bear the load of the superstructure and transmit the load to the foundation. Where sound bedrock is available, the support of the bridge bearing the deck is placed on the rocky surface that serves as a stable foundation.

18.4.1 Abutments and Piers

Abutments are the two terminal supports of a bridge that may be of different designs (Fig. 18.12). They are made of concrete or reinforced concrete, but some abutments, mainly the old types, are made of stone masonry works with lime and sand or clay as cementing materials. For the purpose of smooth passage of road to the bridge, the abutments are designed in various shapes. When an embankment connects the bridge, the beam or girder is directly placed on the embankment by constructing retaining structures for the latter's protection. A single-wing abutment, or a bevelled-wing abutment with footing or with column support is generally designed to provide the seat of the bridge and act as a retaining wall of the embankment.

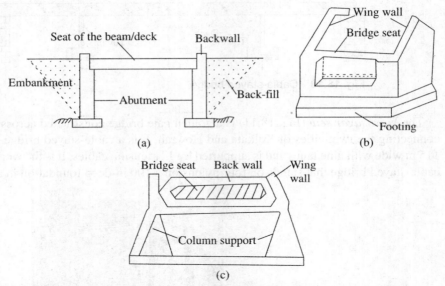

Fig. 18.12 Abutments of different types: (a) a single-wing abutment (with beam or girder seat) showing relation to embankment; (b) a bevelled-wing abutment showing seat of the bridge and footing support; and (c) a bevelled-wing abutment on column support

In addition to two abutments, a medium or long bridge requires intermediate supports provided by the *piers* made of concrete or reinforced concrete. The base or foundation of some of the bridge piers remains under river water at all times, but others may be on land during dry season and under water during rise of river water level (Fig. 18.13). The piers that remain under water are made cylindrical in shape with rounded edge to allow unhindered flow of tides and currents, though piers of other shapes are also prevalent. If the valley spread is large and river water is small, some of the piers may always remain overground. These are land piers or viaduct piers.

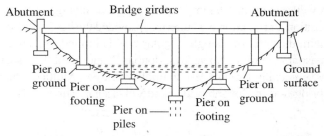

Fig. 18.13 Two end supports (abutments) and the intermediate pier supports of a long bridge in relation to ground surface and water level

The piers take the load of the superstructure to transfer it to the rocky foundation below. In homogeneous and sandy soil, piers rest on spread footings. Spread footing is provided when the pier is founded at shallow depths (Fig. 18.14). However, if the soil is heterogeneous in nature with clay bands and the pier base is very deep, the foundation is laid on piles. The stability of the bridge depends on the stability of the pier foundation. Any deterioration of the pier foundation may cause settlement of the pier and failure of the bridge as experiencedin the case of Chambal bridge in Rajasthan (see Section 18.11.3).

Fig. 18.14 A railway bridge on column and footing supports

18.4.2 Well Foundation for Bridges

Wells are caissons that open at both top and bottom and are made of timber, metal, concrete, or reinforced concrete (Section 6.16). All major bridges in India involving deep foundations are constructed by well foundation. It is difficult and also expensive to construct bridge piers by open excavations in loose deposits of river sections, which necessitates heavy support. It is also difficult to restrict ingress of water. Hence, well foundations are taken up that involves sinking down of the hollow shells (wells) inside the loose deposits by dredging up to firm soil or bedrock. Well foundation is preferred to pile foundation in bridge construction when it is necessary to resist large lateral forces. In a well foundation, the well remains a part of the permanent structure of the bridge. In case of well foundation across a river, the base of the well is always kept below the scour level of the river.

In general, the shape of the well may becircular, twin circular, double-D, dumb-bell shaped, square, or rectangular, but it may also be of other shapes such as hexagonalor octagonal (Fig. 18.15). For very large size piers and abutments of cantilever, suspension, and cable-stayed bridges, large rectangular wells are used. A large-diameter well is advantageous to several small wells to facilitate sinking operation and erection of shuttering, provided the base does not rest on sloping rock surface. The maximum diameter of a single circular well is limited to 12 m for concrete steining and 6 m for brick masonry steining. Nowadays, double-D and dumb-bell shaped wells are used for deep foundations, especially for double-line railway bridges. Multicellular wells are used for founding long-span bridges as in Howrah bridge and Vidyasagar *setu* over the river Ganga.

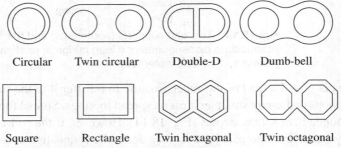

| Circular | Twin circular | Double-D | Dumb-bell |

Square   Rectangle   Twin hexagonal   Twin octagonal

Fig. 18.15   Cross section of wells of various shapes

18.4.3 Components of a Well Foundation

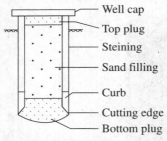

Fig. 18.16   Components of a well foundation

The components of a well foundation include steining, top and bottom plugs, well curb and cutting edge, well cap, and sand filling (Fig. 18.16). The *steining* of a well is made of concrete or masonry work. It is built in a straight line from bottom to top to facilitate tilting of the well at any stage of its construction. The thickness of a steining is designed such that the well can be sunk under its own weight. The thickness will increase with larger wells. For example, the thickness for a well of 3 m outer diameter will be 75 cm and that for a well of 7 m diameter will be 200 cm. The bottom end of the vertical bond of steining is fixed accurately to the cutting edge using nuts or by welding.

The *top plug* of concrete serves as shuttering for laying the well cap. This is provided after filling the well with sand. The *bottom plug* is made of concrete and is used to transfer the load from the steining to the soil below. It is bowl shaped so that it possesses inverted arching action. The *well curb* is made of steel. The outer surface of the well curb is made vertical. The *cutting edge* is made of steel with sharp edge that can penetrate into the ground and withstand stresses from boulders, blows, and so forth. It is anchored into the well curb that partially bears the load of the well at the initial stage when only a part of the cutting edge is in contact with soil. A considerable part of the load coming on cutting edge is taken by skin friction. The framework of the curb is removed after concreting is complete.

The *well cap* made of reinforced concrete is seated over the top of the steining. It is needed to transfer the loads and moments from the pier to the well below. *Sand filling* is done after two or three days of laying the bottom plug. The sand used for filling the well should be clean and free from clay, plant roots, and pebbles.

18.4.4 Sinking of Well to Subsoil

The operations involved in the construction of a well foundation for bridge pier are sinking of the well foundation to subsoil below water, plugging of the bottom, plugging of the top and providing a well cap of concrete, and filling the inside of the well with sand.

The well is sunk by excavating material from inside under the curb. If the ground is dry, initially excavation up to about half a metre above subsoil water level is to be carried out. In case the ground is not dry enough for such excavation, a cofferdam is constructed around the well site so that a dry ground like an island can be obtained to facilitate excavation. The size of the island should be such that there is sufficient space for movement of workers and to operate tools. When the dry area is made ready, the centre point for the well is fixed and the cutting edge is placed. At the initial stage, some wooden sleepers are laid below the cutting edge at regular intervals so as to distribute the load evenly during concreting operation. The excavation of the soil inside the well can be done by workers when it is dry inside. If there is water inside for more than 1 m, the excavation is done by dredgers.

Initially, the well steining is built to a height of only 2 m and later it is restricted to 5 m at a time. Vertical reinforcement of steining shall be bent and tied properly to facilitate the grab movement during steining operation. As the well sinks deeper invariably meeting water, the skin friction on sides progressively increases and the effective self-weight is reduced by buoyancy in the portion of the well below water level, requiring larger steining thickness. To counteract the skin friction and the loss in weight due to buoyancy, additional loading known as Kentledge (comprising iron rail, sand bags, concrete blocks, etc.) can be applied on the well. In sandy strata, the frictional resistance that develops on the periphery of the well is reduced by spraying a jet of water on the entire outer surface of the well.

When the well has been sunk deep enough, pumping out of water from inside the well becomes effective. If the well passes through a clayey stratum, the chances of tilts and shifts are also minimized.

18.4.5 Depth of Well Foundation

A well should be founded on firm bedrock or thick sandy and clayey deposit of sufficient bearing capacity even when available at a deep subsurface below water level. A sandy bed with adequate strength is always preferred to clay bed. However, a thick and stiff clay bed is also suitable for laying the foundation if it has sufficient bearing strength. If thin layers or lenses of clay are present between sandy layers, the well foundation must cross the clay layer and be kept on sandy bed or stiff clay further down (Fig. 18.17).

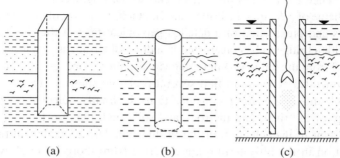

(a) (b) (c)

Fig. 18.17 Depth of well foundation: (a) a well (square-shaped) founded on firm clay bed of high bearing capacity that underlies sandy and clayey layers of low strength; (b) a cylindrical well founded on thick stiff clay deposit of sufficient strength; and (c) a well founded on bedrock occurring at a deeper part below loose river deposits, showing removal of materials from inside by a grab with steel teeth

The most important consideration in well foundation is the maximum scour depth that can be calculated by Lacey's theory (Section18.8). The well should be of sufficient length, called *grip length*, below the scour depth to counteract the overturning moment due to horizontal force acting on the bridge deck and that due to wind and water.

The maximum load on the subsoil under foundation should not exceed the bearing capacity of the subsoil after taking into consideration the scour. The load or total pressure of the well (except the wind and seismic force) should not be more than 25 per cent of the bearing capacity of the subsoil. The effect of the skin friction may be allowed on the portion below the maximum depth of scouring of the river bed material.

Two important criteria are followed in well foundation. First, there should be adequate *grip length* (length inside the bed) below the scour level. The grip length is taken as $1/3D_{max}$ below the scour level according to the code of practice of the Indian Road Congress, where D_{max} is the maximum depth of the well foundation. Second, the well should be based on strata that can resist the load transmitted on it. In general, the bottom of the well is kept on a predetermined (by drill hole) stiff clay or thick sandy zone present in the river deposit whose bearing capacity is adequate in relation to the transferred load. In river with deep water (restricted to 30m depth), *pneumatic caissons* are sunk. Inside the caisson, there is a working chamber filled with compressed air up to 3.5kg/cm^2 that prevents entry of water and permits men to excavate loose materials inside the well and remove them.

18.5 DIFFERENT ASPECTS OF ENGINEERING GEOLOGICAL INVESTIGATION OF A BRIDGE SITE

In the engineering geological investigations of a bridge site, emphasis is laid primarily on the following aspects:

  (i) Selection or evaluation of bridge site in different terrain conditions
  (ii) Exploration of the site by drill holes and laboratory tests of rock and soil samples
  (iii) Evaluation of the foundation condition of abutments and piers
  (iv) Search for concrete aggregates and earth for approach embankments

The engineers make the overall planning of a bridge project and prepare the layout of the bridge site and related road and railway links, and so on. Engineering geologists have limited scope in such a scheme of selecting the bridge site and their work is confined to the evaluation of geological conditions and preparation of a detailed geological map of the site. However, in case of a new project with a bridge to be constructed in a difficult terrain, an engineering geologist's association is required from the initial planning stage to also investigate the road and highway connections.

The general practice is to find several probable sites and the site most suitable from both geological and engineering considerations is finally selected. Topography of the terrain, hydrology of the river, and the rock and soil conditions for the support of the superstructure are important in the selection of the bridge site. The geotechnical problems and approach of investigation in the hilly terrain are different from those in an alluvial plain as discussed in Sections 18.6 and 18.7.

18.6 LOCATING A BRIDGE AT DIFFERENT REACHES OF A RIVER

A river that originates in a high mountainous terrain flows through steep and narrow valleys at the upper reaches and then gradually broadens its valley as it descends down to fan out in the plain. In the course of its journey through the hilly terrain with steep gradients, it forms terraces with thick deposits of boulders, pebbles, and sand at different levels along the banks. In the alluvial plains, the rivers are generally large and carry a huge load of finer sediments in their bed. The following subsections discuss the consideration for a bridge site to be selected in the upper, middle, and lower reaches of a hilly terrain.

18.6.1 Upper Reaches

In the upper reaches of the river, the valley is narrow and V shaped. A single-span bridge supported on two abutments can be constructed in such high-altitude hilly terrain (Fig. 18.1). Rocks are commonly found exposed on both the banks or under a shallow cover of overburden. Terraces are rarely formed and the hill slopes are rocky. With the setting of the rains, jointed rocks of the hill slopes or debris deposited on banks may cause the problem of slide. The site at an upper reach should be selected where hill slopes are stable and rock outcrops exist on both sides of the river. Once the bridge site is chosen, geological mapping is to be done covering the site and its immediate vicinity. The foundation of the area may involve rock cutting. If the site is covered by overburden, initially some pits or trenches can be made to study the nature of the overburden material. In case of thick overburden, each abutment is to be explored by a drill hole made to a depth of 6m in fresh rock or 10 m to 15 m in the unconsolidated materials.

An idea about the hydrology of the river with respect to maximum water level, floodwater discharge, and so on is essential for design purposes and can be obtained from the daily record of measurements in the river. An example of the salient hydrological data required for design of a bridge has been shown in the case study on bridges (Section 18.11). In the high altitudes, meltwater from snow during summer and rains in the monsoon cause an increase in the flow rate and consequent rise in the water level. Huge deposition of bouldery materials may sometimes block the flow path creating sudden rise in water level. The snow-fed Teesta River in Sikkim is an example. A bridge at such a location is, therefore, to be designed for construction at a sufficiently high level. Rock cutting for the abutment foundation and hill slope cutting for road construction near the bridge site may lead to the problem of slide. In such cases, the cut slope is to be stabilized by providing breast wall and making suitable arrangements for drainage.

18.6.2 Middle Reaches

In the middle reaches, the river valley becomes broader. Terraces are formed on the two banks at different levels. A good site for the bridge should possess sound rock for the abutments. It is also to be ensured that vehicles coming from roads along the foothills have sufficient space for turning to move over the bridge. The hill slope along the proposed side should be stable. The hydrological information and rainfall data as stated in the selection of upper reach site should also be taken into consideration in finalizing the site for middle reach. Invaluable information is obtained from the study of the aerial photographs on river basins. The three-dimensional (3D) view will also help to decipher the stability of the slopes. Geological maps are to be prepared covering the proposed site and at least further 100 m on all sides. The mapping throws light on the possible presence of weak zones. Sites having any major weak features such as a fault traversing any pier or abutment are to be avoided.

A bridge at a site in the middle reach with a wider valley will require one or two piers for supporting the superstructure. Bedrock is commonly available at a shallow depth at the pier sites. If the pier and abutment locations consist of deposits free from boulders, some auger holes may be dug at the first instance to find whether bedrock is present at shallow depth. Accordingly, the drill hole programme may be prepared such that each pier site is explored by a drill hole. In case the abutment sites contain thick overburden materials, these are also to be explored by drill holes. The holes should penetrate a depth of 20 m to 30 m at each pier site or 6 m into fresh rock, whichever is less. Water percolation tests are to be carried out in the holes to determine the openness of joints and other planes of discontinuities. Samples of rock cores are tested in the laboratory for compressive strength and other properties.

It is desirable that the piers and abutments are founded on bedrock if it occurs at a reasonably shallow depth below the overburden cover, including the river deposits. The drill holes will help in designing the depths for founding the piers. A geological cross section showing the bedrock along bridge axis may be prepared from drill hole data, which will be a good guide to locate the positions of the piers. The bedrock at the edge of the water and the lowermost terrace is subject to erosion by the stream, resulting in undercutting, which may pose a problem for stability of hill slopes. It may be necessary to protect the hill slopes by providing breast walls with weep holes. Boulder pitching with wire mesh may be done at the water edge to arrest erosion and undercutting by the river.

18.6.3 Lower Reaches

In the lower reach, the river is generally quite broad. It has already attained the matured stage and will debouch into the plains with fan deposits. The bridge is to be located at a sufficiently higher level. The approach roads for the bridge may be taken through the plain and then with a proper gradient to the higher level. This may involve excessive cutting of foothill areas or making tunnels to connect the bridge level. In general, rocks are available for the abutments in the lower reach of a hilly river. The increase in the width of the river will necessitate construction of several piers.

The investigation for a bridge site at the lower reaches may be carried out in the same way as in the middle reaches starting with geological mapping, followed by exploratory drill holes at the indicated pier sites. Before finalizing the design for the pier sites, a bedrock profile of the river along the bridge axis is to be prepared. This profile will show undulations depending upon the extent of erosion in different parts. The piers may be erected at the ridge positions of the undulation to obtain stable bases. This will reduce the heights of the piers and the cost. The troughs of the bedrock profile may be the effect of erosion along weak zones and should be avoided. Under such a situation, span lengths may have to be designed accordingly.

18.7 BRIDGE SITES IN ALLUVIAL PLAINS

Roads and highways in alluvial plains may have to be connected with a number of bridges at the crossings of small streams or big rivers. In a single-span bridge to be constructed across a small stream, the geotechnical investigation is mainly restricted to abutment foundation. In the case of medium bridges requiring construction of one or more piers, the foundation of pier sites, if any, has to be investigated in addition to the abutments. The sites of large bridges over broad rivers, involving construction of many piers including some in the underwater environment, require detailed surface and subsurface geotechnical investigation by exploratory drilling for each of the pier sites and the two abutments.

The landmass in the plains consists of soil or unconsolidated sediments and presence of bedrock is not expected within the depth of pier foundation. Construction of a long bridge over a large river will involve construction of many piers in the river section as well on the banks, which will be under cover of unconsolidated deposits or soil. A geological map is to be prepared showing the area of normal soil, zones of unconsolidated river deposits, and presence of peat or organic soil, if any. Exploration by drill holes will provide information on subsurface condition at the site.

As the bridge will be sufficiently long with several spans and piers, the total load (dead load plus live load) will be many times more than a bridge in hilly region. Hence, the forces transmitted through the supports (abutments and piers) to the foundation will be rather high. The design load considers the structural load and hydraulic factors. Geological investigation has to ensure that the strata on which the piers and abutments are rested have adequate capacity to resist the forces as in the design load.

Pits and trenches are dug at abutments and pier sites close to the banks to find the nature of the soil, and representative samples are tested in the laboratory. Exploratory holes are to be drilled to 30–50 m depth at each pier site, and samples of soils are to be collected from depths by using a special sampler. The logging of the drill holes will reveal the thickness of sand zones and clay zones, whether the soil is of homogeneous or heterogeneous character in its depthwise extension, and whether any organic material or peat bed is present.

Laboratory tests on soil samples of the boreholes are to be conducted for properties such as compression, shear, friction angle, and plasticity. A geological section showing the soil characteristics at depths including laboratory test results of core samples is helpful for designing the foundation of the supports. If the soil is homogeneous and sandy in nature, spread footing is considered in the design. Coarse-and medium-grained sand obtain more strength after compaction. However, if the soil is heterogeneous in nature containing numerous clay stringers or if it is of clayey type, pile foundation is adopted.

The hilly areas, irrespective of the upper or lower reaches, are prone to landslides. The condition may further deteriorate after cutting of the hill slope for construction of bridge abutments and approach roads. Tremors due to earthquake and monsoon rains may also aggravate the situation. Stability of the hill slopes around the proposed bridge sites should, therefore, be studied before finalization of the site, and if required the site may have to be shifted to a safer location. If shifting is not possible, suitable remedial measures are to be suggested.

18.8 BRIDGE FOUNDATION IN SUBSOIL IN RELATION TO DEPTH OF SCOURING

The load of the superstructure of a bridge is transmitted to the foundation material through the supports such as the piers and abutments. Foundation of hard rock possesses sufficient strength to withstand the load or forces transferred to it, but topsoil or subsoil including river deposits cannot bear the superimposed heavy load of the superstructure. For design purpose, the bearing capacity of soil is known from the field load test or laboratory test on undisturbed soil samples collected from drill holes.

The foundation is designed after due consideration of both vertical and horizontal forces acting on the soil at depth. In principle, the shearing stress exerted by the superstructure on the soil should not exceed the shearing strength of the soil. As the designed stress of the structure is in most cases more than the shearing strength of the soil, the foundations of the bridge supports are kept on spread footing or piles. A safety factor of two, three, four, or more is provided in the design of a spread footing. A safety factor of two will mean that the soil is able to withstand twice the designed pressure on the footing. In a spread footing, the total imposed load is distributed over a large area so that the load on unit area (or effective pressure) of footing is decreased. The larger the area of the footing in contact with the soil, the less is the effective pressure on it.

Bridge foundation is also made on piles. In a medium-sized bridge, timber or concrete piles are used. H-beam of steel is used for pile foundation in large bridges. The piles can be most effectively driven if the soil is not of very steep nature. The ideal condition is achieved when bedrock is available at a reasonable depth to drive the pile down to the rock. However, in the absence of bedrock, as is generally the case in bridge construction in plains, the piles are taken down to some gravel or sandy bed at depth. The drill hole data on substrata is the decisive guide to locate the sandy zone down to which the pile should be driven. The idea is to get a firm base for the piles on which the load of the superstructure will be transmitted through the supports such as the piers.

An important consideration in the foundation design of a bridge is the *depth of scouring*. The approximate depth of scouring may be calculated from the Lacey's empirical formula:

$$f = 1.76(m)^{1/2}$$

where m is the mean diameter of the particles in millimetres.

$$D = 1.35(Q^2/f)^{1/3}$$

where D is the normal depth of scour in metres and Q is the discharge of the stream in cubic metres per linear metre of waterways.

The maximum scour depth D_{max} at the nose of the pier is found to be twice the value of Lacey's normal scour depth. D_{max} is measured below the HFL of the stream.

To avoid the problem of scouring, the foundation level is kept below the depth of scouring. Another problem is the underwater erection of piers. In all large bridges, many of the piers are founded in river deposits below the column of water. In fact, the seat of these piers may be far below the unconsolidated deposit under river water. Close to the banks where excavation depth exceeds 5m, piers are constructed by open excavation during dry season when the river water recedes from the banks.

18.9 CONSTRUCTION MATERIALS FOR A BRIDGE

Construction of a bridge requires large quantities of concrete for its various components. The retaining walls on the hill slope and abutment slope protection in the two terminals of a bridge also require concrete. The coarse and fine aggregates for the concrete can be obtained from the rockcuttings and river sands from the bridge site itself. If the passage to the bridge is to be connected by embankments, there will be need for earth. In the plains, plenty of suitable earth will be available, and in the hilly areas, the fill material may be obtained from rock spoils in the hill slopes. In hilly areas, the undercuttings of hill slopes are protected by placing rock slabs, which may be obtained from excavation of the rocky hill slope or collecting big boulders from the rivers bed. To find the 'minute geological defects' in the constituents of the concrete, petrological study is to be carried out in the aggregate material.

Minerals such asopal, chert, chalcedony, and zeolite cause concrete–alkali reaction. There have been instances wherein the piers of a bridge developed serious cracks due to presence of these 'minute' minerals of deleterious effect. While selecting aggregate materials, these deleterious minerals should be avoided. During the construction of the Son bridge in Uttar Pradesh, fairly hard and durable Vindhyan limestone was initially selected for quarrying to meet the requirement of coarse aggregate. However, petrological study indicated the presence of abundant cherts, which react with alkali in cement. The Kajrahat limestone exposed on the right bank of Son River was then used as coarse aggregate by crushing the rock to desired sizes for the bridge construction.

18.10 SALIENT POINTS RELATED TO GEOTECHNICAL EVALUATION OF A BRIDGE SITE

A bridge is the link of roads, rails, and so on on the two sides of a river. Hence, the bridge site will have to be selected as a part of the overall plan to accommodate the approaches from either side. The structure should be cost-effective. A narrow river section reduces the length and hence the construction cost is also reduced, but it may involve long diversion of existing roads for entering the bridge. A site that facilitates easy connections without long deviation from the roads should be chosen.

The hydrological study of a river provides information on the maximum water level, river discharge, and related aspects. Study of aerial photographs provides data on drainage pattern, bank erosion, terrace deposits, bed load, and so on of the river. All these pieces of information are required to locate the bridge site on safe ground and for taking measures against adverse geological situations.

The mere presence of rock is not a sufficient condition for the bridge supports. The rock should have the capacity to resist the designed loads on the supports. Laboratory test provides information on bearing capacity of the rock. In-situ load test is the best way to confirm the load-bearing capacity of foundation rock. In hilly sections, rocks are generally exposed or available under shallow overburden cover. In the investigation of bridge sites in such areas, exploratory drill holes indicate the depth to bedrock. A bedrock profile of the river section along the bridge axis will help to locate the piers and abutments on rock foundation.

In the alluvial plains, a bridge site should be explored by drill holes for all the pier and abutment sites. Sandy zone in the river deposit at depth may be chosen for placement of the pier base. Soil samples from drill holes are tested in the laboratory for their strength parameters. The pier and abutment foundations are designed on spread footings or piles so that the load transferred by these supports to the soil is less than the bearing strength of the soil. Materials for concrete aggregate required for constructing a bridge may be found from the rock cutting and river sand. The sand should be free from minerals that react with cement to cause concrete–alkali reaction.

A bridge fails due to settlement of supports under deteriorating condition of the foundation strata. A study of such bridges is helpful to understand the problem of settlement and probable cause of deterioration of the foundation strata of the concerned bridge. The knowledge can be applied in the construction of other bridges under similar geological set-up.

18.11 CASE STUDIES ON BRIDGES INCLUDING A COLLAPSED BRIDGE

A bridge construction usually encounters several types of problems that are related to the adverse geology of the bridge site, especially in relation to the weak foundation conditions of the piers. Geotechnical investigation helps to identify the weakness of the site, if any, and specify correct treatment. The following case studies are presented to give practical examples of the problems faced at the bridge sites of varied geological set-ups and the geotechnical work conducted at the site that helped in safe construction of the bridge.

18.11.1 Aqueduct-cum-road Bridge in Madhya Pradesh

A 35.5 m-high and 940 m-long aqueduct-cum-road bridge over the river Narmada near Tilwaraghat connects Jabalpur with Nagpur by National Highway 44 in Madhya Pradesh. Of the total span of the structure, the length of each of the 18 main ducts is 18m and that of the 11 viaducts is 20 m. An earlier bridge at its upstream part remained submerged by floodwater for quite some time and this new one was then constructed. The following is the salient hydrological data of the river:

(i) Catchment area: 17863 km^2
(ii) Maximum flood discharge: 28,316.50 m^3
(iii) Longitudinal slope: 1:3000
(iv) Highest flood level: 385.47 m
(v) Lowest water level: 360.02 m
(vi) River bed level (lowest): 358.00 m
(vii) Maximum mean velocity: 3.05 m/s

It is a 29-span bridge with 30 piers of which 19 piers have been designed for open-well foundations, nine piers for sink-well types, and the remaing two abutment piers were founded on rafts. The site is represented by different varieties of schists such asmica schist, chlorite schist, and sericite schist, which are jointed to various extents and contain shear zones. Geotechnical investigation was conducted by geological mapping (Fig. 18.18) including exploratory work and 3D log of the pier sites.

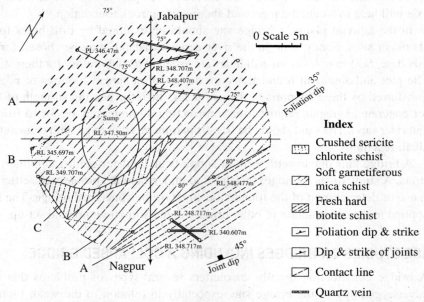

Fig. 18.18 Geological map of foundation of pier 9 of Narmada aqueduct (Chibber, Chepe, and Ramachandra 1991)

The pier sites were drilled by 19 large-diameter (calyx-type) boreholes upto 25–30 m depth, which gave only chips of rocks. Later, diamond core drilling was undertaken in ninepier sites. The cores were tested for crushing strengths. The designed load was 6–10 kg/cm^2. The bedrock (schists) obtained at depths showed crushing strengths of 80–170 kg/cm^2. The drilling proved the presence of firm rock in most places within a depth of 5–11 m and piers were founded on rock except for sixpiers whose bases were laid on stiff clay–*kankar* at depths of 5 m from the ground. Geotechnical study helped to find stable foundations for the piers at reduced depths as compared to the earlier design and thus saved substantial amount of expenditure in project constrction (Chibber, Chepe, and Ramachandra 1991).

18.11.2 Tikira High-level Road Bridge in Orissa

The 243m-long Tikira high-level road bridge in Angul district of Orissa was constructed across a tributary (Tikira *nala*) of the river Brahmani. It connects Talchir with Rengali dam project over the river Brahmani. The area covering the bridge site is a part of the Talchir coalfield under Gondwana basin. The site lies over a wide boundary fault between the Archaeans and the Gondwanas. Partly weathered, greenish, and gritty-to-pebbly Talchir sandstones of the Gondwanas occur on the right bank, whereas Archaean pegmatite is found on the left bank and quarzite occupies the central part of the bridge site.

Originally, 11 piers were proposed for the bridge site, but in the final design seven piers were provided. The span length between two pier varies between 21.5m and 37m. Drilling at the seven pier sites indicated very low core recovery (only chips) that caused difficulty in fixing the depths for pier foundation. Later excavation revealed the presence of rock (schist, quartzite, and pegmatite) at the foundation levels of four pier sites but in a shattered or sheared condition. From geological consideration, consolidation grouting for 6m to 9 m was suggested in shattered rocks of all the pier sites and this was adopted to strengthen the foundation. As per geotechnical suggestion, the three pier sites located on highly fractured and brecciated rocks were specially treated by means of consolidation grouting, spread footing, and deep anchoring as a measure against possible differential settlement. The bridge has been functioning without any trouble since its construction in 2000 (Mukherjee and Ghosh 2004).

18.11.3 Failure of a National Highway Bridge over Chambal River in Rajasthan

The failure of the 742 m-long bridge over Chambal in Dholpur district of Rajasthan on National Highway No.3 is attributed to incompetent foundation strata. Four of the arch spans of the bridge collapsed due to the subsidence of a single pier in 1978, fourteen years after construction. With a view to finding the reasons for the failure, the foundation area was geotechnically explored with the help of drill holes at the 22 pier sites of the collapsed bridge.

Geotechnical study by exploratory drilling revealed that certain piers founded on jointed limestone got affected due to the presence of solution cavities and fissures in the rock. The foundation of other piers consisted of soft claymatrix conglomerate. At the pier foundation that was affected by settlement, the drill holes indicated that the clay matrix of the conglomerate was altered to 1 m-thick residual clay. Boreholes at the remaining sites of piers encountered limematrix conglomerate, which was proved to have an adequate bearing strength, 10kg/cm². Artesian conditions were observed in all the drill holes. Based on the geotechnical findings of the subsurface condition, seven new piers had to be constructed by taking the foundation to deeper parts in bedrock that could safely withstand the imposed load (Mathur 1983).

18.11.4 Foundation of Banas Bridge in Rajasthan

The experience gained from the investigation on the failure of the Chambal bridge was well utilized in selecting the foundation for another bridge in Rajasthan over the river Banas near Sawaimadhopur town. The Banas bridge consists of 12 piers. The span between piers 1 and 2 in the right bank is 26 m. However, the length of each of the fourspans covering piers 3 to 7 is 27 m. The area in and around the bridge is covered by alluvial soil with few outcrops of limestone with thin shale layers.

Geotechnical study of the bridge site included subsurface exploration by eightdrill holes. In the river section, bedrock was encountered below 7.5 m to 9 m of overburden material. The overburden thickness on the left bank was found to be only 1 m to 2.5 m. Groundwater was observed to occur at a depth of 3 m to 5m below the surface. Surface mapping and exploratory drill holes helped to take a decision on the foundation depth. The foundation was placedon dolomitic limestone below the depth of the scouring and also below the zone of weathering along joints. Out of the twelvepiers, three were founded on bored piles and nine on open foundation, resting on rock. The foundation was placed on dolomitic limestone after ensuring that it had adequate bearing strength to withstand the imposed load of about 15 kg/cm² (Mathur 1983).

18.11.5 Distressed Railway Bridge of Bhagalpur District, Bihar

An old railway bridge over Sahabgunj loop in Bhagalpur district of Bihar was severely affected through settlement and subsidence of its right abutment and one of its four piers (pier P4) in November 2007, resulting in twisting of the rail line (Fig. 18.19) and causing suspension of service for a long time. Deep and wide cracks encircled the distressed abutment, and subsidence cracks in two or three steps along the banks of the river and see page of water through the cracks were noticed. Study of available exploration data at the bridge site indicated that the abutments and the piers of the bridge were founded on 10 m to 11m thick silty clay layer having low bearing capacity, which was responsible for the settlement and subsidence.

Fig. 18.19 Twisted rail line over a distressed railway bridge in Bihar caused by the subsidence of abutment and a pier (Pal, Roy, and Mishra 2009)

Further investigation revealed that the distress in the abutment and pier has been caused by the combined effect of scouring below the foundation, plastic deformation of foundation materials, swelling pressure, neo-tectonic activity, and piping action. In order to rectify the defect, stabilization measure of the foundation was taken up by the engineers. This included sheet piling encircling the distressed right abutment. From geotechnical consideration, it was also recommended to provide large number of relief wells to create see page path to relieve the hydrostatic pressure. Ground cracks were sealed by sand mixed with coal tar. On completing these measures on the bridge pier and abutment, the operations on rail line over the bridge were restored (Pal, Roy, and Mishra 2009).

SUMMARY

- The main components of a bridge are girders, abutments, and piers. Piers are made of concrete columns, which act as support of the bridge and carry the girders or beam on them to pave the path for movement. Abutments are the two end supports of a bridge made of concrete or masonry works.
- Girder bridge, cantilever bridge, arch bridge, and suspension bridge are the major types of bridges. In a girder bridge the load is distributed vertically to the abutments and piers, whereas in a cantilever bridge the load is transmitted vertically only to the two end piers. In an arch bridge, both vertical and lateral thrusts operate on the supports. A suspension bridge requires to be anchored to firm rock or a big concrete block.
- A bridge site, related to a railway or highway, is generally selected from engineering consideration to suit the overall project layout. Engineering geological investigation at a bridge site is carried out by geological mapping and exploratory drilling at the pier sites and abutments.
- The main aspects of study for selecting a bridge site include topographic features, land and slope stability, rock types, rock structures, thickness of overburden materials including river deposit, and foundation conditions for the abutments and pier sites. A geological section is prepared along the bridge axis showing depths to bedrock for the purpose of engineering design of the bridge components and fixing locations for the piers.
- In a hilly terrain, availability of rocks at the river section provides a good condition for pier foundations, but if terrace deposits are present at the bridge site, these may lead to problems in stability.
- If a bridge needs to be constructed in the downstream reaches of a river or in alluvial country where the river is wide, the site will generally be covered with thick unconsolidated deposits.
- Exploratory drill holes in the bridge site are planned in such a way that one drill hole is taken up for each pier site, extending down to the maximum scour depth of the river or until sound rock is available. If firm bedrock is not available at a reasonable depth, the piers are founded on undisturbed soil below the scour depth of the river on spread footing or pile foundation.
- Bridge site investigation involves searching for aggregate materials required for construction work of the abutments and piers and retaining structures. The river deposits, terrace materials, and crushing of rocks obtained from hill slope cutting generally provide suitable quality concrete aggregates.
- In the material investigation, it is to be seen that the aggregate materials selected are free from alkali-reactive minerals as these may cause cracks in concrete piers.
- Case studies on some Indian bridges suggest that a bridge may collapse by settlement of piers due to weak foundation. Adverse geological features such as faults, shear zones, joints with infilling, and clay seams pose foundation problems necessitating adequate treatments before foundation is laid on such soft and structurally weak rocks.
- In a railway bridge, it was seen that the pier was founded on silty material of low competence, causing settlement of pier and development of cracks. Extensive stabilization measures such as sheet piling, providing relief wells to combat hydrostatic pressure, and grouting of cracks were fruitful to stabilize the abutment and pier of the distressed bridge.

EXERCISES

Multiple Choice Questions

Choose the correct answer from the choices given:

1. In a bridge site, engineering geological investigation is carried out by:
   (a) study of aerial photographs to select possible site followed by site visits
   (b) exploratory drill holes to know about the subsurface condition
   (c) both (a) and (b)
2. Materials generally used in the construction of a bridge include:
   (a) timber and steel
   (b) concrete and steel
   (c) any one or a combination of timber, steel, and concrete

3. In a single-span girder bridge:
   (a) the two abutments take the load
   (b) no pier is necessary
   (c) both (a) and (b)
4. In a multi-span bridge, load is transmitted:
   (a) on abutments and piers to be taken by the ground
   (b) only on the piers
   (c) only on the abutments
5. In a suspension bridge, the load is taken by:
   (a) the bridge anchored to bedrock
   (b) large concrete blocks on two sides that provide stability
   (c) both (a) and (b)
6. In a cantilever bridge:
   (a) the total weight is transferred vertically to two piers to bear the load
   (b) no load is imposed on abutments
   (c) both (a) and (b)
7. In an arch bridge:
   (a) the ground on two abutment sides where the arches meet take the load
   (b) the two abutments bear the horizontal thrust
   (c) both (a) and (b)
8. In concrete for bridge piers, the harmful aggregate materials include:
   (a) opal, chalcedony, zeolite, and excessive mica
   (b) cherty limestone fragments
   (c) both (a) and (b)
9. The scour depth required for design of bridge foundation is calculated from:
   (a) Lacey's formula
   (b) Laplace's formula
   (c) both (a) and (b)
10. A bridge along the national highway over the river Chambal in Rajasthan failed due to:
    (a) overtopping
    (b) settlement of several piers
    (c) wrong choice of foundation depth for piers
11. Subsidence of abutments and pier causing twisting of rails in a rail bridge in Bihar was due to:
    (a) presence of a clay layer that swelled and exerted thrust
    (b) neo-tectonic activity and piping
    (c) both (a) and (b)

Review Questions

1. What is a bridge? What are the forces that act on a bridge? Write notes on girders, abutments, and piers of a bridge.
2. Give a short account on the major types of bridges and the forces acting on their support system.
3. Describe the foundation and support systems of bridges. What is superstructure and substructure of a bridge? State how piers of a bridge are founded on river bed materials.
4. What is well foundation of a bridge? Illustrate and describe the different components of a well foundation.
5. What is the approach in sinking a well to subsoil? What are the considerations required to decide on the depth of well foundation?
6. Discuss the different aspects of geological investigation of a bridge site and the manner of their performance.
7. Give a short account on the construction of bridges in the upper, middle, and lower reaches of a river in relation to its flow pattern, erosion, deposition, and transportation characteristics.
8. Describe the nature of geological investigation and construction approach taken up in selecting a bridge site when the river flows through an alluvial plain.
9. State the method of testing to know the load-bearing capacity of soil and the type of foundation designed in soil for constructing a bridge. What will be the approach to found the bridge pier below the scour depth of a river?
10. List the types of materials required for the construction of a bridge. Where do you expect to find such materials? Explain the type and percentage of deleterious minerals whose presence in concrete is harmful in bridge construction and add a note on how the problem can be solved.
11. From the study of case history of different bridges, what do you infer regarding the foundation problems that may lead to partial damage and even failure of a bridge? How can such problems be solved?

Answers to Multiple Choice Questions							
1. (c)	2. (c)	3. (c)	4. (a)	5. (c)	6. (c)	7. (c)	8. (c)
9. (a)	10. (b) and (c)	11. (c)					

19 Highways, Runways, Canals, Power Channels, and Flumes

●　●　●　●　●　● LEARNING OBJECTIVES ●　●　●　●　●　●

After studying this chapter, the reader will be familiar with the following:

- Investigation approaches of highway alignment, runway construction, and power channel alignment
- Different types of pavements and materials used for pavement and runway construction
- Design aspects of a canal and problems encountered in canal constructions

- Geological and engineering consideration for power channels and flumes
- Utilities and functions of a flume
- Case study of an aqueduct-cum-bridge over the river Narmada
- An ancient aqueduct used for irrigation and water supply

19.1 INTRODUCTION

This chapter deals with the geological approach in the selection of a highway alignment. The main factor that requires attention in highway construction is the pavement, which must bear the load of moving vehicles. Another important aspect is the materials used, including concrete aggregate. This chapter presents the specification of the Indian Bureau of Standards for concrete aggregate and also highlights the procedure for locating other pavement material.

19.2 HIGHWAYS

Highways pass through plains and land slopes extending over various types of soil and rocks. Engineering geological investigation of a highway alignment is taken up for identifying its geological problems and determining suitable materials for its construction. The work approach in highway investigation is described below.

19.2.1 Site Investigation for Highways

In recent times, the need for geotechnical investigation for highway has become essential as most of the highways are designed as two-way roads varying in width from 6 m to 8 m and having 2–3 m shouldering on both the sides (Fig. 19.1).

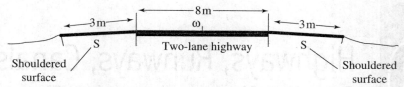

Fig. 19.1 Highway with two-way 8 m-wide roads and 3 m-wide shouldered surface on both the sides

The principal works involved in highway investigation include the following:

(i) The entire alignment is studied to assess the geotechnical problems that may arise along the highway, determine whether realignment is necessary in the problematic sections, and/or identify the type of treatment necessary to rectify the defects.

(ii) The volumes of cutting and filling required in different parts of the alignment are estimated, the availability of materials such as soil and rocks in the nearby areas are determined, and sites for borrow areas for soil and rock quarry are located.

(iii) Study along the alignment also includes measurement of soil slope and that of rocky sections to ascertain the inclination of stable cut slopes and measures to be taken in the design for the unstable slopes.

(iv) Soil samples are collected at every 200–300 m intervals to conduct laboratory analyses, which provide knowledge regarding the variations of the soil types encountered along the alignment. The ultimate aim is to find the suitability of the soil for subgrade of the highway.

(v) An important objective of the site visit is to assess whether the ground conditions of the different sections are safe to carry the dead and live loads of the automobiles.

(vi) At the outset, aerial photographic study is taken up covering the proposed alignment with special emphasis to locate the presence of peat, swampy land, and similar weak features along the alignment and, if present, to ascertain their effect when pavement is laid.

(vii) It will also be necessary to determine the amount of pitting, trenching, and drilling involved in investigating the subsoil condition and the cost involved.

(viii) The most important part of the investigation is the preparation of a geological strip map covering the areas along the alignment and its neighbouring parts, plotting soil cover, rock outcrops with rock attitudes, and morphological features.

A report is prepared on the basis of the investigation and submitted with the geological map and test data on soil and rock to help the engineering design and cost estimation of the highway construction. The sampling of soil is done by pitting and trenching from 1 m or 2 m below the surface. The tests conducted in the laboratory for soil include water content, dry density, and consistency limits (Section 6.3). The report submitted also includes the intensity of the seismicity of the area with specific mentioning of the areas of instability along the alignments and recommendation for measures to be adopted to stabilize these sections. In general, the rocky sections remain stable even with a cut slope of 60°, but in soil, the inclination of cut slope should be 30° or less; however, this general rule may vary in individual sections depending on soil or rock condition and attitude of rocks. Slopewise inclined beds, even if they are low dipping and jointed, may require retaining structures to prevent slope failures.

19.2.2 Placement of Pavement Material and Drainage

The *pavement* is the most important part of a highway construction. It should be a stable and waterproof surface that can withstand the wear and tear resulting from the impact of wheels. It is to be so designed that the dead and live loads of automobile movement are safely distributed to the pavement without exceeding its bearing capacity. The pavements are constituted of certain layers of materials placed on the original ground called *subgrade*. The subgrade is the base on which concrete slabs are directly laid for a thickness of 5–15 cm, or concrete slabs are placed with some materials of crushed stones and sand as base course to make *rigid pavement* (Fig. 19.2).

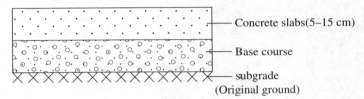

Concrete slabs(5–15 cm)
Base course
subgrade
(Original ground)

Fig. 19.2 Rigid pavement showing the constituent layers

Several layers of material need to be placed above the subgrade for *flexible pavement* as shown in Fig. 19.3. The sub-base layer that lies above the subgrade is made of soil taken from local pits. This is followed by a 15–20 cm thick layer of crushed rocks called *base course*. The base course materials are properly compacted and then another layer of 5–10 cm thick asphalt concrete is placed on its top. In some roads, even a thin coat of asphalt powder is provided to make the topmost surface fairly smooth. The crushed rocks of the base course are generally mixed with gravel, sand, and clay with cement or asphalt acting as binders. The layer of compacted base course materials of flexible pavement laid on subgrade acts as the most protective bed to carry the load of vehicle movement.

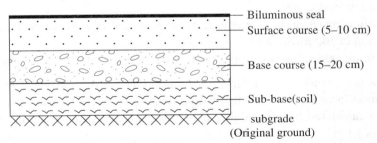

Biluminous seal
Surface course (5–10 cm)
Base course (15–20 cm)
Sub-base(soil)
subgrade
(Original ground)

Fig. 19.3 Flexible pavement showing the constituent layers

Alternate cooling and heating due to fluctuating day and night temperatures may cause heaving and cracking of pavement, especially in summer months. In general, to prevent such cracking of pavement, expansion joints are provided with tie rods between the pavement slabs. Deterioration of the base course material with time may cause erosion of pavement, showing signs of heaving at places. The main cause of such deterioration is the entry of water through the surface to the material below. During rainfall, water enters into the subgrade and cracks

Fig. 19.4 Cracks developed due to passage of water into the subgrade

are developed. Figure 19.4 shows the cracks developed in a pavement by passage of water into the subgrade; the material of base course are scattered with parts of the road section being broken. Proper drainage is necessary to avoid such breakage. Rain or snow-melt water in hilly areas entering into the subgrade from top, below, or sides may cause further deterioration of the pavement and its side shoulders. To avoid such deterioration and cracking of road pavement, *under-drains* are provided, which prevent the entry of water into the subgrade. These drains are connected to gutters in the uphill side. In the plains, reinforced concrete pipes are placed below the saturated soil to drain out water to the underlying impervious layer.

19.2.3 Quality of Aggregates and Their Functions

Raw materials such as crushed rocks, gravel, sand, and clay are used as concrete aggregate for road constructions. The following guidelines are specified by Indian Standard (IS) (387–1970) for use of aggregate in concrete (see also Table 12.3):

(i) Specific gravity not less than 2.6
(ii) Water absorption not to exceed 5 per cent
(iii) Deleterious materials (coal, clay, soft shale, micas, opal, etc.) not to exceed 5 per cent by weight of total aggregate
(iv) Aggregate crushing value not to exceed 30 per cent in fresh sample and 45 per cent in weathered sample
(v) Abrasion value (Los Angeles test) not to exceed 10 per cent in 100 revolutions and 40 per cent in 500 revolutions
(vi) Soundness value (sodium sulphate test) for five cycles not more than 12 per cent

The test procedures for laboratory determination of values for crushing, abrasion, and soundness have already been described in Section 12.4. The percentage of deleterious materials can be calculated by determining the content of deleterious materials in the aggregate (see Section 12.7).

Toughness is the most significant aspect of aggregates. The top of the pavement is hardened and smoothened by concrete or asphalt. The upper part is made impervious. The lower part made of aggregate should also be made impervious; otherwise, it may heave and the pavement might develop cracks. In general, aggregate materials for highways are made of angular fragments, which are resistant to cold and hot temperature conditions. It should also have high resistance to thrust caused by moving vehicles, and hence, trap rocks such as basalt are considered to be the most desirable material for aggregate.

The hard variety of sandstone with siliceous matrix when crushed may also serve as good quality aggregate. The size fractions of the materials should be such that the void or pore spaces are filled up by fines and the binder provides good bonding of the aggregates. Materials such as clay, powder of dolomite, and limestone with some chemicals are used as binders. The void-filling fines may be made from dust of silica, limestone, slate, or soapstone. The aggregate should be inert and not undergo any chemical reaction. It should have very low coefficient of expansion so that it does not undergo degradation such as erosion and weathering by natural processes.

19.3 RUNWAYS

The most important aspect of runway construction that is discussed in this section is the laying of pavement, including its construction materials. Drainage problems and the load-bearing capacity of the pavement are the other important aspects discussed along with seepage problem and its solution.

19.3.1 Plan of a Runway

Runway is one of the important components of an airport, which extends over a large area, accommodating the passenger building, hangars, apron, and taxiways (Fig. 19.5). It is built on a rectangular area in the airport land to facilitate landing and take-off of the aircraft. Aircraft weighing 90 tonnes need an area of 30–40 m width for a runway and 12 m width for taxiways with the length being at least 1800 m. Larger aircraft may require about 2500 m-long runways, and close to sea level a stretch of 3000 m might be needed. The runway is a man-made structure constructed of either concrete or asphalt, or a mixture of both. However, a natural surface covered by gravel, dirt, grass, or ice is also utilized as runway for small and light aircraft, provided the ground is plain and has no humps or uneven features.

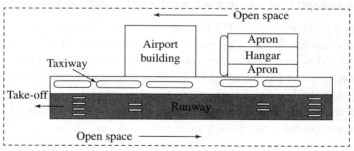

Fig. 19.5 Plan of a runway and other components of an airport

19.3.2 Investigation Approach

The main consideration in the runway investigation is the topography and the drainage. The location to be chosen for a runway must be an even ground. The area surrounding the runway must be free from any large construction so that the view of the runway from a distance or height is not blocked. Once the layout of the runway is fixed, a programme for exploratory drill holes is planned. Depending upon the nature of the ground, the drill holes may be spaced 100 m to 150 m apart through the central line of the runway and the depth to penetrate will be 10 m to 12 m. On the basis of the exploratory work, a cross section may be prepared showing variations in the subsurface strata for the design of the runway pavement. Undisturbed samples of different types of soil are collected from the drill holes and tested in the laboratory for size fractions of particles and properties such as density, moisture content, and consistency limits as well as for expansion nature for swelling clay, if any.

If the area is composed of homogeneous type of soil without any change in the soil type at depth, auger holes spaced at longer intervals penetrating to a depth of about 6 m may be taken up instead of drill holes. Investigation for foundation of an airport is the same as that explained for building foundation (Sections 6.13–6.16). Groundwater condition should be ascertained from the drill holes with respect to water table from the drill holes or from any springs if available. Areas with old fill material or where groundwater level is close to the surface require special consideration with respect to settlement problems and their remedy.

19.3.3 Runway Pavement

The materials to be used in the construction of the pavement for a runway are decided based on the local ground conditions and the type of aircraft to be used on the runway. In general, the materials required and types of pavements are the same as that described in Section 19.2.9 (see Figs 19.2 and 19.3). Where ground condition permits, a rigid concrete-type pavement is considered to be the most favourable. A concrete pavement is very stable and requires minimum maintenance for a long term. However, if the study of topographical and drainage condition indicates the possibility of settlement problems, it is preferable to install an asphalt–concrete pavement. It is easy to undertake patchwork in such type of pavement as and when required or on a periodic basis. If the runway is used only by small and light aircraft with very low traffic, it may even be possible to use sod surface. In all cases, the specification given for aggregate materials in Section 19.2.3 may be followed.

19.3.4 Seepage Problem in Pavement and Corrective Measures

Runway pavement surface is prepared and maintained to withstand maximum friction for wheel movement. Water may seep into the pavement flowing from side slope or upslope and may even seep below the subgrade. Meteoric water such as rain or melting snow may flow into the pavement entering through cracks or concrete joints. In areas with high groundwater table, water may enter the pavement by capillary action. All these factors increase the moisture content of subgrade materials, thereby changing the soil characteristics. This causes decrease in stability of the subgrade when its water content exceeds the plastic limit.

As a measure to reduce the water content in the subgrade or make it free from moisture, pumping is undertaken to drain the accumulated water and adequate drainage arrangements are made. The drainage arranged may be opened or covered ditches (sub-drains). In covered drains, the possibility of clogging by muddy suspensions should be considered. Excessive entry of water softens the flexible pavement, which is then easily cut by heavily loaded axles. To minimize the entry of rainwater into the pavement surface, the surface is finely grooved so that water flows through surface film into the grooves, thereby minimizing the chance of slip of aircraft tyre.

In order to stabilize and take measures against heaving of expansive soil in the pavement, quicklime and anhydrous lime products are used. Fly ash mixed with lime forms a synthetic product that can retard the expansion. Transverse cracks may develop due to difference in day and night temperatures, especially in summer and winter months. Spraying of sufficient quantity of water on the pavement during dry spells may prevent formation of cracks and solve the problem of water seepage.

19.4 CANALS

This section describes the investigation approach needed for selection of canal alignment. The excavation or filling involved in canal construction requires special study. The inclination of cut slope and the requirement of anti-seepage lining and protective coating for a canal to be constructed in soil and rock get special consideration in design. This section describes all these aspects and the problem of canal construction through hill slopes and its right solution with illustrative figures.

In the discussion of canal, the status of cut slope plays an important part. In the uncovered section, lining needs to be provided to arrest seeping of water and stop consequent drainage. Geotechnical investigation aids in selecting the power channel alignment and flume route so that the stretches of unstable area can be identified and stabilization measures can be adopted.

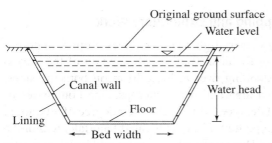

Fig. 19.6 Nomenclature of different parts of a canal

19.4.1 Basic Aspects

Canals are artificial waterways that are mainly used for irrigation and navigation by boat. The different parts of a canal are shown in Fig. 19.6. A long canal is directly fed by reservoir water impounded by the dams or connected with the tailwater of hydroelectric projects. A network of such canals is used to distribute water to the farmers for irrigation purpose. For example, the tailwater of the Bhakra dam is used for the irrigation of vast areas of Punjab, Haryana, and Rajasthan by a network of 174 km-long main line canal and 1110 km-long branch canals.

Engineering geological investigations for canals include the following aspects:

(i) Geological mapping and exploration by drill holes along canal alignment
(ii) Estimation of cutting involved in soil and rock
(iii) Stability of cut slopes in soil and rock and support requirement
(iv) Settlement of soil and problem of weak rock in canal construction
(v) Groundwater and its possible influence on canal excavation
(vi) Lining requirement of a canal for stretches in soil and rock

Ideally, the layout of the canal should be straight and it will have small gradient along flow direction. A contour plan and cross section along the canal alignment are generally obtained from the project authority for engineering geological investigation. Realignment of the canal may be required in certain stretches having adverse geology or important religious or historical structures. A long canal commonly passes through an undulating terrain consisting of soil as well as rock. To find the extent of cutting involved in soil or rock, it is necessary to measure the horizontal disposition as well as the vertical extent of rock or soil up to the canal level. A canal constructed in soil or rock may be rectangular, semicircular, trapezoidal, or polygonal in cross section (Fig. 19.7).

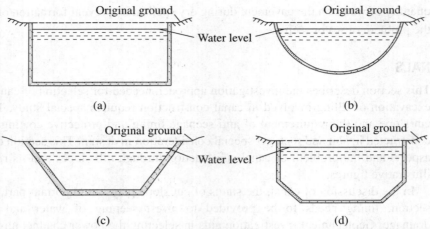

Fig. 19.7 Canals of different cross sections: (a) rectangular; (b) semicircular; (c) trapezoidal; and (d) polygonal

19.4.2 Site Investigation by Geological Mapping and Exploratory Work

At the initial stage of site investigation, study of aerial photographs gives a quick appraisal of the morphological and geological features of the area. Geological mapping of the area is then undertaken along the canal alignment extending over at least 50 m on both the sides. The geological map will show the soil-covered area and rock outcrops. The boundaries of different types of soil are to be demarcated, and the types of rock and their degree of alteration, attitudes, and weak structural features are to be plotted. If any cultivated land falls within the canal alignment, it is to be delineated in the map. During mapping, care should be taken to plot the slide-prone areas where slope stability measures may be necessary. The areas of tectonic disturbances, pervious grounds, boggy areas, quicksand, and subsiding grounds are also to be plotted, and care should be taken to avoid such unfavourable grounds wherever practicable. The geological map will provide the base for further exploratory works by trial pits and drill holes for the canal alignment.

Pits may be dug where the canal grade is available within a couple of metres below the ground level. The type, in-situ density, and field permeability of the soil are measured in the pits. Drill holes are made at 300–500 m intervals up to a depth of 3 m below the estimated level at canal invert. However, if fresh rock is available at the upper horizon and geological knowledge indicates continuation of the same rock at depth, the drill hole should be restricted to only 3 m in fresh rock. If there are variations in soil or rock types, additional holes, even at intervals of 50 m, may be drilled to cover the varied types of strata.

Diamond core bits are required for obtaining cores for study of rock features at depth and laboratory tests for rock properties. While drilling in soil, samplers may be used to collect undisturbed soil samples from depths down to canal level. The soil samples obtained from canal grade from pits and drill holes are tested for various parameters, especially those related to soil stability analyses. If there is any anticipated major weak feature (e.g., fault) continued at depth for a long stretch, seismic survey is carried out to determine how far it has affected the canal media. The seismic survey reveals the thickness of overburden and bedrock profile. Electrical resistivity survey helps in the detection of pervious strata such as sandy and silty beds.

The logging of soil or rock cores will provide data on the soil or rock type and the condition of rock including the thickness of soil zone formed by in-situ decomposed or alteration of bedrock and the dip of bedding, intensity of joints, and weak zones such as a fault. The samples of rock and soil obtained from drilling are sent to the laboratory for quantitative data related to their strength parameters and other engineering properties. The swelling index and clay minerals present in the canal bed are determined in the laboratory. A cross section along the alignment is prepared based on surface mapping and drill hole data to portray the subsurface rock or soil condition up to a depth of 3 m below the canal floor. The approximate extent or volume of rock or soil cutting required for the canal can be estimated from this geological profile. The data on soil and rock cuttings helps in the estimation of cost for contractual purpose. The more the bore holes, the more will be the accuracy of the estimate. The geological section and data of subsurface explorations are needed for the design of the canal.

19.4.3 Excavation and Filling Involved in Canals

Depending upon the terrain morphology and geology, a canal construction may involve entirely excavating or entirely filling, or both cut and fill (Fig. 19.8). The canals are commonly rectangular or semicircular in cross section when construction involves cutting and/or filling of steep soil or blasting of rock. Canals that are trapezoidal or polygonal in cross section are made in clayey soil ground, attributing better stability and operational facility. The major problems in a canal excavation pertain to providing stable side slope of the canal and that of the hill side

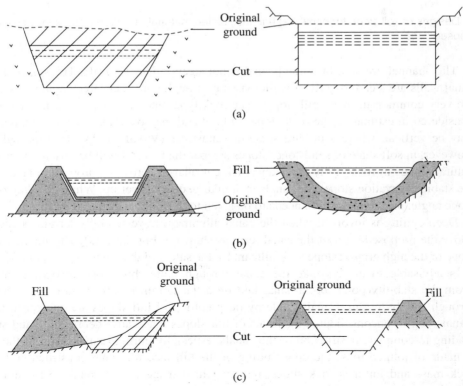

Fig. 19.8 Canal construction: (a) entirely in cut; (b) entirely in fill; and (c) both cut and fill

traversed by the canal; settlement, scouring, heaving, and cracking of ground of canal bed; and seepage from the soil or fractured rocks traversed by the canal. It is a favourable feature if the canal passes through a terrain having ground elevation coinciding with the top level of the canal. This is because the cutting involved in such case will be only for the canal dimension.

19.4.4 Design Aspects of Soil Slope and Water Depth of a Canal

In the design of a canal, the side slopes provided for the underwater portion of the canal are slightly steeper than the portions over water (in berms). The side slope in unconsolidated material is kept such that the ratio of vertical-to-horizontal distance is always less than 1:1 (Fig. 19.9). The approximate design slopes of vertical (V) to horizontal (H) for a 5 m-deep canal for different types of deposits are as follows:

  (i) Fine sand 1:2
 (ii) Pebbly and sandy soil and also in clayey loam 1:1.25
(iii) Decomposed peat 1:2.5
 (iv) Medium-to-coarse sands and also in sandy loam 1:1.5

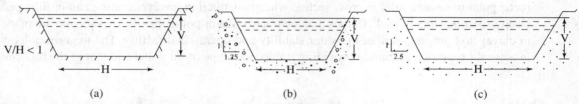

Fig. 19.9 Designs of soil slope for canal: (a) unconsolidated material; (b) pebbly and sandy soil; and (c) decomposed peat

The channel gradient of a lined canal is not steeper than 1V:1.5H. The slopes provided in canal walls are steep to vertical while passing through hard rock. Rectangular canals as such are very common in rocky hill slopes. The rock types and their degree of alteration are to be considered in estimating the rock slope. In hard and massive rocks such as granite, the slope may be vertical, whereas in hard shale or sandstone 1V:0.5H to 1V:1H slope will be safer. However, in soft shale or sandstone, slopes steeper than 1V:1H will be unsafe. A survey of the natural slopes in rocky hilly faces and rock cut quarries provides proper guidance regarding the stable excavation slopes for canals in similar geological set-up. While deciding on the safe slope angle, due consideration needs to be given to the heights of the slope faces.

Deep cutting is involved when the canal alignment traverses hilly terrain or high ground above the proposed level of the canal level as shown in Fig. 19.10(a). The stability of the hill slope or the high ground slope is significant for the safety of the tunnel and needs consideration. It is advisable to always take the canal through flatter hill slope involving shallow cut. From the stability point too, a canal laid on a flat terrain is always safer than that passing through a hilly terrain. Canals made by deep cutting of hill slopes are subject to landslides. Canals dug in unconsolidated materials of hill slopes may cause debris slide and subsidence leading to complete destruction of the canals, especially during rainy season. If the hill slope is made of jointed rock, the undercutting of the hill face for canal construction disturbs the rock mass and initiates rock slides, resulting in damage of the canals. Structural study of the rock types traversed by canal alignment with respect to faults, joints, bedding dip, and so on helps to ascertain the failure mechanism of rock slopes. The laboratory test results on shear, cohesion, internal friction, and so on of soil mass are used to decide the safe angle in soil slope.

Remedial measures for protection against slides of hill slopes should be taken up for the safety of the canal structure. If the hill slopes are devoid of sound rock, benches of 1 m or 2 m width are provided with side drains in the upper reaches. This is needed to ease the overlying load, prevent rainwater flow to the canal, and arrest rock fall or debris slide. If the pathways of the canal involve cutting of unstable hill slopes, retaining structures are provided, which include breast wall or other structures with side drainage as shown in Fig. 19.10 (b) and (d). In soft and pervious strata, the measures adopted include compaction of soil followed by lining. If the soil is of swelling type, which creates problems of heaving and cracking of the canal bed and walls, the expansive clay materials are removed and backfilled by normal earth followed by lining to the affected part (Fig. 19.10c).

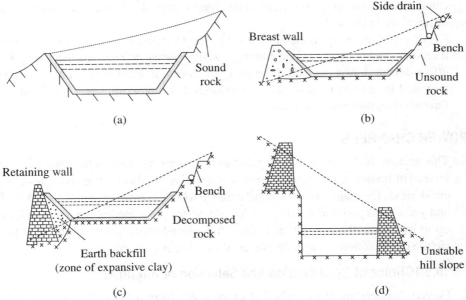

Fig. 19.10 Canals traversing steep hill side: (a) deep cutting of hill slope in sound rock; (b) protective measures in unsound rock such as benching with side drain and breast wall in the downhill; (c) removal of expansive clay, backfilling with normal earth, and benching of uphill slope and downhill retaining wall; and (d) protection of uphill and downhill slopes by breast walls

Groundwater condition of canal areas is known by measuring the seasonal variations in water levels through the piezometers inserted into the completed boreholes. If the water table lies near the surface, special drainage measures will be necessary. The design should provide proper drainage arrangement for draining surface water for the quantity of water that enters into the canal during rains. Similarly, the drainage measures should ensure stability of the canal slope and prevent groundwater hazards.

Canals passing through loess type of soil may cause subsidence and deformation with the consequent creation of fissures through which water flows out. Peaty soil is also not good for canal construction. Vegetation cover is provided by the sides of the canals to protect them from slide and slumping. The canal water carries certain percentage of particles in suspension, which are carried away and cannot settle down in the canal if the water moves under certain velocities. Settlement of suspended materials may cause an increase in water level and consequent siding in canal wall.

Some amount of water in the canal is lost by evaporation and seepage through the canal bed or walls. *Lining* is provided to the canals to prevent water percolation or seepage and to avoid scouring of canal when it passes through loose sandy or clayey soil. It also reduces silting and arrests slides in unstable areas. The velocity of water flowing in canals is designed such that there will be no scouring and siltation. This non-scouring velocity is dependent on the depth of canal water and the average diameter of the bed material. The non-scouring velocity of a 5 m-deep canal is 11 m/s in sedimentary rock and 17 m/s in a concrete-lined canal.

Depending upon the materials used, linings are of two types—*anti-seepage* and *protective* linings. Anti-seepage lining uses compact clay, asphalt, and bituminous coating in addition to concrete or reinforced concrete. Protective linings are given to prevent erosion and scouring of canal bed by paving stone slabs and pitching with reinforced concrete slab. If the soil of the canal contains swelling clay, reinforced lining is provided to prevent heaving and avoid creation of fractures in the soil.

All types of lining in canals are subjected to pressure when the water table is high. It is, therefore, necessary to make drainage arrangements underneath the lining to avoid adverse effect during drawdown and dewatering of canals. In hard rock lining is not needed, but it is provided in uneven rock surface for smooth flow and complete prevention of water outflow through the pores and fractures.

19.5 POWER CHANNELS

This section deals with the geological and engineering reasons for opting for a power channel instead of tunnel construction. The case study given in this section indicates that the stability of cut slopes is the main problem encountered in taking a power channel through varied topography and geological conditions. This section elucidates the remedial measures that need to be taken up to overcome the construction difficulties. In addition, it also highlights the importance of siphon construction to carry the water of canal or power channel.

19.5.1 Choice of Construction and Selection of Alignment

Power channels are the canals that carry water from a natural lake, run-of-river scheme, or storage reservoir for the sole purpose of power generation. In general, in a hydroelectric project, power channel is a part of the water conductor system that comprises tunnel and also covers duct. Decision regarding whether to construct a power tunnel or a channel is made depending on the cost factor involved in the project as per the engineer's design as well as on geological conditions such as stability of hill slope, type of constituent material or rock including its permeability, and geological structures along the power channel alignment. In hilly terrain with soft and shattered rocks, tunnel construction may be a difficult proposition involving expensive treatment, whereas a channel construction in such terrain will be more economical and safe.

Power channels traverse mostly rugged and hilly terrains. During investigation for selecting the alignment, it should be examined whether the power channel route passes through stable lands. Geological situation plays an important role in the consideration of protective measures against sliding or collapse of channel walls. The channel while crossing the hill slopes with soft rock or soil is commonly lined by concrete to permit undisturbed flow of water. However, in intact and impervious rock free from joints, the channel may be kept unlined to reduce the cost. Concrete capping of power channel is required where fall of debris is expected from the hill slope into the channel water.

19.5.2 Case Study on Power Channel

The 2.27 km-long, 18 m-wide power channel of the Loktak project of Manipur is aligned through lake sediments comprising clay, silt, and pebbles with occasional vegetable matters. Its bed slope ratio is 1V:1.5H to 1V:3H. The power channel experienced problems of subsidence of banks and heaving of floor in several places and artisan water flow after it began its operation. Figure 19.11 shows the geological profile across the centre line of the power channel having adverse features such as heaving, black humus clay, and ground cracking. Remedial measures such as drainage through sand filter holes, grouting of bed, loading at places by means of concrete blocks, and providing cut and cover ducts in certain sections could restore the operation of the power channel (Chowdhury 1990).

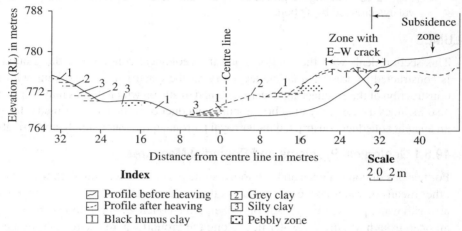

Index

▱ Profile before heaving		② Grey clay	
⊡ Profile after heaving		③ Silty clay	
⊓ Black humus clay		⊡ Pebbly zone	

Fig. 19.11 Geological profile across the centre line of the power channel of Loktak hydroelectric project (Chowdhury 1990)

19.5.3 Siphon to Carry Water of Canal or Power Channel

A *siphon* that looks like an upturned 'U' is made of concrete or steel pipe. It is used to carry the canal or power channel water from one slope to another across a valley (Fig. 19.12). Large inverted siphons are used to convey water being carried by canals or flumes across the field for irrigation. Any settlement of foundation of the siphon may damage the connected structures making the entire canal system inoperative. Protection of siphon foundation against settlement is, therefore, necessary. Hence, the geological condition of foundation site of proposed siphon structure needs to be thoroughly evaluated and protective measures suggested where necessary.

The Gumti hydroelectric project of Tripura is a run-of-river scheme with a diversion dam. The stored water is diverted to the powerhouse through a 2.42 km-long power channel that includes mostly open canal with cover duct at places. The power channel traverses highly undulating terrain with several low valleys, involving construction of several siphons to cross small valleys including lowlands, locally called *charas* (Fig. 19.13). The area consists of sandstone and shale, which are altered and structurally disturbed. The canal was partly damaged by slides of hill slopes and debris flow. Geological study was carried out in the slide-prone areas including siphon areas. Remedial measures such as slope moderation by unloading the loose sandstone blocks, terracing of

Fig. 19.12 Siphon carrying canal and power channel water from one slope to another

overburden material, constructing retaining structures and catchwater drains, and afforestation were taken up as proposed after the geological investigation of the area (Awasthi and Singh 1992).

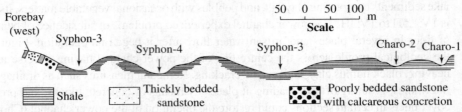

Fig. 19.13 Geological profile along the Gumti power channel showing several low valleys crossed by siphons

19.6 FLUMES

This section deals with the landform and the geological condition of the ground when a flume is constructed to cross a small valley, carrying the canal water. It explains with illustration the construction of the engineering structure named trestle on which flumes may be placed. The section also includes a case study on a bridge-cum-aqueduct over the river Narmada. It further describes an ancient aqueduct constructed under royal patronage for water supply and irrigation purposes.

19.6.1 Geological Problems and Remedial Measures

For reasons of convenience and economy, some portions of a canal alignment may be crossed by other means of water conveyance such as flumes, aqueducts, siphons, and pipelines. *Flumes* are artificial water passageways made of timber, steel, concrete, or reinforced concrete in the form of an open trough usually for water flow along the ground where excavation for a canal is difficult or impracticable. They may be placed on the ground slope or above the ground on support. Similar to a canal, they may be of various shapes such as semicircular, trapezoidal, or rectangular. Undulated landform, steep ground slopes, and hilly terrain may be found unfavourable for canal construction. In a topography that does not favour construction of a canal, a flume that is generally of trapezoidal or rectangular shape can be constructed.

In areas that involve huge filling, instead of a canal, a flume is generally provided to reduce the cost. At places of lowland, the flume is taken above ground by placing it on a trestle. When the flume is laid on a continuous stretch, it can be constructed as a cast-in-place concrete monolithic structure or it may be a pre-cast concrete or reinforced concrete structure. In the latter case, the walls are separated from the inverts by joints to function as breast walls. The trestle-elevated flumes are of two types. In one type, the entire load is carried by the span between the supports, see Fig. 19.14 (a), and in the other type, longitudinal beams between the trestle carry the load, as shown in Fig. 19.14 (b). The mean flow velocity in flumes is about 1.5 m/s to 2.0 m/s when the difference in levels between the entry point of water to and the exit point of water from the flume is 0.1 m to 0.15 m.

A long *flume* has been constructed in the Jaldhaka hydroelectric project, Stage-II, of West Bengal, which passes through the hilly terrain to conduct uninterrupted flow of water. Excavation of the flume bench extending over a stretch

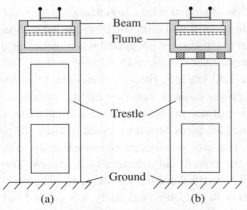

Fig. 19.14 Reinforced concrete flume placed on trestle: (a) load carried directly by support; and (b) load carried by longitudinal beams

of 30 m revealed heaving and subsidence cracks in the ground. The affected zone consisted of sheared and pulverized phyllite in a decomposed state. As remedial measures, the materials were removed up to certain depth and backfilled by graded boulder, pebbles, and sausage filters of 1 m thickness. After carrying out these measures, no further problem was observed at the flume path (Chatterjee 1983).

19.6.2 Aqueducts to Carry Water for Canal or Flume

An *aqueduct* is a water supply or navigable channel (conduit) constructed to convey water. In modern engineering, the term aqueduct is used for a bridge-like structure supporting a water conduit including canals or flumes. The structure is built to cross rivers, gullies, or other natural depression. In Fig. 19.15, an aqueduct supported and elevated on trestles is connected with the canal or channel portion by a flaring flume transition laid on the ground. A hinge provides watertight sealing to the flume at its entrance to and exit from the aqueduct. Sometimes, an aqueduct-cum-bridge is built over large rivers for flow of traffic as well as conveying water. The aqueduct-cum-road bridge over the river Narmada connecting National Highway 44 between Nagpur and Jabalpur is an example.

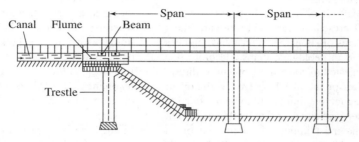

Fig. 19.15 Aqueduct built to cross a valley carrying trestle-elevated flume with the longitudinal beam in between supports carrying the load

The Narmada aqueduct-cum-bridge forms a part of the Bargi canal system of the left bank. An 11.8 m-wide canal carries 127 cumecs of water through the aqueducts to supply 54.28 million cubic metres of water to Jabalpur city for drinking purpose and for irrigation of land around Jabalpur and Narsingpur districts of Madhya Pradesh. The 940 m-long aqueduct (length of the 18 main ducts being 38 m) is made of pre-stressed concrete with carriage way width 7.5 m, footpath 1.5 m, and slope 1:400 (for more details, see Section 18.11.1). The structure resting on a large number of piers is stated to be the highest and unique of its kind in India (Chibber, Chepe, and Ramachandra 1991).

It may be of interest to the readers to know that though particularly associated with the Romans, aqueducts were devised much earlier in the Indian subcontinent. Evidence can be found at the site of present day Hampi, in Karnataka (Fig. 19.16). The massive aqueducts near the river Tungabhadra supplying irrigation water were 15 miles (24 km) long. The waterways also supplied water to royal bath houses (Wikipedia 2012).

Fig. 19.16 A 24 km-long aqueduct of ancient India in Hampi, Karnataka (Wikipedia 2012)

- A canal carries water for irrigating vast areas of farming land. The area along the canal alignment is investigated by geological mapping followed by pitting, trenching, and at places drilling to record the surface and subsurface geology, morphology, and stability of land slopes.
- Representative rock and soil samples of the canal line are tested in the laboratory to determine their engineering properties for design purposes.
- With the help of a contoured map and data of exploratory works, the volume of cutting and filling involved in soil and rock along the canal alignment is calculated and a cross section is prepared showing the nature and thickness of soil and rock in different parts of the canal alignment.
- Loess, organic soil, swelling clay, and soft and weak rocks if present in the canal route are likely to pose problems of slides. Such unstable stretches are demarcated in geological maps and protective measures suggested to contain the potential slides.
- In the stretches of the canal covered with soil or porous rocks, concrete linings are to be provided to arrest water seepage and scouring by flowing water.
- Runway is the most important component of an airport area used for landing and take-off of aircraft. It may extend to a length of 1800 m to 3000 m and is about 30 m wide depending on the nature of aircraft using it.
- During geotechnical investigation, the subsoil condition along the runway is known from exploratory drill holes and laboratory test data.
- The pavement of a runway should preferably be made of concrete, but flexible asphalt pavement is also suitably made depending upon the type of aircraft using it and the availability of material in the area.
- Adequate drainage arrangement is necessary to prevent entry of meteoric water, especially rainwater, into the pavement to prevent subsidence and ensure stability of the runway.

- A power channel is a part of a hydroelectric project where water is conducted through a tunnel or cover duct and partly through power channel to powerhouse for generation of electricity.
- Compared to a tunnel, construction of a power channel is safer and more economical if the area does not have sufficient rock cover or if the rock is soft or fissured, involving expensive treatment.
- The investigation approach for power channel is the same as that of canal and includes geological mapping and exploratory work by trenching and drilling.
- A power channel aligned through rugged hilly terrain may involve deep cutting in many stretches. The stability of the hill slope of these stretches is of important consideration during the investigation of the area, and stabilization measures are proposed wherever needed. Except in areas with firm rock, lining with proper capping is provided along the entire power channel.
- In a steep valley where a canal faces steep ground, involving large volume of filling, a flume supported on a trestle serves better for the purpose of water conductor.
- Flume is made of cast-in-place or pre-cast concrete and may be circular, semicircular, or rectangular in cross section. At places where a valley or a small stream needs to be crossed, the flume carrying running water is placed on an aqueduct or bridge, the two ends of which remain connected with a canal or a power channel.
- Geological investigation for a flume alignment is mainly aimed at studying the stability of the ground where the flume is placed or aqueduct is constructed.
- There are instances of subsidence and development of cracks in the ground of flume bench due to the presence of soft and decomposed rocks. The stabilization measures of such affected zone include removal of soft rocks and backfilling by suitable materials such as graded boulders and sausage filters.

EXERCISES

Multiple Choice Questions

Choose the correct answer from the following:

1. Choose the correct answer.
   (a) For stability, the side slope in a canal, should be 1:2 in sandy soil.

   (b) The side slope for pebbly, sandy, and clay soil should be 1:1.25.
   (c) Both (a) and (b) are correct.
2. Choose the correct answer
   (a) Lining is provided in a canal to prevent water seepage.

(b) Lining is provided in a canal to avoid scouring of side materials.

(c) Both (a) and (b) are correct.

3. Choose the correct answer.

(a) Power channels are preferred to tunnels if the geological conditions are not suitable.

(b) Channel construction is very economical in shattered soft rock.

(c) Both (a) and (b) are correct.

4. Choose the correct answer.

(a) Flumes are made of steel, concrete, or timber and can be semicircular, circular, triangular, or polygonal in shape.

(b) When the topography does not allow low constructions such as a canal, a flume can be constructed.

(c) Both (a) and (b) are correct.

Review Questions

1. What are the types of geological investigation taken up in a highway alignment? Draw a section of road pavement and label the different parts.

2. State the types and sources of construction materials necessary for highway construction.

3. What are the standards specified by Indian Standards for aggregate material to be used in a highway construction?

4. What are the different components of an airport? What should be the length of an airport if it is to be constructed near sea level? Describe the pavement of a runway giving your opinion regarding the type of pavement suitable for a long runway.

5. Give a short account on seepage problem in runway pavement and the measures to be taken to rectify the problem.

6. What methods need to be adopted to retard expansion and stabilize expansive soil in runway pavement? How are cracks developed in a runway pavement and how can such development be prevented?

7. Give an account of the principle aspects of engineering geological investigation in canal construction. Draw sketches and explain the different types of canals.

8. Enumerate the approximate design and shape of canals in relation to their depths. What are the designs of canal slopes in different soils for a 5 m-deep canal?

9. Describe the method of canal excavation through hill slopes. What are the stabilization measures for soil slope and avoiding leakage through soil?

10. What is a power channel and what are the factors to be considered in constructing a power channel instead of a tunnel?

11. Describe the importance of morphological study of terrain condition when selecting alignment for a power channel.

12. Describe the types of geological problems mentioned in the case study on power channel and measures taken to solve the problems.

13. Write a short note on the utility and function of siphons in carrying water of canals and power channels. What type of measure is necessary for founding a siphon structure?

14. What is a flume? State the shape of a flume and the reason of its construction over other means of water conductors.

15. Explain the two types of trestle-elevated flumes in relation to the support requirement.

16. What is an aqueduct? What is the purpose of constructing an aqueduct and an aqueduct-cum-bridge?

17. Name the place where an aqueduct was constructed during ancient India and state its purpose. Name the river from which water was drawn for supply to people and royal baths.

Answers to Multiple Choice Questions

1. (c) 2. (c) 3. (c) 4. (c)

20

Natural Hazards

20.1 INTRODUCTION

This chapter explains the theory of continental drift and plate movement in relation to origin of earthquakes. It also explains the various aspects of seismic waves and the methods of determining the origin of an earthquake. The chapter also highlights the use of Richter scale for measuring magnitude and Mercalli scale for measuring intensity of earthquakes. Geological investigation in any project site should include the study of seismicity of the area from past earthquakes known from earthquake zoning map, and suitable factor is to be incorporated in the design of the structure to protect the engineering structure from damage of seismic shock. This chapter also describes the available methods of forecasting an impending earthquake including the Global Positioning System (GPS).

20.2 EARTHQUAKE

Earthquake is the rapid vibration of ground caused by the sudden fracture and movement of large segments of rocks in the earth's crust. The vibration may cause extensive damage to man-made structures such as buildings, dams, and bridges and trigger landslides, avalanche, and flash flood resulting in loss of property and lives. This section discusses the causative factors, different types of seismic waves, method of finding the origin of earthquake, and the nature of the earth's interior.

20.2.1 Causative Factors

Earthquakes are caused by sudden release of strain energy that rocks had accumulated over prolonged time of many thousands of years. Rupture of the earth's crust along some faults results in earthquakes. The rocks around the faults are weak and hence they are broken easily. Powerful forces compress the rocks along a fault for a prolonged period. These forces will make them to break apart and move. The sudden movement shakes all the rocks around leading to an earthquake.

About 90 per cent of all earthquakes have depths less than 100 km (generally within 70 km), but some have more than 100 km. Faults or ruptures restricted to a depth of one or two kilometres produce minor earthquakes. The crust of the earth is composed of brittle rocks. The breakdown of the crust occurs when elastic strain built up in the crust exceeds the effective strength of the rocks in that part. Earthquakes associated with ruptures on normal faults commonly reach the earth's surface, but when hidden inside the crust, as in the case of low-angle reverse faults (thrust), only seismographs can detect them.

20.2.2 Seismic Waves and Other Earthquake-related Terminology

The point below the earth's surface where the rocks first start moving is the *focus* or *hypocentre* of an earthquake. The location on the surface of the earth vertically above the focus is the *epicentre*. The depth of the focus from the epicentre is called *focal depth*. The distance of the epicentre from any place of interest (say, a seismograph station) is called the *epicentre distance*. The movement of a large section of rocks in the crust creates vibrations called *seismic waves* that travel outwards from the focus in all directions through the earth. Figure 20.1 shows the seismic waves propagating in all directions from the focus due to an earthquake created by the movement of the blocks of rocks along the plane of rupture (fault).

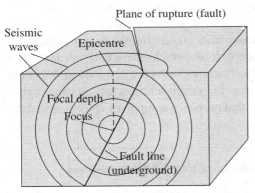

Fig. 20.1 Propagation of seismic waves in all directions from the focus of an earthquake

There are three types of seismic waves, two of which are *body waves* and propagate through the earth's interior. The third type is the *surface wave* that travels through the top or surface layer of the crust. The two body waves— which are called so because they travel through the body of the earth— are termed primary wave (*P-wave*) and secondary wave (*S-wave*). The surface wave is known as *Love wave* and moves from side to side.

There is another type of surface wave known as *Rayleigh wave*, which rolls like the ripples in lake water. The P-wave is a *compressional wave* and is the fastest of the two body waves. This wave is generated when the rock is pushed or pulled forwards or backwards. During the propagation of compressional wave, particles of materials move back and forth parallel to the direction in which the wave is moving. The S-wave is the *shear wave* caused when rock is shaken from side to side like the wave motion. During the propagation of shear waves, particles of material move up and down perpendicular to the direction in which the wave is moving (Fig. 20.2).

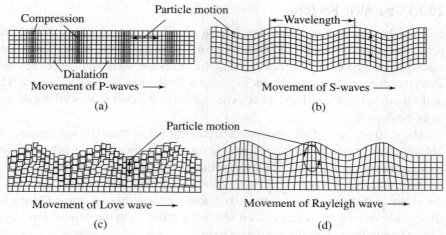

Fig. 20.2 Types of seismic waves: (a) primary wave; (b) secondary wave; (c) Love wave; and (d) Rayleigh wave (→ indicates directions of wave movement)

20.2.3 Locating an Earthquake

Earthquake waves are recorded by a device known as *seismograph* and the record is called *seismogram* (Fig. 20.3). In the simplest form, a seismograph carries a heavy weight tied to a horizontal rod that hangs from a fixed point and moves freely in the lateral direction when the ground vibrates due to earthquake shocks. The rod carries a writing pen at its other end and a sheet of paper rolled around a cylinder. As the cylinder rotates, the pen continuously draws lines indicative of ground motion during earthquakes. A seismologist reads the pattern of lines in the seismogram and measures the different parameters of the earthquake. In its present form, the seismographs are very sophisticated and sensitive. There are strong motion seismographs that can detect and record highly disturbing earthquakes near the source but other seismographs are designed to detect and record even very small earthquakes from a long distance. In modern seismographs, seismic signals are recorded on film or magnetic tape in digital form that enables a computer to process the data.

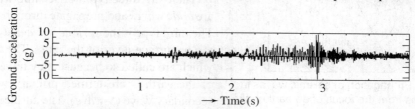

Fig. 20.3 Seismogram of earthquake that occurred in Bhuj, Gujarat, in January 2001

A seismograph records a series of waves that appear like waves on a sea with several peaks and troughs. The height of the peak from zero position is the *wave amplitude* and the time taken to complete the wave from one peak to the next (one cycle of motion) is the *wave period*. The number of cycles of motion of wave per second is the *frequency* of a wave, which is measured in hertz. The amplitude that is measured from the seismogram curve is many times more (may be a thousand times) than the actual, and the amplifying factor of the seismograph device needs to be taken in to account to get the correct value. Human ear can hear sound if its wave frequency is 15 Hz or more, even in thousands. However, ground vibration of an earthquake has frequencies of 20 Hz to less than 1 Hz. Hence, though in general an earthquake is only felt, sometimes it may be heard to.

Prior to the invention of seismograph, earthquake effects were measured by visits to the places affected by the earthquakes. The locality of maximum damage was then traced as the epicentre. The classic work of R. D. Oldham (1899) of Geological Survey of India (GSI) on the Assam earthquake of 12 June 1987 was based fully on field visits that demarcated the epicentre from the place of maximum disturbances and also worked out the source mechanism by adopting a state-of-the-art technique. After the invention of the seismograph, the distance of the earthquake focus from the seismograph stations can be measured from the travel times of the P- and S-waves. The seismograph records the exact time of arrival of both P- and S-waves from the starting place of disturbance. Since the P-waves travel faster than the S-waves, they arrive first and then the S-waves arrive. The difference of arrival times of the two waves increases with the increase of distance of the seismograph station from the earthquake disturbance. This time difference is a vital parameter in determining the epicentre distance of an earthquake (Fig. 20.4).

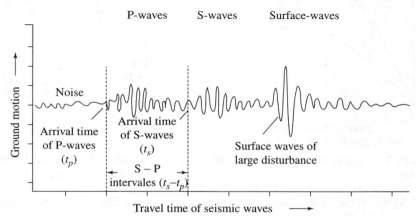

Fig. 20.4 Seismogram of an earthquake showing relation of arrival times of different seismic waves

The various methods of locating the focus of an earthquake and epicentre are nearly the same in principle. All methods basically depend on one factor, that is, the travel time of P- and S-waves from the source (focus) to the seismographic station, which enables direct measurement of the distance between these two points. Seismologists from their experience of studying the wave patterns in seismograms have been able to determine by trial and error the average travel time of P- and S-waves for any specified distance. The time versus distance data can be plotted in a graph and entered in a table. The appropriate distance between the seismograph station and earthquake focus can be read out from this graph or from the table by comparing the arrival times with those that were actually measured from an earthquake. The distance of an earthquake from at least three seismograph stations when plotted by triangulation gives the location of the epicentre, and the time of earthquake disturbance at the source can also be calculated.

Figure 20.5 illustrates the determination of epicentre distances d_1, d_2, and d_3 to seismograph stations A, B, and C, respectively, from the plotting of time intervals of S- and P-waves in a graph. Thereafter, taking these distances as radii, a minimum of three circles are drawn to find the geographical location of the epicentre from their meeting place (Fig. 20.6). It should be noted that the circles do not actually intersect at one point, but the actual location is to be obtained from the interpolation of overlapping arcs. Today, seismograph stations spread out in the different parts of a country and even in other countries can record the seismic waves generated even from a moderate-size earthquake in any part of the world and can precisely locate the epicentre with the help of computers.

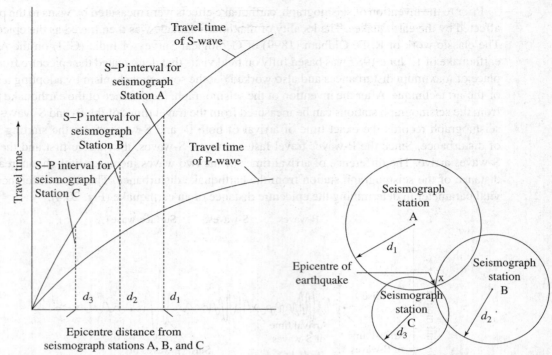

Fig. 20.5 Time versus distance curve of seismic waves

Fig. 20.6 Locating an earthquake epicentre from three seismograph stations

Again on the basis of past experience, seismologists have prepared a ready reference chart for the average distance of an earthquake epicentre from a seismograph station corresponding to the difference of time intervals of P- and S-waves in reaching the station. Thus, the distance of epicentre can be obtained from consulting this table when the arrival times of S- and P-waves are known from a seismogram. Though the distance is known, the geographical location of the epicentre is not known. Taking the distance as radius and the seismograph station as the centre, a circle is drawn. The meeting point of a minimum of three circles drawn from three different stations gives the geographical location (latitude and longitude) of the earthquake epicentre as shown in Fig. 20.6.

20.3 EARTH'S INTERIOR AND PROPAGATION OF SEISMIC WAVES

Knowledge of the earth's interior and its crust is necessary to understand the propagation of seismic waves inside the earth. The total depth from surface to core of the earth (i.e., the radius of the earth) is 6371 km. The crust of the earth varies between 5 km and 75 km (Fig. 20.7), of which the undulating land area of the crust is only one-fourth of the total area, the remaining three-fourths being covered by sea of shallow depth. The temperature of the earth's interior increases slowly towards the interior and is above 2000°C at a depth of 400 km, which is the upper mantle (Fig. 20.8). Further down, towards the lower mantle, there is a rapid rise in temperature and the rocks are in a semi-liquid state.

In fact, the earth's crust is floating on a semi-liquid layer of molten rock (magma) and down below is the liquid outer core, whereas the inner core is in a solid state extending beyond a radius of 3400 km. This solid inner mantle of the earth is composed of dense metal—iron and nickel. The earth was formed approximately 4.5 billion (4.5×10^9) years ago during which time

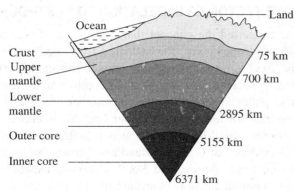

Fig. 20.7 Interior of the earth showing its cores, upper mantle, and crust with depths (Wikipedia 2011)

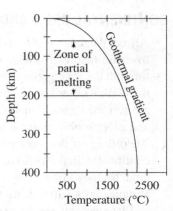

Fig. 20.8 Geothermal gradient

the denser substance sank towards the core by a process of planetary differentiation, whereas the less-dense materials migrated towards the upper part of the earth. It is argued that the solid inner core would be in a giant crystal form. In a recent research (August 2011), Prof. Kei Hirose of Japan placed a sample of iron–nickel alloy in the grip of the tip of two diamonds and then heated it to a temperature of 4000 K with a laser beam. Later, the sample was observed with X-rays; the observation strongly supported the theory of the earth's inner core being in the form of a giant crystal running north to south (Wikipedia 2012).

Seismic waves show properties of reflection and refraction similar to light and sound waves. They change direction and speed as they travel through different densities of the earth's rocks. The P-waves can propagate through solids such as rock and also through liquid and air. When it passes through air, some sound may be produced. S-waves cannot travel through liquid or air. Hence, a person will not feel it while travelling by air or sea. P-waves arrive a few seconds earlier to S-waves. Surface waves travel along the surface of the crust and are the slowest of the seismic waves. They arrive last but usually cause more damage than the P- and S-waves.

As soon as an earthquake is generated in rocks in the interior of the earth, the seismic waves spread out in all directions and take only a few seconds to travel long distances. When these waves reach the surface, the ground shakes and an earthquake is experienced. A number of earthquake shocks also take place before and after the main shock and are termed *foreshocks* and *aftershocks,* respectively. In general, the propagation of velocity of the seismic waves depends on the density and elasticity of the medium.

Velocity tends to increase with depth and ranges from approximately two to eight km/s in the earth's crust and up to 13 km/s in the deep mantle. In the absence of any resistance on the way, the P- and S-waves of all major earthquakes travel quickly through the middle part of the earth to reach even the opposite side of the earth (Fig. 20.9). For instance, the tremor of an earthquake generated in the North Pole may be feebly felt in the South Pole. However, this may not always be true as by the time it reaches the opposite side, the waves become very weak and can be detected only by a powerful seismograph.

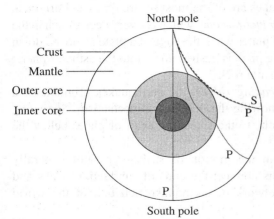

Fig. 20.9 Movement of P- and S-waves

20.4 CONTINENTAL DRIFT AND PLATE TECTONICS IN RELATION TO EARTHQUAKES

Alfred Wegener of Germany first asserted the concept that light rocks of continents resting over the heavy crustal materials of deeper parts slowly drifted from their original positions through the long geological time. According to the modern accepted concept of earth scientists, the earth's outer crust is made up of several huge fragments of continental dimensions, which are termed tectonic plates. These plates move relative to one another above a hotter, deeper, and more mobile zone at average rates of few centimetres per year due to convection current. According to this concept of *continental drift*, the earth was initially made up of a single continent called *Pangaea*. Fossil evidence of Gondwanaland also proves that once there were land bridges connecting Australia, India, Africa, and South America. However, eventually Pangaea broke up to form the seven continents of the current time.

Plate tectonic theory explains the processes responsible for global and regional earthquakes. Bolt (1999) has discussed in detail the theory of plate tectonics in relation to the earthquake processes. According to this theory, the *lithosphere*, which is the crust and outer mantle of the earth ranging from 80 km to 200 km in thickness, consists of several large and stable segments of rocks called *plates*. These plates are, in fact, oceanic and continental parts of huge dimensions. The plates have taken the current pattern by fragmentation of the ancient supercontinent *Pangaea* over the past 200 million years. Below the lithosphere is the *asthenosphere* composed of rocks very close to melting point. The asthenosphere occasionally rises to the lithosphere through some weak places as a *hot spot*. Each lithospheric plate moves horizontally relative to the neighbouring plate on the softer rock of asthenosphere in the lower part. The energy for the driving mechanisms of the drifting plates comes from the heat produced by the decay of the radioactive elements in rocks inside the earth.

20.4.1 Possibilities of Plate Movement and Resultant Earthquakes

There are three main possibilities of movement of the plates. First, the plates move together causing ruptures of the crust at the edges of the plates. Second, the adjacent plates may move apart from each other creating a widening gap between them. Finally, the plates may slide alongside each other on *transform faults*. Thus, there are three environments in earthquakes caused by plate movements, namely compressional (or convergence), extensional (or divergence), and transform:

(i) At the convergent boundaries, the earthquakes are of the most severe types and originate from shallow depths to deep parts of the *subduction zone*, which is a very deep trench in the earth's crust. There may be convergence of plates when the edge of a plate is moved down and forced into subduction. The subducting plate is pushed down into the asthenosphere, where it is heated and absorbed into the mantle rock.

(ii) At the divergent boundaries such as *spreading ridges,* the earthquakes originate at a shallow depth along the spreading line, showing the divergence mechanism. Spreading ridges or *mid-oceanic ridges* are formed due to the spreading apart of plates below the ocean.

(iii) At *transform*, earthquakes originate from the motion of strike–slip fault generally within a depth of 25 km. Transform faults connect the ends of subduction zones and mid-oceanic ridges into a continuous network of earthquake-rich belt of the world (Fig. 20.10).

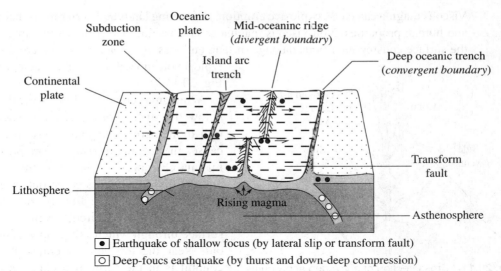

Fig. 20.10 Diagram illustrating plate tectonics and the three types of plate boundaries

Plate geometry is not a permanent feature but it goes on changing. In the subduction zone, cool materials of the earth's crust plunge inside and melt because of the high temperature inside the earth. As the density of the lava increases, it becomes more buoyant and tries to come out to the surface. However, material from the crust moves to the core and vice versa leading to an earthquake or a volcano. In *mid-oceanic ridge*, there is constant ejection of materials from the earth's surface to form new crust. This material is deep seated on both sides of the plates and is responsible for breaking apart or unification of the plates. The processes arising from the growth of the plates produce earthquakes along the mid-oceanic ridges.

20.4.2 Interplate and Intraplate Earthquakes

The earth contains several major lithospheric plates and their boundaries are seismically active. There are many other minor plates too, but it is along the boundaries of the major plates that earthquakes and volcanic eruptions take place. Earthquakes that occur at the contact of two interacting plates are called *interplate earthquakes*. All the major earthquakes of the Himalayan region are interplate earthquakes and originate in the collision zone of the Eurasian plate and the Indian plate.

The earthquakes that occur within the plates are called *intraplate earthquakes*. Such relatively low-intensity earthquakes arise from the localized system of forces in the continental region. Stress due to compression is responsible for the generation of tectonic movement related to intraplate seismic events that take place along intersecting fractures or a single plane of rupture. Frictional forces along one of the fracture surfaces initiate the earthquake. The earthquakes of peninsular India are of intraplate type.

20.5 VOLCANO AND ITS ACTIVITY RELATED TO EARTHQUAKE AND OTHER EFFECTS

Earthquakes may also occur due to volcanism, which is a phenomenon connected with volcanic activity. Volcanoes have many effects on the earth's crust. The molten rock or magma erupting from the earth's interior due to volcanism behaves differently depending upon its viscosity.

Viscous magma can cause explosive eruptions destroying lands and burn forests, habitats, crops, and human properties. Non-viscous magma as lava issuing out of the earth's interior flows to the surface causing enormous damage of land and loss of human lives. The cross section of a typical volcano will reveal the features as shown in Fig. 20.11.

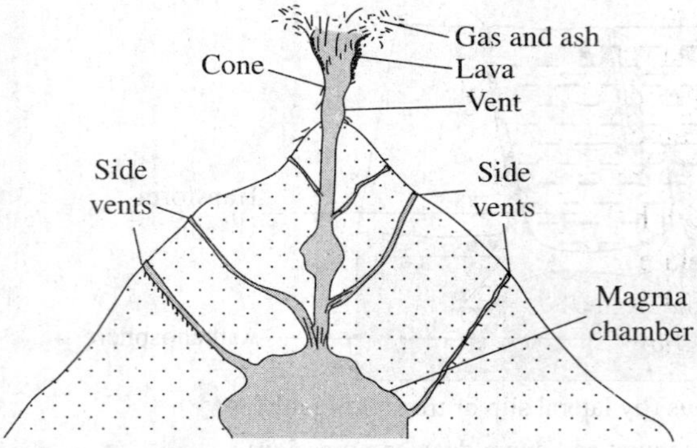

A magma chamber lies several kilometres below the surface at the interior where the gas-rich molten rock (magma) exerts pressure. This causes the magma to rise up. The pressure in the magma chamber is reduced as the magma rises and the dissolved gases come out of the solution as expanding bubbles. Finally, the force of the gases blasts open a circular vent on the earth's surface through which ash, cinders, and lava flow out. Near the top vent or crater, the volcano's conduit is in the form of a cone. In addition to the main vent, a volcano may release lava and ash from side vents creating subsidiary cones.

Fig. 20.11 Cross section of a volcano showing its different parts

The surface features or landform originated by a volcano may be of different types from towering peaks to vast sheets of lava. The features of volcanoes vary with the type of eruption, nature of lava and other erupted materials, and subsequent erosion. There are four major types of volcanoes, namely lava cones or shield volcanoes, cinder cones, composite cones, and lava plateaus.

A *lava cone* volcano is usually formed by slow upwelling of viscous lava from steep-sided volcano (Fig. 20.12) as seen in parts of California. It is also called a *shield volcano* as it resembles a warrior's shield. The Mauna Loa volcano in Hawaii is a classic example of this type of volcano. It rises from the sea floor nearly 10,000 m deep to a height of nearly 4100 m above the sea level.

A *cinder cone* volcano is composed of loose pyroclastic materials. The volcano mainly ejects rock fragments of solid lava. Large quantities of ash are also released by this type of volcano (Fig. 20.13). The eruption from the volcano is very explosive and the erupted materials generally do not rise to more than 300–400 m above the surface. The slope of the volcanic cone may be above 40°.

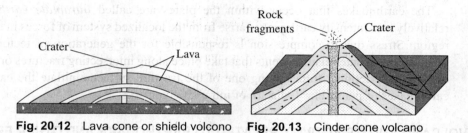

Fig. 20.12 Lava cone or shield volcono   **Fig. 20.13** Cinder cone volcano

A *composite cone* volcano possesses concave cone-shaped sides and erupts ash and lava alternatively (Fig. 20.14). Mt Fujiyama in Japan and Mt Vesuvius in Italy are examples of this type of volcano. It is also known as *stratovolcano* because of its very large and symmetrical structure. It is composed of lava flows and pyroclastic materials. In AD 79 the entire city of

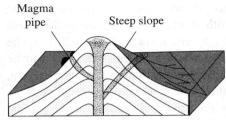

Fig. 20.14 Composite cone type volcano

Pompeii was buried in three days by ash and pyroclastic material that erupted from this type of volcano, which was dormant through centuries. Sometimes solid lava seals the main pipes to the crater and the top part blasts off to release the pent-up gases. The summit may collapse leading to the formation of a very big cavity called caldera. The Ngorongoro crater in Tanzania is the world's largest crater created by such blast of volcanic summit.

In the *lava plateau* type of volcano, the magma from the interior comes to the surface as lava flow through fissures and after solidification forms a huge blanket of basalt (Fig. 20.15). The Deccan plateau basalt of India (Fig. 20.16) is an example where the thickness of basalt is as high as 2199 m and the basaltic rock covers an area of about 650,000 km². Another such lava plateau is the Columbia plateau, also known as Columbia Basin, located in the Pacific Northwest.

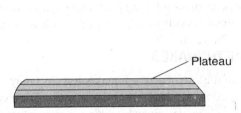

Fig. 20.15 Lava plateau type volcano

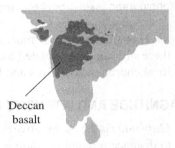

Fig. 20.16 Deccan plateau basalt, India

Earthquake related to volcanism can cause ground deformation and destruction of buildings and other man-made structures. It can also cause ground subsidence and large ground cracks. When the injection of magma from volcano is sustained, it may produce a number of earthquakes, called volcanic tremors. A volcanic tremor is thus an indicator of an impending earthquake. Movement of molten rock from the earth's inner mantle by thermal convention current with gravitational effect and by plate tectonic motion can result in volcanism. Magma intrusion by such volcanism may form large intrusive bodies such as batholiths, lopoliths, and laccoliths and also dykes and sills (see Section 2.2).

When water interacts with volcanism, it acts as geothermal energy to create geysers, fumaroles, hot springs, and even mudflows. On the earth, volcanism occurs in distinct geological settings, generally associated with boundaries of enormous rigid plates that make the crust's upper mantle and crust. Nearly 80 per cent of the earth's volcanic activity takes place where two plates converge and one overrides the other. The island arc trench at the subduction zone is an example of such volcanic activity. Another site of volcanism is the oceanic ridge system where plates move apart on both sides of the ridge and magma flows out to the ocean floor creating a new landform. The Hawaii Islands, in which active volcanoes still exist, were thus formed in the Pacific Ocean. An example is the Kilauea volcano through the crater of which lava still flows out (Fig. 20.17). With the movement of the Pacific plate, it is believed that Kilauea will move farther away from the hot spot under the plate.

Among the chain of several volcanoes in the Hawaiian Islands, Kilauea is geologically the youngest and it is currently the most active volcano. It still exhibits its upwelling basaltic

Fig. 20.17 Upwelling viscous lava with fire in Kilauea volcano (photograph taken from helicopter)

lava, which extends over a wide area of the surface with glow of fire as the molten lava flows through certain channels (Fig. 20.17). The magma chamber of Kilauea is estimated to be nearly 60 km below its main vent to the surface. The volcano showed its vigorous eruptive force with upwelling lava and shaking of the ground throughout the nineteenth century, and continued to erupt many times in the twentieth century and is still active. The frequent volcanic erupts of Kilauea have made it a centre of study on volcanism for volcanologists.

Local people believe that Pele, the goddess of fire, lives inside this volcano. This belief and several tales associated with its eruptive history have made it a spiritual site of significance for the Hawaiian people. Amidst the burnt forests and villages damaged by the fire from Kilauea, there still stands an isolated hamlet, where, it is said, long back lived a pious priest couple of the local church, whose lives and homes were saved by the mercy of Goddess Pele.

20.6 MAGNITUDE AND INTENSITY OF EARTHQUAKES

Magnitude is the quantitative measure of the size of an earthquake. Its effect may vary with respect to distance, ground conditions, construction standard, and other factors. Each earthquake has a unique amount of energy, but the magnitude values given by different seismological observatories for an event may differ. The scale for magnitude of earthquakes is not linear. The energy released by an earthquake of a particular magnitude may be equal to the energy released by 30 earthquakes of the previous magnitude. For example, the energy released by an earthquake of magnitude 6 is about 39 times that released by an earthquake of magnitude 5 and is about 900 times that released by an earthquake of magnitude 4. The energy released by an earthquake of magnitude 6.3 is equivalent to the energy released by the explosion of the atom bomb in Hiroshima, Japan in 1945.

20.6.1 Richter Magnitude Scale

Depending upon the size, nature, and location of earthquakes, seismologists use different methods to estimate the magnitude. The widely used scale for measuring seismic magnitude is the Richter scale developed in 1935 by Prof. Charles Richter of California. Richter found that at the same distance, *seismograms* (records of earthquake ground vibrations) of larger earthquakes have larger wave amplitudes than those of smaller earthquakes and that for a given earthquake the seismographs at farther distances have smaller wave amplitudes than those at close distance. It means that larger the intrinsic energy of the earthquakes, greater is the amplitude of ground motion at a given distance. On this observation, he calibrated the scale of magnitude using the measured maximum amplitude of waveform (shear wave) with periods of about one second. The record had to be obtained from a specific type of seismograph called Wood–Anderson Seismometer. Later seismologists have developed the scale factors to extend the Richter magnitude scale to all types of seismographs.

The Richter magnitude (M) is given by the following expression:

$$M = \text{Log}_{10} A + 3 \log_{10} (8\Delta t) - 2.92$$

Here, *A* is the amplitude in millimetres measured directly from the photographic paper of seismogram and Δt is the difference in seconds between the times of P- and S-waves in the wave curve.

The Richter scale in whole number varies from 1 to 9. It is such that the magnitude of the quake is 10 times greater than the previous whole number. For example, an earthquake of 6 has 10 times the force of that with a magnitude of 5, or an earthquake of magnitude 7 has 100 times the intensity of that of magnitude 5. The typical effects of earthquakes in Richter scale are stated in Table 20.1 for different magnitudes.

Table 20.1   Richter scale of earthquake magnitude*

Magnitude	Description	Earthquake effect
< 2.0	Micro	Micro earthquake, not felt
2.0–2.9	Minor	Generally felt, but rarely causes any damage
3.0–3.9	Minor	Often felt, but rarely causes damage
4.0–4.9	Light	Noticeable shaking of indoor items
5.0–5.9	Moderate	Can cause major damage to poorly constructed buildings over a small region; at most, slight damage to well-designed buildings
6.0–6.9	Strong	Can be destructive in areas up to about 160 km across in populated areas
7.0–7.9	Major	Can cause serious damage over larger areas
8.0–8.9	Great	Can cause serious damage in areas several hundred kilometres across
9-0–9.9	Great	Devastating in areas several thousand kilometres across

*Based on US Geological Survey (USGS) document

20.6.2 Mercalli Intensity Scale

Intensity is a quantitative measurement of an earthquake based on its impact on people, land, and building. It is denoted using Roman capital letters. The intensity varies with distance from the focus of an earthquake. In a map of an earthquake-affected area, the curved line showing the locations of equal seismic intensities is called the *isoseismal line* and the map is called an *isoseismal map* (Fig. 20.18).

The seismic intensity of a locality depends on the magnitude of an earthquake, location from epicentre, and nature of paths followed by seismic waves and their final destination. There are several scales for measuring seismic intensity. The intensity scale used for many years was the Rossi–Forel scale, devised in 1883, with 10 divisions designated from I–X. The two common intensity scales currently used are the Modified Mercalli (MM) scale and the Medvedev–Sponheuer–Karnik (MSK), which is a macroseismic scale. These two scales are similar in many ways, each having 12 divisions ranging from I to XII. Table 20.2 provides the description of the different intensity levels according to these scales.

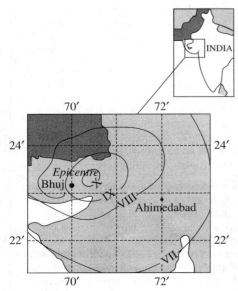

Fig. 20.18   Isoseismal map of Bhuj (Gujarat) earthquake of 26 January 2001 with seismic intensity X (MM/MSK scale)

Table 20.2 Description of MM and MSK levels of intensity scales

Level	MM intensity scale *	MSK intensity scale**
I	Not felt except by a very few persons under especially favourable conditions.	Not felt by human senses, but recorded only by seismographs.
II	Felt only by a few persons at rest, especially on upper floors of buildings.	Noticed only by some people at rest, especially on upper floors of buildings.
III	Felt quite noticeably by persons indoors, especially on upper floors of buildings, but many do not recognize it as an earthquake. Standing motor cars may rock slightly. Vibration similar to passing of a truck.	Noticed by some people inside buildings. The vibration felt is similar to passing of a light truck. Very observant individuals may notice objects swinging slightly, especially on upper floors of buildings.
IV	During the day, felt indoors by many, outdoors by few. At night, some are awakened. Dishes, windows, doors are disturbed and walls make creaking sound. Sensation is similar to a heavy truck striking a building. Standing motor cars rock noticeably.	Noticed by people inside buildings and by some outside. Some people wake up. The vibration is similar to a heavily laden truck passing by. Floors and walls develop cracks. Furniture may move. Liquids in open container rock.
V	Felt by nearly everyone, and many are awakened. Some dishes and windows may be broken. Unstable objects are overturned. Pendulum clocks may stop.	Noticed by most people inside buildings and many outside. Many people who are asleep wake up. Animals become nervous. Buildings shake and hanging objects swing. Pictures are hit against walls. Pendulum clocks may stop. Light objects move. Open doors or windows swing violently. Small amount of liquids spill from containers. Buildings of mud or stone–clay walls (type A buildings) may damage slightly. Flow of springs may alter.
VI	Felt by all. Many are frightened and run outdoors, or walk unsteadily. Windows, dishes, glassware are broken, Books fall off shelves. Some heavy furniture move or overturn. There may be few instances of fallen plaster and slight damage.	Felt by majority of people both inside and outside buildings. Many are frightened and go out. Domestic animals flee from their shelters. On some occasions, crockery and glassware break, books fall from shelves, pictures move, and unstable objects overturn. Heavy furniture may move. There may be moderate damage to type A buildings and slight damage to buildings made of brick, masonry, mortar, or timber frame (type B building).
VII	Noticed by persons driving motor cars. It will be difficult to stand and furniture may be broken. Damage will be negligible in buildings of good design and construction, slight to moderate in well-built ordinary structures, and considerable in poorly built and badly designed structures. Some chimneys may be broken.	Majority of people are frightened and run out to the street. Many find it difficult to stand up. People driving cars feel the vibrations. Large bells ring. Many type A buildings suffer serious damage and type B buildings suffer moderate damage. Buildings with metal or reinforced concrete structure (type C buildings) suffer slight damage. Landslides may occur on hilly roads. Waves appear in lakes. Water levels in wells and springs change.
VIII	The damage will be slight in specially-designed structures, considerable in ordinary substantial buildings with partial collapse, and heavy in poorly built structures. Chimneys, factory stacks, columns, monuments, and walls may fall. Heavy furniture may be overturned.	There will be fear and general panic among people, including those driving cars. Branches of trees may break off. Heavy furniture move and turn over. Many type A buildings are destroyed and some collapse. Many type B buildings suffer serious damage and some are destroyed. Many type C buildings suffer moderate damage and some serious damage. Stone walls collapse. Small landslides may occur on slopes and cuttings. Cracks that are several centimetres wide are developed on the ground. New springs appear. Existing wells become dry and dry wells get filled with water.

(Contd)

Table 20.2 (*Contd*)

IX	There will be general panic. The damage is considerable in specially-designed structures, and well-designed frame structures are thrown out of plumb. The damage is heavy in substantial buildings, with partial collapse. Buildings may be shifted off foundations.	There will be general panic and significant damage to property. Animals run in confusion and make strange noises. Many type A buildings collapse. Many type B buildings are destroyed and some collapse. Many type C buildings suffer serious damage and some are destroyed. Monuments and columns fall and underground pipes are partially cracked. Railway tracks are bent. Jets of water and mud are observed in saturated lands. Numerous cracks up to 10 cm in width appear in the ground. Many landslides occur and rock falls. Large waves appear in lakes and reservoirs. Dry wells get filled and existing wells dry up.
X	Some well-built wooden structures are destroyed, and most masonry and frame structures are destroyed along with foundation. Rails are bent.	Majority of type A buildings and many type B buildings collapse. Many type C buildings are destroyed. Railway tracks are altered. Underground pipes are twisted. Steel surfaces form undulations. Grounds develop cracks that are more than 10 cm wide. Landslides occur in hill slopes and steep river banks. There will be changes in the water levels in wells. New lakes are formed.
XI	Few, if any, masonry structures remain standing. Bridges are destroyed. Rails are greatly bent.	Significant damage to all types of buildings, bridges, dams, rails, and roads. Underground pipes are destroyed. Land is pushed out of shape by landslide and rock falls.
XII	Damage is total. Lines of sight and level distorted. Objects thrown upwards into the air.	All structures, including underground structures, are destroyed or severely damaged. The topography changes. There are large cracks on the ground and significant horizontal and vertical landslips are observed. Rock falls occur. Waterfalls appear and rivers are diverted.

*USGS pamphlet (1986): *The severity of an earthquake*
**IAG Information and Dissemination, 2004

20.7 SEISMIC ZONING AND CODES FOR EARTHQUAKE RESISTANCE

20.7.1 Seismic Zoning Map of India

Seismic zoning map of a region is prepared based on the extent of damage suffered by the region due to the effect of past earthquakes. The seismic zoning map of India showing five zones (zones I–V) was first prepared by Geological Survey of India (GSI) in 1935 based on the damages suffered in a particular region due to past earthquakes in the areas with an intensity higher than Rossi–Forel intensity VII (equivalent to MM intensity VIII). This gave a design criterion of the severity of damage that each zone is liable to suffer in case of future earthquakes (Table 20.3). Since then several modifications have been made by Bureau of Indian Standards (BIS).

Table 20.3 Seismic zones in different states and Union Territories of India

State/Union Territory	*Seismic zone*	*State/Union Territory*	*Seismic zone*
Arunachal Pradesh	V	Lakshadweep	II
Assam	V	Madhya Pradesh	I and III
Andaman and Nicobar Islands	V	Mizoram	V
Andhra Pradesh	II and III	Meghalaya	V

(*Contd*)

Table 20.3 *Contd*

Bihar	V	Manipur	V
Chandigarh	II, III, IV, and V	Maharashtra	II, III, and IV
Daman and Diu	III	Nagaland	V
Dadra and Nagar Haveli	III	Orissa	II and III
Delhi	IV	Punjab	II, III, and IV
Goa	II and III	Rajasthan	II, III, and IV
Gujarat	II, III, and IV	Pondicherry	II and III
Haryana	II, III, and IV	Sikkim	IV
Himachal Pradesh	IV and V	Tripura	V
Jammu and Kashmir	IV and V	Tamil Nadu	II and III
Karnataka	II and III	Uttar Pradesh	II, III, IV, and V
Kerala	II and III	West Bengal	II, III, IV, and V

Nearly 60 per cent of the areas in India fall under the zones III, IV, and V. The Himalayan region is within zones IV and V. Peninsular India was earlier considered to be a seismically stable landmass but after the Killari earthquake of 1993 and Jabalpur earthquake of 1997 (see the table in Section 20.9), modification has been made as regards the seismicity of the peninsular region. In the revision of seismic zones in 2000, zone I (*very low damage risk zone*; maximum intensity of V in MM or MSK) has been merged with zone II by the BIS seismic zoning committee. Thus, there are now four zones, numbered II, III, IV, and V, in the earthquake zoning map of India (Fig. 20.19). The intensity and probable earthquake damage that may be suffered by these zones are as follows:

Zone II : The maximum intensity in MM and MSK scales is estimated to be VI. This zone is termed as *low damage risk zone*.

Zone III: The probable intensity in MM and MSK scales is VII. This is termed as *moderate damage risk zone*.

Zone IV: The area under this zone is equivalent to seismic intensity VIII in MM and MSK scales. This zone is the *high damage risk zone*.

Zone V : This covers the area liable to seismic intensity IX or more (MM and MSK scales). This is the *maximum damage risk zone*.

20.7.2 Codes for Design of Earthquake-resistant Engineering Structures

Deformation or damage of structures is caused by earthquake forces and ground shaking. Seismic codes are prepared as guidelines for design of earthquake-resistant structures. When an earthquake occurs, the ground moves in lateral sides and also in the upward and downward directions. The forces that cause the movement of the ground can be resisted by the structure if it is constructed following the guidelines provided in the seismic codes. The codes are based on the probable maximum shocks that a structure may experience during its lifetime. The probable maximum intensity of earthquake of an area is considered from the seismic zoning map, which has been prepared from the past earthquake records of the area.

Throughout the world, codes for aseismic design are prepared to save man-made structures from the destructive effects of earthquakes. In fact, the structural design is codified to respond to earthquake vibrations without damage in case of moderate earthquakes and save the structure

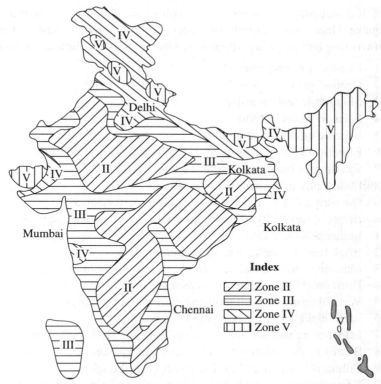

Fig. 20.19 Earthquake zoning map of India showing the four zones

from total collapse during high-intensity earthquakes. In 1962, BIS published the code IS 1893 for earthquake-resistant structures, which include structures such as buildings, dams, and retaining walls. In the latest revision of BIS in 2002, this code has been separated into five parts: (i) part I for retaining tanks, (ii) part II for buildings, (iii) part III for buildings and retaining walls, (iv) part IV for industrial structures, and (v) part V for dams and embankments. In addition, in 2000, codes for bridges over roads and railways were made in separate codes by IS 1984 and also by Indian Road Congress IRC 6. The codes take care of the following criteria in the design:

  (i) The structure can resist the lateral forces so as to prevent the collapse of structures.
 (ii) The load-bearing capacity (size, shape, and structural system) of the structure is kept in such a way that the flow of inertia to the ground takes place smoothly.
(iii) The lateral load-resistant system is made in such a way that under low-to-intermediate earthquake shocks, there will be no damage to structures.
(iv) During high-intensity earthquakes, the structure will be saved by providing ductility.

Since the Bhuj (Gujarat) earthquake of 2001, it is mandatory to design structures (e.g., buildings) following the code of earthquake-resistant structures in areas of seismic zones III, IV, and V.

20.7.3 Tips for Earthquake-resistant Design and Construction

The catastrophic Bhuj earthquake has also initiated the Indian Institute of Technology Kanpur (IITK), and the Building Materials and Technology Promotion Council (BMTPM) to release 24 tips for dissemination of knowledge on various aspects of earthquake, especially for construction of buildings. It is to be remembered that an earthquake does not directly kill a person, but loss of life and property occurs as a result of the collapse of building structures during an earthquake. Hence, it is desired that students should know the basics of earthquakes and codes of constructing buildings in earthquake-prone areas as highlighted in the following tips:

Tip 1   Causes of earthquakes
Tip 2   Ground shakes
Tip 3   Magnitude and intensity
Tip 4   Seismic zones in India
Tip 5   Seismic effects on structures
Tip 6   Effects of earthquakes on architectural features in buildings
Tip 7   Twisting of buildings during earthquakes
Tip 8   Philosophy of aseismic design of buildings
Tip 9   Ductility of buildings for good seismic performance
Tip 10   Effect of flexibility of buildings on their earthquake response
Tip 11   Indian seismic codes
Tip 12   Brick masonry behaviour during earthquake
Tip 13   Masonry buildings having simple structural configuration
Tip 14   Horizontal bands needed in masonry buildings
Tip 15   Vertical reinforcement required in masonry buildings
Tip 16   Earthquake-resistant stone masonry buildings
Tip 17   Effect of earthquake on reinforced concrete (RC) buildings
Tip 18   Beams in RC buildings for resisting earthquakes
Tip 19   Columns in RC buildings for resisting earthquakes
Tip 20   Beam–column joints in RC buildings for resisting earthquakes
Tip 21   Open ground-storey buildings vulnerable to earthquakes
Tip 22   More damage in short columns due to earthquakes
Tip 23   Preference for buildings with shear walls in seismic regions
Tip 24   Reduce effect of earthquake on buildings

The tips cover topics such as basic introduction to earthquakes and terminology including magnitude and intensity as well as earthquake zoning and the codes of aseismic design. These basic aspects have been fully dealt with in the preceding sections, especially in Sections 20.6 and 20.7. The details are provided with the belief that readers will find the tips useful when constructing buildings in earthquake-prone areas and will consult experts for finalizing their design and construction detail. For details regarding each tip, consult the website of National Information Centre of Earthquake Engineering (www.nicee.org).

20.8 RESERVOIR-INDUCED SEISMICITY

It has been found that under water-filled condition, some large reservoirs trigger earthquakes known as *reservoir-induced seismicity* (RIS). The initial activity, generally of low magnitudes, results from the instantaneous effect of loading and delayed effect of pore pressure diffusion. Following this initial activity, major incidents of high-magnitude earthquakes take place several years after impounding water in the reservoir to maximum water level. The delay in the large event depends on the permeability of reservoir rocks, geological structures, and volume of water. Widespread seismicity is observed in the periphery and deeper parts. The maximum intensity in RIS is followed by decay in the activity continued for months and years. In Koyna of India, even after three decades of impoundment, seismic intensity of lesser extent has been recorded.

20.8.1 Incidents of Reservoir-induced Seismicity in Different Parts of the World

Reservoir-induced seismicity has been reported from as many as 120 reservoir projects of the world as mentioned in the International Symposium on RIS held in Beijing, China, in 1995. In the 1960s, several severe RIS occurrence of magnitudes more than 6.0 on the Richter scale were recorded from reservoirs of different countries such as Koyna of India, Xinfengjiang of China, Kariba in Zimbabwe, and Krembasta of Greece. Study of reservoir loading in relation to RIS has indicated that in most cases seismicity is associated with the initial impoundment of the reservoirs. However, there are several cases in which RIS has been found to occur after some years of initial impoundment. In the case of Aswan dam of Egypt, it was after 17 years of impoundment that the reservoir triggered an earthquake (Table 20.4).

Table 20.4 RIS of magnitude more than 4.5 on Richter scale (Gupta 2002)

Country	Name of dam	Height of dam (m)	Reservoir volume (m³ × 10⁶)	Year of reservoir filling	Year of maximum RIS	Magnitude
India	Koyna	103	2780	1962	1967	6.3
Zambia	Kariba	128	175,000	1958	1963	6.2
Greece	Kremasta	160	4750	1965	1966	6.2
China	Xinfengjiang	105	14,000	1959	1962	6.1
Thailand	Srinakarin	140	17,745	1977	1983	5.9
Greece	Marathon	67	41	1929	1938	5.7
USA	Oroville	236	4400	1967	1975	5.7
Egypt	Aswan	111	164,000	1964	1981	5.6
New Zealand	Benmore	110	2040	1964	1966	5.0
Australia	Eucumbene	116	4761	1957	1959	5.0
USA	Hoover	221	36,703	1935	1939	5.0
India	Bhatsa	88	947	1981	1983	4.9
USA	Kerr	60	1505	1958	1971	4.9
Japan	Kurobe	186	149	1960	1961	4.9
France	Monteynard	155	275	1952	1963	4.9
China	Shenwo	50	540	1972	1974	4.8
Ghana	Akosombo	134	148,000	1964	1964	4.7
Spain	Canelles	150	678	1960	1962	4.7
China	Danjiangkou	97	16,000	1967	1973	4.7
France	Grandval	88	292	1959	1963	4.7
Greece	Kastraki	96	1000	1968	1969	4.6
New Zealand	Lake Pukaki	106	9000	1976	1978	4.6
Tajikistan	Nurek	317	10,500	1972	1972	4.6

It may appear that RIS is the result of additional weight that a reservoir experiences after being filled with water. Calculations have shown that the additional pressure or stress due to the impounded water is only a fraction of already existing tectonic stress in the rocks below the reservoir covering several kilometres. The reason for the earthquake is explained to be the extra pressure resulting from the impoundment of water, which spreads out as pressure waves or pulses through the pores or fractures of the rocks (Bolt 1999). Detailed field study of the reservoir areas has led to the conclusion that the factors responsible for RIS include ambient stress field condition, availability of fractures, hydro-mechanical properties of the underlying rocks, geology of the area together with dimensions of the reservoir, and the fluctuations in lake level.

20.8.2 Conditions Required for Generation of Reservoir-induced Seismicity

The following conditions generally need to be fulfilled for generation of RIS, which is a shallow focus earthquake with considerable damage potential over only a limited area.

  (i) Water body of a reservoir depth of more than 100 m can produce RIS.
 (ii) Access of reservoir water to the potential focal zone through interconnecting rock cracks is necessary for RIS.
(iii) Optimum structural set-up (under gravity faulting and not thrust faulting environment) very close to a limiting condition is required, so that a relatively meagre water load may act as triggering effect. As already mentioned (Section 15.9), there is no evidence of RIS in the Himalayan reservoirs mainly due to their location under thrust fault environment.

20.9 SEISMOTECTONIC FRAMEWORK OF INDIA

Seismotectonics deals with the tectonic features of the earth related to origin of earthquakes. Understanding of tectonic features of a terrain helps to delineate seismically active zones and design earthquake-resistant structures for such zones. The Himalayan region and the peninsular plateau of India have been suffering from frequent earthquakes of varying intensities related to tectonic history of these regions. The Himalayan region has been subjected to several tectonic movements since Tertiary time. The maximum number of high-magnitude seismic events in India has been recorded in this region. Earthquake records of the past 100 years indicate that the Indian subcontinent has experienced over 600 earthquakes of magnitudes 5 and above and most of the high-magnitude earthquakes (magnitude > 7) took place in the Himalayan terrain whereas peninsular India experienced mostly low-to-medium earthquakes.

The most prominent tectonic event of the Himalayas is the collision of the Indian plate with the Eurasian plate producing large-scale seismic activity in the plate boundary. The plate boundary was at the Indus suture zone when collision began in the Tertiary time. In subsequent time, it shifted southwards along the Main Central Thrust (MCT) and further later along the Main Boundary Fault (MBF). The southern shifting is related to the convergence of the Indian plate. The detachment surface represents the upper part of the Indian plate, underlying the entire Himalayas. The thrust sheets (MCT and MBF) merge with the detachment surface at depth. High-magnitude earthquakes located within this 50 km block between MCT and MBF result from differential movements of discrete tectonic blocks. These blocks are the source of neo-tectonic activity and geothermal manifestation (Narula 1999).

According to the tectonic model for the western Himalayas outlined by Dhar (1991), 'the Himalayan thrust belt is composed of two distinct segments: (i) a shallow portion, dipping gently northward designated as "detached" and (ii) the steeper portion situated down-deep from the detachment separates the basement of Indian shield from the overlapping sedimentary wedge composed of Siwaliks of the lesser Himalaya. The basement thrust, on the other hand, juxtaposed the basement rocks of the Indian shield in the foothill walls with similar rocks in the hanging wall side.' As per this model, all the great earthquakes (magnitude > 7.8) are produced at long intervals in the detachment side whereas relatively frequent earthquakes of smaller magnitudes take place in the basement thrust area.

During the past 100 years, eastern Himalayas has experienced many earthquakes of magnitudes above 7 including the great Assam earthquake of 1897 of magnitude 8.7 (Table 20.5) having their epicentres concentrated along the MCT. Some lineaments running parallel to MCT are responsible for the Shillong earthquake of 1950 that devastated large areas of Assam. Maximum seismic activity of the eastern Himalayas has been recorded from the subduction zone where Indian plate has thrust below the Burmese plate. In the Shillong plateau, high-magnitude seismicity has been developed from the northerly subduction zone of Indian plate. Some major earthquakes have originated from faults parallel to the MCT. Regional tectonic features such as Bomdila and Kopili lineaments across the eastern Himalayas have experienced several seismic events along them.

Table 20.5 Major (magnitude > 7) earthquakes of Himalayan terrain (Ramchandran, Pradhan, and Dhanota 1981)

Date	Magnitude	Location
10 January 1869	7.5	Cachar, Assam
30 May 1885	7.0	Sopor, Jammu & Kashmir
12 June 1897	8.7	Shillong plateau
4 April 1905	8.0	Kangra, Himachal Pradesh
31 August 1906	7.0	North-east India
12 December 1908	7.5	Assam
12 May 1912	8.0	Assam and Bihar
8 July 1918	7.6	Srimangal, Assam
9 September 1923	7.1	Assam
3 July 1923	7.1	Dhubri, Assam
2 July 1930	7.1	Dhubri, Assam
27 January 1931	7.5	Assam
14 August 1932	7.0	Assam and Burma
15 January 1934	8.4	Bihar and Nepal
16 August 1938	7.2	Chin Hills and Assam
21 January 1941	7.1	North Assam
23 October 1943	7.3	North and South Assam
12 September 1946	7.8	Mandalay, Burma
29 July 1947	8.6	Assam, Bengal, and Bihar
15 August 1950	8.6	Assam and Arunachal Pradesh
21 March 1954	7.2	Assam
16 July 1956	7.0	Epicentre near Burma
6 August 1988	7.3	Manipur–Burma border

Based on the synthesis of various works on tectonic framework of the eastern Himalayas, the region was divided into five tectonic domains (Gupta 2003):

(i) Main Himalayan belt (collision zone) trending E–W to NE–SW
(ii) NNW–SSE trending Abor–Mishmi Complex (the syntaxis zone of the Himalayan and the Burmese arcs)
(iii) NNE–SSW to N–S trending Arakan Yoma belt (subduction Zone) that joins at the south with the high seismicity zone of Sunda arc
(iv) Shillong plateau and Mikir massif marking plate boundary with Assam valley
(v) Bengal Basin and plate boundary zone of Tripura folded belt area

In the western Himalayas, the isoseismals drawn for a number of major earthquakes are found to follow the regional trends (Narula and Shome 1992), but the seismicity trend in the eastern Himalayas indicates alignment along transverse features such as Bomdila and Kopili lineaments. The Tista lineament, which is oblique to the Himalayan belt, defines the western limit of seismicity of the eastern Himalayas.

The peninsular region was once considered to be a seismically stable landform, but earthquake hazard study of the area in the past two to three decades have indicated that the region has been subjected to several high-magnitude (> 6) earthquakes such as the Latur earthquake of 30 September 1993 and the Jabalpur earthquake of 22 May 1997 with magnitudes 6.3 and 6.0, respectively, and MSK intensities IX and VIII, respectively. Of the earthquakes recorded in the peninsular region of India, the Kutch earthquake of 1819 had the highest magnitude of 7.8 and intensity XI. Among the several earthquakes experienced by peninsular India and recorded so far, fifty-six are of magnitudes 5 to 6 and nine are of magnitude more than 6, of which two are of magnitudes 7 and above (Table 20.6). The depths of focus of these seismic events are shallow and confined within 20 km from the earth's surface.

Table 20.6 Earthquakes (M > 5.5) of peninsular India (Pande 1999)

Date	Location	Magnitude
16 June 1819	Kutch, Gujarat	7.8
1 April 1843	Bellary, Karnataka	5.8
19 April 1845	Lakhpat, Gujarat	5.7
19 June 1845	Lakhpat, Gujarat	6.3
27 May 1846	Narmada, Madhya Pradesh	6.5
26 April 1848	Mount Abu	6.0
25 December 1856	Dahanu, Maharashtra	5.7
18 November 1863	Nimar, Madhya Pradesh	5.7
3 July 1867	Chennai, Tamil Nadu	5.7
2 June 1927	Jabalpur, Madhya Pradesh	6.5
14 March 1938	Indore, Madhya Pradesh	6.3
26 June 1938	Paliyad, Gujarat	6.8
21 July 1956	Anjar, Gujarat	7.0
10 December 1967	Koyna, Maharashtra	6.5
14 April 1969	Bhadrachalam, Andhra Pradesh	6.0

(Contd)

Table 20.6 *(Contd)*

14 April 1969	Kothagudem, Andhra Pradesh	5.7
30 September 1993	Latur Osmanabad, Maharashtra	6.3
22 May 1997	Jabalpur, Madhya Pradesh	6.0
26 January 2001	Bhuj, Gujarat	6.9

Seismotectonic evaluation of peninsular India indicates that the region is characterized by intraplate types of earthquakes (Fig. 20.20). The region has varied rock types that include Archaean crystalline, Cuddapah sedimentary metamorphics, Deccan Trap, and Tertiary sedimentaries. The terrains have experienced more than 1000 earthquakes, mostly of magnitudes ranging from 3 to 5, in the past six centuries. The tectonic movement was restricted to two areas, which are the relatively stable craton with large number of seismic events and the relatively active craton characterized by structures such as rifts and graven.

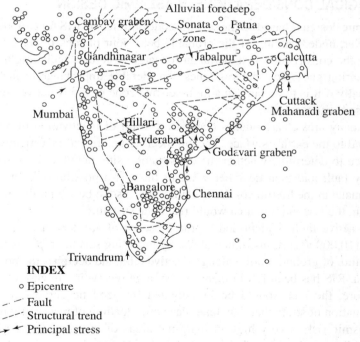

Fig. 20.20 Seismotectonic map of peninsular India (Pande 1999)

The work of Pande (1999) has indicated that from seismotectonic angle, the region can be divided into two units:

(i) Higher seismicity domain indicative of very active and extensive heat sources resulting from strong tectonism in the intracrustal zones

(ii) Relatively stable tectonic domain where events of damaging earthquakes are large and attributed to the compressional region along basement factors. For example, the Jabalpur earthquake took place along the Narmada rift indicative of a relatively deep source.

One interesting feature of some low-intensity seismic events of peninsular India is the rumbling sound. During the five decades from 1950 to 1999, 13 places of this region have experienced microseismicity, which is characterized by tremors known as *swarm earthquakes*, which are sometimes accompanied by a sound. These earthquakes of low intensity and low magnitude occur generally at shallow focal depth, and ground shaking lasts for a very small period of time in seconds. Swarm earthquakes occur as a result of stress adjustment within a fault system where several structural planes intersect and where enough stress cannot accumulate to create even a moderate earthquake.

In Khandwa region of Madhya Pradesh, earthquakes with acoustic phenomenon were observed for 25 years until 1999 and had continued even afterwards. Indian Meteorological Department (IMD) and GSI monitored these earthquakes for five months and recorded 700 microseisms with the peak magnitude of 3.1. The source of these earthquakes is within the soft and saturated red mole of Deccan basalt at a depth of 1–10 km. The occurrence of such swarm seismicity for a continuous period at a place suggests that the area is under the process of stress building that may be a precursor of a major earthquake in future.

20.10 GEOLOGICAL CONSIDERATIONS IN ASEISMIC DESIGN

Aseismic design is based on the seismic code outlined by BIS for earthquake-resistant structures. However, understanding of the probable behaviour of rock under a seismic event is required before the construction is undertaken for large structures including high dams. In this regard, engineering geological data assumes considerable importance in the design of the structure, especially if it is to be built in a high-seismic zone or if any active fault is present close to the proposed structural site.

In many areas, earthquakes occur as a result of reactivation of ancient faults that are not exposed to the earth's surface. The engineering geological investigation should aim at the first instance to determine whether the construction site including its surrounding area is affected by any fault and then decipher whether there is a possibility of its reactivation to generate earthquake in the lifetime of the structure. It is also to be seen if the motion in case of a seismic event in the surrounding area would reach the structure.

An active fault is identified by the presence of any tectonic effect in recent or Holocene times (10,000 years). Instrumental test by installing seismographs is to be taken up to measure the nature of ground motion along an active fault and decipher the possible earthquake ground motion. RIS has been felt in many of the large reservoirs. In case of a large reservoir project, therefore, the area should be investigated by geodetic survey to determine the nature of deformation of strata that may take place after loading of reservoir with water.

Seismic risk is very high in marginal areas of the continental plates. Geological inputs of possible earthquake effects are taken into account in structural design of this part. The Himalayan region with complex tectonic features, especially the terrain in and around MBF and MCT, is highly susceptible to seismic hazards. The 246 m-high dam of Tehri hydropower project is located within a high-seismic risk zone of the Himalayas. Based on the data of detailed geotechnical study on seismic effect, a number of defensive features were incorporated in the design of this dam including liberal free board and flatter slopes, and even the type of the dam was changed from the earlier straight concrete dam to clay-core rock-fill dam. By model test, 'the safety of the dam was checked in actual accelerogram on the severe earthquake parameter with peak ground acceleration value of 1.38 g vertical and 0.72 g horizontal both acting simultaneously and the dam was found to be safe against this earthquake' (Gupta 1999).

An aseismic design is aimed to mitigate or resist the disastrous effect of lateral movements and vibrations of ground associated with an earthquake shock. The parameter of peak ground acceleration or maximum ground velocity is of main consideration in the aseismic design. In practice, the designer takes in to account the magnitude of earlier-recorded earthquakes and the expected maximum magnitude for estimating ground motion parameter, so that in case of any seismic event there will be no damage to the structure. The peak ground acceleration for the various earthquakes located within a distance of 100 km from the focus can be calculated on the basis of attenuation relationship as follows (McGuire 1974):

$$a_m = (472) \times (10^{0.278M}) / (R + 25)^{1.3}$$

Here, a_m is the maximum ground acceleration (cm/sec^2), R is the distance from earthquake focus, and M is the magnitude of earthquake.

Peak ground acceleration up to 0.19 for subsurface structures (e.g., tunnel) located within 80 km distance from the focus of an earthquake of magnitude 8 is considered safe. However, this may not be true for surface structures for which the induced stresses due to seismicity should be considered in the design for safety of the structure. In fact, earthquake causes damage in three ways—ground shaking, slip along a fault if present in the vicinity of a structure, and induced landslide. Geological investigation is taken up to evaluate all these possibilities for consideration in the design of the structure. The normal approach to resist earthquake shocks is to tie the structure firmly with foundation ground. All ground movements are transferred to the structure that is designed to survive the inertial forces of the ground motion. In major structures (say, a large building), these inertial forces can exceed the strength of the structure. The design should ensure that the building is constructed of ductile material so that it may deform but not collapse. Moment-resistant steel frame structures are good for the purpose.

20.11 FORECASTING EARTHQUAKES

Forecasting of an earthquake has not yet reached a very accurate stage. The phenomenon of earthquake prediction involves innumerable unknown parameters that cannot be detected with technical precision. Earthquake scientists have forwarded several suggestions towards finding clues of possible earthquakes as stated in the following subsections.

20.11.1 Measurement by Global Positioning System

The following are the various approaches for forecasting an impending earthquake:

(i) Studies in China and Japan have revealed that certain animals and fish behave very strangely before the occurrence of an earthquake. However, its practical application in forecasting an earthquake is found to be a difficult proposition as it involves keeping a constant watch on such animals.

(ii) Continuous monitoring of the behaviour of the ground around an active fault zone for a sufficient period may provide some fruitful results. If the site of a large structure such as a dam is traversed by a fault, precision survey by repeated levelling is required to determine the possibility of earthquakes. If the survey shows minor displacement or level difference between the two sides of a fault, an earthquake is expected to occur but the time cannot be predicted.

(iii) It has been found that prior to an earthquake, water levels in lakes and wells suddenly go down. Well water may completely disappear. Spring water may become turbid and ground temperature rises before an earthquake. Such observations with respect to water level in wells and springs sometimes give clues of an impending earthquake.

(iv) Prior to a seismic event, due to change in rock property, the velocity of P-waves may change up to 10 per cent. Fluctuation in the speed of P-waves in the record of seismograph installed close to an active fault is, therefore, considered suggestive of a possible earthquake. If the active fault occurs in a large reservoir area, there is every chance of generating an earthquake (RIS) under full reservoir condition.

(v) Geophysical survey by magnetic and electrical conductivity methods in high seismic zones may provide indication of the possibility of earthquakes. In tectonically disturbed terrain or faulted rocks, if magnetic and electrical conductivity tests show anomalies, an earthquake is expected in the area. However, it is argued that the geophysical survey rather proves the presence of a fault and in an earthquake-prone area a fault may be reactive and hence causes earthquake hazards.

(vi) Radon, a radioactive gas, comes out to the surface in higher concentration through an active fault or deep well prior to an earthquake. Nine days before the 1995 Kobe earthquake, a tenfold increase in radon concentration was detected from a place 30 km away from the epicentre (Bolt 1999). The measurement of this increased concentration of radon provides clues of impending earthquakes. However, continuous measurement of radon release in a fault zone for uncertain period of time is not a realistic approach.

(vii) The detection of strain in the rocks of earth's crust by geodetic survey, statistics of frequency of occurrence of earthquakes in time and space, and monitoring of the foreshocks by precision seismographs may provide important clues of earthquake.

(viii) In recent times, measurement of low build-up of tectonic strain in the earth with the help of orbiting satellite enables forecasting of earthquake in specific tectonic belts. Seismologists can also measure movement along major faults using *Global Positioning System* (GPS) satellites to track the relative movement of the rocky crust by a few centimetres each year. This information may help predict earthquakes. Scientists in the US have been able to measure the relative motion of plates by observing changes in earth's magnetic field preserved in ocean crust. 'These rates of movement of plates has been confirmed by direct measurement using space satellite through the GPS and by the relative motion of radio telescope with respect to quasar signals from outermost space. All our information about relative plate motion can be fed into a computer model that tells us the motion of any given plate with respect to any other' (Yeats 2001). This provides a promising approach in earthquake prediction.

(ix) In recent times, an earthquake prediction programme funded by the National Aeronautics and Space Administration (NASA) has accurately predicted 15 of California's 16 largest earthquakes since January 2000. The scorecard uses records of past earthquakes to predict California's most likely locations and time to have quakes of magnitude 5 or above in every 10 years. The 'scorecard' is in fact one component of NASA's QuakeSim project, which develops tools for quake forecasting. It integrates historic data, geological information, and satellite data including high-precision, space-based measurements from GPS satellites and Interferometric Synthesis Aperture Radar (InSAR) with numerical simulations and pattern recognition techniques for updated forecasts of quakes (USGS Website 2004: Earthquake Activity).

20.11.2 Earthquake Disaster Mitigation

It is not possible to prevent an earthquake but its disastrous effects can be mitigated. The impact of collapsed buildings or houses during ground shaking by earthquakes is the main cause of loss of lives. It was found that the Latur earthquake of 1993 had collapsed 90 per cent of houses (mostly brick built) in that area. It is thus the design of the house, including the construction material used, that causes significant damage during an earthquake. Earthquake-resistant structures are constructed in earthquake-prone areas to resist earthquake shakings. The main purpose of an earthquake-resistant house is to protect people inside the structure from the disastrous effects of an earthquake.

Primarily, an earthquake-resistant structure is firmly attached to the ground so that the ground motion is transferred to the structure, which is designed to survive the inertial forces of ground movement. Wherever available, light materials such as wood and plaster are used in the construction of houses. Hollow concrete bricks are also used. BIS has formulated building codes for earthquake-resistant structures for the expected seismic intensity of the terrain as given by the seismic zoning map. It is mandatory to follow these building codes while constructing any structure in areas under seismic zones III and IV.

In seismically active terrain, in addition to constructing earthquake-resistant structures, the place should be connected by good communication system so that in case of a seismic event, amenities can reach in a short time. A disaster management body for educating people about the effects of earthquakes and making advance plans regarding earthquake survival should be in place. A network of instant warning systems is the first and foremost requirement to transmit the news of an incoming destructive earthquake to vulnerable places by fast-moving microwaves.

20.12 CASE STUDY ON A DEVASTATING EARTHQUAKE

The 8 October 2005 earthquake known as the *Kashmir earthquake* was a major one with a death toll of 70,000 and 106,000 were injured. It is said that the earthquake was related to the major tectonic event of collision of Eurasian and Indian tectonic plates. Figure 20.21 shows a map depicting the Eurasian and Indian plates along with the Arabian and African plates. The epicentre of the Kashmir earthquake is shown by the dark solid circle.

The Kashmir earthquake is considered to be the seventeenth deadliest earthquake of all times. The United States Geologic Survey (USGS) measured its magnitude as a minimum of 7.6 in the moment magnitude scale and Japan Meteorological Agency measured it as 7.8 with its epicentre falling at 34°29/35//N 73°37/44//E, nearly 19 km northeast of Muzaffarabad in Pakistan-administered Kashmir. The earthquake affected India, Pakistan, and Afghanistan. Such a high-intensity earthquake is comparable to the major earthquakes of the earth such as the 9.1 magnitude of Indian Ocean earthquake of 2004 (see Section 20.13), 7.5 magnitude of Quetta earthquake of 1935, 7.7 magnitude of Gujarat earthquake of 2001 (see Fig. 20.18), and Sumatra earthquake of 2009.

A total of 147 aftershocks were registered in the first day after the initial quake, with one of magnitude 6.2. As on 27 October, after 19 days of the earthquake, a total of 978 aftershocks were registered. The severity of the damage of the earthquake is attributed to its origin being connected with the severe upthrust and poor construction of buildings and houses. Measurements from satellite have shown the rise of the Himalayas by a few metres as a consequence of this earthquake, giving ample proof that the uplift of the Himalayas is still continuing (Wikipedia http://wikipedia.org/wiki/2005_Kashmir_earthquake).

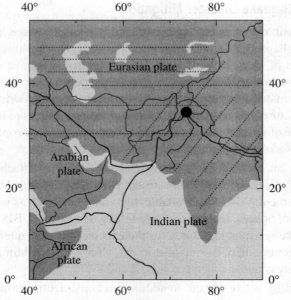

Fig. 20.21 Map depicting the tectonic plates

20.13 TSUNAMIS

Tsunami is a natural phenomenon that causes immense damage of property and lives of people living in the coastal areas of oceans. It has not been a regular feature in India, but in 2004 it affected costal India causing extensive damage. This section discusses the effect of the tsunami of 2004. It also highlights the methods of coastal protection as well as measures to be taken to prevent damages caused by tsunamis.

20.13.1 Origin of a Tsunami

A tsunami is indicative of the destruction that can be caused by ocean waves. The word *tsunami* is the combination of two Japanese words *tsu* (meaning harbour) and *nami* (meaning wave). This natural hazard in the form of large waves of ocean causes destruction in the coastal area. Tsunami is different from tidal waves, which result from the attraction of the sun and the moon. Tsunami waves have larger wavelength, higher amplitude, and greater velocity than tidal waves.

Tsunami originates from an earthquake, which has its focus in deep ocean floor. An undersea volcanic eruption or seabed slide may also create tsunami. In the ocean floor, an earthquake is generated in the subduction zone due to the sliding of one lithospheric plate against another. The energy released by an undersea earthquake of a large magnitude of 7 or above on the Richter scale may cause rupture of subduction zone and sudden rebound of ocean floor giving rise to tsunami.

'A typical tsunami wave has about 10–20 m height, 200 km wavelength, and a speed equivalent to that of an airplane, i.e., 700–900 km/hour in deep waters. Tsunamis contain about 1 % to 10% of the total energy of the earthquake that in case of a great event may release energy equivalent to 1023 ergs equivalent to four million atom bombs detonating simultaneously. Since the velocity of the Tsunami wave is a function of depth ($c = \sqrt{gD}$ where c is the velocity, g acceleration due to gravity, and D the water depth) the wave speed decreases on approaching the shallow continental coasts. As the period remains constant and the velocity decreases, the wavelength gets shortened and the wave amplitude increases from a metre in the deep sea to as much as 30 m in the shores.' (Pande 2005)

20.13.2 Destructive Action of Tsunami

The damaging effects of a tsunami depend on the geomorphic feature of the coastal belt and the density of people and their properties. A tsunami that comes as waves in succession of crests and troughs can cause destruction in two ways. First, the impact of the waves damages everything in their path and large areas bordering the ocean including agricultural lands are inundated by saline water making the land unproductive. Second, the retreat of the waves with considerable speed destructs everything on their path including buildings, household materials, livestock, and even people and ultimately dragging them into sea water. Low-lying lands remain fully inundated by water for a long time even after the ocean water has receded.

The tsunami waves originating in the deep ocean travel at a very high speed. Though the speed of the waves decreases as they reach the coastal land, the wave height increases many times. These high waves hit the coastal areas as troughs and crests in succession within a duration of few minutes. The wave coming in as a trough causes sea water to recede, exposing and scouring the seabed for a long stretch, but within minutes the high wave reaches the shore causing severe destruction of property and life.

The rough configuration of the ocean floor, especially the Pacific Ocean, is susceptible to tsunami. Most of the tsunamis take place in the Pacific affecting the coastal areas of countries such as Japan, Indonesia, Malaysia, Mexico, South America, Philippines, New Zealand, and Hawaii. In 1883, an undersea volcanic eruption of Krakatau Island caused tsunami waves 35 m high, which devastated localities in East Indies killing more than 36,000 people. Japan is the country that is mostly affected by tsunamis. It has been recorded that in 1933 a tsunami in Japan caused the waves to rise to an extent of 30 m killing 299 inhabitants.

The greatest destruction recorded in the history of Japan was caused by the 11 March 2011 earthquake with a magnitude of 9.0 on the Richter scale and the consequent tsunami. The earthquake triggered a powerful tsunami that swept over homes and farmlands in the northern part of Japan. This resulted in the death of many human lives, and nearly 20,000 people went missing. It also set off a nuclear crisis forcing thousands of people to evacuate their homeland.

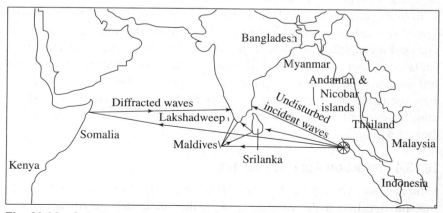

Fig. 20.22 Schematic diagram showing diffracted wave directions of December 2004 tsunami (Jayabalan, Durairaj, and Muraleedharan 2009)

20.13.3 December 2004 Tsunami on Coastal India

On 26 December 2004, the collision of the Indian Ocean plate and the South-east Asian plate (Burma plate) caused displacement rupture of 1200 km length along the Andaman–Sumatra trench creating an earthquake that threw up the ocean floor by 6 m and triggered a tsunami. The earthquake with its focus 30 km beneath the ocean bed released energy equivalent to that of 100 atomic bombs of the type used in Hiroshima. The magnitude of the earthquake was 8.9 on the Richter scale.

The Sumatra earthquake was so destructive that the landmass of Sumatra was uplifted to different extents and shifted in southwest direction by nearly 8 m. The waves of the tsunami travelled at a speed of 500 km/h. The height of two waves nearly 150 m apart in the mid-ocean was only one or two metres. As they approached the shallow coastal area, the waves became as high as 30 m. The giant waves dashed against the coastal land as a high wall and extended to a distance of 1.0–1.5 km inside the land causing devastating effect on people and properties.

The tsunami that swept through the coastal areas of Indian Ocean on that day was the worst in the recorded history of tsunamis of the world. Nearly two lakh people of many countries of Asia and Africa bordering the Indian Ocean died and many more suffered from its destructive effects. Of the 11 countries affected by the disturbances of the tsunami (Fig. 20.22), the death toll recorded includes 122,232 in Indonesia, 30,974 in Sri Lanka, 10,776 in India, and 5395 in Thailand. In addition, thousands of people remain untraced. The other countries that suffered from the effects of this tsunami include Somalia, Maldives, Malaysia, Myanmar, Tanzania, Kenya, and Bangladesh. Since 1767, Indian coastal areas have experienced 15 tsunamis that originated in the Indian Ocean prior to the December 2004 tsunami, but those were not so destructive.

In India, Andaman and Nicobar Islands and the coastal parts of Tamil Nadu, Andhra Pradesh, and Kerala were affected by this tsunami that caused huge loss of property and human lives. The calamity was so severe in the coastal areas of South India that it caused complete collapse or damage of thousands of houses and agricultural lands and the death of more than 6000 people. Many of them were fishermen who were engaged in fishing in the ocean, far away from their coastal habitats. One strange feature of the tsunami is the change in the marine environment caused by the fleeing of fish and other marine species from South Indian coast to other areas. The drastic decrease in fish in the post-tsunami ocean water has become a constant source of worry for thousands of fishermen of the coastal south whose livelihood has been threatened. Another tsunami-created feature is the covering of the rocky ocean bed of the coastal belt by sand and silt deposits of enormous thickness.

In Andaman and Nicobar Islands, the giant waves of the tsunami moved with great speed and deeply scoured the coastal landform and also brought about a change in the sea floor profile. The land was uplifted or subsided at places. Eruption of mud volcanoes and reactivation of old faults were other features observed during the tsunami. About 1800 people of these islands died and another 5500 persons remain untraced. Large stretches of agricultural land along the shore were submerged by sea water as a result of the high waves. The ocean water receded after some time, but even after two years the land was not usable for agricultural purpose as the salinity of the soil was high.

20.13.4 Protection Against Tsunami

India had not experienced the destructive effects of a tsunami prior to the 26 December 2004 tsunami that lashed the coastal states of the south and the Andaman and Nicobar Islands bordering the Indian Ocean. This has led to the realization that tsunami is a natural hazard that should not be ignored as India is prone to further tsunamis and may have to experience its destructive effects again. The gigantic trench (fault) below the Indian Ocean created during this tsunami extends up to the Andaman Islands. This fault may be reactivated or a new rupture can take place in future to cause an ocean bed earthquake that triggers another devastating tsunami.

It needs to be understood that the important factor is not the source of the tsunami but the steps taken to protect life and property against its destructive effects. Whether it will occur again is not of significance but getting warned about it on time is. Observation of the condition of waves in mid-ocean during a tsunami can predict the danger of an incoming tsunami. Japan, whose coastal region is frequently affected by tsunamis, has placed several seismograph systems on the ocean bottom covering the shelf and deep ocean parts for a distance of 200 km from the coast to get warning signals of the advancement of a tsunami before it strikes the land.

When any earthquake is generated in the ocean bed, the tsunami centre at the shore station instantly records the shaking, and in case of an impending tsunami immediately informs coastal inhabitants through a well-grid warning system. India has gained technical knowledge in this respect and is in the process of placing seismographs and warning systems to monitor round-the-clock activity of the tsunami waves and storm surges in the Bay of Bengal and Indian Ocean and provide warning to the coastal inhabitants.

Protection from tsunami should be taken up along with coastal management (see Section 8.9). Construction of buildings and other engineering structures and human habitation should not be allowed in the coastal areas as tsunamis may cause extensive destruction of life and properties of people in such areas.

SUMMARY

- All major earthquakes are associated with the movement of lithospheric plates that create rupture in rocks in the interior of the earth and consequent shaking of the ground. The earthquakes are detected in seismograms that record the seismic waves generated from the place of its origin called focus. The place on the earth's surface vertically above the focus is the epicentre of an earthquake.

- There are three types of seismic waves, namely primary (P) wave, secondary (S) wave, and surface (L) wave. The P- and S-waves are body waves that move through the interior part of the earth, whereas the L-waves travel through the top layer. The wave curves of a seismogram help to calculate the distance of the epicentre from a seismograph station. The actual location of the epicentre is known from the plotting of epicentre distances from at least three seismograph stations.

- The magnitude of an earthquake is a measure of its released energy expressed by the Richter scale. A human being can feel an earthquake of magnitude 3.5; however, damage is caused by an earthquake of magnitude 5.5 and above and an earthquake of magnitude 8 or above may bring about severe damage covering areas of 100 km around the epicentre.

- The intensity of an earthquake measures its impact on people, land, and man-made structures. The intensity is expressed in Modified Mercalli (MM) scale or Medvedev–Sponheuer–Karnik (MSK) scale, both having levels I–XII in increasing order of intensity. The major disturbance starts from intensity level IV culminating in the total damage of all structures in intensity level XII of the MM and MSK scales.

- An earthquake may be generated due to the stress imposed by the impounded water of a dam on the reservoir floor. This is known as RIS. In India, in the Koyna project of Maharashtra, an RIS of magnitude 6.3 on the Richter scale shook the dam and the grounds around. In case of a fault traversing the reservoir area, it may also be reactivated under full reservoir condition and generate RIS.

- Based on the past earthquakes, Indian landform has been grouped under four earthquake zones from zones II to V in the order of increasing seismic disturbances. BIS has formulated codes for earthquake-resistant structures for these zones.

- In the Himalayan terrain, most of the major earthquakes today occur by movement of rocks along faults and thrust of the region. The areas bounded between MCT and MBF are the most susceptible to earthquakes.

- The peninsular plateau earlier considered to be a stable landmass has been affected by several earthquakes, some of which have magnitudes as high as 6.
- Seismotectonic evaluation of the Himalayas and peninsular plateau gives an insight of the origin of earthquakes in relation to tectonic features and helps designing earthquake-resistant structures in these terrains.
- Since it is not yet possible to predict in advance the occurrence of an earthquake, it is necessary to take steps towards mitigating the destructive effects of an earthquake by constructing structures following the earthquake-resistant codes recommended by BIS in high-seismic zones.
- Tsunami is a natural hazard created by an undersea earthquake or volcanic eruption or by a large slide of marine bed.
- Collision of lithospheric plates causes rupture of subduction zone and generates earthquakes of high magnitude. The energy released by the earthquake causes rebound of ocean floor giving rise to tsunami in the form of high waves with great speed that lash the coastal landform leading to severe destruction of property and loss of lives.

- An earthquake of magnitude 8.9, originated in the Indian Ocean by the collision of the Indian plate with South-east Asian plate, triggered the tsunami of 26 December 2004. It lashed 11 coastal territories bordering the Indian Ocean and ruptured the Indian Ocean floor creating a trench 1200 km long.
- The Andaman and Nicobar Islands and coastal parts of India were severely affected by the tsunami that destructed property and caused the death of thousands of people.
- Protection of coastal landform is necessary to prevent the destructive impacts of tsunami. This can be effected by the settlement of human habitation away from areas very close to the coast and protecting the coastal landform by constructing concrete sea walls or use of groynes.
- Measures that act as a warning system against an incoming tsunami include installing a series of seismographs in the Indian Ocean and the Bay of Bengal from the shelf to deeper parts of the ocean and devising systems to instantly transmit news of an incoming tsunami to the coastal areas.
- India is already in the process of placing seismographs in oceans and taking measures to protect the coastal inhabitants from the destructive effects of tsunamis.

EXERCISES

Multiple Choice Questions

Choose the correct answer from the choices given:

1. Of the following, the wave that moves like a ripple in water surface is:
   (a) P-wave
   (b) S-wave
   (c) Rayleigh wave
2. Choose the correct earthquake magnitude in Richter scale (from the given numbers in bracket) for the conditions of earthquake shocks as mentioned in (i), (ii), (iii), and (iv):
   (i) Well-built buildings were slightly affected but poorly constructed buildings are seriously damaged (Richter scale 3.5, 4.5, 5.5, and 6.5).
   (ii) Serious damage of buildings was caused by earthquake shocks covering populated areas of about 100 km across (Richter scale 5, 6, 7, and 8).

   (iii) The earthquake shock caused rattling noise but not significant damage (Richter scale 2.9, 3.9, 4.9, and 5.9).
   (iv) An area covering nearly 500 km across was seriously damaged (Richter scale 7.2, 7.9, 8.0, and 8.7)
3. Reservoir-induced seismicity is related to:
   (a) plate tectonic event
   (b) compression of water of large reservoir water under full reservoir condition
   (c) additional pressure or stress due to impounded water
4. An Indian project where RIS was strongly felt within a few years of reservoir filling is:
   (a) Bhakra dam-reservoir project of Punjab
   (b) Koyna dam-reservoir project of Maharashtra
   (c) Tehri dam in the Himalayas

5. The condition responsible for generation of RIS is:
   (a) water body of reservoir of more than 100 m depth
   (b) excessive leakage of reservoir water
   (c) Tectonic history of the region
6. RIS is not observed in the reservoirs projects of Himalayan areas because of their location in:
   (a) a thrust fault area
   (b) in Siwalik rocks
   (c) a seismic zone
7. A tsunami can be generated by:
   (a) flow of volcanic lava over the ground surface in the sea coast
   (b) underwater volcanic eruption
   (c) underwater earthquake
8. The great tsunami of the Indian Ocean swept the coast on:
   (a) 15 August 2004
   (b) 1 January 2004
   (c) 26 December 2004
9. The effect of tsunami can be prevented by:
   (a) settlement of habitation away from the coastal area
   (b) protection of coastal land
   (c) placing seismographs in ocean
10. Beach protection can be effected by:
    (a) construction of concrete wall along coast
    (b) providing groins
    (c) providing sand dunes

Review Questions

1. Enumerate the causative factors of an earthquake. Illustrate and explain different types of seismic waves, focus, epicentre, and focal depth of an earthquake caused by block movement.
2. Draw a diagram and show the different parts of the earth's interior with respect to rock condition and thickness of different parts. State the variation of the earth's temperature with depth.
3. What is a seismograph? Explain how an earthquake is located by a seismograph. What was the method of measuring the earthquake effects prior to the invention of the seismograph?
4. What is a seismogram? Illustrate and explain how the epicentre of an earthquake can be located from the seismogram data.
5. State briefly the concept of continental drift. Define plate tectonics and discuss how plate tectonic theory explains the different processes of earthquakes.
6. Enumerate the basic differences between the Richter and MM scales. How many divisions are there in each of these earthquake scales? At what intensities of earthquake in MM scale are the masonry structures damaged and underground pipes twisted?
7. What is seismic zoning? How many seismic zones are shown in India's seismic zoning map? Highlight the earthquake intensity of each of these zones. State the seismic zones that cover the Himalayan terrain.
8. Write a short account of the seismotectonic framework of India. Name some major earthquakes of Himalayan belt with their corresponding magnitudes.
9. What is the current status in earthquake fore-casting? Explain GPS and its role in earthquake prediction.
10. State the measures that are necessary to adopt safety measures against an earthquake's disastrous effects.
11. What does the term 'tsunami' mean? How is tsunami generated in the ocean?
12. Write an account of the destructive action of a tsunami with reference to that of 26 December 2004.
13. State the types of measures required for sea coast protection to save life and property from the danger of tsunami.
14. What is the warning system related to tsunami? How will the system help to save the life of people?

Answers to Multiple Choice Questions

1. (c) 2. (i) 5.5, (ii) 6, (iii) 4.9, and (iv) 8.0 3. (b) 4. (b) 5. (a) 6. (a)
7. (b) and (c) 8. (a) and (c) 9. (a) and (b) 10. (a), (b), and (c)

21 Landslide Evaluation and Mitigation

● ● ● ● ● ● ● **LEARNING OBJECTIVES** ● ● ● ● ● ● ●

After studying this chapter, the reader will be familiar with the following:

- Different types of landslides and their characteristics
- Geological, man-made, and natural causes of landslides
- Methods of investigation of landslide-affected areas and unstable regions
- Preparation of landslide hazard zonation map and mitigation measures
- Major landslides of India with reference to their causes and effects

21.1 INTRODUCTION

Landslides bring about immense loss of property and even loss of human lives. Geological and human actions and to a certain extent natural causes are responsible for causing landslides in soil mass and rocks, especially in hill slopes. Investigation methods of landslides include searching for their causes, identifying measures to restore the damaged structures, and suggesting measures to ensure future safety. Landslide zonation map is prepared to highlight the areas of instability where there is a chance of occurrence of slide. This chapter describes all these aspects and explains with illustrative figures landslide mitigation measures such as modification of slope geometry, drainage arrangement, and reinforcement of slope.

21.2 HAZARDS OF LANDSLIDES

Landslide is a geological phenomenon involving downward movement of large quantities of materials such as rock, earth, sand, and clay in any combination (Fig. 21.1). The movement of materials may be very slow from a few millimetres to a few centimetres per year, when it is called *creep*. In a *flow* type of landslide, the materials may move very fast, even over 15 km/h. The collapse of the masses in a landslide may also be very sudden as in an *avalanche*, especially in steep hill slopes of high mountainous terrain.

Landslides bring about changes in the landform and affect people by causing massive damage of buildings and loss of crop and livestock and even human life. Landslides may disrupt the functioning of powerhouses, waterways, and

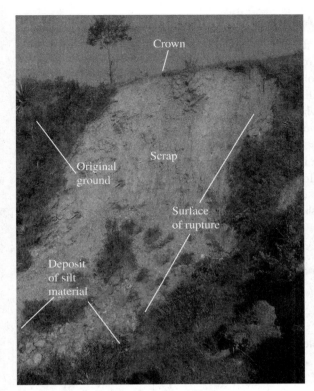

Fig. 21.1 Landslide of a steep hill slope

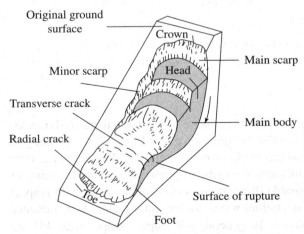

Fig. 21.2 Diagram of a landslide (rotational slide in soil) illustrating different parts (cf. Fig. 21.1)

sewage disposal systems and cause destruction of railways, roadways, and even dams. Apart from affecting people, landslides inflict enormous economic loss to the general public and government organizations by way of defraying cost in restoration, reconstruction, repairing, and related post-landslide works and also treatment of affected people. The different parts of a landslide with commonly accepted nomenclature is illustrated in Fig. 21.2 by means of a soil slide along a hill slope. In hilly areas with steep natural slopes or cut slopes made for construction of roads and highways, such slides are very common. An entire road surface along the hill slope may collapse, or the materials from the crest or sides of the hill face may fall on the road causing road blockage.

A hill slope may also slide downhill where road formations are undercut by flowing streams. The slope materials consisting of unconsolidated or loose earth and jointed rocks are more prone to landslides than the consolidated dense soil and competent rock. Apart from hilly areas, landslides also occur in the plains or low relief areas after excessive rains, especially in places such as quarry sites, open cut mines, river banks, dumping grounds of mine wastes, and roads constructed on cut-and-fill materials.

The effect of a landslide cannot be judged from its severity alone. A catastrophic landslide in a remote area may dislodge and bring down huge mass of rock and soil from the hill slope but without any loss of human life, whereas landslides of not-so-severe nature in a populated area may lead to disaster. Mudflows and debris avalanche, which are common in mountainous regions, also cause damage to life and property.

Landslides accompanied by heavy rains may be of catastrophic nature, moving at a high speed and flowing with debris, devastating natural and man-made structures that are on the way. An earthquake-triggered landslide in hilly areas frequently causes blockage of road by the slide materials causing transport disruption. A road cutting through faulted and fissured rocks is liable to cause huge rock falls, damaging road formations and causing extreme difficulty to resident people as well as travelling public.

21.3 LANDSLIDE TYPES—CLASSIFICATION AND DESCRIPTION

Landslides are classified in different ways depending on the various essential factors involved in the slide mechanism such as the nature of materials, depths of sliding, speed of mass movement, and nature of the failure plane. The most obvious factor is the nature of the masses or materials such as rock, debris, and earth that are displaced from the parent body due to the landslide. The mechanism of material movement, which includes *falling*, *toppling*, *sliding*, or *flowing*, is also a very important factor in the landslide process. Combining these two factors—the types of materials and the mode of movements—the United States Geological Society (USGS) has provided a classification (USGS 2004) that is widely used in landslide investigation. This scheme of classification, provided in Table 21.1, is followed in describing the different types of landslides.

Table 21.1 Classification of landslides (Cruden and Varnes 1996)

Type of movement		*Type of material*		
		Bedrock	*Engineering soils*	
			Predominantly coarse	*Predominantly fine*
Falls		Rock fall	Debris fall	Earth fall
Topples		Rock topple	Debris topple	Earth topple
Slides	Rotational	Rock slide	Debris slide	Earth slide
	Translational			
Lateral spreads		Rock spread	Debris spread	Earth spread
Flows		Rock flow (Deep creep)	Debris flow (Soil creep)	Earth flow
Complex		Combination of two or more principal types of movement		

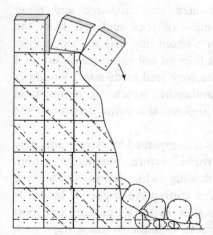

Fig. 21.3 Fall—free fall of detached blocks of rock from steep slope

21.3.1 Falls

Falls are abrupt movement of masses of geological materials (rocks, soils, and debris) that become separated from the parent body in steep slopes and fall freely or bounce and roll down under gravity (Fig. 21.3). Differential weathering, erosion and undercutting of slope, unfavourable disposition of divisional planes with respect to slope, and interstitial water are the main factors that influence the fall of masses. In general, hill slopes steeper than 45° are generally involved in falls. The failure surface in falls is planar, wedge shaped, or vertical (Fig. 21.4).

An example is the well-jointed metadolorite that remained exposed in one of the high walls of Barapani (Meghalaya) tunnel after its roof collapse covering a wide area and repeatedly caused rock falls.

Another example is the high rock falls that occurred along the Sikkim highway maintained by the Border Road Authority, which

Fig. 21.4 Rock fall of jointed rocks, the failure surfaces being mostly vertical

interrupted traffic for quite some time. Here, the rock fall was triggered by the blasting for road extension in highly jointed arenaceous phyllite (see Fig. 21.17). The devastating Malpa landslide of 1998 in Uttaranchal was mainly a rock slide.

21.3.2 Topples

Toppling of the soil or rock blocks takes place along planes of discontinuities by forward rotation with respect to the point below the sliding mass. The toppling failure is characterized by end-over-end motion of rocks (Fig. 21.5). The rocks involved in a topple move by action of gravity or forces exerted by adjacent blocks. When the centre of gravity (CG) of rock blocks fall outside the rock mass, then there is a possibility of toppling. The forces for topple of blocks may be created by water or ice in cracks in the mass. The initial failure surface may be single or multiple (Fig. 21.6).

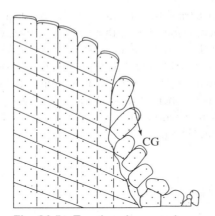

Fig. 21.5 Topple—downward movement of rocks by forces exerted by adjacent blocks

Fig. 21.6 Topple of rock blocks from a dam abutment along multiple joint planes

Examples of topple-type slide of jointed rock blocks are seen in joint locks in parts of road cuts (highway) from Jammu and Kashmir where snow penetrating into the joint openings widens the gaps and causes toppling due to gravity. The toppling of decomposed rocks shown in Fig. 21.6 was observed in one of the abutments of a Bihar dam project after the dam and abutments were excavated.

21.3.3 Slides

The slide type of failure may take place in varied types of materials including rock, soil, and debris. Slides in rock always occur along well-defined surfaces with down slope movements. Faults, shear zones, joints, and bedding or foliation planes of rock are usually the planes of

Fig. 21.7   Rock slide

weaknesses along which the masses slide. The two main types of landslides under this category are rotational slide and translational slide.

Rotational slide

In this type of slide, which basically occurs in soil zone, the surface of rupture is curved concavely upward, and the movement is caused by rotation of the slid material about an axis parallel to the ground slope and transverse across the slide direction. Rotational slide usually shows steep scarp at the upslope end and bulging of toe of the slid material at the bottom (Fig. 21.1). The materials may creep slowly, or there may be a sudden collapse or *avalanche* of large quantities of materials that move over long distances. Figure 21.7 shows a rock slide in parts of Sikkim, where a huge mass of rock collapsed from a scarp face and rolled down a long distance.

Translational slide

During transitional slide movement of materials takes place along a more or less flat (planar) surface, sometimes with a slight backward tilting (Fig. 21.8). The surfaces of ruptures are usually the contact planes between firm rock and overlying loose earth, bedding planes of layered rock, and other weak planes in rock or earth deposit. A translational slide may be a *block slide* involving down slope movement of the sliding mass as a single unit or few closely related units as a coherent mass. Block slides are usually seen in bedded rocks in which large blocks of rock move along the dipping surfaces of beds or along divisional planes such as faults and joints (Fig. 21.9).

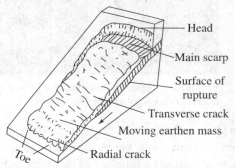

Fig. 21.8   Translational slide with movement of materials along a nearly flat surface

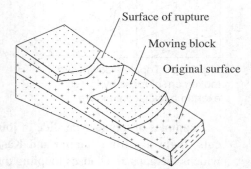

Fig. 21.9   Block slide of rock mass parallel to bedding plane

21.3.4 Lateral Spreads

Lateral spread type of mass movement occurs in flat or low relief areas in saturated soil consisting of loose sediments (sand and silt) that fail by liquefaction. Loose sediments underlying and overlying firm and coherent mass (dense soil or rock) may liquefy when fractures are developed in the overlying coherent materials due to shear and tensile stresses. The materials accompanied by shear and tensile fractures then fail by slumping and move in a lateral direction and hence it is termed *lateral spread* (Fig. 21.10). The failure is triggered by ground motion due to earthquakes or induced vibration.

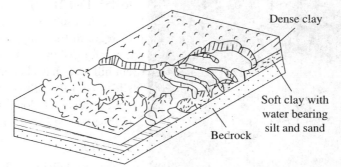

Fig. 21.10 Lateral spread showing lateral movement of the materials by liquefaction

21.3.5 Flows

The movement of masses is like the flow of fluid plastic in viscous state. The movement may be a complete run out of the source materials of catastrophic nature. Depending upon the material involved and the rate of movement, flows are classified into four groups, namely rock flow, debris flow, earth or mudflow, and creep.

Rock flow

'Flow movements in bedrock include deformations that are distributed among many large or small fractures, or even micro fractures, without concentration of displacement along a through-going fracture' observed Varnes 1978. Hungr et al. (2001) described the rock flow as extremely rapid, massive, flow-like motion of fragmented rock from a large rock slide or rock fall. The rock slope involved in rock flow may be 45° to 90°.

Debris flow

Debris flow is the most destructive type of landslide and results after heavy rains or extensive melting of snow in the source area due to supersaturation of the ground with water. The materials involved in the debris flow include large quantities of rock fragments or boulders and loose soil with sand, silt, clay, and organic matter, the fine particles (2 mm or less) being less than 50 per cent. The masses in the debris containing large quantity of water moves along the hill slope generally within 10–20 km/h, but it may move at a faster rate depending upon the water content in the slurry as well as the terrain morphology.

The debris flow is commonly initiated at the steeper part of the hill slope, and with the flow continuing downhill, the volume of materials increases by addition of water, earth, boulders, and vegetation including trees. As it enters flat ground, the materials of debris flow spreads up as a fan encompassing a large track of land (Fig. 21.11). Debris flow may even dam a stream and block flow of water. The bursting of such a dam creates flash floods leading to further problem of ground slope failure. The variety of debris flow in which the movement of the material is sudden and very rapid (> 5 m/s) is called

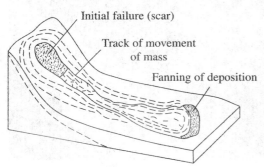

Fig. 21.11 Debris flow showing pattern of movement of materials

Fig. 21.12 Debris avalanche of loose overburden materials from a high hill

debris avalanche. Debris avalanche is common in hilly terrain with slopes having thick deposit of loose overburden materials and inclination of 20° to 45° (Fig. 21.12).

As a result of cutting of the hills consisting of regolith (decomposed rocks) with overburden cover for making highways and roads, many of the cut hill faces have become unstable. Hence, many slides (mostly debris slide or debris flow type) are encountered along the Sikkim and Darjeeling highways (see Fig. 21.12 and Section 21.8).

Earthflow

Heavy rainfall in the hilly terrain of a moderate slope constituted of soil or earth deposit is the immediate cause of earthflow. The fissures developed by the movement of earth materials help infiltration of water into the ground. The ingress of large quantity of water increases the pore water pressure and reduces the shear strength of the soil deposit, resulting in sliding of the slope materials as slurry of earth or mud with water (Fig. 21.13). The materials involved in earthflow are mainly fine-grained particles (clay, silt, and fine sand), but certain percentage of coarser particles (debris) are usually present in the flowing mass.

The difference between debris flow and earthflow depends upon the content of fines in the flowing masses. In earthflow, more than 20 per cent of the materials by weight contain particles finer than 2 mm in size (Fig. 21.14). If the earthflow contains materials as high as 50 per cent of fines, it is normally called *mudflow*. The rate of material movement in earthflow is slow to rapid (> 1.8 m/s or nearly 6 km/h or more), whereas in the mudflow it is rapid to very rapid (> 5 m/s or over 18 km/h). Compared to earthflow, mudflow contains significantly greater water content to facilitate faster flow (Hungr et al. 2001).

Examples of mudflow are many as seen in Fig. 21.14. In Norway, in earlier years mudflow in hill slopes used to be initiated due to the presence of quick clay (sensitive clay) band or lens in soil. Any

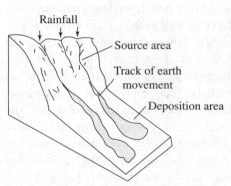

Fig. 21.13 Earthflow with movement of the material as slurry of earth

Fig. 21.14 Earthflow mixed with debris after heavy rains in Darjeeling

vibration will cause this clay to expand and the soil mass moved as flow over long distances along with overlying trees or man-made structures including houses. New techniques of soil protection by electro-osmosis have prevented such slides.

Creep

In creep, the slope materials move down slope at a very slow rate (few millimetres to few centimetres per year). The shear stress in the materials induces the deformation responsible for the slow movement but it cannot cause shear failure. Depending upon the operative process, the creep type of mass movement may be continuous, progressive, or seasonal in nature.

Creep in a slope of rock or earth can be recognized from bending of trees, tilted electric poles, and retaining walls or fences out of alignment (Figs 21.15 and 21.16). Layered bedrock below the soil slope also bends downwards due to creep resulting in ripple-like features in the soil.

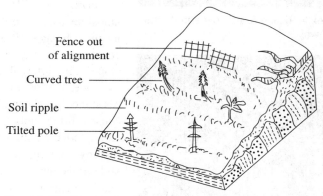

Fence out of alignment

Curved tree

Soil ripple

Tilted pole

Fig. 21.15   Creep of soil cover landmass (cf. Fig. 21.16)

21.3.6 Complex

Severe landslides may take place by the combination of two or more types of material movements, namely falls, topples, slides, flows, and lateral spreads. This combined effect of mass movement is termed *complex*. Such slides of combined nature are distinctly visible in many highways.

Some of the slides investigated in Sikkim, Darjeeling, and Kashmir highways are of complex type being debris slide combined with earth slide or rock slide. Figures 21.17 and 21.18 show the disturbingly famous B-2 and Pagla slides of Sikkim and Darjeeling highways, respectively, where landslides take place almost every year with the onset of monsoon. See the figure in Section 21.4.2 for a slide in an engineering construction site, which is mainly of man-made origin.

Fig. 21.16   Soil cover of hill slope under creep showing bending of trees

21.4 CAUSES OF LANDSLIDES

Two types of forces operate landslides, namely driving forces and resisting forces. The driving forces (shear stresses) act to cause movement of the masses whereas resisting forces (shear strength) tend to resist the slides. Landslides occur when combined driving forces within

Fig. 21.17 Complex landslide (earthflow and debris slide) in Sikkim highway

Fig. 21.18 Complex landslide (rock and debris slide) in Pagla Jhora in Darjeeling highway

a land slope exceed the forces of resistance of the materials that constitute the slope. Gravity is the principal driving force of landslide movement.

The other forces involved in the movement of masses include earthquake, water filtration, freezing and thawing action, and volcanic eruption. These forces aid gravity to overcome the resistance offered by the ground material and initiate its movement. Most damaging landslides are related to natural causes whereas some are due to human actions.

Terzaghi (1950a) described the causes of a landslide as *external* and *internal* with respect to the changes in slope morphology and its constituent materials enforcing instability of slope. The internal causes are responsible for the modification of materials and reduction of resistance to shear stress, whereas the external causes bring about a change in the ground slope inducing an increase in shear stress and thereby making the slope body vulnerable to slide movement. In fact, the internal and external causes mentioned by Terzaghi that are responsible for land instability are geological or geology dependent.

In the overall consideration, the causes of landslides can be grouped under three major categories, namely geological, man-made, and natural.

21.4.1 Geological Processes

Geological processes are primarily responsible for causing morphological change of land slope and instability of landmass in different ways as follows:

(i) Erosion brings about a change in morphological features including steepening of land that are prone to slides. Erosion also helps removal of cement and fines from materials that weaken the strength of slope materials resulting in slope failures.

(ii) Weathering of rock changes the physical and chemical properties that bring about reduction of shear strength, thereby promoting landslides.

(iii) Divisional planes in rocks such as adversely oriented faults, joints, and other planes of discontinuities often cause rock fall type of landslides.

(iv) Swelling and expansion are also associated with upheaval resulting in landslides.

(v) Shearing, jointing, and fissuring are other geological factors that weaken the strength of materials and promote landslides.

(vi) Leaching of certain ingredients from the earth creates sensitive clays (e.g., quick clay), which may cause serious landslides (Section 5.9).

21.4.2 Human Actions

In many cases, landslides are created by human intervention as evident from the following examples:

(i) Cutting of ground slope for various constructions such as buildings, roads, railways, and other utility purposes, especially in hilly areas, leads to slope failure and movement of slope-forming materials including fractured and weathered rocks (Fig. 21.19).

(ii) Construction for human settlement in unstable areas is responsible for many landslides.

(iii) Blasting during mining or rock quarrying weakens the stability of land slopes and triggers their failure.

(iv) Vibration of machineries during engineering construction is another reason to initiate failure of unconsolidated materials.

(v) Dumping of rocks or soil in a place after digging of ground or stockpiling of ore develops stress in weak underlying materials and initiates their slide movement.

Fig. 21.19 Rock-cum-debris slide caused after slope cutting for construction purpose

(vi) Vegetation and tree roots help in the binding of shallow soil deposit with the underlying bedrock. Cutting of trees and removal of vegetation cover are responsible for sliding of soil deposit over the bedrock.

(vii) Overgrazing is also a common cause of slide movement of unconsolidated soil deposits.

(viii) Water leakage from utilities is another factor responsible for slides.

21.4.3 Natural Causes

The following are the natural causes responsible for many of the landslides:

(i) Landslides are mostly associated with heavy rainfall. Saturation of ground by heavy rains or rapid snowmelt is responsible for the mudflows or debris flow including debris avalanche.

(ii) Erosion or undercutting of slopes by rivers results in slope failures in river valley areas. Wildfire may also cause denudation of hill slopes and lead to failure of slope consisting of soil and regolith.

(iii) Earthquakes induce stresses that bring about failure of weak slopes sometimes as liquefaction when material moves like a liquid.

(iv) Filtration of excessive water to ground causes groundwater pressure that often acts to destabilize the ground slope.

(v) Volcanic eruption that causes accumulation of loose materials and ash later generates extensive debris flow with water pressure due to rains.

(vi) Ocean waves may also create oversteepening of seashore and bring about coastal slide of the shore materials.

(vii) Freezing and thawing processes in high-altitude regions may generate loose slope materials.

(viii) Sometimes, the action of thunder may trigger failure of unconsolidated ground materials.

21.5 INVESTIGATION OF AREAS AFFECTED BY LANDSLIDES AND SLIDE-PRONE AREAS

In the investigation of the slide-prone areas and areas where landslide has taken place recently, aerial photographs and satellite imageries are studied at the first instance followed by visits to the area for ground checking. The type of landslide or the mechanism responsible for slides can be decided only after ground survey. The ground inspection provides signatures of 'minor details' that are not recognizable from the photographs. For example, ground cracks or defined line of ground shifting covered by plants or bushes that bear testimony of landslides may be difficult to visualize from the study of photographs.

A ground with some lateral shifting is indicative of old landslide effect. There may be some downward movement or subsidence of ground that delineates the boundary of the sliding surface of old landslides. In the study of the geomorphology, the disturbances are clearly brought out including features such as the subsidence of the top and heaving of the toe of a landslide surface. The slip surface when exposed is studied in relation to subsidence and the presence of slickenside in the planes of the landslide-affected mass. Slickenside in the ground within the boundary of a landslide is a definite proof of mass movement. Slipped block at the head is a clear clue of slide caused by rotation of ground. Anomalous vegetation cover provides the clues for the area being affected by old slides. Old slide scars or cracked grounds bear testimony to the area being affected by landslides in the past. An area of old landslide may be covered by plants but features such as inclined trees, tilting of fencings, and electrical poles bear evidence of landslide and creep. Land slope with extensive toe cutting is an area prone to probable landslide. Movement of talus material or sand fills exposed in the hill slopes are indicative of instability.

The general approach in field investigation of landslide is the preparation of a contour plan and plotting of geological features including rock structures covering the area affected by the landslide. The causes of the landslide are evaluated and treatment of the slide-affected area for reconstruction is proposed or remedial measures to contain future slides are suggested to the engineers. The proposed measures are generally based on experience. It is, however, necessary to carry out comprehensive field investigation that includes study of the landslide pattern, delineation of the boundary of slide-affected zone, and revelation of the slip surface of the slide. The ground fissures, displacement or shifting of the strata, springs with oozing water, and so on that are created by landslides are to be traced and recorded in the geological map. It is to be ascertained whether the landslide was triggered by an earthquake, and if so, the intensity of the earthquake is to be known.

The slide mechanism with respect to shearing strength of the materials before and after the slide, pore water pressure, and friction angle and other parameters of the material involved in the slide are also to be studied. The shear strength parameters and other engineering properties of rock or soil of the area subjected to slide are to be estimated by laboratory and in-situ tests for designing effective measures by the engineers. The sampling should be representative for determination of strength and other parameters by laboratory tests. Today, several instruments are available to carry out in-situ tests. The estimate of shear strength parameters and other test results are required for stability analysis. Depending upon the analysis, the corrective measures are designed with the appropriate safety factor to contain the slide and stabilize the affected area.

21.6 Landslide Hazard Zonation Mapping

The preparation of landslide hazard zonation (LHZ) map is an important step towards delineating areas susceptible to landslides. Different parts of the country have been affected by large numbers of slides in the past, and several other parts are prone to landslides. Identification of such areas affected by landslides in the past and those that may be affected in future is the objective of LHZ mapping. These maps can help the community to construct houses and other important structures such as schools, hospitals, and powerhouses away from areas of potential landslide risk. The landslide zonation map also helps in engineering planning for mitigation means against the disastrous effects of landslide and in taking decisions towards development work in the unstable areas under a strict regulatory board.

Several methods are available for evaluating the landslide susceptibility parameters of an area and preparation of LHZ maps based on these parameters. The guidelines given by the Bureau of Indian Standards (BIS) code of 1998 and modified BIS code of 2004 are mostly followed in preparing LHZ maps of our country (see BIS code through IS 14496 Pt II). The guidelines in BIS code are based on six causative factors related to lithology, structure, slope morphology, relief, land use and land cover, and hydrology. The value 1 or 2 is assigned as landslide hazard evaluation factor (LHEF) to each of these causative factors in BIS code, the total LHEF being 10 (Table 21.2). This BIS code modified (2004) by GSI includes 10 causative factors with 14 estimated total LHEF (Table 21.2). The scale used in preparing LHZ map depends on the requirement and is generally 1:50,000 to 1:25,000 in macro LZS map and 1:10,000 to 1:5,000 in meso or micro LHZ map.

Table 21.2 Causative factors and LHEF for BIS and modified BIS codes

Causative factors (BIS)	LHEF	Causative factors (modified BIS)	LHEF
Lithology	2	Lithology	2
Structure	2	Structure	2
Slope morphology	2	Slope morphology	2
Relative relief	1	Relative relief	1
Land use and land cover	2	Land use	1
Hydrological conditions	1	Land cover	1
		Hydrological conditions	1
		Rainfall	1
		Landslide incidences	2
		Slope erosion	1
Total	10	Total	14

The values of total estimated hazards considered for differentiating the intensity of five landslide zones in BIS and modified BIS methods are shown in Table 21.3. The practical approach in preparing LHZ map is to divide the area into several facets marking the total estimated LHEF on each facet. Depending upon the total estimated hazard value of each facet, the entire area is classified into five hazard zones starting from very low hazard to very high hazard.

Table 21.3 Landslide zones for total estimated hazards in BIS and modified BIS methods

Total estimated hazards in BIS code	Total estimated hazards in modified BIS code	Description of zones
< 3.5	< 4.9	I. Very low hazard
3.5–5.0	4.91–7.0	II. Low hazard

(Contd)

Table 21.3 (*Contd*)

Total estimated hazards in BIS code	Total estimated hazards in modified BIS code	Description of zones
5.1–6.0	7.1–8.4	III. Moderate hazard
6.1–7.5	8.41–10.50	IV. High Hazard
> 7.5	> 10.50	V. Very high hazard

Figures 21.20 (a) and (b) show LHZ of two areas prepared following the guidelines of BIS code (refer BIS code number for LHS, IS-IS14496, in Pt II, 1998, modified 2004).

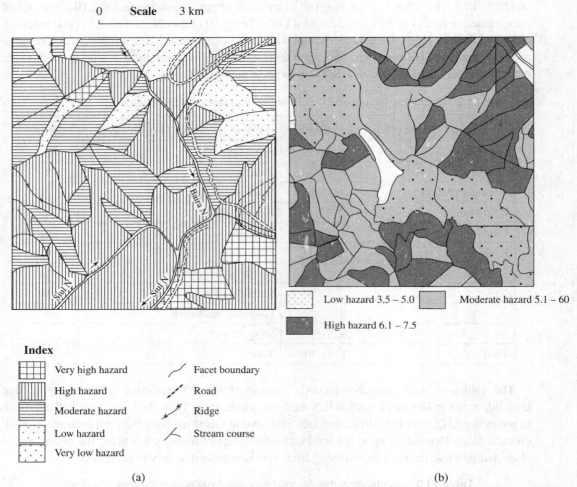

Index

▦ Very high hazard	╱ Facet boundary
▥ High hazard	╱ Road
▤ Moderate hazard	⚹ Ridge
⋮ Low hazard	╱ Stream course
⋰ Very low hazard	

(a)   (b)

Low hazard 3.5 – 5.0   Moderate hazard 5.1 – 60

High hazard 6.1 – 7.5

Fig. 21.20 LHZ maps: (a) dam project site in Chamba district, Himachal Pradesh (Deva and Srivastava 2006); and (b) Bhimtal Naukuchiatal area, Nainital district, Uttaranchal (Gupta 2006) (also see Anbalagan, Chakrabarty, and Kohli 2008)

In India, GSI is the nodal organization carrying out landslide zonation mapping from the study of the geology, geomorphology, structure, and tectonic features of landslide-prone areas and suggesting landslide mitigation measures. The states are in link with GSI to share information regarding areas of instability to supplement GSI's database of ongoing assessment of landslide-prone areas and carry out zonation mapping on Survey of India's contoured maps (topo-sheets). GSI has already covered large parts of the country affected by landslides and areas of instability by hazard zonation mapping giving quantitative overview of distribution of landslide hazards in India (Fig. 21.21).

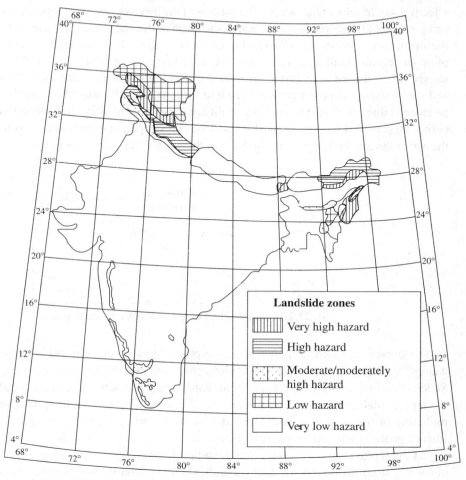

Fig. 21.21 GSI map showing quantitative overview of landslide hazards in India

21.7 LANDSLIDE HAZARD MITIGATION

Landslide hazard mitigation measures are taken up for correction of existing landslides or prevention of potential slope failures. In principle, the measures involve stabilization of the soil or rock mass by increasing the resistive forces and/or reducing the driving forces. Five types of methods are generally adopted in landslide mitigation, which are afforestation, modification of slope geometry, drainage arrangements, slope reinforcement, and provision of retaining structures.

21.7.1 Bioengineering or Afforestation

Trees largely increase the stability of the underlying ground by increasing root strength, intercepting the direct effect of precipitation, and reducing pore pressure by evapotransportation. Therefore, bioengineering methods including large-scale afforestation and protection of existing vegetation cover need to be adopted in the landslide-prone areas. The selection of suitable plant species should be such that they can withstand the existing hydrological conditions of the terrain.

21.7.2 Modification of Slope Geometry and Prevention of Land Erosion

The geometric modification of hill slopes or changing the shape of the slope surface is a cost-effective way of stabilizing a slope. The success of the method, however, depends on the size, shape, and position of the slope. In this measure, the angle of the soil slope is reduced by cutting a part of the top surface or depositing fill materials on the foot of the slope. A combination of both reduction of upper part and infilling of the foot of the slope may also be done (Fig. 21.22). In very high hill slopes, filling the toe is preferred, as reducing slope from the top involves excavation machinery and is expensive. While excavating a place for bringing the fill material to stabilize a slope, it is to be ensured that this does not create instability of the excavated place. In high and steep soil slopes, berms are provided by cutting the slope surface into stepped segments so as to place the weight of the soil at the toe. In doing so, it is to be seen that such cutting does not initiate creep movement.

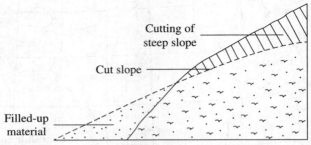

Fig. 21.22   Stabilization of ground slope by cutting and filling

The slope after re-profiling needs to be protected against erosion and infiltration of meteoric water. This is done by providing impervious cover and making adequate drainage arrangements. Several types of anti-erosion and environment-friendly biomats and bionets (synthetic products) are now available, which can be spread over the re-profiled surface to prevent erosion. Asphalt mulching of the surface is also adopted as a protective measure against erosion. Grassing of surface protects erosion by binding the soil with long roots. Catch drains (contour drains) are dug at the top to drain out surface water, and perforated pipes are inserted into the soil to drive out water from inside the soil mass. Landslide mitigation by modification of slope geometry cannot be applied to stabilize a slope affected by a long translation slide. In case of debris flow type of landslide, check dams with soil cement are sometimes constructed in a stream, which can yield much sediment from large landslide-prone areas at the upstream part.

In rock slopes, geometric modification is done by the removal of highly weathered, extremely fissured, and overhanging rock blocks from the rocky surface. Vegetation or plants that widen the roots to rock joints are also removed. The sole idea is to remove the shallow unstable rock mass from the upper surface and expose the firm and sound rock from beneath to obtain a stable bedrock surface. The measure prevents rock falls or topples from the hills slopes, which damage habitation structures and endanger people at the foothill region. The topographic modification of hill slopes can be done by pickaxes; when explosives are used, only controlled blasting is to be undertaken to avoid development of cracks in the rock mass.

21.7.3 Drainage Arrangements in Relation to Groundwater Management

Drainage is the widely used and most effective method in stabilizing ground, especially when used in combination with modification of slope geometry. Slope stabilization by drainage is also less expensive than other landslide mitigation measures. An integrated drainage arrangement for the re-profiled slope includes diverting surface run-off from the unstable area by shallow surface channels and lowering of groundwater level by deep drains and trenches. If groundwater occurs at a shallow depth, generally a combination of vertical boreholes connected with sub-horizontal (10°–15°) boreholes oriented uphill are used for drainage by gravity. Shallow unbroken drainage trenches filled with permeable granular material and connected by perforated pipes at their bottom parts are also very effective in removing water from the soil mass at a shallow depth (Fig. 21.23).

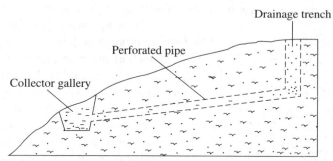

Fig. 21.23 Stabilization of ground slope by drainage arrangement

In order to provide deep drainage, isolated wells are developed covering the area to be drained. This system needs constant operation of drainage pump for each of the wells. In case of drainage of water from the failure surface, deep drains or trenches are to be dug extending below the slip surface or shear plane of the landslide. In the sites affected by deep landslides, tunnels or galleries are excavated into the intact soil mass below the slid mass linked with a series of vertical wells sunk from the ground surface. This drainage measure, though expensive, is very effective in removing water from the soil and lowering the groundwater level. If the groundwater level is very deep, a combination of vertical drainage wells connected with a network of sub-vertical borehole drains is used to remove groundwater from the unstable area. Landslide mitigation measure by drainage arrangement needs regular maintenance of the drains so that they do not become non-functional by clogging within a short time.

In rocky hillsides, the geometry of the slope and the geological structure of the rock play the deciding role in planning drainage network. Water enters mainly through the joints or fissures in rocks and develops pressure, initiating rock slides along weak structural planes. Hoek and Bray (1981) have proposed placing selective shallow and sub-shallow drainages in the hillside to prevent entry of water through the cracks in rocks and reduce water pressure in the immediate vicinity of the potential breakage surface. Shallow surface drains, especially at the top part of hill slopes, help to divert surface run-off and infiltration through the rock cracks. In general, wide drains are dug to collect surface water, which is then removed to a distance from the unstable area and is connected with other smaller drains. Other protective measures towards slope protection may also be provided in combination with drainage arrangement (Fig. 21.24). In the case of lowering groundwater level in fissured rocks, deep vertical trenches or holes are dug and water is pumped out from sub-surface rocks.

Fig. 21.24 Stabilization of slide-prone hill slope by drainage (lined surface drain) combined with re-profiling of slope by berms

21.7.4 Slope Reinforcement

The work approach in soil reinforcement is basically done in two ways, which involve insertion of reinforcement elements such as *anchoring* and *nailing* in the ground. It is then followed by improvement of the mechanical character of slope material by consolidation or bonding by the process of grouting with varied admixtures.

The anchoring technique is adopted to stabilize soil mass of failed slope or ground of impending failure. The anchors increase the resistive force of the unstable ground mass to prevent failure by driving forces. The ground slope is first treated by reinforced concrete for grip of the anchors and the anchors are then inserted fairly deep into the ground slope (Fig. 21.25). Nailing technique is applied to mobilize the cohesion and angle of internal friction of soil for stabilization purposes. This measure is very effective to contain the potential failure surface. Nails are introduced in the ground by means of injecting cement mortar at high speed. In another method, steel nails are inserted perpendicular to the unstable slope by screwing or percussion process. Steel plates fashioned as nail heads are fixed inside with expanded bulb by means of injected mortar.

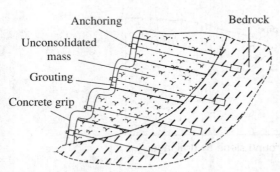

Fig. 21.25 Reinforcement of soil mass by anchoring

Grouting of the slope material is often done successfully to stabilize the ground of unconsolidated materials. The grout mixture is chosen depending upon the nature of the ground materials. In general, a grout of cement–water–bentonite with admixture of some chemicals is injected into the ground mass at a high speed. The injected grout after setting consolidates the slope providing high strength and prevents failure.

In rock slopes, the common methods of reinforcement of rock slopes are *bolting* and *shotcreting* (Fig. 21.26). Bolting enhances binding and increases the shear strength of the fissured rock. Shotcreting is done in rock surface by a hose at high speed with a view to increasing the strength of the rock and thereby preventing rock falls.

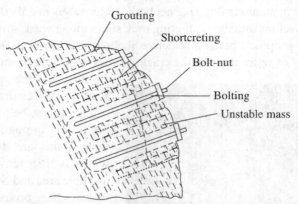

Fig. 21.26 Reinforcement of rocky slope by bolting

21.7.5 Retaining Structures

By far, the common practice of protecting slope failures is to provide some kind of a retaining structure to the slope surface. The nature of these structures depends on the resistive force to the load imposed on the soil slope. Gravity retention structures used include masonry or stacked masonry wall, concrete wall, rubble filled masonry wall, and rock-filled gabion wall (Fig. 21.27). Of these, the stacked masonry walls of dimension stones (Fig. 21.28) at places with wire wrapping and concrete walls are common. Weep holes are provided in many of the retaining walls to reduce water pressure caused by infiltration of rainwater through the slope surface.

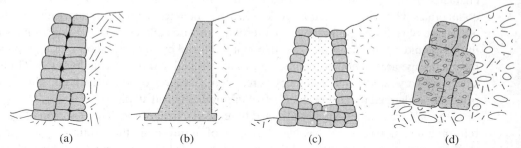

(a)   (b)   (c)   (d)

Fig. 21.27   Retaining structures: (a) stacked masonry wall; (b) cast in-situ reinforced concrete wall; (c) rubble filled masonry wall; and (d) rock-filled gabion wall

Fig. 21.28   Stabilization by stacked masonry wall

Instead of massive retaining structures to protect unstable soil slopes, cast in-situ reinforced concrete piers and sheet H-piles are also provided. In the case of rock slopes that may affect human habitation including man-made slopes due to rock fall or rock slides, trenches are dug at the foothill or walls, and fencing is constructed to trap fallen rock blocks and boulders at the bottom part of rock slopes. Boulders wrapped in wire nets are also placed to create barriers to rock slides.

21.7.6 Other Methods of Soil Stabilization

In addition to the methods discussed here, *thermal* or *freezing* treatment and *electro-osmosis* techniques are applied as stabilization measures in clayey earthen ground.

In the method of thermal or freezing treatment, the heating or cooling of the clayey soil increases the shear strength of the material that helps to prevent surface creep. The success of this method, however, depends on the nature of the materials and the topography including the ground slope. The thermal treatment method has been successfully undertaken in foreign countries such as Russia and Romania. The process involves burning of a liquid or gas with air introduced at depth in soil through some boreholes. This causes solidification of soil and decrease in its compressibility and increase in strength. The high temperature (around 500°C) also changes the sensitive character of clay minerals by making the soil non-expansive and thus helps the process of stabilization.

The *electro-osmosis* method involves dewatering of soil mass of the ground. The method involves inserting two electrodes and passing current to enforce movement of the hydrous ions to cathode from anode and thus improve the resistive force of soil by the removal of water (Section 5.9.2). This method is, however, applicable only in homogenous clayey soil. In fact, the electro-osmosis method of soil treatment brings about changes in the structure of the soil. The process of introducing electric current is done by providing electrodes of certain electro-chemical composition that causes deposition of the metal salt in the soil pores. Use of aluminium as anode gives good results in the process of soil stabilization and the process is also known as electro-chemical hardening.

The modern methods used in land slope stabilization, namely nailing, use of geogrids, cellular faces, coir geotextiles, and steel wire mesh, are important.

The fundamental principle of the soil *nailing* technique is mobilizing the intrinsic mechanical character of the ground such as cohesion and angle of internal friction so that ground activity collaborates with the stabilization work. Nailing, similar to anchoring, induces normal stress to the advantage of stability. Nailing by rapid response diffuse called CLOUJET causes nails to embed in the ground by means of an expanded bulb obtained by injecting mortar at high pressure into the anchorage area. Another method known as soil nail and root technology (SNRT) introduces steel nails very rapidly into a slope by percussion, vibration, or screw methods. This geotechnical principle is claimed to have been effective in stabilizing soil slopes.

Geogrids are synthetic materials used to reinforce the ground. The insertion of geosynthetic reinforcements, therefore, has the function of reinforcing the ground, conferring greater stiffness and stability upon it and the capacity to be subjected to greater deformations without reaching fracture points.

Cellular faces, also known as *crib faces*, are special supporting walls realized by means of head grids prefabricated in reinforcement concrete or in wood. Compacted granular material is inserted in the spaces of the grid. The characteristic modularity of the system confers notable flexibility of use in terms of adaptability to the morphology of the ground and because it does not require any deep foundation other than laying a plane of lean concrete. Hence, this method is useful in stabilizing the ground.

Coir geotextile is another method universally used for bioengineering and slope stabilization because of the mechanical strength necessary to hold soil together. It has been proven that coir geotextile lasts for only three to five years, and when it degrades, it is converted to humus that enriches soil.

Steel wire mesh is used for soil and rock slope stabilization. After levelling, the surface is covered by a steel wire mesh, which is fastened to the slope and tensioned. It is a cost-effective method of stabilization.

21.8 INSTANCES OF MAJOR LANDSLIDES IN INDIA

The Himalayan terrain being highly fragile, is a perennial source of landslides of all possible descriptions, small to disastrous, very old to very recent. Other parts of the country such as the north-eastern hilly region, the Western Ghats, the Nilgiris, the Eastern Ghats, and certain areas of the Vindhyas are affected by intensive landslides almost every year, especially after heavy precipitation during monsoon. This section discusses some among the large number of severe landslides that affected various parts of the country in the past few decades.

With the onset of monsoon, heavy rains bring about large-scale landslides affecting many localities of Sikkim and Darjeeling, including the national highway passing through these Himalayan terrains (Fig. 21.29). The 1968 flood in parts of Darjeeling district in West Bengal unleashed about 20,000 landslides covering vast areas of Sikkim and Darjeeling and killing thousands of people. This incident of innumerable landslides spread over a period of three days with precipitation of 500 mm to 1000 mm was an event of a 100-year return period. The effect was so catastrophic that the 60 km-long stretch of the national highway passing through these areas was seriously damaged by 92 major landslides and communication with the other parts of the country was completely disrupted (Bhandari 2006).

Fig. 21.29 Disruption of highway in Darjeeling by a huge landslide of hill slope after heavy rains and snow

The hilly areas of the north-eastern states other than Sikkim and West Bengal, namely Assam, Meghalaya, Tripura, Nagaland, and Arunachal Pradesh, have also suffered serious landslide damages.

Excessive precipitation is responsible for increase in water levels in lakes, streams, canals, and reservoirs and results in floods that change the groundwater condition and bring about extensive landslides. At places, *landslide dams* (dams created by the landslide itself by depositing its debris) are created in the river valley, blocking the water flow by the debris of the flood-induced landslides. Many old slides are reactivated by heavy and prolonged rains.

The Teesta valley in Sikkim and Darjeeling experiences a mean annual rainfall of 5000 mm and suffers from repeated landslides almost every year. In Sikkim, the rainfall is associated with cloudbursts. Cloudbursts may last for nearly three hours at a stretch causing devastating landslides. Cloudbursts of intensities exceeding 1000 mm in 24 hours inevitably will trigger large-scale mass movement.

In the South, landslides of severe nature are common during the rainy season along the steep slopes of Western Ghats overlooking the Konkan valley. Landslides are also frequent in the Nilgiris where many of the slides occur as debris avalanches consisting of soil and boulders. The major landslides in Nilgiris are the Runnymede slide, the Glenmore slide, the Coonoor slide, and the Karadipallan slide. These landslides caused enormous damage of land and property, and even loss of lives. In the Nilgiris landslide during October–November 1978, as many as 90 people died. Kerala experienced several large landslides such as the Amboori landslide of 2001, which resulted in severe damage and the death of 23 people.

A landslide of unprecedented dimension and unique character was the Alakananda tragedy of 1970 in Uttaranchal, which was due to the massive floods in the river Alakananda and the breach of the landslide dam in its confluence with the river Patalganga. In addition, a 300 km-long road consisting of calcareous quartzite and slate passing through the Alakananda valley of Garhwal, east of Srinagar, has been badly affected by the landslides. In 1998, the Malpa rock avalanche wiped out the entire Malpa village of Uttaranchal in Kumaun Himalaya causing immense destruction, including killing of 220 people (BMTPC 2005).

Of all the causes responsible for the landslides and mass movement in India, the main reasons include ground saturation by heavy rainfall, steep land slope, unfavourable rock attitudes, and loose soil and bouldery deposits covering the hill slopes. Almost all major landslides occur during monsoons (SW and NE monsoon) in the western flank of Western Ghats and during occasional cyclonic events in the eastern flank indicating that the main triggering mechanism is the oversaturation of overburden caused by heavy rains. Cloudburst phenomenon that brings about heavy downpour within a short spell is another main cause of many of the catastrophic landslides.

Human activities are also responsible for landslides in many places. In hilly areas, slope terracing is adopted mainly to prevent soil erosion and to enhance percolation during dry season for crop cultivation. This causes blockage of natural drainage lines on slope surface for drainage of excess storm water during heavy rains. This leads to oversaturation of soil and development of pore pressure, resulting in slides of soil mass. Indiscriminate cutting of trees for timber and domestic use for fuel as well as overgrazing are responsible for deforestation. Deforestation in combination with a fragile geological environment is responsible for a large number of landslides. Bhandari (2006) observed that vast areas of Sikkim, Kumaon, Garhwal, Himachal Pradesh, Kashmir, and several other hilly regions have been robbed of the protective vegetal cover to less than 30 per cent as against twice as much needed. The slopes without the vegetation cover cannot hold soil cover, resulting in widespread erosion and initiating landslides.

21.9 CASE STUDY ON LANDSLIDES IN INDIA IN RECENT PAST

Landslides of devastating and destructive forms have been identified in different parts of India in places such as the north-eastern, eastern, and western Himalayas and the states of Tamil Nadu, Assam, Sikkim, West Bengal, and Jammu and Kashmir.

Hilly areas with slopes containing weathered and decomposed rocks (regolith) and soil mass are vulnerable to landslides. The mountainous areas of the Himalayas, Nilgiris, and Western Ghats and also the hills of Sikkim and Darjeeling are subjected to considerable loss of human lives in addition to loss of road and highways, landed properties, agricultural fields, and forests. The following are some instances of devastating landslides of the recent past:

(i) 11–12 July 1996—A massive landslide in Jaldhaka valley and South Kalimpong hills caused huge damage to properties and loss of lives.

(ii) 9 June 1997—Widespread devastation in Gangtok town disrupted communication along the national highway in addition to causing damage of hill slopes and lands.

(iii) 18 August 1998—The major landslide of Malpa, Uttaranchal, led to the death of 220 people. The entire hill slope with heterogeneous rock blocks moved down slope causing large-scale destruction.

(iv) 7 July 1999—Kurseong town of North Bengal was devastated by severe landslide causing heavy damage to the township.

(v) 8 October 2005—The Kashmir earthquake (described in Section 20.12) triggered 2400 landslides causing immense destruction and loss of lives. However, in addition to failed slopes, numerous hill slopes developed features that facilitated water percolation leading to landslides in subsequent years too.

(vi) 14–15 August 2007—The hilly areas of Himachal Pradesh were destroyed by a massive landslide due to monsoon rains. In this catastrophic event, at least 62 people died in Dhara village, 185 km from Shimla, due to monsoon cloudbursts. The calamity struck without any warning when a huge mass of slush and rubble started moving down from uphill at an altitude of 2700 m towards Dhara village at a terrific speed, wiping out everything on its way and left a trail of misery. More than two-thirds of the population perished.

21.10 CASE STUDY ON MIZORAM LANDSLIDES (MUKHERJEE AND BHAGWAN 2009)

The landslide activity in Mizoram in north-eastern India, especially during monsoon, not only creates frequent disturbances to the traffic on National Highway 54 passing through this hilly state, but also causes damage to buildings, culverts, roads, and cultivable and non-cultivated hill slopes. Since 1996, a stretch of NH 54 has been facing continuous problems of subsidence, creep, and landslides. In the same year, about 120 km stretch of the highway was severely damaged by landslides and subsidence. It not only created problems for the moving vehicles, but also enhanced further risk of failure of the hill slopes of the area.

Hence, a detailed investigation related to the causes of the slides and subsidence was carried out by the scientists of the Central Road Research Institute and suggestions were forwarded for their remediation. This extract is based on their work and suggestions for remedial measures.

The major portions of the road including the affected areas of slide and subsidence consist of alternate layers of slate, shale, and sandstone. Severe tectonic activities on the rock formations caused intense fracturing of these rocks. The fragile and fragmented rocks are visible almost everywhere. Because of the highly weak and unstable condition of the strata, landslides are very frequent along the road and its neighbouring areas. In 1996, even after taking up certain corrective measures, the situation of slide disturbance did not improve. In 2003, in addition to the highway, many shops and houses by the roadside were completely damaged during the monsoon. The problems of sinking of road and sliding of hill slopes were so intensive that traffic movement along the entire highway had to be stopped.

After thorough geological investigation of the highway and the surrounding areas, it was found that the finer components of the rock formations are composed of silt and clay. Highly decomposed shale is the dominating rock type in the area, which formed clay after weathering and decomposition. Joint planes are highly sheared and have no cohesion. In many places, the sedimentary beds are folded and fragmented forming innumerable planes of discontinuity, which have enhanced the chance of slides. The rocks are actually subjected to wedge-type failures. Huge piles of fragmented and disintegrated blocks resulted from the failures accumulated in different parts of the roads. Shear testing of the materials under slide indicated very low shear strength (35 kPa) and angle of internal friction (30°).

After thorough geotechnical investigation of the landslide and subsidence situation, the following remedial measures have been mainly suggested to prevent the recurrence of the problems:

(i) Plugging of all existing potholes on the shoulder portions of the road
(ii) Sealing of the tension cracks
(iii) Construction of side drains along the roads and culverts
(iv) Provision of proper arrangements for surface drains and feeder drains
(v) Provision of step-like chutes by the roadside besides the drains
(vi) Construction of heavy and light retaining structures such as masonry walls, wire crate walls, drum retaining structures, bamboo crate walls, perforated tin sheets, and log retaining structures to protect the unstable and affected locations
(vii) Piling of bamboo, installing of lime-column benching, and erosion control by vegetation using jute or coir geogrid mantle
(viii) Proper maintenance of the retaining structures

SUMMARY

- Landslides bring about disastrous effects by destruction of habitation and properties, and even cause loss of lives. The country suffers immense economic loss in the reconstruction of public utilities damaged by landslides.
- A combination of the types of mass movement (fall, topple, slide, or flow) and materials involved (rock, earth, and debris) are used in classifying landslides.

- In *fall* type of movement, the rock, earth, or debris falls freely from the parent body, whereas in *topple* type, the materials move under gravity.
- In *slide* type, the failure takes place along well-defined surfaces such as bedding, joint, fissure, or other weak structural planes. The nature of movement involved in the earth slide may be rotational or translational. The *flow* type of mass

movement is common in hill slopes covered by debris and earth that contain large amount of water to facilitate the flow at a significantly high speed.

- In low-relief areas, liquefaction of saturated sandy and silty materials initiates landslide by movement of the mass as *lateral spread*. Combined mass movement that results from two or more of the landslides types (e.g., falls, topples, slide, and flow or spread) is termed *complex*.

- Geological processes, action of man, and natural causes are mainly responsible for landslides. Erosion and weathering bring about constant changes in the morphology of the ground, which may result in instability and even failure of the ground slope. Fault, shearing, and jointing and other unfavourable geological structures are responsible for most of the rock falls or rock slides.

- Human actions such as cutting of hill slopes, chopping of trees, and removing of vegetation from ground surface destroy the stability of landform and initiate landslides. Earthquake triggers large-scale landslides in many areas. In addition, undercutting by rivers and infiltration of rainwater, which increases water pressure in earth and debris, may result in slope failure.

- Large parts of India, especially the mountainous regions of the Himalayas, Western Ghats, Eastern Ghats, north-eastern states, and Nilgiris suffer from landslide hazards almost every year.

- Geotechnical investigation of landslide-affected areas is carried out to find causes of the slide and suggest measures to prevent sliding of unstable areas. Preparation of LHZ maps helps in the planning of protection measures against this natural hazard. GSI has already completed the zonation mapping of the country by taking the help of the states who provide information regarding slide-affected and slide-prone areas.

- Landslide remediation measures are arranged under four groups, which are modification of slope geometry, drainage arrangement, reinforcement measures, and/or protection of slopes by retaining structure. Geometric modification is done by cutting and filling of ground slope by suitable fill material to reduce the slope angle and increase the shear strength.

- Drainage arrangement is the cheapest of all the hazard mitigation measures. Shallow surface drains and deep drains are provided to drive out surface water and underground water from unstable ground or from areas of potential slope failure. The barren surface of the slope is covered by impervious material or mats to arrest infiltration of rainwater.

- Plantation is also undertaken to stabilize the soil slope. The reinforcement measures of rock include anchoring or rock bolting, shotcreting, nailing, and grouting.

- Protection of unstable slope is undertaken by different types of gravity-retaining walls made of masonry blocks, gabion, buttress, piles, piers, and cast in-situ reinforcement concrete walls. Rock traps, trenches, and fenced walls are provided at foothills for stopping rock fall or failure of rock slope.

- Modern methods applied for effective slope stabilization use crib facing composed of granular materials, coir geotextiles, and steel wire mesh, which are found to be cost-effective methods of slope protection.

EXERCISES

Multiple Choice Questions

Choose the correct answer from the choices given:

1. Debris covering a land slope that moves at a speed of more than 5 m/s is:
   (a) debris flow
   (b) debris avalanche
   (c) topples

2. From a hill slope of jointed rocks, large quantities of rock blocks fall down by rotation with force of gravity, the initial force applied by the adjacent blocks being:
   (a) complex
   (b) falls
   (c) topples

3. After heavy rainfall, the water-soaked materials of a soil slope containing nearly 50 per cent of fines moves at a speed of nearly 20 km/h is:
   (a) lateral spread
   (b) earthflow
   (c) debris flow

4. Liquefaction causes failure of a low-relief area covered with saturated soil and the materials involved in the landslide move sideways is:
   (a) earthflow
   (b) translation slide
   (c) lateral spread
5. Landslides in India mostly occur during:
   (a) monsoon, after rainfall
   (b) cloudbursts (in hilly areas)
   (c) summer (in the plains)
6. Landslide remediation methods include taking measures such as:
   (a) modification of slope geometry
   (b) drainage arrangement
   (c) providing retaining structure
7. Soil stabilization against landslides is done by adopting measures such as:
   (a) planting trees with extended roots
   (b) nailing of soil mass
   (c) reinforcement by bolting
8. Rock slopes are reinforced to protect from landslide by:
   (a) adopting rock bolting
   (b) taking up grouting
   (c) shotcreting and nailing
9. Modern methods of slope protection uses:
   (a) coir textiles
   (b) still wire
   (c) crib facing of granular materials
10. Sensitive soil is stabilized by:
    (a) applying the method of electro-osmosis
    (b) heating the soil by thermal process
    (c) elaborate grouting

Review Questions

1. How does a landslide cause damage or destruction of human lives and property? Give an account of how landslides inflict economic loss to the general public and government organizations during and after the enormous damage from catastrophic landslides.

2. Draw a schematic diagram to show the following parts of a landslide: crest, toe, surface of rupture, scarp face, radial cracks, and transverse cracks.
3. Explain the classification of landslides based on the types of materials such as rocks, soils, and debris and their movement involved in the landslides. How will you differentiate rotational slide from translational slide?
4. What is a creep? What may be the rate of movement of the materials in a creep? How do you recognize the development of creep in a landmass that may result in a landslide in future? Describe the flow types of landslides with reference to debris flow and mudflow.
5. Give an account of the geological and natural causes of landslides. How can human action cause landslides and slope failures? Explain with examples.
6. What will be your approach to investigate an area affected by landslides to identify the causes of the slides?
7. What are the main features of landslide zonation mapping as specified by BIS? State the approach followed by GSI in preparing landslide zonation map of India. How does aerial photographs and remote sensing help in recognizing the areas of instability?
8. What are the steps needed to be taken to mitigate landslide hazards? Give a brief account of each of these steps to bring stability of landmass and avoid landslide hazards.
9. In case of clayey earthen ground, what are the special methods applied to stabilize the ground from landslides? What is the basic concept of electro-osmosis method to improve soil stability and in what type of soil is it applicable?
10. What are the main regions of India that are mostly affected by landslides every year? Give an account of the prime causes and triggering mechanisms responsible for major landslides in India.

Answers to Multiple Choice Questions

1. (b)	2. (c)	3. (b)	4. (c)	5. (a) and (b)	6. (a), (b), and (c)
7. (a) and (b),	8. (a), (b), and (c)	9. (a), (b), and (c)	10. (a)		

22 Karstic Terrain Investigation

22.1 INTRODUCTION

Dam and reservoir projects in limestone might face the problem of competency of reservoir due to the formation of large surface and underground caverns through which the entire reservoir water may drain to a lower valley. Chemical and radioactive studies can evaluate whether the reservoir water under storage condition is likely to cause large-scale leakage. This chapter deals with such practical problems related to karstic limestone reservoir and their solution, which includes providing deep grout curtain along the periphery of the reservoirs. It also provides a short case history of such problems in different countries of the world and describes the extent of grout curtain provided to arrest leakage.

22.2 SOLUBILITY OF LIMESTONE AND FORMATION OF KARST

The word *karst* is derived from the 'Karst' region located in the Adriatic coast of Slovenia where the landform is characterized by excessive erosion and the presence of numerous surface and subsurface caverns in limestone. Karst topography generally develops in humid climate and in areas consisting of limestone and other soluble rocks. A combination of erosion of country rock by acidic groundwater and mass wasting by sinkholes (deep caves) is responsible for creation of karst topography.

The sinkholes and caverns are formed by the solution action of water on limestone. Limestone is only slightly soluble in pure water, but when the water is charged with carbon dioxide, solubility increases manifold. Rainwater when it absorbs some carbon dioxide from the atmosphere forms carbonic acid. When the acidic water falls on a terrain consisting of limestone, it can dissolve the rock effectively because of chemical reaction between calcium carbonate of the bedrock (limestone) and carbonic acid of meteoric water. The chemical reaction can be expressed as follows:

$$CO_2 + H2_O \rightleftharpoons H_2CO_3 \text{ (Carbonic acid)}$$

$$H_2CO_3 + CaCO_3 \rightleftharpoons Ca(HCO_3)_2 \text{ (Calcium bicarbonate)}$$

The dissolved portion of limestone forms calcium bicarbonate and the remaining insoluble material in limestone will be carried away by the moving water and deposited elsewhere. The rainwater that percolates down and then moves as underground water is generally charged with carbon dioxide derived from the uppermost soil zone. Carbon dioxide in the soil zone exerts on the water passing through it a pressure that is several hundred times more than that exerted by carbon dioxide in the atmospheric air. The solubility of limestone by groundwater subjected to such conditions will be increased manifold. As a result, a network of underground channels and caves is formed that contributes enormously to the subterranean drainage system.

Solution is not the only factor in the formation of limestone cavities. Corrosion and mechanical erosion also play a significant role. The debris accumulated in the cavity floor by mechanical erosion or corrosion of the carbonate rock still contains many soluble salts. The rainwater entering through the openings in limestone widens the openings by dissolving the side walls and then in course of its onward movement through the cavity floor leaches the soluble salts in the accumulated debris. As the process of solution, corrosion, and leaching continues simultaneously, the openings enlarge to give rise to big solution cavities.

The solubility of limestone is dependent on its chemical purity, but in many limestone reservoirs, the chemical purity of limestone is found to be unimportant for the creation of caves. More cavities may be formed in impure limestone as the insoluble content goes into suspension after dissolution begins. Extensive development of caverns in many of the project sites containing dense and massive limestone is not uncommon. The huge limestone mass hinders infiltration of rainwater except through the joints, faults, and other fracture planes. As a result, the bedrock of limestone, characterized by joints and fractures that provide path to downward moving water, goes into solution with progressive development of large caverns. If limestone formation undergoes a lack of confining beds such as impervious shale and claystone at depths, but mainly intersected by joints and faults, a network of deep and long cavities may be developed in the subsurface region by passage of water through these weak planes, hastening the process of solution of limestone (Fig. 22.1).

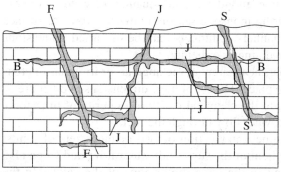

Fig. 22.1 A network of caverns formed in limestone along weak planes (J–J and B–B, joints; F–F, fault; and S–S, shear zones)

22.3 SURFACE AND SUBSURFACE FORMS OF LIMESTONE CAVERNS

Caverns or caves in limestone are found to occur in various sizes and shapes on the earth's surface or at depths. In general, it is possible to easily enter into these caves. On the basis of their location, the limestone caverns may be classified into two groups, namely surface caverns and subsurface caverns. Surface caverns exist in the exposed parts of limestone formation and extend downwards such that one can easily enter them. Subsurface caverns occur in the underground region of karstic limestone extending for long distances such as tunnels or channels and have outlets at the ground surface.

22.3.1 Surface Caverns—Sinkholes and Swallow Holes

Fig. 22.2 Large sinkholes in limestone connected with subsurface solution channels

Limestone caverns observed on the surface, commonly called *sinkholes or swallow holes*, provide path for rainwater to move downwards. These are generally formed along the joints, especially at the junction of two intersecting joints. Most of the surface forms of caverns are seen to be nearly vertical or inclined openings and extend very deep to a long distance in the subsurface region with the outlet at the surface (Fig. 22.2). These sinkholes or cavities are partially filled with clay or debris carried from the top by the percolating water. They form networks of deep cavities by widening the openings of intersecting sets of joints. In karstic limestone terrain, large circular to semicircular cavities that are several tens of metres in diameter and even giant valley-shaped depressions may be formed. The giant cavities and depressions are generally the result of collapse or subsidence of roof of underground cavities and later erosion by percolating surface water.

22.3.2 Subsurface Caverns—Solution Channels or Tunnels

Subsurface caverns occur as conduits or channels and are termed *solution channel*, *solution tunnel*, or *subterranean channel*. During summer months, a small quantity of water may flow through them or they may remain dry, but during monsoon season, rainwater from the surface flows into the subsurface channels. These channels or tunnels may be of varied dimensions from a fraction of a metre to tens of metres in diameter or height.

The caverns may be triangular, semicircular, near-circular, rectangular, or irregular in cross section. Many of them are constricted at some places but are of huge dimensions in other areas. These variations in dimensions are attributed to the chemical purity or impurity of limestone, structural features, and velocity and extent of water flow through the rock.

Presence of argillaceous matter in limestone, intersection of two or more sets of joints, occurrence of shear or fault zones, steep gradient of the surface of water flow, and so on also add to the cavity formation activity resulting in the huge dimensions of the cavity. In the Kopili project of Meghalaya, cavities in limestone vary from a fraction of a metre to as large as 20 m in diameter. During construction of a dam project (Kruscica project) of Croatia in limestone, a cavity as large as 50 m long, 30 m high, and 25 m wide was found at a depth of 50 m below the ground surface.

22.3.3 Stalactites and Stalagmites

Stalactite and stalagmite are two typical features found in many subsurface limestone caverns. Dissolution of limestone by percolating water through faults, joints, and fractures in rock and its dripping and deposition in the form of crystalline calcium carbonate are responsible for the formation of these features (Fig. 22.3). If calcium carbonate crystals hang like icicles from the roof of the cavern ('St' in Fig. 22.3), they are called *stalactites* (Fig. 22.4). The pillar-, conical, or hump-shaped mass of calcium carbonate that grows as a ridge after deposition of the mineral on the floor of the cavern is called *stalagmite* ('Sm' in Fig. 22.3). The subsurface limestone channels that occur like tunnels following *zigzag* paths are formed by solvent action along the joints or other weak planes such as faults.

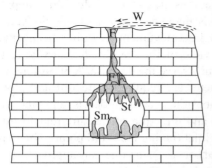

Fig. 22.3 Stalactites (St) and stalagmites (Sm) formed in a limestone cavern (W, water passage and F–F, fault plane)

Fig. 22.4 A cluster of stalactites in a large cavern in limestone formation of Kopili project

Some authors believe that solution cavities may be formed below the water table or piezometric surface of groundwater (Davis and Dewist 1966). For instance, the Hales Bar dam in the US that failed after construction was found to have several large cavities in the zone of saturation below the dam foundation. Though the overall observations suggest that cavity formation may take place at any level, in most of the projects, maximal karstification is found to be limited to the upper 30 m to 60 m zones.

A channel or tunnel type of subsurface cavity may be several kilometres long. These are mostly connected by many subsidiary channels similar to the surface streams having several tributaries. Some subsurface cavities may also form labyrinths, which contain complicated series of paths through which it is very difficult to find the way. So, while moving along a long tunnel or channel to investigate it, some markings by paints are to be made for identification of the return route. The floors of the subsurface solution channels generally contain deposits of pebbles, sand, silt, and clay of varying thicknesses.

22.4 SPELEOLOGICAL AND OTHER METHODS OF EVALUATION OF KARSTIC CONDITION

A karst aquifer is a carbonate aquifer where groundwater flow mainly occurs through bedding strata, fractures, conduits, and caverns created by and/or enlarged by limestone dissolution. It is important to study the scientific nature, geology, and biological aspects of caverns. Subsurface information can also be obtained by exploratory drilling which gives the ideas of joints, faults, and shear zones. The following sub-sections details out some common methodologies.

22.4.1 Speleological Study

Speleology is the scientific study of physical, geological, and biological aspects of caverns. The investigation of surface and subsurface caverns in karst area including mapping and exploration of subterranean caves and solution channels comes under the purview of speleology. In a project area with karst topography, aerial photographic study is undertaken at the first instance to find the accurate locations and characteristics of the various cavities or sinkholes of the project area. Individual cavities are then studied by field traverses. A large-scale geological map of the area is prepared plotting all the cavities, giving numbers to each of them, and recording their dimensions and other characteristics. Many of the observed caverns of the surface are extended to subsurface region and these are in fact inlets or outlets of big underground channel- or tunnel-type cavities.

The most strenuous but important part of speleologic study for engineering geological purposes is the underground survey of the cavities including geological mapping of the subsurface solution channels. In most cases, this involves clearing of the clogged materials or debris in the interior parts of the cavities for free movement. Lighting arrangement in the underground caverns is necessary for a clear view of the cavities for survey. A torchlight powered by at least four cells each of 2.5 volts may serve the purpose. A 50 m-long measuring survey tape and a Brunton compass or clinometer are required to record and plot the rock types, joints including their attitudes, cavity dimensions, nature of debris, and other subsurface features along the channel route. The distance and direction of the different features of the solution channel are measured starting from the inlet (taking it as zero point) when proceeding along the channel until its outlet end is reached, or as far as one can approach. Preparing a sketch of subsurface geological features and noting the height or diameter of the cavity at certain intervals are to be done in the field diary. Care should be taken to ensure that the sketch and writing in the notebook are not spoiled by water dripping from the roof of the cave.

A geological map and geological section are then prepared for the subsurface channel from the data obtained from subsurface survey of the cavities (Figs 22.5 and 22.6). In a reservoir project of karstic area, such study of the large subterranean solution channels is extremely helpful to find their pattern and identify possible leakage path under impounded condition of the reservoir. While working in the subsurface cavities, it may be interesting to note that insects and small fish, living inside the caves or solution channels where light cannot penetrate, are blind with protruding eyes (*personal experience of this author*). Many subsurface cavities may have a stinking smell as they become the abode of a large number of bats.

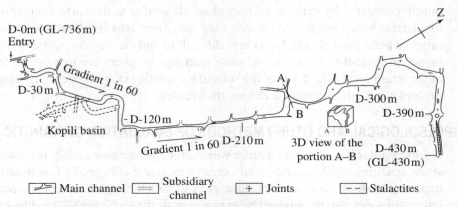

Fig. 22.5 Speleologic map of a subsurface solution channel of Kopili project, Meghalaya; distances are marked from entry (D-0 m) to exit (D-430 m)

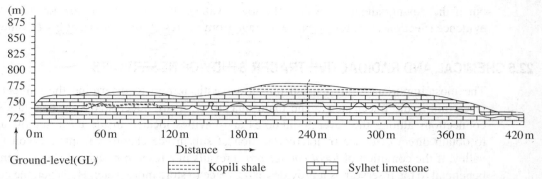

Ground-level(GL)

Kopili shale   Sylhet limestone

Fig. 22.6 Geological cross section along the subsurface solution channel of Fig. 22.5 showing its varying dimensions and ground profile above the channel

22.4.2 Geophysical Study

Geophysical study is an effective method to determine the subsurface nature of limestone and presence of cavities, if any, in a reservoir area of karstic terrain. In general, *electrical resistivity* method is followed to find the 'anomalies' that may be due to the presence of cavities. *Gravity survey* method is also applied to investigate the limestone terrain for detecting subsurface cavities by comparing density anomalies of rock at different depths. A few test holes are, however, required to be drilled at the places of anomalies detected through these two geophysical methods of investigation (Chapter 11) to prove their presence in the subsurface.

Temperature logging conducted through some boreholes with the help of sensitive temperature measuring instrument is another geophysical method of finding the possible presence of subsurface cavity. A fall of temperature from one level to another in the subsurface region is indicative of differential movement of water in limestone formation, which is possible only from circulation of cooler water in comparison to surrounding areas with connection to surface water through subsurface cavities.

22.4.3 Exploration by Drilling

Most of the important subsurface information is obtained from exploratory drill holes covering the limestone strata inside and along the rim of a reservoir. The rock cores obtained from drilling give an idea of the nature of joints, faults, and shear zones including their frequency of occurrence, presence of scouring action, and cavity formation along the fracture planes. Percolation test is carried out through each of the holes drilled at every 3 m intervals. Any excessive leakage or non-development of pressure after using packer is indicative of the presence of void space or cavities through which leakage takes place. Sometimes, an inclined drill hole, instead of a vertical drill hole, provides more chance of cutting across subsurface cavities. Engineering geologists should apply their knowledge of surface and subsurface conditions to identify the locations of drill holes in a limestone terrain.

The drill holes are used for measurement of water level throughout the year to provide a critical view of the fluctuating water level during dry and wet seasons. Like other sedimentary rocks (e.g., sandstone), limestone may not contain groundwater with regular slope. The piezometric surface (generally undulating in nature) is prepared plotting the water level data to obtain the overall downgradient direction of the subsurface flow. It is to be remembered that the topography or dip of limestone beds is not a criterion to find the down slope direction of piezometric surface. In a reservoir project, an intrabasin down slope direction is not harmful, but if the downgradient is towards the lower valley, the fact is to be considered with other evidence (Section 15.6) for probable leakage from the reservoir to adjacent lower valley.

22.5 CHEMICAL AND RADIOACTIVE TRACER STUDY OF RESERVOIRS

The most important study carried out in a karstic terrain to determine the competency of reservoirs is the tracer study. The main objectives of the tracer study are to establish local connections among the several sinkholes and cavities located inside the reservoir area and to obtain direct evidence of interconnection of subsurface channels from reservoir to lower valley. If the condition of *local connections* prevails in a reservoir basin, it is considered to be beneficial to the reservoir as it provides extra space to store more water. However, the condition of *interconnection to lower valley* causes leakage when water is impounded in the reservoir basin. If the tracer study indicates that the clusters of locally connected cavities are joined with the interconnected subsurface channels with lower valley, it poses the serious problem of leakage of reservoir water under the water-impounded condition.

The material used as tracer may be a chemical or radioactive substance. The chemicals are of varied nature, the most common being the common salt. In addition, a fluorescent colour dye or a chemical such as potassium dichromate (Geological Survey of India's preparation) is also used.

22.5.1 Chemical Tracer Study

A solution of common salt, or fluorescent colour dye (e.g., rhodamine, a commercial product), or potassium dichromate is introduced through the cavities and also through the drill holes of the reservoir areas soon after some rain in the area when the cavities are filled with plenty of water. The quantity of the substance to be used needs to be judged based on the length of the travel path. From actual application, it can be said that the quantity of chemicals to be used for a 1–5 km travel path should be 40–100 kg of common salt, or 2–10 kg of the dye, or 3–30 kg of potassium dichromate.

The chemicals are to be diluted with water before pouring into the cavity or drill holes. After introducing the chemical tracers, samples are to be collected at every 6–12 h intervals from selected points in the lower valley (or from cavities and drill holes) for a period of 15 days to 2 months. The samples are to be tested both qualitatively and quantitatively for the common salt and only quantitatively for potassium dichromate. Even if traces of the latter, as low as 0.001ppm, are present in water, it exhibits a pink colour in the presence of diphenylcarbazide dissolved in acetone. Rhodamine will exhibit a reddish tint when seen with naked eye and fluorescence under a fluoroscope even after dilution to a great extent.

22.5.2 Radioactive Tracer Study

When the chemical tracer test does not provide any fruitful result, radioactive tracer study is undertaken, which includes the use of environmental tritium in groundwater and injecting tritium through drill holes. It is postulated that tests of thermonuclear weapons create a source of natural tritium in groundwater. If groundwater mixed with tritium moves from a higher valley to a lower valley, the older water will be found in the lower valley and the recent water in the upper valley. Thus, testing of the environmental tritium in the groundwater provides a clue to understand the possibility of leakage from a reservoir.

In the other method, a very small amount of tritium-loaded heavy water is injected into the boreholes within the reservoir area. Sampling of water is done at regular intervals from the drill holes in the reservoir periphery and from the springs and seepages in rocks of lower valley adjacent to the reservoir. The samples are tested using sophisticated spectrometer to identify whether they contain tritium-bearing water, and if present, the travelling speed is calculated from the time of sampling and that of injection. The data provides information on the possibility of free flow of reservoir water including the rate of movement. The radioactive tracer study was undertaken in India in collaboration with the atomic scientists at the Tata Institute of Fundamental Research (Gangopadhyay 1970).

22.6 CONSTRUCTION OF DAMS AND RESERVOIRS IN KARSTIC LIMESTONE

The engineering geologist working on dam and reservoir projects in a karstic terrain needs to take up a thorough investigation with respect to the competency of the reservoir. The mere presence of cavities in limestone does not mean that the project cannot be undertaken. In fact, many dam and reservoir projects could be constructed on karstic limestone. However, there are also instances of many failures. The geological conditions may be different in different places. Available record of projects constructed on limestone reveals that most of them experienced the threat of leakage, and very extensive and expensive treatment had to be undertaken. In many projects, the dam is located on cavernous limestone wherein the problem of treatment is restricted to the dam site and abutments. However, if the reservoir area consists of cavernous limestone likely to cause leakage, it may involve expensive treatment to contain the leakage.

It is the duty of the engineering geologist to reveal all the features of the caverns and study their behaviour to assess the seriousness of the problems that may arise under reservoir condition. In many cases, the cavities that are vulnerable to leakage of reservoir water can be treated successfully to prevent leakage. The main method of treatment is the grouting of the cavernous limestone along the reservoir rim to arrest the path of leakage to the lower valley. One or two rows of drill holes for grouting purposes are taken up at close intervals covering the entire cavernous limestone bed at depth. The purpose is to seal the voids in limestone. Efficacy of grouting is examined by post-grouting pumping test. If the test indicates that the percolation of water is negligible, the grouting undertaken is considered satisfactory to contain leakage.

Case histories of dams and reservoir projects in cavernous limestone show that most of these projects in different parts of the world have faced utmost difficulty during their construction on account of adverse geological situations related to the presence of subsurface cavities. Extensive treatment had to be taken up to overcome the difficulties. In several cases, the dam and reservoir sites had to be shifted in favour of geologically better sites, and in some cases, the reservoir projects had to be abandoned due to the threat of uncontrollable leakage. Section 22.7 presents a brief account of the karstic limestone problems faced by several countries and the treatment undertaken to contain the leakage.

22.7 INSTANCES OF KARSTIC LIMESTONE PROBLEMS AND REMEDIAL MEASURES

This section discusses the problems associated with karstic limestone in some projects undertaken in India and other countries and their remedial measures.

22.7.1 Projects in India

This subsection discusses a few projects undertaken in karstic limestone terrain in India.

Kopili hydroelectric project, Meghalaya

The Kopili hydroelectric project in eastern India consists of tertiary limestone capped by shale and underlain by sandstone deposited on Precambrian gneissic basement. A vast area covering the Kopili and Umrong reservoir basins and the lowermost Langlai valley is characterized by exposed limestone. The project is aimed at harnessing the power potential of the river Kopili by constructing a large dam across it, storing water in the Kopili and Umrong basins, and carrying the water through a long tunnel to obtain a drop of nearly 400 m.

During the investigation stage, study of aerial photographs revealed the presence of nearly 200 and 150 sinkholes in the Kopili and Umrong reservoir basins, respectively. Exploratory drill holes in the two reservoir areas followed by water percolation tests proved the presence of caverns at varying depths, the top 40 m of limestone formation being the worst-affected part with large solution channels. Chemical and radioactive tracer studies conducted in the reservoir basins indicated possible leakage under storage condition. From the overall investigation result, it was concluded that the limestone of the reservoir area is highly cavernous and liable to cause leakage. Hence, an altogether new project scheme had to be planned changing the earlier scheme, which involved shifting the dam site from limestone to granite terrain nearly one kilometre downstream and proposing to construct the project in two phases to avoid the risk of reservoir leakage.

Obra dam project, Uttar Pradesh

The 23 m-high Obra dam in Uttar Pradesh constructed for hydroelectric power generation rests on Kajrahat limestone and shale of Lower Vindhyan formation. A part of the reservoir in the vicinity of the dam consisting of Kajrahat limestone (bed L-1) had posed the problem of leakage from the reservoir. Another highly cavernous limestone bed (bed L-2) that occurs in many places of the reservoir spread was also prone to leakage of reservoir water. Structurally, the limestone bed L-1 being in the synclinal part of the regional fold had enhanced the problem of leakage.

In the dam and the river bed section, leakage of the reservoir water through the cavernous limestone was controlled by providing positive cut-off extended to the underlying impervious shale bed-2. The cut-off alignment passes through the carbonaceous shale bed-1 after intercepting the limestone bed L-2, which was treated with concrete plugs followed by curtain grouting. In addition, two lines of concrete diaphragm were provided at 3 m intervals for a length of 365 m through the river deposit. Curtain grouting was also undertaken for a length of 475 m along the cut-off alignment and below the concrete diaphragm to seal the leakage path through the cavernous limestone including its faulted contact with Archaean basement.

The cavernous character of the limestone bed L-1 was treated by curtain grouting over its entire thickness and the consumption of cement–bentonite mix used in the grout was of the order of 900 kg/m. In the treatment of the clay-filled caverns of the spillway and powerhouse sites, the clay was removed and the cavities were then filled with concrete followed by consolidation grouting in the limestone foundation zone to a depth of 3–4 m. In addition, curtain grouting was carried out to a depth of 15 m extending up to the impervious shale bed so as to seal the leakage path in cavernous limestone. The grout intake was up to 184 kg/m.

22.7.2 Projects in Karstic Limestone in Other Parts of the World

This section discusses a few projects undertaken in karstic limestone in various parts of the world.

USA

Logan Martin dam   This is a multipurpose project developing part of the river Coosa in Alabama. The dam site consists of Precambrian dolomite limestone with numerous caverns. This involved elaborate treatment by grouting. The total depth taken together of the grout holes were 46,200 m and consumed as much as 84,400 m³ of cement grout.

The Hondo reservoir   This reservoir located in New Mexico is situated on cavernous limestone capped by thick soil. The water table lies at a depth of about 60 m. After the reservoir was constructed, large cavities (voids) were developed in the soil through which large-scale leakage started rapidly. The loss of water was so great that the reservoir was abandoned.

Hales Bar project   Of all the projects constructed under the Tennessee Valley Authority in the US, the Hales Bar project constructed on cavernous limestone caused the maximum trouble. On account of excessive leakage beneath the dam, it took eight years to complete the project. Soon after the dam completion, heavy leakage started. It took almost 30 years to contain the leakage to a considerable extent by elaborate and expensive treatment.

Norris dam   The foundation and abutments of this dam are constituted of cavernous limestone. The caverns posed serious leakage hazards and necessitated extensive and systematic treatment (by grouting) covering a stretch of 930 m of the abutments and foundation area. Each of the grout holes penetrated as deep as 100 m to 120 m.

Canada

Manitoba hydroelectric project   The project area including the reservoir is characterized by cavernous limestone, which posed the problem of leakage of stored water after the construction of the project. To prevent the leakage, 30 km-long grout curtain along the entire reservoir periphery had to be provided at a prohibitive cost. The estimated cost of providing grout curtain was as much as $140 million in the 1960s.

Croatia

In Yugoslavia, several dam projects, namely Liverovice, Peruca, Moste, Zvornik, Senj, Trebisnjica, Buskobalato, and Kruscica, were built on karstic limestone with the help of extensive grouting. During construction of the Kruscica project, a big *ponor* (cavity) 55 m long, 30 m high, and 25 m wide was discovered about 50 m below the right abutment. This involved a change in the design of the dam. The 70 m-high proposed arch dam had to be changed to a rock-fill dam.

France

The Chambon dam on Triassic limestone, the Santest dam on Lias limestone, the St. Pierre Cognet dam on Upper Lias limestone, the Castilian dam on Jurassic limestone, the Chaudonune dam on Lias limestone, and the Genissiat dam on Cretaceous limestone encountered problems of leakage due to the cavernous character of the limestone. The caverns were treated by very expensive grouting. The grout intake in different dam projects was as much as 137 to 520 kg/m.

Turkey

May dam founded on karstic Mesozoic limestone and Neogene marl was completed in 1960 but it did not hold water. The reservoir area is covered with 15 to 20 m-thick alluvium. The reservoir water seeps into the alluvium and disappears in the karstic limestone. To arrest leakage, the observed sinkholes and caverns in limestone were packed with impermeable materials, but several new caverns were then developed and the leakage could not be fully contained.

Iraq

Docan dam and reservoir   The foundation area consists of Cretaceous limestone, marl, and shale. Extensive curtain grouting had to be carried out along the reservoir rim and in the dam foundation. The grout curtain covered 420,665 m³ of limestone with the total length of the boreholes 193,185 m. The total cement consumption was 45,000 tonnes and cost £240,000 in 1959.

Italy

Val Gallina dam is located on Upper Trias dolomite limestone. To prevent leakage, large-scale grouting was undertaken, which resulted in a grout intake of 950 kg/m. Fedaia dam was constructed in 1956 on limestone and tuff and extensive grouting had to be undertaken. Even so, excessive leakage through the caverns in limestone continued for decades.

Algeria

In Fodda dam and reservoir project spread over the Jurassic limestone, the dam foundation was found to be highly fissured and full of caverns. It necessitated shifting of the site to a different place. In the new location too, very expensive treatment had to be carried out in the fissured limestone. In the Beni–Bahdel dam project, the foundation area consisted of Jurassic limestone. With the completion of the project, leakage started at rates between 500 and 1000 l/s. Extensive grouting had to be undertaken to control the leakage to sufficient extent.

SUMMARY

- Karst is the rugged landform of limestone terrain developed by numerous caverns and mass wasting and erosion of ground surface. The rainwater charged with atmospheric carbon dioxide and groundwater charged with carbon dioxide derived from soil make the water slightly acidic. This causes dissolution of limestone and formation of surface and subsurface caverns.
- Construction of storage structures in karstic terrain might encounter problems of stability of structure and leakage of stored water. A thorough investigation is necessary to build engineering structures, especially for storage purposes in karstic terrain.
- At the preliminary stage of investigation, aerial photographic study followed by ground checking of a project area in karstic terrain helps to locate the surface cavities and evaluate their characteristics. Detailed geological mapping is then taken up by extensive field traverses to plot and record the dimensions and characteristics of all the observed cavities and their underground links.
- Geophysical survey and exploratory drilling including permeability tests of rocks through the drill holes are important aids to find the locations and extents of caverns and solution channels in the subsurface region. The drill holes are also used as piezometers to measure groundwater levels and study the flow pattern of underground water through cavernous rocks.
- Speleologic study that involves detailed investigation of the limestone caverns including mapping of the subsurface solution channels is immensely helpful in the evaluation of foundation stability of storage structures and competency of reservoir in karstic terrain.
- Tracer study is also undertaken in karstic reservoir sites using a saturated solution of common salt or fluorescent dye and also radioactive tracer (such as tritium), which is injected into the groundwater through a drill hole in the reservoir area.
- An observation drill hole in the lower valley side is used to collect samples of groundwater at regular intervals and test the water sample. Presence of

tracer in the test sample is suggestive of possible leakage under reservoir condition.

- Many of the storage projects of the world in karstic limestone either failed or involved very expensive treatment. A deep grout curtain is required to be provided along the vulnerable stretches of the reservoir periphery to seal the possible leakage path.

- In karstic limestone, presence of cavities in dam foundation may pose the problem of stability as well as large-scale leakage. In general, such cavities remain filled with loose river deposits. In their treatment, these cavities are first filled and compacted with inert substance and then treated by grouting to ensure the stability of the structure and to arrest leakage.

EXERCISES

Multiple Choice Questions

Choose the correct answer from the choices given:

1. The karstic condition in a terrain is developed if the area consists of:
   (a) soft impermeable shale
   (b) soluble limestone
   (c) both (a) and (b)
2. In a reservoir project, leakage can take place through cavities in limestone formed:
   (a) only by solution in pure limestone
   (b) in impure limestone
   (c) in both pure and impure limestone
3. Limestone cavities are formed by:
   (a) only surface water
   (b) only underground water
   (c) both surface and groundwater
4. Geophysical study by the process of temperature logging in a karstic terrain at depth indicates the presence of karstic condition of cavity formation at depth by the:
   (a) fall of temperature from one level to another
   (b) high temperature at all levels
   (c) low temperature at all levels
5. Stalactites are formed from solution of limestone and deposition of calcium carbonate:
   (a) on the floor of a surface cavity
   (b) in the roof of an underground cavity
   (c) on the floor of an underground cavity
6. Stalagmites are formed from deposition of calcium carbonate:
   (a) on the floor of a surface cavity
   (b) on the floor of a subsurface cavity
   (c) in the roof of a subsurface cavity
7. Speleologic study is conducted for limestone cavity:
   (a) from surface by exploratory drill holes
   (b) from surface by geophysical method
   (c) entering subsurface and studying physical, geological, and biological aspects
8. Tracer study is taken up to investigate:
   (a) dimension of underground cavity
   (b) whether there is connection of cavities within a reservoir basin
   (c) whether cavities of a reservoir basin are connected with those of lower valley
9. Tracer study uses substances such as:
   (a) common salt
   (b) fluorescent dye
   (c) radioactive substance
10. There are several dam and reservoir projects taken up in the world in cavernous limestone threatening large-scale leakage in which:
    (a) the project had to be abandoned
    (b) the dam and reservoir sites had to be shifted to save the situation
    (c) very expensive grouting was undertaken to treat the cavernous limestone
11. The Hales Bar dam and reservoir project of the US constructed in karstic limestone faced problem of excessive leakage after completion of the project and it required:
    (a) rejection of the project
    (b) 30 years to contain the leakage
    (c) taking extensive and expensive treatment

Review Questions

1. What is karst? What type of rock is responsible for creation of karst condition? Describe the processes of formation of karst topography.
2. Describe the chemical reaction of limestone with acidic water. Explain how limestone forms solution and creates caverns. Write short notes on stalactite and stalagmite.

3. Explain how large caverns and solution channels are formed in pure and impure limestone.

4. Give an account of the formation of cavities of various sizes and forms in surface and subsurface regions comprising faulted, sheared, and jointed limestone.

5. Name the approach followed in the investigation of karst terrain. Define speleology. Explain how speleologic investigation helps in the preparation of subsurface maps of large solution channels of karst areas.

6. Give an account of geophysical and drilling explorations that provide aids to detect subsurface caverns and obtain subsurface information of karstic terrain.

7. Name the different types of tracers used to prove the possibility of leakage of reservoir water under full storage condition. Write briefly the procedure of chemical as well as radioactive tracer study.

8. What are your inferences about the construction of a reservoir project in limestone from the case studies of limestone projects in other countries? State the nature of remedial treatment undertaken in different parts of the world to seal the probable leakage paths from a limestone reservoir to lower valley.

Answers to Multiple Choice Questions

1. (b)	2. (c)	3. (c)	4. (a)	5. (b)	6. (b)
7. (c)	8. (b) and (c)	9. (a), (b), and (c)	10. (b) and (c)	11. (b) and (c)	

23 Guidelines for Writing an Engineering Geology Report

23.1 INTRODUCTION

A professional, either from the discipline of geology or engineering, after the completion of engineering geology work in a project site, submits a report on the work to the project authority. This chapter discusses the objective of writing such a report and the data that needs to be incorporated in it so that it serves the best purpose of planning and design of the engineering project. The report generally starts with a brief account of the project features followed by the morphology and geology of the area with special emphasis on the geological problems of the site and their solutions. The site plan showing the project features, geological map, and cross sections along with the laboratory test results on the engineering properties of materials including rocks of the site need to be submitted with the report.

23.2 OBJECTIVE OF AN ENGINEERING GEOLOGY REPORT

The objective of preparing an engineering geology report is to communicate the geological condition of a site to the concerned authority of the state or central government or to the private concern that deployed the services of engineering geologists to investigate the site. Engineering geology reports are mostly required for civil engineering projects with the object of providing information on the surface and subsurface geological conditions of the project sites. The information given with recommendation in a report helps in the preparation of design of the engineering structure and its safe construction.

In addition, engineering geology reports are required on natural hazards such as earthquakes, tsunamis, and landslides to reveal their causes and recommend remediation methods to the government authorities. When a stretch of a highway or a part of a communication system is affected by landslides, an engineering geology report is of immediate importance for suggesting measures for arresting the slides and restoring the communication system at the earliest.

The importance of preparing an engineering geology report with comprehensive data cannot be overstressed. In fact, engineering geologists are involved in investigating various types of geology-dependent problems and reporting to the concerned authorities, providing suggestions to solve the problems. The report carries specific information on the practical aspects of geology covering the stability of site and weak geological features and recommendation for improvement of site conditions. Engineers need the report for proper understanding of geological defects so that the design can be made with cost estimate for remedial measures.

23.3 BASIC ASPECTS

In the book *Application of Geology to Engineering Practice* (Berkeley 1950), special stress was laid on the importance of writing engineering geology reports in simple language so that the report is easily comprehensible to engineers. This holds good even today.

The work of an engineering geologist will not be appreciated if it is not easily understood by engineers because of the use of high-sounding geological terms or its manner of presentation. Attewell and Farmer (1976) also observed that 'the most important part of the investigation report (engineering geological) will be the actual results *clearly and objectively presented.*'

In fact, report writing is the most vital part of the activities of an engineering geologist. The report must fulfil the requirement of the engineers. The data in the report should be exhaustive and the manner of reporting should be such that it can serve the desired purpose of the engineers without any further inquiry.

The engineering geology report may be written using engineering terms wherever applicable. In principle, the report should be clear and concise, but at the same time it should carry all essential geotechnical data related to the work or project site. Depending upon the engineering need, the report may carry a short description of the morphology of an area, but detailed description of the geology and comprehensive data on soil and rock properties in quantitative terms are necessary for cost-effective design and safe construction of the project.

An engineering geologist may investigate a project site by a single visit working in the field for a short time and then report on specific geological problems. However, all major engineering works are done in different phases such as preliminary phase, design phase, and construction phase, and the engineering geologist's association with the work of a major engineering project is required from its beginning until the completion of the construction. At the end of the field investigation for an engineering project, the geologist submits a report to fulfil the engineering needs for each phase of the work.

23.4 GEOLOGICAL INPUTS

The *title* of the report should reflect whether it is of preliminary, feasibility, or construction stage work. The type of the project, whether dam site, tunnel alignment, and so on, should also be known from the title. Next to the title comes the *abstract*, which highlights the nature of the project under investigation and the engineering geological work undertaken.

After the abstract, the report starts with an *introduction* which states the location and salient features of the project, the aims and objectives, or the purpose, and the nature of investigation.

The introduction of the report should clearly bring out the stages or phases of investigation for all major engineering works. This is followed by the main aspects, namely the descriptive parts of site geology.

The *description* of any matter comes under a meaningful heading. For example, under *subsurface geology*, the geological features interpreted or studied from exploratory drilling or geophysical investigations are described in detail. In the very early stage of work involving selection of a site out of two or more alternative sites given by the engineers, the merits and demerits of all the given sites are described in the report and then the best possible site for engineering construction based on geotechnical consideration is suggested.

The salient features of the alternative sites are also presented in a tabular form highlighting the superiority of the selected site over the others. A report ends with one or two paragraphs dealing on *conclusion and recommendation*.

The presence of any *adverse geological condition* at the site under investigation needs to be described in sufficient detail to give a clear understanding of its role in the foundation of the engineering structure and the measures to be adopted to overcome the adverse geology. Suppose a fault traverses the foundation of a heavy structure; mere reporting about its presence in the foundation is not sufficient for an engineering design. The report must clearly bring out its significance mentioning that it is a weak feature and is likely to create a problem of subsidence or differential settlement as may be the case under stress of the structure.

The report should also discuss the details of such a weak feature such as the length, width, and depth of affected stretch due to the fault including its position and orientation with respect to the engineering structure and the nature of *treatment necessary*. Such details help engineering design to include the cost of treatment.

It is very important to present the field observations in an intelligent way. All observational facts mentioned in the report should be meaningful and related to the problems of the engineering structure proposed for construction. If presented with maps and tables, the observed features and data of field measurements become more meaningful than when provided mere description. Such data must be presented in the report along with *maps, cross section, graphs*, and *diagrams* to facilitate engineers to prepare their report on the design of the structure.

When provided with illustrative maps, sections, and tables, the subject matter of the report is easily understandable and can be utilized by the engineer for structural design. For example, contoured geological map and geological cross section, submitted with an engineering geology report, are essential for the engineering design of structures such as dams, tunnels, and powerhouses.

As far as practicable, the descriptions of field observations should be supported by factual evidence and quantitative data. The results of soil or rock *laboratory tests* are to be presented in a tabular form and graphs with a site plan showing sampling sites. While mentioning about subsoil or rock condition, it is necessary to submit a cross section showing the geological features at depths. In the description of subsurface *hydrology*, providing simple numerical data in case of *water table* may not give the true picture of its place of existence. It becomes more meaningful if it is shown in a geological cross section drawn along the boreholes used to measure water table.

The *core logs* of drill holes are also to be presented schematically in log charts (Section 11.7, Figs 11.21 and 11.11). Submission of a large sheet of field data is not desirable. Instead, graphs, cross sections, and so on prepared from the bulk data with suitable scale may be submitted with the report.

The broad guidelines followed by engineering geologists of GSI in writing reports for different stages of civil engineering works including the types of information to be furnished in these reports are enumerated in Section 23.5.

23.5 REPORTS FOR DIFFERENT PHASES OF SITE INVESTIGATION

Reports are mainly created during the three phases—Planning phase, Design phase, and Construction phase. Depending on the project features, physiology and feasibility of the project, planning phase reports cover the suitability of the proposed site from geotechnical considerations. It is always advisable to include a recommendation while preparing planning stage reports as it helps with decision-making. If the project is in design phase, details of expansive soil including the area of its occurrence are elaborated. For construction phase reports, notes written by the resident geologist on the geological features should highlight immediate attention of the engineers for construction purposes involving treatment of weak zones. It is to be noted that the details of geological defects observed and suggestions given for their corrective measures are the most important aspects to be included in any construction stage report.

23.5.1 Planning Phase Report

The following are the points to be noted when submitting an engineering geology report in the planning phase:

- State briefly the project features supplemented by a contour plan showing the location of the site or different sites under the project.
- Provide a short description of the physiology, regional geology, and site geology including major geological structures of the area.
- Include available small-scale geological map of the area or that made from aerial photographic study.
- Mention the earthquake status with respect to seismic zone of the project area.
- Mention the probable environmental impact due to the project construction. Also, mention whether the construction of a reservoir project may lead to loss of any archaeological treasure or will endanger the wildlife in that area.
- In *conclusion and recommendation* part at the end of the report, express opinion on general feasibility of the project and suitability of the proposed site from geotechnical consideration. In case of alternative sites, recommend the one with the best geological condition. Recommendation may also be given if a better location is sited from field observation than the one or those proposed by the engineers. The programme of further geological work necessary including exploratory work and laboratory tests of soil or rock for a project site is also outlined in the recommendation.

23.5.2 Design Phase Report

The following aspects need to be included in the report submitted in the design phase:

- Provide detailed project features and include a map showing the single or different sites or alignments depending upon the nature of the project.
- Report the general topographical features of the project area and detailed morphology of the site observed from aerial photographic interpretation, followed by ground checking

in and around the structural sites. Describe the main features to be observed at the site including rock, soil, and overburden conditions; morphology, landslope, erosional feature, and drainage pattern; and stability status of the site.

- Describe site geology in detail as revealed from investigation of the site by large-scale geological mapping. Include geological map of the site showing rock types with attitudes and structural details.
- Describe the subsurface geology including foundation grade and thickness of overburden from results of exploratory work. Enclose a geological cross section showing the bedrock and fresh rock levels.
- State groundwater condition and its influence in the project area. This is especially important while submitting report on the investigation of tunnel alignment or sites for subsurface structures.
- Mention the earthquakes that have affected the project area under investigation consulting old earthquake records and seismic zonation map of India. The information helps the design engineer to decide about the seismic factor to be incorporated in the design of the project structure.
- Discuss the problems that may arise due to adverse geological structures such as a fault or shear zone in the foundation and suggest the remedy.
- Provide information regarding quality and quantity of construction materials available in and around the project area for the construction of the proposed structure corroborating laboratory and field test data for their engineering properties.
- Give details of expansive soil, if present, including the area of its occurrence and its possible adverse role on stability and the relevant remediation measures.
- For design purposes, mention the depth to sound rock suitable for founding the structure and provide engineering properties such as strength, compressibility, and permeability of the foundation rocks in a tabular form together with in-situ stress and groundwater condition when available.
- Provide a summarized account of the laboratory or field test results on rock and soil in the body of the report in a tabular form and the details as appendix of the report. Submit the logs of drill holes in proper log charts.
- Depending upon the type of the project, provide the foundation condition in *conclusion and recommendation* in clear terms, including the defects of the rock or soil in any part of the structural site and the method of treatment for their remedy.

23.5.3 Construction Phase Report

The following are the points to be noted when submitting an engineering geological report in the construction phase:

- A resident engineering geologist in the construction stage provides short reports or notes based on the site inspection. The geologist meets the engineers to discuss geotechnical problems observed after excavation of the site and suggests remedy whenever required. The observations, suggestions, and discussions are reflected in these reports.
- In-field reports or notes are written by the resident geologist on the geological features that require immediate attention of the engineers for construction purposes involving treatment of weak zones. Such notes sometimes require approval of the supervising officer and may be routed through headquarters to the engineers in a short time.

- Some geological 'surprises' may appear in the foundation after excavation at the construction stage. The resident geologist studies the geological features related to such surprises and submits a report explaining the significance of these earlier unknown geological features and the associated construction problems, if any. The engineering design is generally kept flexible to cater to the need of such changed circumstances due to unknown geology.

- A detailed report of construction stage work is submitted to headquarters covering all geotechnical works carried out from the site excavation until the completion of construction. The details of geological defects observed and suggestions given on their corrective measures are the most important aspects to be included in the construction stage report. This report will also include large-scale geological maps of foundation and documents of other survey works to serve as valuable guide for taking measures on any post-construction foundation problem.

23.6 REPORT FOR SPECIAL INVESTIGATIONS

The services of engineering geologists are also utilized for short-time work related to specific geological investigation such as search for railway ballasts or to ascertain the stability of a habitation site or a hill slope traversed by a road or rail line, subsoil condition of airport, and foundation of bridge piers. In all these cases, engineering geologists are involved in field investigations including exploratory works to find subsurface soil or rock condition. The report of such investigations contains the morphology, details of rock or soil of the site, and especially adverse geological conditions, if any, and the remedial measures of geological weaknesses.

This report, like any engineering geology report, first states the features investigated and the problems inferred and then recommends, whenever necessary, for further geotechnical work at the site by exploration to reveal the subsurface condition of rock or soil. When the exploration is completed, the engineering geologist revisits the site and submits a detailed report describing the substrata condition and outlining the remedial measures in case of adverse geological condition of the site. The report also contains geological map and records of subsurface exploratory works.

23.7 IN-FIELD PREPARATION OF WRITE-UP FOR A REPORT

The following suggestions help in the preparation of a comprehensive report with maximum inputs of field observations:

At the end of a day's fieldwork or after investigation for a couple of days, the geologist should prepare a detailed write-up on the observation in the field itself when it will be possible to remember every single detail of the field observation. As the work progresses, this daily or time-to-time prepared write-up can be further updated, so that it finally takes the shape of a draft report. If field testing is done on rock or soil, the geologist can write up the methods followed, the result obtained, and any other related matter in the same form to be submitted to the headquarters.

In the field, the geologist can also prepare a fresh copy of geological map, geological section, drill hole log sheets, and other data chart or tables, as may be required to be submitted with the report as supporting evidence. In this way, it is possible to document all the field observations in the form of a draft report. If these details are not updated as and when the observations are made, and the geologist waits until returning to the headquarters to write the report, minute details and even some vital observations on geological problems may be forgotten or left out. In the headquarters, the final report can be submitted after adding laboratory test results on rock or soil and making the necessary minor modifications after consulting other professional reports or relevant literature on earlier work in the area.

23.8 ILLUSTRATIVE EXAMPLE OF AN ENGINEERING GEOLOGY REPORT

Selected portions of a report titled *Reconnaissance Investigation of Three Alternative Alignments for a Diversion Tunnel in Eastern Nepal (1965–1968)* written by the renowned engineering geologist Dr J.B. Auden are provided in this section as an illustrative example of an engineering geology report. The full text of the report (20 pages) may be read from the publication (Auden 1971).

It may be noted from Dr Auden's report that the title of the report itself gives a clear idea regarding the type and location of the engineering project and that it is a preliminary stage report. This report is illustrated by seven geological maps with cross sections of subsurface strata and four tables bearing quantitative data. The report starts with an abstract followed by introduction and then description of the relevant matters under the following eight headings:

1. Introduction
2. Question of feasibility
3. Topographical survey
4. Geological environment and structure
5. The three alignments
6. Geological conditions and seismicity
7. Formations along the alternative alignments
8. Conclusion

Abstract

Studies were made by the author about the feasibility of a project for utilizing the waters of the river Sun Kosi in Eastern Nepal for the combined purposes of irrigation of the Terai region and generation of hydroelectric power. The project would require the construction of diversion barrage, pick-up weir, tunnel, and powerhouse in the Sun Kosi basin. The complex geological and tectonic set-up of the area has been discussed through which three alternative tunnel alignments varying in length from 6,220 m to 14,000 m are proposed within a distance of 85 km along the river.

To satisfy the topographic and engineering requirements, the alternative tunnel alignments are to traverse some of the formations such as Nathens, Daling, and Darjeeling gneiss. Out of the three alternatives, two alignments are to intersect the main boundary fault, Daling megathrust, and a few fault planes. All the three alignments fall within the high-intensity seismic belt close to the Gangetic downwarp in front of the Himalayan chain.

A classification of the rock group liable to major or minor overbreaks within the tunnel section has been enumerated on the basis of the physical conditions and inherent strength of the different rock groups to be encountered within the alignments.

Besides the geological considerations, the report deals with other aspects requiring dovetailing of the tunnel diversion project with other developmental river valley projects along the main path of the river Kosi.

Introduction

While a member of the Food and Agricultural Organization (FAO), Rome, the author made a brief reconnaissance investigation in 1965, 1966, and 1968 of the alternative alignments for a diversion tunnel between the river Sun Kosi and the terrain south of the Mahabharat Lekh. The project involved the diversion of water from the river for irrigation in the Terai plains and the generation of hydroelectric power. The work was part of the feasibility study undertaken by FAO first with its own staff and subsequently under contract with the firm of consulting engineer Nippon Koei of Tokyo.

In all, three alignments were studied, now termed 'A', 'B', and 'C', while a fourth alignment 'D' was also later considered by Messrs Nippon Koei. The ESE–WNW distance between the alternatives A and C is 85 km from coordinates 26°50′:86°49′ to 27°14′:86°05′. It was assumed for purpose of initial reconnaissance that the internal diameter of the diversion tunnel would be 7.8 m, with a gradient of 2.18 m/km, and that the pressure tunnel would deliver 140 m³/s.

The project would require the construction of a diversion barrage on the Sun Kosi, a tunnel under the Mahabharat Lekh, a power station at the tunnel outlet, a pick-up weir on either rivers Kamala or Trijuga, and a system of canals in the Terai.

Comparative analysis of tunnel alignments

Parameters	Tunnel C (Goltar)	Tunnel B (Chiptar Ghat)	Tunnel A (Khampu Ghat)
Intake elevation (m)	451	340	183
Catchment area (km²)	10,100	16,250	17,600
Length of tunnel (m)	14,000	13,800	6,220
Outlet of tunnel	Chandaha Khola	Tawa Khola	Trijuga Khola
Annual discharge (m³)	380	510	550
Available water (m³)	80	130	140
Width of Kosi river bed at diversion barrage (m)	170	180	180
Geological condition	Complex	Complex	Relatively simple
Geological formations along tunnel alignment	Nahan, Gondwana, Daling, and Darjeeling gneiss	Nahan, Gondwana, Daling, and Darjeeling gneiss	Nahan only
Thrust plane	Main boundary Daling megathrust	Main boundary Daling thrust	None observed
Faults	Not studied in detail	Hunter and Sarang fault complex	No important fault observed
Seismicity	The regional seismicity must be regarded as strong and as equally affecting all three alternative projects.		

Conclusion

Only a few of the problems concerning the alternative alignments for a tunnel diversion project have been discussed in this report. There is no doubt that from the point of view of geological conditions, the least complex alignment would be that between Khampu Ghat and Deoghar (alignment A). This tunnel would be driven entirely within the lower Siwaliks and would avoid the two major regional overthrusts, which would be present in B and C alignments. The tunnel would be half the length of B and C tunnels and the available water from the catchment of 17,600 km^2 would be much more than that further upstream.

The geological and hydrological characteristics of each tunnel alignment are not, however, the only factor that needs to be considered. This assignment of water potential, making full use of the local topography with subsidiary structure and tunnels, power production, different irrigable areas, and case of canal distribution, is complex. It is understood that the different factors have been the subject of computer analysis in Tokyo since 1968. There is further problem of dovetailing the tunnel diversion project with the possible construction of a high dam on the Sapta Kosi and a reservoir that would extend far up the Sun Kosi tributary. Finally, it may well be that the most economical project for the irrigation of the Terai will be by utilization of the groundwater, without recourse to tunnel diversion. Whatever the ultimate outcome, the work undertaken along Sun Kosi and Mahabharat Lekh will be of some value from a geological and engineering perspective.

SUMMARY

- An engineering geology report of a project site is written in a manner that can be clearly understood by the engineers. The report reveals the geological condition of the project site and helps the engineers in planning and design of the structures.
- The information required to be furnished in an engineering geology report varies depending upon the type of the project or the nature of the engineering construction.
- An engineering geology report must have a title depicting the nature and place of work. The report is then written under different heads beginning with an abstract of the work done, followed by information on preliminary aspects such as project features and aims and objective of the work undertaken.
- The main geotechnical aspects such as the physiography, geology, and geohydrology of the area are then described in detail. Geological maps, geological cross sections, drill hole logs and their diagrammatic representation, and other exploratory works prepared in tabular form are included with the report to illustrate the descriptions.
- At the end, all reports contain a *conclusion and recommendation* section where an opinion is expressed by the author on the feasibility of the project highlighting the adverse geological features and suggesting their remedial measures.
- Engineering geological work on a project is carried out in different stages or phases of engineering works. The report of the planning stage work describes the morphology and general geology of the project area and also gives considered views about the suitability of the structural sites and overall feasibility of the project.
- When several alternatives sites are investigated, the report gives a comparative statement of geotechnical aspects of all the sites, mentioning the most favourable one.
- The design stage report contains all details of surface and subsurface geology revealed from site exploration by drilling or geophysical studies. Detailed geological maps, cross sections, tables, graphs, and so on required for design purposes are submitted with the report.
- The construction stage final report includes all works carried out beginning from site excavation until the completion of construction and contains several maps, sections, and other documents prepared during the entire construction work.

Multiple Choice Questions

Choose the correct answer from the choices given:

1. The objective of writing an engineering geology report is:
   (a) to communicate geological condition of engineering project sites for planning and design of project
   (b) as a routine office work to inform the geologist's activity to the higher authority
   (c) for office record of engineers to find if any adverse condition will happen in future to the project

2. The stage of work-activity at which an engineering geology report is to be submitted is after:
   (a) in-office study of site condition from remote sensing or aerial photographs
   (b) completion of field or site investigation in planning stage and also after completion of design stage wok, but time to time during construction stage when the construction work progresses
   (c) engineering construction faces any trouble due to geological defect

3. The initial or planning stage report contains:
   (a) description of the geology with a geological map of the site selected by the project engineer
   (b) selection of a site out of several alternative sites, which is considered favourable from both geological and engineering consideration
   (c) description of geological and structural features especially weak features, if any, and their significance, seismicity status, and suggestion for exploratory works showing location of drill holes

4. The content of the design stage report will include:
   (a) large-scale geological map and detailed description of structural features and recommendation for exploratory work
   (b) geological map, geological cross section, record of drill core study (core log), groundwater level, result of laboratory test results of subsurface rock or soil, and in some cases field test results, source of construction materials, adverse geological features, and measures for their rectification

   (c) site plan showing location for alternative site or sites with description stating why the site (selected by project engineer) is not found geologically favourable

5. The general expectation from a construction stage report and the time when it is submitted is:
   (a) any weak geological features such as a fault, a shear zone, or excessive joints observed at the site after excavation and suggestion for remedial measures for the weak zones are reported as and when observed. A comprehensive report is submitted also at the end of construction with all data collected from fieldwork and laboratory tests
   (b) a comprehensive report is submitted after completion of project containing details of engineering construction.

6. In the investigation of an area affected by natural hazards such as landslides a special-purpose engineering geological report should contain:
   (a) an estimate of the loss of public and private properties and recommendation for monetary help from the government
   (b) the causes and effects of the landslide and the measures required to stabilize the affected area and also suggestion on the type of construction activity for houses and engineering structures in the area on the basis of zones of seismicity and BIS guidelines.

Review Questions

1. Explain the types of work to be covered in the initial and planning stages of investigation in the project sites.
2. List the sequence of topics to be written in geotechnical reports.
3. What is the information to be furnished during the planning and design stages of report on the work-activities in an engineering project?
4. What are the subjects (field and laboratory data) that would go as enclosure with the report for assisting design purposes?
5. What is the data to be included in the final stage of the report on the work-activity in an engineering project?

Answers to Multiple Choice Questions

1. (a)   2. (b)   3. (b) and (c)   4. (b)   5. (a)   6.(b)

24 Physiography, Stratigraphy, and Ores and Minerals of India

● ● ● ● ● ● **LEARNING OBJECTIVES** ● ● ● ● ● ●

After studying this chapter, the reader will be familiar with the following:

- Physiographic features of India
- Rock formations of India and their occurrence
- Important ores and their occurrence
- Precious and semiprecious stones of India

24.1 INTRODUCTION

The physiographic features of India are related with the varied geology of the country. The three distinct physiographic forms include the peninsular region of the south, the Himalayan terrain of the north, and the Indo-Gangetic plains in the central part washed by the rivers Indus and Ganges and their tributaries. India has a diverse geological set-up starting from the oldest rock formations to the younger alluvium. Different regions of the country are constituted of rock types of different geological periods such as the Precambrian gneiss, schist, and granites in the peninsula and eastern India, sedimentary metamorphic rocks of the Dharwar and the Cuddapah of the South, the Vindhyan formation of central India with good quality building stones, the Gondwana formation of Damodar valley with best coal seams, and the Deccan trap area of the Peninsula with vast sheet of lava flow hardened to basalt. Gold, iron, copper, aluminium, chromium, and other ores and many minerals of economic importance including uranium also occur in the rocks of the country. All these aspects form the subject matter of discussion in this chapter.

24.2 PHYSIOGRAPHIC FEATURES OF INDIA

The physiographic form of the Indian subcontinent today is the result of the vast geological formations of this land mass being a part of the primordial crust of the earth. India can be divided into three distinct physiographic divisions:

(i) The peninsular plateau
(ii) The extra-peninsular region comprising the Himalayas
(iii) The Indo-Gangetic plains

It can also be stated that the Indian peninsula with its oldest rock formations was a part of the ancient Gondwanaland, which drifted northwards and striking with the Central Asiatic plates rose up to form the Himalayas out of Tethys Sea. The Indus and the Ganges originated from the Himalayas and with their tributaries deposited piles of sediments through geological times to form the vast tract of alluvial plains.

24.2.1 The Peninsular Plateau

The peninsular plateau occupies the southern part of India fronting on two great arms of the Indian Ocean—the Arabian Sea on the west and the Bay of Bengal on the east. The peninsula is also called peninsula *shield* as it is the most stable land mass of the earth's crust composed of ancient rocks. The land mass is roughly triangular in shape and consists of a vast tableland broken by the river valleys. The land slopes gently downwards towards the east.

The peninsular area is washed by the rivers Narmada, Tapti, Godavari, Krishna, and Cauvery, the old rivers of India. The Aravalli range, which is the eroded land mass of a very old mountain range of the geological past, covers the north-western part of the peninsula. The Vindhyan range (Fig. 24.1) and the Satpura range are the other mountain systems occupying the north and central parts of this ancient land mass.

The east and west coasts of India that border the eastern and western parts of the peninsula are of uniform and regular nature. A large portion of the coasts is sandy and the gently shelving coastal strip is washed by shallow sea. There are raised beaches with lands extending over areas of both the coasts. The beaches lie above the level of highest denudation by the tides. There is evidence that the western coast is slightly higher in level than the eastern coast.

Fig. 24.1 The Vindhyan hill in Madhya Pradesh comprising horizontally-to-sub-horizontally bedded red sandstone

There is a scanty margin of alluvial deposit on the western coast except in Gujarat, whereas there is a wide belt of riverborne alluvium on the east coast in addition to the great deltaic deposits of the rivers Mahanadi, Godavari, Krishna, and Cauvery. Such deltaic deposit is not found in the rivers flowing in the west, namely the river Narmada and the river Tapti. The other significant feature is that whereas all other rivers flow to the east, these two rivers flow towards the west—a feature explained by the fact that these two rivers do not flow in channels formed of their own erosion, but run along fault-controlled tectonic valleys.

24.2.2 The Extra-peninsular Area

The extra-peninsular region occupies the northern parts of India, extending over nearly 2000 km from Arunachal Pradesh in the east to Jammu and Kashmir in the west and being governed by the Himalayas (Fig. 24.2). The Himalayas is classified into three parallel or longitudinal zones, each having distinct well-marked orographic features. The following are these zones:

(i) The Great Himalayas is the innermost line of high ranges, rising above the limit of the peripheral snow. The average height extends to 6500 m. High peaks such as Mount Everest, K2, Kanchenjunga, and Nanga Parbat are situated here.

(ii) The Lesser Himalayas or the middle ranges is a series of ranges that is closely related to the Greater Himalayas but is of lower elevation; they do not rise above 4000–5000 m. The Lesser Himalayas forms an intricate system of ranges whose average width is about 80 km.

(iii) The Outer Himalayas or the Siwalik ranges lies between the Lesser Himalayas and the plains. It is formed of a group of low hills that are about 1000–1300 m high.

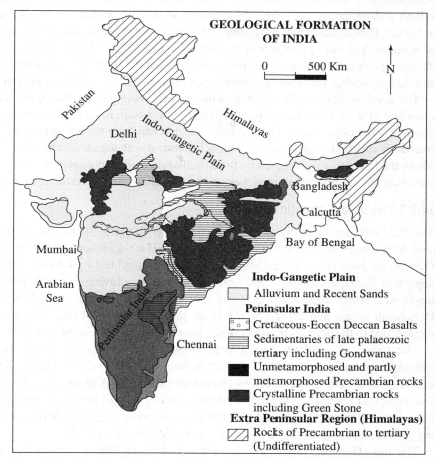

Fig. 24.2 Simplified geological map of India showing major geological formations and three physiographic divisions

Geologically, the northern or Tibetan zone, lying beyond the line of highest elevation, is composed of a continuous series of fossiliferous marine sedimentary rocks, ranging in age from the Palaeozoic to the Eocene age. The Central or Himalayan zone, comprising the Lesser or Middle Himalayas, is composed of crystalline or metamorphic rocks such as granite and gneisses with unfossiliferous sedimentary deposits of ancient age. The Outer or Sub-Himalayan zone corresponding to the Siwalik ranges is composed of Tertiary, especially upper Tertiary, sedimentary river deposits.

Several high ridges intersected with deep valleys are the characteristic features of the Himalayan mountainous region. The three most famous rivers of India, namely the Indus, the

Ganges, and the Brahmaputra, originate from the snow-clad peaks of the Himalayas. The Great Himalayas and the Trans-Himalayas in the further northern parts are composed of metamorphic rocks such as quartzite and phyllite, at places with intrusive granite. The extra-peninsular region has undergone several tectonic events causing extensive folding, faulting, and thrusting of rock formations of this Himalayan terrain.

The prominent features of the extra-peninsula, the mountain border of India formed by the upheaval of the crust during the late Tertiary time, are the internal structures made up of series of broad anticlines and synclines. The anticlines are faulted in their southern limbs lying in juxtaposition with their younger limbs. The outer ranges are separated by narrow longitudinal valleys or depressions. The reverse faults are the most common characteristic feature in the tectonics of the sub-Himalayan ranges. Overfolds, inversions, and thrust planes assume an increasing degree of intensity. The most prominent is the *Main Boundary Fault* (MBF) at the southern boundary of the extra-peninsula. Several of the mountain ranges of the Himalayas are separated by deep defiles and valleys, having a number of watersheds, peaks, and passes and other rugged features.

The northern flank of the Himalayas hinges with the gigantic Tibetan plateau. The south-east extension of the Himalayas is characterized by the Assam ranges, which have somewhat unique physiographic expressions. The extremely rugged and lofty central ranges of the Himalayas are constantly capped by snow and ice. This is in contrast with the mountains of the Lesser Himalayas where the land mass is occupied by the Siwalik ranges. The conspicuous features of this part are the series of escarpments, deep slopes and broad valleys, which result in a peneplain landform.

24.2.3 The Indo-Gangetic Plains

The third physiographic division, namely the Indo-Gangetic plains, encompasses the extensive area from Assam in the east to Punjab in the west. The region is named after the rivers Indus and Ganges, the twin river systems that drain it. Geophysical test by GSI has estimated a sediment deposit of thickness 100 m to 5000 m in the Gangetic alluvium. The Gangetic plains owe their origin to a crustal depression that was later filled by sediments created by the erosion of the Himalayan terrain and carried by the rivers through geological time. The persistent flatness is entirely due to the aggrading works of the rivers of the Indus–Ganga system.

The Indus, which flows with a large amount of water at the western part, has five main tributaries, namely Sutlej, Beas, Ravi, Chenab, and Jhelum. The Ganges in its downstream course from the Himalayas is fed by tributaries such as the Alakananda, Bhagirathi, Mandakini, Jhanhabi, and Yamuna. 'These alluvial plains of the Ganges are of absorbing interest in human history, being thickly populated, and the scene of many important developments and events in the cultural and social history of India', observed Krishnan (1949).

The Indo-Gangetic plains, also known as the 'Great Plains', run parallel to the Himalayan mountains from Jammu and Kashmir in the west to Assam in the east, draining most parts of northern and eastern India. The plains encompass an area of 700,000 km^2. The area is divided into two distinct drainage basins by the Delhi Ridge. The eastern part consists of the Ganga–Brahmaputra drainage systems and the western part consists of the Punjab plain and the Haryana plain. This divide is only 300 m above the sea level. The Ganges in its lower part flows into the ocean and forms the Gangetic delta.

The large area covered by the Indo-Gangetic plains is classified into the following four divisions:

(i) The *Bhabar* belt is at the foothills of the Himalayas. It consists mainly of boulders and pebbles and is generally narrow, about 7–15 km in width. Here, the streams flow below the porous detritus.

(ii) The *Terai* belt lies next to the Bhabar region and consists of newer alluvium. The rivers flowing subsurface in the Bhabar region reappear in this zone, which receives heavy rainfall throughout the year. The area is thickly forested and populated by wildlife.

(iii) The *Bhangar* belt consists of older alluvium and forms the alluvial terraces of the flood plains. In parts of the Gangetic plains, the low upland areas are covered by laterites.

(iv) The *Khadar* belt lies in the lowlands, covering areas of West Bengal. It is made up of freshly deposited newer alluvium, which are piles of sediment carried by the rivers flowing down the plain.

The Indo-Gangetic plain is considered to be the world's most extensive expanse of uninterrupted alluvium, formed by the deposition of the numerous rivers. The plains are conducive for irrigation through canals, and the areas are extensively irrigated mainly for crops such as rice and wheat as well as maize and sugarcane. Indo-Gangetic plains are one of the world's most densely populated areas.

24.3 STRATIGRAPHY OF INDIA

India possesses rock formations from geologically very old to very recent times. A simplified classification of Indian rocks has been given in Table 24.1 following Wadia (1966) and Krishnan (1949).

Table 24.1 Geological formations of India

Era/period	*Period/epoch*	*Major rock formations of India*
Quaternary period	Holocene or recent	Newer alluvium
	Pleistocene	Older alluvium and Karewas of Kashmir
Tertiary period	Pliocene Miocene	Siwalik system and Assam sedimentaries
	Oligocene	Murree and Barail series
	Eocene Palaeocene	Deccan trap
Mesozoic era	Cretaceous	South coast, Assam, and Himalayan system
	Jurassic Triassic	Gondwana system
Palaeozoic era	Permian	Himalayan sedimentaries
	Carboniferous Devonian Silurian Ordovician	Devonian limestone of Hazara and Chitral Kashmir sedimentaries, Muth quartzites, and Shimla slates
	Cambrian	Cambrian of Salt Range and Cambrian of Spiti and Kashmir
Precambrian	Proterozoic	Vindhyan system and Cuddapah system
	Archaean System	Dharwar system, Aravalli series, Shillong series Gondite series, and Kodurite series Gneissic complex, Charnockite series
(Archaean Complex)	Gneisses and Granites	Bengal gneisses, and Bundelkhand gneisses

24.4 THE ARCHAEAN COMPLEX

'The Archaean complex possesses the oldest rocks of the crust forming the core of the mountain chains of the world. They are thoroughly crystalline, extremely contorted and faulted and intruded by plutonic rocks and generally have deformed and foliated structures and as such named fundamental complex or the basement complex', observed Wadia (1966).

24.4.1 Gneisses and Granites

The Archaean complex of India belongs to Precambrian fundamental basement complex intruded by several plutonic rocks such as the Bengal gneisses (Fig. 24.3), charnockites, and Bundelkhand gneisses. The highly schistose sediments, the para-gneisses and schists such as Dharwars and Aravallis, are younger than the gneisses.

Fig. 24.3 Gneissic rock of Archaean complex with contorted structure (Wadia 1966)

Fig. 24.4 A barren hill top of charnockite with a spheroidally weathered rock mass in Tamil Nadu

Bengal gneisses These rocks belong to the oldest member of the Archaean complex. They are highly foliated and are, at places, schistose in nature. They are found in West Bengal, Bihar, Orissa, and Karnataka. The rocks are also found to contain bands of limestone, dolomite, hornblende schist, and corundum rocks. The rocks have formed a dome-shaped structure in the northern part of West Bengal and are called dome gneisses.

Bundelkhand gneiss This gneiss is sparsely foliated and occurs as pink granite. Hornblende and chlorite schist are associated with the gneiss. The rocks are traversed by dykes and sills and at places by pegmatite. The gneissic rocks are also found in parts of the Peninsula and are known as Balaghat gneiss, Hosur gneiss, and so on. They were used in building many temples of south India.

Charnockites series Charnockites occur in the peninsular region, especially in Tamil Nadu (Fig. 24.4) and Nilgiri hills. The rocks have a distinct petrological province that is different from any other occurrences of the gneissic rocks. It was first observed in the hills in Tamil Nadu and was named by Sir T.H. Holland, the then Head of GSI, as charnockite after Job Charnock, whose tomb in Kolkata is made of this rock. Petrological composition of charnockite includes hypersthene, blue quartz, plagioclase, hornblende, and biotite and accessory minerals such as zircon, iron oxide, and graphite. The rocks are formed by magmatic differentiation and segregation and show fine-grained basic veins.

24.4.2 Precambrian Sedimentaries

There are several sedimentary sequences (series) in India which show few or no metamorphism. Among them the most ancient is the Bijawar and Gwalior series, older than 1800 million years, and the Cuddapah series which is around 1700 million years old. The sedimentation in the Vindhyan series ranges from 1400 to 500 million years. The following subsections give details of each of these series.

Dharwar system

The Dharwar rocks are of sedimentary origin but are unfossiliferous and exist occupying a synclinal basin. Metamorphism of the rocks has influenced obliteration of the original forms, and presently the rocks possess crystalline and schistose characteristics and are intruded by granite bosses, veins, and sheets and by dolerite dykes. The rocks have diverse characteristics—clastic sediments, volcanic, and plutonic rocks metamorphosed to different extents. They contain iron ores associated with hematite and magnetite schists. The Dharwars also contain volcanic rocks such as rhyolite and andesite as well as crystalline limestone.

Dharwar system is of great economic importance as it contains many metallic ores such as gold, manganese, iron, copper, and even tungsten and lead. The manganese of Dharwars occurs in Gondites of Balaghat and Nagpur and also Kodurites of Orissa and Andhra Pradesh.

Fig. 24.5 Contorted and folded Aravalli sedimentaries of Rajasthan

Fig. 24.6 Aravalli sandstone metamorphosed to quartzite

Aravalli series

Dharwarian rocks are exposed in a large area covering the Aravalli range of Rajasthan. The Aravalli range is the most ancient mountain range of India and came into existence during the Dharwar era, when the sediments that were deposited in the seas were lifted up forming mountains. The metamorphism of the sedimentaries of the Aravallis gave rise to folded, extremely contorted, and foliated rocks (Fig. 24.5). The Aravallis are composed of quartzite (Fig. 24.6), mica schist, hornblende schist, and other feldspathic schists and gneisses.

The *Shillong series* of Assam developed at the same time as the Aravallis. These rocks consist of thick deposits of quartzite, arenaceous phyllite, slates, and schists with the intrusion of granite and basic interbedded traps.

Cuddapah system

The unfossiliferous sedimentary rocks of the Cuddapah system were formed in Proterozoic period after the Dharwar time when there were earth movements creating wrinkles in the Aravallis. The Cuddapah system is composed of compacted shales, slates, quartzite, and limestone with secondary minerals such as mica, staurolite, andalusite, and garnet. Volcanic eruption took place during the lower half of the system with the lava flow forming traps and tuff beds. The Cuddapah rocks are classified into two divisions, an upper and a lower. They contain several well-defined series that may be homotaxial as shown in Table 24.2.

Table 24.2 Classification of Cuddapah rocks

Upper Cuddapah	Nallamala series
	Krishna series
	Kaladgi series
Unconformity	
Lower Cuddapah	Bijwar series
	Cheyair series
	Gwalior series
	Basic volcanic series
	Papaghani series

Vindhyan system

The Vindhyan system (Fig. 24.1) of the Proterozoic period is a vast stratified formation consisting of sandstone, shale, and limestone of 4000 m thickness. These stratified rocks show unique uniform characteristics, having horizontal beds and red to buff colour (Fig. 24.1). The Lower Vindhyan rocks, however, experienced igneous intrusion containing granite and rhyolite.

The Vindhyan rocks are classified under two divisions. The entire rock succession of the two divisions is formed with an unconformity as shown in Table 24.3.

Table 24.3 Classification of Vindhyan rocks

Upper Vindhyan	Bhander series	Sandstone, shale, and limestone
	Rewa series	Sandstone and shale
	Kaimur series	Sandstone, conglomerate, and shale
Unconformity		
Lower Vindhyan	Kurnool series	Limestone and calcareous shale
	Bhim series	Limestone and shale
	Malani series	Rhyolite and tuff
	Siwana and Jalore	Granite boss

Extensive occurrences of the Vindhyan rocks in high hills are found in Madhya Pradesh and also in Chhattisgarh, Uttar Pradesh, and Rajasthan. Throughout their thickness, the Vindhyan rocks show shallow water deposits with ripple marks and sun cracks However, the vast piles of sediments show no evidence of presence of any fossil. Vindhyan sandstones are excellent building stones and have been used in the construction of royal palaces and various monuments of India from ancient times.

24.5 PALAEOZOIC ERA

The Palaeozoic Era constitutes rocks from Cambrian to Permian ages. These rocks occur in a number of areas covering the Salt Range, Spiti and Kashmir areas of the Himalayas, the coastal areas of the South, and parts of Assam. The rocks of these places are described in this section with respect to the geological periods starting from the Cambrian age.

24.5.1 Cambrian System

Marine fossiliferous rocks of the Cambrian period were developed in Kashmir and Spiti areas of the Himalayas in Kumaon, north of Shimla, and in the Salt Range, northwest of Punjab. These rocks contain well-preserved fossils that help in fixing the age of rock formations. The lowermost Cambrian bed occurs in the Salt Range as Salt marl, overlying which is a shallow water deposit containing fossils of brachiopod Neobolus and other fossils, namely *Schizopolis*, *Lingula*, *Orthis*, *Redlichia* (a trilobite), and *Nyolithes* (mollusc). The brachiopod and trilobite resemble the Cambrian of Europe, which stamps the rocks as of Cambrian age. Fossiliferous sedimentary rocks of Lower Cambrian time are found in the Salt Range of Punjab and in the Spiti areas of Central Himalayas. The succession of fossiliferous Lower Cambrian Salt Range rocks is shown in Table 24.4.

Table 24.4   Fossiliferous rocks of Lower Cambrian Salt Range

Formation	Series	Thickness (m)
Salt Range (Lower Cambrian)	Purple sandstone	70–122
	Unfossiliferous	
	Neobolus shale	30
	Magnesian sandstone	83
	Dolomites	137
	Sandstone	

24.5.2 Silurian and Devonian

The best development of Silurian and Devonian rocks is found in the Spiti area of the Himalayas where the rocks overlie the Haimanta and underlie the red quartzite and grits (Muth quartzite) of Devonian period as shown in Table 24.5.

Table 24.5   Types of Silurian and Devonian rocks

Devonian	Muth Quartzite
Upper Silurian	Siliceous limestone, coral limestone, and shaley limestone with brachiopod and gastropod
Lower Silurian or Ordovician	Shales and flaggy sandstone, dolomite, and limestone with trilobite and brachiopod, pink quartzite and gritty conglomerate
Cambrian	Haimanta black shales and slates

The rocks of the Silurian period include limestone, flaggy shale, quartzite, sandstone, and conglomerate possessing characteristic Silurian fauna. Such Silurian rocks also occur in Kashmir. Overlying the siliceous and coral limestone of the Silurian, there is the typical Ordovician of white-coloured Muth quartzite, which is of considerable thickness (nearly 100 m).

In Chitral area of Kashmir, stratified sequence also developed resembling the Devonian rocks of Spiti. In addition, the Hazara area of Kashmir contains conglomerate followed by red sandstone and shale overlain by limestone which are 600 m thickness. The rocks of Chitral area of Kashmir are from Devonian period followed by upper Silurian of Spiti.

24.5.3 Carboniferous and Permian

The Permo-Carboniferous rocks occur in the Salt Range, which is a thick and highly fossiliferous stratum. The rocks are divided into two groups. The lower group is speckled sandstone containing glaciated boulders and mottled sandstone with fossils of upper Carboniferous age. Following these strata, there is the Productus limestone (150 m thick) of Permian age. It consists of calcareous sandstone followed by arenaceous limestone, cherty limestone, and sandstone with typical Permian fossils and the basement bed of Permo-Carboniferous boulder conglomerate.

In the Central Himalayas of Chitral province, trilobite and brachiopod fossils are found in calcareous sandstone of *Lipak series* (lower Carboniferous) and the *Po series* (upper Carboniferous) having Fenestella shales interbedded between them. The Lipak series is equivalent to the Syringothyris limestone of Kashmir. In some places, an agglomerate bed of volcanic origin overlies the Carboniferous rocks. The Permo-Triassic rocks possess the typical fossiliferous bed *Productus limestone* of marine origin.

24.6 GONDWANA FORMATION

Gondwana rock formation is the most important in Indian stratigraphy as good variety coal seams are available in this formation. The deposition of the Gondwana sedimentaries started from the Permo-Carboniferous time and continued until the Jurassic time. This formation is divided into three groups as shown in Table 24.6.

Table 24.6   Classification of Gondwana formation

Formation	Series	Stage
Upper Gondwana (Jurassic Age)	Umia series	
	Jabalpur series	
	Rajmahal series	
Middle Gondwana (Triassic)	Maleri series (Upper Trias)	
	Kamthi series (Middle Trias)	
	Panchet series (Lower Trias)	
Lower Gondwana (Permo-Carboniferous)	Damuda series	Ranigunj stage
		Ironstone stage (ironstone shale)
		Barakar stage (coal bearing)
	Talchir series	Karharbari stage
		Talchir stage (boulder bed)

The coal-bearing rocks of the Gondwana system are found mainly in the Damodar valley and also in parts of the Mahanadi and Godavari valleys. The best occurrence of coal is in the Damodar valley covering the Ranigunj, Asansol, and Jharia areas of West Bengal and also in parts of Bihar. The Gondwana rocks also contain marks of climatic change from arctic cold to tropical and desert conditions. Among the Talchir rocks, the *Talchir tillite* shows evidence of some glacial deposition. Figure 24.7 shows a geological section of Gondwana system rock formation of Satpura with faulted contact of basement Archaean with lowermost Talchir series followed by Damuda, Pachmarhi, Kamthi, and Jabalpur with a faulted contact of Deccan trap (Wadia 1966; original drawing by Meddlicot of GSI (1872)).

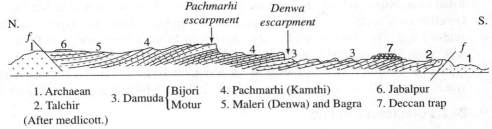

Fig. 24.7 Generalized section through the Gondwana basin of the Satpura region (Wadia 1966)

The Gondwana rocks occur extending over a wide area from the Ranigunj hill to Nagpur. The rock formations are also found in other places such as Satpura hills (Fig. 24.7), Godavari valley, Jabalpur, and the east coast. The component rocks are mainly well-bedded shale and sandstones. The lower Gondwana rocks are about 300 m thick, whereas the thickness of middle Gondwana is over 1000 m. The Gondwana sedimentaries contain a number of plant fossils such as *Gangamopteris*, *Glossopteris*, and *Vertebraria* with some insects and worm tracts in the lower stages. The majority of the plant fossils are found in the middle Gondwana rocks and vertebrate fossils including reptiles in the upper Gondwana rocks (Panchet stage). The Damuda coal-bearing strata at places are traversed by igneous intrusion such as dykes, destroying the coals.

24.7 MESOZOIC ERA

As described in Section 24.6, the deposits of Gondwana sedimentaries continued covering Triassic and Jurassic times. The Mesozoic rocks representing Triassic, Jurassic, and Cretaceous periods have been developed in different parts of India such as the Himalayas, Kashmir, the south coast, and Assam.

24.7.1 Triassic Period

The Triassic formation is well developed in the northern parts of the Himalayas and the adjoining Garwal and Kumaon. The component members of the system are mainly limestone and dolomite with intercalations of blue colour shale. The rocks are richly fossiliferous containing Triassic marine fossil of the genera *Trilobites*, *Ceratites*, *Flemingites*, and *Spirifer*.

In Kashmir, the development of Trias is also very large. A thick bed of compact blue limestone and dolomite is displayed in many places. In addition, the Triassic rocks occur with an unconformity over the Devonian rock, the basal parts having a 30 m-thick fossiliferous limestone bed of Triassic fossil.

In the Salt Range area, Trias is developed in the western part over the *Productus limestone* with ammonite fossils of *Ceratites* of Lower Trias in flaggy limestone. The Trias of Salt Range are named *Ceratites bed* on account of the abundance of the fossil ammonite ceratite. This bed consists of calcareous sandstone and arenaceous limestone.

24.7.2 Jurassic Period

The Jurassic strata of the Himalayas has two distinct rock units, namely *Kiato limestone* of 610 m to 915 m thickness of Lower Jurassic and the *Spiti black shale* of Upper Jurassic overlying it and extending over almost the entire distance of the Himalayas. The Spiti shale contains several fossil ammonites such as *Phylloceras*, *Litoceras*, and *Hoploceros* and also a large number of lamellibranch such as *Belemnites*, *Pecten*, and *Corbula*, which are typical indicators of the Jurassic age. The Hazara area has also developed Jurassic formation with fossils of the Spiti area types. Moreover, the Jurassic beds are developed in the Salt Range area containing Jurassic-type fossil *Ceratites*. Marine transgression during the Jurassic period took place in coastal India and in parts of Kutch and Tiruchirappalli and in the low-lying flat areas of Rajasthan.

24.7.3 Cretaceous Period

The Cretaceous system of India is marked by extensive occurrences in South India where the *Niniyur*, the *Ariyalur*, the *Tiruchirappalli*, and the *Utatur* stages of sedimentary rock formations occur in succession as lower to upper beds. The Utatur stage is very famous for its phosphatic

nodules, which are an important source of phosphate in the country. In the central provinces, there is the well-developed *Lameta bed*, which contains fossils that helped to ascertain the late Cretaceous or Palaeocene age of the Deccan traps.

Towards the end of the cretaceous period, there was a great geological event in India when a volcanic eruption caused lava flow extending over an area of 500,000 km^2, which formed the basaltic rocks of the Deccan trap (Fig. 24.8). The volcanic activity continued until the Tertiary period and even the Eocene time. The basalts of Deccan trap are used in various engineering

Fig. 24.8 Deccan trap basalt (also called trap rock) formed of volcanic lava flow

constructions such as railway ballasts and building stones, and the famous black soil well suited for cotton growth is derived from this rock.

24.8 TERTIARY ROCK FORMATIONS

The Tertiary period is the most important in Indian stratigraphy and tectonic history. The volcanic activity of the Deccan is associated with two great events: the break-up of the Gondwana land mass and the Himalayan orogeny related to the uplift out of Tethys Sea began at this period and the volcanism of the Deccan traps continued. The basaltic rocks of the Deccan traps (Upper Cretaceous to Eocene) are found over an extensive area of Maharashtra and other parts of the Deccan (see Fig. 24.2).

The rocks of the Tertiary system are found mostly in the Himalayas. In the peninsula, they occur in the coastal areas of Gujarat, Kerala, and Tamil Nadu. In the Shimla area, the Tertiary sedimentaries are divided into three series: the Subathu series consisting of grey and red shales followed upwards by the Dagshai series comprising red claystone and the Kasauli series composed of sandstone. Table 24.7 gives the Tertiary succession of Kashmir Himalayas along with the thickness of rock formations and the rock types:

Table 24.7 Tertiary Succession of Himalayas

Siwalik system (5300 m)	Fluvial deposits of clays, sand-rocks and conglomerate with mammalian fossils	Upper Siwalik—clays and conglomerate Middle Siwalik—massive sandstone Lower Siwalik—sandstone and shale

(Contd)

Table 24.7 *(Contd)*

Sirmur system (3000 m)	Murre series—freshwater sandstones and clays 2600 m thick
	Subathu series—marine nummulites with
	Laki series of shale and coal seams

In the east in Assam, nummulitic limestone occurs in Khasi hills. Oil is associated with these rocks of Oligocene–Miocene age. In the Salt Range area, nummulitic limestone representing Eocene is extensively developed. The Middle Tertiary is absent but Upper Siwalik rocks of Pliocene age are present as shown in Table 24.8.

Table 24.8 Tertiary rock formations in the east

Siwalik system (6500 m)	Upper— boulder conglomerate	Pliocene
	Middle—sandstone and clays	to
	Lower—sandstone and nodular clays	Miocene
Unconformity		
Kirthar series (1700 m)	Massive nummulitic limestone	Middle and Upper Eocene
Laki series (30m)	Clays with pyritous and bituminous shales and coal seams	

Along the foothills of the Himalayas, Siwalik molasse consisting of conglomerate, sandstones, and shales with a height of nearly 6100 m are traceable for a long distance (Fig. 24.9). The Siwalik system ranges in age from the Middle to the Late Tertiary period. The Siwalik is famous for vertebrate fauna apart from *echinoids*, *gastropods*, *foraminifera*, and *corals*. Though the strength of the rocks is not high, a number of large dams including the 225 m-high concrete Bhakra of Punjab, one of the largest dams in the world, were successfully constructed in the Siwalik rocks due to ingenious engineering design.

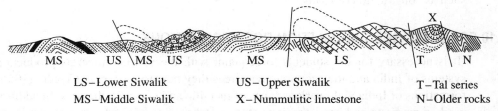

LS–Lower Siwalik   US–Upper Siwalik   T–Tal series
MS–Middle Siwalik   X–Nummulitic limestone   N–Older rocks

Fig. 24.9 A Geological section of Siwalik system showing the sub-Himalayan zone east of the river Ganges (Wadia 1966; original drawing by Middlemiss of GSI (1885))

24.9 QUATERNARY PERIOD

The Quaternary period is marked by glacial condition in parts of the Himalayas showing indication of Ice age as in Europe. The enormous large terminal moraines and scattered blocks of smooth hummocky hills of low elevation with striated marks bear testimony of an Ice age in Pleistocene. Equally, the irregularities in the drainage flows bear evidence of such glaciated condition with ice movement. The Quaternary Ice age deposits are found in Kashmir.

The Indo-Gangetic plains comprising the thick piles of sediments covering the vast area from Punjab to Assam belong to the Quaternary period. The plains were formed mainly by the aggrading action of Himalayas through its rivers and monsoon rains. The alluvial deposits consist of clay, silt, loam, and so on. The alluvium is divided into the following groups as shown in Table 24.9.

Table 24.9 Deltaic deposits of the Indus and the Ganges of recent time

Newer alluvium	*Khadar* of Punjab
Older alluvium	*Bhangar* of the Ganga
Unconformity	
Rocks of unknown ages of Archaean and Gondwana	

The older alluvium or *Bhangar* occurs on the ground above the flood level whereas the *Khadar* is confined to the river channel and their flood plains. The Bhangar confined to West Bengal and Uttar Pradesh is composed completely of Pleistocene deposit, whereas Khadar gradually passes to recent time, occupying the higher level. The Khadar contains comparatively less *kankars* (coarse sediments). The deltas of the Ganges and the Brahmaputra have khadar deposits. Similarly, the delta deposit of the Indus contains sediments covering up to recent time. The alluvial plains bear the most fertile soils of the country, with new deposition of silt every year by the rivers during monsoon time.

The Indo-Gangetic alluvium is a great source of freshwater. At all places of the alluvium deposit, groundwater is available at a shallow depth by dug wells or tube wells. The term *Bhabar* denotes gravel talus and steep slopes. *Terai* is the densely forested marshy ground at a lower level of the *Bhabar*. The term *Bhur* indicates elevated lands along the banks of the river Ganges. The efflorescence products of alluvium *Reh* or *Kallar*, which covers some areas of the alluvial deposits, are detrimental for irrigation purposes as they destroy agricultural fertility.

In addition to the alluvium, the Pleistocene is also the creator of the desert of Rajasthan, namely the *Thar*, formed of windborne sand, stretching from the west of the Aravallis to the Indus basin and southern confines of Punjab. The Pleistocene formed the laterites found in various places of India by sub-aerial alteration of bauxites, which are used for many purposes such as construction of roads.

24.10 ECONOMIC RESOURCES FROM ROCK FORMATIONS

'It is necessary for the students to acquaint with the various minerals products of the rock systems of India and the economic resources they possess', observed Wadia (1966). The rock formations of India yield various types of metallic and non-metallic ores. In addition, minerals used as precious and semi-precious stones, radioactive minerals, rare minerals, and so forth and also coal and petroleum are available in the rock formations of India. A review of all these ores and minerals and also of coal, petroleum, and building stones with respect to the rocks and location of their occurrence is provided in this section.

24.10.1 Metallic Minerals

India has a large number of economically useful minerals and they constitute one-fourth of the world's known mineral resources. The Archaean rocks of India are rich in ore deposits. Rocks of other types also yield economic minerals. However, coal and petroleum occur only in sedimentary rock formations.

Gold   *Gold* occurs as native gold associated with quartz vein and also as placer deposit in sands. The principal gold-bearing rock is found in Kolar district of Mysore where gold occurs in quartz veins in hornblende schist of Dharwar rock formation. Here, the auriferous veins contain gold as minute particles. After mining the rock, it is crushed and the crushed materials are dissolved by the process of amalgamation to extract gold. The Hatti area of Dharwar schist in Karnataka and Ramgiri in Andhra Pradesh also produce gold to a lesser extent. Placer gold to a small extent is available from some rivers such as the Subarnarekha of Bihar.

Iron   *Iron* occurs in a large number of deposits in the form of iron oxides (mainly hematite and magnetite). Iron ores occur in Dharwar and Cuddapah systems of rock formations, the common rock types being hematite–magnetite–quartz schists. High-grade hematite iron ores are available in the banded hematite–quartzite of South Singhbhum district of Bihar and Maurbhanj and Keonjhar of Orissa. In fact, about two-thirds of the country's iron deposit lies in a belt along the border of Bihar and Orissa. Other hematite iron ore deposits are found in Madhya Pradesh, Karnataka, Maharashtra, and Goa. Magnetite iron ore is found in Tamil Nadu, Bihar, and Himachal Pradesh. The Damuda series of rocks in West Bengal also holds valuable deposits of bedded or precipitated iron ores in the ironstone shale.

Copper   Ores of *copper* occur in Nellore of Tamil Nadu, Khetri and Dariba of Rajasthan, Singhbhum of Bihar, and in parts of Sikkim and Karnataka. Copper ores are also found in Madhya Pradesh as veins or dissemination in Dharwar schists and phyllite. The Sikkim deposit has got prominence and geological similarity with that of the Singhbhum copper ores, being associated with the Daling schist and phyllite. Presently, copper ores are mined from Mosaboni, Rakha, and Ghatsila of Bihar. They are found mainly as sulphides (chalcopyrite, malachite, azurite, and cuprite) in schistose rocks and phyllites. The Khetry deposits of Rajasthan and Malanjkhand of Madhya Pradesh are also mined giving good productions.

Chromium   *Chromium* ores are available in Singhbhum in Bihar, Cuttack district in Orissa, Krishna district in Andhra Pradesh, and Mysore and Hassan in Karnataka. Chromium occurs by magmatic differentiation in n ultrabasic rocks such as dunites, peridotites, and serpentines. Some chromites occur in Chalk hills (magnesite veins) near Salem. It is used as furnace lining and also as alloy with steel for chrome steel used in armour plates.

Manganese   India has large deposits of *manganese* next to Russia. The chief centres of manganese mining by quarrying method are Balaghat, Bhandar, Jabalpur, and Nagpur of Central India. Visakhapatnam and Gangapur of Orissa and Panchmahal of Maharashtra also yield manganese. Gondite and Kodurite of Andhra Pradesh and Orissa contain manganese ore minerals psilomelane, braunite, and pyrolusite. The Dharwar system of rocks (sedimentary metamorphic) found in Karnataka and Madhya Pradesh also contains manganese ores. Present manganese mining areas include Madhya Pradesh, Maharashtra, Bihar, and Orissa

Cobalt and nickel   Cobalt and nickel are not obtained in India in sufficient quantities. These metals are found in copper mines of Khetri and Jaipur of Rajasthan. Nickeliferous pyrrhotite and chalcopyrite occur in some places in South India but in sparse amount. *Nickel ores* are also reported from Cuttack and Maurbhanj in Orissa.

Titanium   *Ilmenite* is the main source of titanium. It occurs as an accessory mineral in igneous and metamorphic rocks as well as mineral sands. Ilmenite reserves occur in Kerala and along the sandy beaches of both the east and west coast.

Lead, silver, and zinc ores   Galena is the main source of lead and also contains up to one per cent silver. Besides, anglesite, which is an oxidation product of galena, occurs as lead spar, white in colour. The ore also contains zinc to some extent. The rock containing lead ores is crushed and concentrated by floatation process. The top is skimmed off as lead whereas the leftover part contains significant content of silver. Zinc, which is usually present with lead and silver, is non-soluble and remains in a separate layer and is easily removed. Rajasthan is the leading producer of lead of the country. Lead is mined mainly in Zawar, Bhilwara, Rajpura-Dariba, and Rajsamand areas of Rajasthan. Besides, Surgipalli area in Orissa and Agnigundala in Andhra Pradesh also mine lead ores. The lead ore galena is found in many places of India as crystalline schists in parts of Himalayas, Tamil Nadu, and West Bengal, and as veins and pockets in Vindhyan limestone. Lead ores of Hazaribagh and other places of Bihar and in parts of Madhya Pradesh are often mixed with a small content of silver (few ounces per tonne).

Aluminium   The prime source of aluminium is *bauxite*, which is its hydrate. Aluminium has various applications in modern industries such as manufacture of utensils, electrical appliances, and aeronautical parts. With the discovery of bauxite in the east coast hilltop covering Orissa and Andhra Pradesh, which has 74 per cent of the total reserve, India is now very rich in aluminium ore. The estimated reserve in 1970 was 345 million tonnes, which has now come up to 2000 million tonnes. Prior to finding it in the east coast, bauxite as laterite caps was available in Katni of Jabalpur and the hills of Balaghat in Madhya Pradesh. Other important occurrences are Kalahandi, Mahabaleshwar, Bhopal, and the Palani hills and some parts of Tamil Nadu and the Western Ghats.

Tin   Occurrence of *tin* has been reported from gneissic rocks and pegmatite in Palanpur of Hazaribagh district of Bihar where it is present as cassiterite crystals.

Tungsten   *Tungsten* (*wolfram*) and antimony ores occur in Nagpur, Tiruchirappalli, and Rajasthan in gneissic rocks. Tungsten steel is largely used in the manufacture of munitions and armour plates of heavy guns. Thin wires of tungsten are suitable for and manufactured to use in electric lamps.

24.10.2 Non-metallic Minerals

Mica   India produces three-quarters of the world's mica (muscovite). Among the non-metallic minerals, mica is found abundantly in Koderma, Gaya, and Hazaribagh districts of Bihar. High-quality mica also occurs in Andhra Pradesh and Rajasthan.

Kyanite and sillimanite   *Kyanite* and *sillimanite* are used for refractory purpose as these minerals can withstand high temperature. These minerals are also used for lining of kilns and furnaces and for some other purposes in the chemical industry. India has the richest deposits of kyanite and sillimanite. Kyanite occurs in Lapsa Buru (Kharsawan of Bihar), the largest deposit in the world. Sillimanite occurs in Sonapahar in Meghalaya and Pipra in Madhya Pradesh in sillimanite–quartz–schist of Archaean complex.

Graphite   *Graphite* is used for crucibles, blast furnace lining, refractory, and many other purposes. It occurs in the crystalline metamorphic rocks gneisses and schists as *disseminate* flakes, needles, and grains and also as veins. It is found in various parts of the country such as Jharkhand, Karnataka, Kerala, Orissa, Rajasthan, and Tamil Nadu. In Orissa, it occurs as an essential constituent of khondalite.

Asbestos   Amphibolite and its fibrous variety chrystabolite are used as asbestos. The asbestos minerals have been found in the pockets of gneisses and solution cavities of soluble rocks in many places such as Saraikela in Singhbhum. One rich occurrence is in the Archaean rocks of Ghattihosahalli in Karnataka.

Steatite This mineral is known for its softness and uniform nature. It is soapy to touch and is also called soap-stone. It has a hardness of one in Mohs scale of hardness. It is used for manufacturing plates, bowls, and, pots. It occurs in many places in the Archaean and Dharwar rocks of the peninsula. Workable deposits are found in Bihar, Jabalpur, Salem, and Jaipur.

Barite *Barites* occur in many places of India as veins and beds of shale in sufficient quantities. The chief localities of barite occurrence are Salem in Tamil Nadu and Sleemanasad in Jabalpur. It also occurs in parts of the Cuddapah basin and it is mined from the Mangampet deposit of Andhra Pradesh. Barite is used as a pigment for mixing with white lead and as a flux in the smelting of iron and manganese in paper manufacturing and pottery glaze industries.

Uranium *Uranium* ores are available in Bihar where it is mined from Jaduguda. Pitchblende occurs in pegmatite veins in the gneisses and schists of the Sengar mica mines in Gaya district of Bihar. It is associated with uranium minerals. The Gaya pitchblende has also radium content.

Rare minerals The pegmatite veins of the crystalline rocks of India contain a few of what are called 'rare minerals' as accessory minerals. These rare minerals have extended use in the nuclear industry in the manufacture of special steel and other highly specialized use. *Samarskite*, a very rare mineral, is found in the mica-bearing pegmatite of Nellore and in parts of Mysore. The most common of rare minerals such as monazite, columbite, niobates, and tantalites occur in the pegmatites of Gaya, Hazaribagh, and Nellore and also parts of Rajasthan and Travancore.

24.10.3 Coal and Petroleum

The Gondwana system is the *coal*-bearing rock formation of India, which has the world's largest deposits of coal. High-grade coal occurs in Jharia and Bokaro of Bihar and Ranigunj coal fields of West Bengal covering the Damodar valley area. Lignite coals are found in Neyveli in Tamil Nadu. Coal also occurs in the Tertiary sedimentary formations in parts of Assam.

Petroleum deposits are found in Assam and Gujarat. Fresh reserves are located off Mumbai and petroleum is being raised by offshore drilling from the coast of Mumbai. Tertiary sedimentary formation is the potential source of petroleum in India found in Assam, Tripura, and Manipur. Sedimentaries of Punjab, Himachal Pradesh, Kutch, and Andaman Islands are also probable sources of petroleum. A deep-seated deposit of petroleum is expected from the Krishna and Godavari valley.

24.10.4 Precious and Semi-precious Stones

The Panna *diamond* belt, which produces diamonds for the country, covers the districts of Panna, Chhatarpur, and Satna in Madhya Pradesh and some parts of Banda in Uttar Pradesh. The diamond occurs in the kimberlite rocks in old volcanic pipe.

Crystallized and transparent varieties of corundum of red colour form the highly valued gem ruby and those of light blue tint form sapphire. These precious stones are available in the pegmatite of Bihar. They are also found in the terrace debris of Kashmir.

Beryl when transparent and possesses good colour and lustre serves as a high quality gemstone. The colour varies and has shades of green, blue, and even yellow. The green variety is the highly priced *emerald* and the blue is *aquamarine*. These occur in pegmatite veins in the

Archaean gneiss of Bihar and Nellore. These gemstones are also reported from Shigar valley area of Kashmir. In addition, *garnets* and *zircon*, which are used as semiprecious stones, are also available in Bihar.

Precious and semiprecious stones such as *garnet*, *zircon*, *topaz*, and *moonstone* occur in the Archaean rocks of Kashmir and Rajasthan to a limited extent.

Rock crystals as crystallized transparent quartz are cut for ornamental objects. The principal places where crystalline quartz of requisite purity and transparency occur are Thanjavur, Kashmir, and Jabalpur. *Amethyst*, which is a rose or purple variety of quartz, is found in Jabalpur and has a high demand for ornamental use. It occurs mainly in the geodes filling up lava cavities.

Various forms of chalcedonic silica, agates, carnelian, bloodstone, onyx, jasper, and so on are known in the common name of *akik* (agate) in India. These are available from the amygdaloidal basalts of the Deccan where these occur as infilling of the cavities in lavas.

24.10.5 Building Stones

Building stones, which are suitable for engineering constructions, should have some indispensable qualities. Rocks that can bear the ravages of weather through the ages, possess the requisite strength and in some cases attractive appearance and tint, and can be dressed without difficulties are considered as suitable building stones. Several rock formations of the country comprise rock types that are quite suitable for building purposes, that is, for use as rock mass for foundation of engineering structures and as rock fragments for construction materials.

Granites, gneiss, basalt, marble, limestone, and sandstone are well utilized for building purposes. Granites and sparsely foliated gneisses are very good building stones possessing high strength and durability. Indian pink granite and gneisses are highly valued building stones that have found market in foreign countries also as these can be quarried in large blocks and used for decorative purposes for wall and floor. Many varieties of peninsular gneiss and granite of the Archaean age are very attractive building stones.

The marble of Rajasthan, especially the milk white marble mined from Makran, is used for exquisite construction purposes. The Taj Mahal is constructed of this marble. Good quality marble is also available in Ajmer, Jaipur, Alwar, and few other places of Rajasthan. Dolomite of Jabalpur is also suitable for use as building stone.

The limestones of Cuddapah, Bijwar, and Aravalli formations are used as building stones. In addition, good quality limestone for constructive purposes is available in the Vindhyan formation. The Cretaceous limestone of Tiruchirappalli includes ornamental variety.

The red-coloured Vindhyan sandstone is considered to be very good for architectural purposes. It was used in the construction of many ancient monuments and royal buildings of the past. In addition, upper Gondwana sandstone available in Orissa and some other places are also good building stones.

The basalt of Deccan traps are used in many of the engineering constructions, especially for highways and railways as detailed in Chapter 19. The laterites of India spread over many places have been used for roads and other construction purposes for a long time. The slates and arenaceous phyllites, which split into big slabs, are used in constructing floors and walls of buildings and houses.

SUMMARY

- The three distinct physiographic features of India include peninsular India, extra-peninsular India, and the Indo-Gangetic plains. The peninsula is constituted of very old rock formations and lava flow forming the Deccan plateau. Washing the peninsular region, the rivers Godavari, Krishna, and Cauvery flow towards the east and the rivers Narmada and Tapti flow towards the west. India has long coastlines along the east and the west facing the Bay of Bengal and the Arabian Sea, respectively.
- India has a diverse geological set-up constituting very old to younger rock formations. The Archaean rock system consisting of gneisses and granites are found in the peninsula and also in parts of Rajasthan, Chhattisgarh, Bihar, and Jharkhand.
- The Dharwar system of rocks, the earliest formations of sedimentary rocks later metamorphosed, are found mainly in Karnataka, Jharkhand, Rajasthan, and Madhya Pradesh. The rocks are mainly schists, marbles, slates, quartzites, and granites. The Cuddapah rock formations occur in Tamil Nadu, Andhra Pradesh, and Madhya Pradesh comprising slates and marble.
- The Vindhyan system of rocks comprising sandstone, shale, and limestone are found mainly in Madhya Pradesh, Chhattisgarh, and Uttar Pradesh. The red colour Vindhyan sandstones have been used as building stones from very ancient time in the architectural works of royal buildings and monuments.
- The Gondwana formation contains sedimentary rocks having the best coal seams of India. These rocks are found in eastern India in the Damodar river valley area and also in the Mahanadi and Godavari valley areas of the peninsula.
- The Deccan trap systems of rocks are volcanic in nature and found in vast areas of south India including Maharashtra. Volcanic activity caused the break-up of the ancient Gondwana land and uplift of the Himalayas.
- The Tertiary rock formations are sedimentary sequence of rocks found mostly in the Himalayas and the two coasts of the peninsula covering Gujarat and Kerala. The rocks are mainly sandstone and shale with some brown coal deposits.
- The Quaternary system is characterized by the Ice age deposits of Kashmir and the formation of the alluvial plains of North India and the desert of Rajasthan.
- The rock formations of India yield various types of metallic and non-metallic ores including gold, iron, copper, lead, aluminium, manganese, and chromium and salt, alum, gypsum, borax, asbestos, and phosphatic minerals. Mica, uranium, thorium, and other rare elements are also present in India. High-grade coal seams occur in the Gondwana rocks. Petroleum occurs offshore of Mumbai and in the Tertiary rocks of Assam and Tripura.

EXERCISES

Multiple Choice Questions

Choose the correct answer from the choices given:

1. Rocks of primordial crust are available in India in:
   - (a) Kashmir
   - (b) Himalayas
   - (c) the peninsula
   - (d) none of the above
2. The Deccan trap rocks were formed by:
   - (a) deposition of sediments
   - (b) volcanic eruption
   - (c) uplift of land mass
   - (d) igneous intrusion
3. The Himalayas is characterized by:
   - (a) sedimentary rocks
   - (b) faulted and folded rocks
   - (c) ice-capped ranges
   - (d) all of the above
4. Bhangar is the name of the:
   - (a) younger alluvium
   - (b) older alluvium
   - (c) deltaic deposit
   - (d) both (a) and (b)
5. Alakananda, Bhagirathi, Mandakini, Jhanhabi, and Yamuna are the tributaries of the riiver:
   - (a) Ganges
   - (b) Indus
   - (c) Brahmaputra
   - (d) both (a) and (b)
6. Aravalli range consists of:
   - (a) sedimentary rocks
   - (b) metamorphic rocks

(c) igneous rocks

(d) sedimentary rocks later metamorphosed

7. Volcanic activity formed the Deccan traps in the:
   (a) Tertiary period
   (b) Precambrian period
   (c) Cambrian period
   (d) Mesozoic period

8. Main Boundary Fault lies:
   (a) to the north of the Himalayas
   (b) to the south of the Himalayas
   (c) along the river Indus
   (d) along the river Ganges

9. The best coal deposit of India is found in:
   (a) Barakar stage
   (b) Ranigunj stage
   (c) Ironstone stage
   (d) both (a) and (b)

10. Manganese-bearing Gondite and Kodurite rocks are found in:
    (a) Bihar
    (b) Andhra Pradesh
    (c) Kashmir
    (d) both (a) and (b)

11. Petroleum deposit is found in:
    (a) Assam
    (b) offshore of Mumbai coast
    (c) Kashmir
    (d) both (a) and (b)

12. Iron ores occur as hematite in:
    (a) Singhbhum district of Bihar
    (b) Keonjhar and Maurbhanj of Orissa

(c) Tamil Nadu

(d) both (a) and (b)

Review Questions

1. What are the main physiographic divisions of India? Name them.

2. Name the east and west flowing rivers of the Peninsula.

3. How was the Deccan plateau formed? What is its main constituent rock?

4. State the formation of the Himalayas and its main tectonic features.

5. What is the extent of the Himalayas in the extra-peninsula? Name the states at its eastern and western ends.

6. How were Indo-Gangetic alluvial plains created? What is the main use of the alluvial soil?

7. What are the two divisions of Indo-Gangetic alluvium? Name them and state their characteristic features.

8. In which part of India will you find the rocks of primordial crust? What are these rock types?

9. Name two Precambrian rock formations of India, their occurrence, and the rock types.

10. Which part of the Gondwana system is coal bearing? Name the coal-bearing stages and their place of occurrence.

11. From which rock formation did we first get trilobite fossil and what is its age?

12. What is the age of the Siwalik system? Where do we get the occurrence of Siwalik formation?

Answers to Multiple Choice Questions

1. (c)	2. (b)	3. (d)	4. (b)	5. (a)	6. (d)	7. (a)	8. (b)	9. (d)
10. (b)	11. (d)	12. (d)						

APPENDIX

A

Geotechnical Problems of Dams and Their Solutions

The subject matter of this presentation may be read with Chapter 11 to gain practical knowledge on geotechnical problems encountered during the construction of some dams in India and the measures adopted to solve the problems. Instances of failure of a few dams with causes of failure are also provided. The names of the authors and their publications consulted in preparing this appendix have been added to 'References' at the end of the book.

A.1 DAMS FOUNDED ON IGNEOUS ROCKS

This section discusses some dam projects that have been founded on igneous rocks.

A.1.1 Koyna Dam and Reservoir Project, Maharashtra

The Koyna hydroelectric project of Maharashtra includes a 104 m-high and 853 m-long rubble concrete dam constructed across the river Koyna, a tributary of the river Krishna. It is a reservoir of capacity about 2890 million cubic metres and consists of a head race tunnel, a tail race tunnel, and a powerhouse to generate 540 MW power. The dam site consists of the basalt of the Deccan trap overlain by nearly 30 m thick cover of laterite and lateritic soil. The basalt of the dam site represents three distinct massive lava beds separated by tuff breccia, vesicular basalt, ash, and red bole, which contributed to the weakness of the bedrock. The basalt is also jointed at places and shows presence of several shear zones. All these weak zones posed problems of settlement, seepage, and uplift of the dam.

The shear zones and clay seams in the basalt of the foundation were scooped out and backfilled by concrete. Anchor rods were embedded as grout buttons along the walls of the weak zones. In the right abutment, the red bole layers and the tuff breccia were treated by a cut-off followed by effective grouting. Consolidation grouting was done to a depth of 7.6 m in the weak patches of foundation followed by curtain grouting to arrest problems of settlement, uplift, and seepage.

The reservoir area the Koyna project lies within the peninsular shield, which is considered to be a stable landmass. However, since 1962—when water was impounded in the reservoir—the project had experienced frequent low-intensity earthquakes. In 1967, major earthquakes of intensity measuring 5.0 to 6.3 in the Richter scale damaged parts of the Koyna township and underground power plant, though the dam, designed with a seismic factor of $0.05g$, could withstand the shock. The dam was strengthened by buttress, cabling, and grouting to protect it from high-intensity tremors (Srinivasan 1975).

A.1.2 Ukai Multipurpose Dam Project, Gujarat

A 68 m-high and 4928 m-long composite dam (earth-cum-masonry) has been constructed across the river Tapti under Ukai project to irrigate a vast area of Gujarat and to generate 300 MW power besides flood moderation. The dam site comprises Deccan basalt intruded by dykes of dolerite and gabbro, which occur under thick overburden materials of sand, gravel, and boulders of 10 m thickness in the river section. In the *earth dam* portion, two rows of grout curtains were extended to a depth of 15 m below the cut-off trench. The unconsolidated material of the river section was treated by a positive cut-off to a depth of 25 m by placing concrete diaphragm. The shear zones in the rocks of the earth dam foundation were grouted, the average grout intake being 57 kg/m.

In the 905 m-long *masonry dam* section, the main defects of the foundation rock included sheeted joints, shear zones, and deep weathering of dolerite. Consolidation grouting was undertaken to strengthen these weak zones. The two red bole zones were excavated to a depth of 3.4 m and backfilled with concrete. In the spillway portion, the rocks were grouted with deep holes to strengthen the contact zones of dykes with basalt. In addition, grout curtains were undertaken from the inspection gallery, and drain holes were provided to relieve the undue uplift pressure. In the design of the dam, a seismic coefficient of $0.10g$ was adopted to withstand earthquake shocks. The Ukai dam encountered the problem of scouring due to huge flood discharge in 1969 and again in the subsequent years. The dam was then anchored by steel grip rods embedded in concrete to protect the structure from scouring (Mahendra and Mathur 1975).

A.2 DAMS BUILT ON SEDIMENTARY ROCKS

Some of the dams built on sedimentary rocks are briefly described in this section.

A.2.1 Rana Pratap Sagar Dam, Rajasthan

The Rana Pratap Sagar project utilizes water from the river Chambal by means of a 54 m-high and 1143 m-long masonry dam constructed across this river, a powerhouse at the toe of the dam, and other ancillary structures for power generation and irrigation of large areas of Rajasthan and Madhya Pradesh. The dam site consists of horizontally bedded Vindhyan sandstone and shale. The rocks are folded and the fold axis runs across the dam alignment. Several shear zones of thickness 2.5–3 cm traverse the foundation rocks. In the left bank, slump features are seen in shale and glide cracks in sandstone.

The horizontal disposition of the sandstone and shale with clay seams, presence of thin shear zones along the bedding, block jointing in sandstone, and susceptibility of the downstream rocks to scouring posed the problems of differential settlement, leakage, and toe erosion. The clay seams occurring up to 3.3 m below the foundation were treated by two cut-off trenches, one in the toe and another in the heel of the dam, followed by intensive grouting between the two trenches.

In the spillway foundation, two rows of anchor rods spaced at 1.2 m intervals were embedded 3 m inside the foundation rock to resist shearing along the dam length. In addition, a key trench was dug at the toe of the dam. In the downstream bucket portion, a reinforced cement concrete (RCC) raft was tied to the underlying sandstone to prevent scouring from the thrust of the spillway water. As remedial measures against slumps and cracks of the left bank and against slide, the steep ground of this part was flattened and a clay blanket was provided, and in some places retaining walls were constructed. In addition, consolidation grouting was adopted up to a depth of 9 m in the foundation (Sanganeria 1975).

A.2.2. Srisailam Dam, Andhra Pradesh

The Srisailam dam in Andhra Pradesh is a 139 m-high masonry construction across the river Krishna. The dam site is a classic example of the problems encountered in construction on sedimentary rocks. A narrow gorge is created due to the presence of soft shale that undergoes easy weathering. Nearly horizontal beds of quartzite and shale are the main rock types of the dam site. The 0.4–2 m-thick massive quartzite bed possesses high strength; however, the shale in the abutment is of very low strength and the undersaturated condition causes it to disintegrate when even a low pressure is applied. Underlying the massive quartzite of the river section, there are fissile quartzite and sandy shale having clay seams along the bedding planes.

The main foundation defects were low strength of the abutment shale and a 3 m-thick highly weathered quartzite bed in the right abutment, block jointing of the quartzite, and sliding due to the low dip (3°–7°) of the beds having several clay seams. As a remedial measure, a 60 m-long drift was driven, and the soft shale beds were mined out and later filled with concrete. An impervious clay blanket was provided in the upstream parts. The block joints were treated by stage drilling and grouting over the entire foundation at 3 m intervals. A 30 m-wide RCC anchor block was provided at the toe and the dam was constructed as a single monolith structure to arrest any possible slides (Rao and Narasimham 1975).

A.3 DAMS FOUNDED ON METAMORPHIC ROCKS

This section deals with some dams that were founded on metamorphic rocks.

A.3.1 Idukki Dam, Kerala

The 167 m-high Idukki thin concrete arch dam with crest length of 365 m is only one of its kind in India and is one of the largest arch dams in the world. The dam is constructed across the river Periyar in Kerala in a 'V'-shaped gorge section where the river takes an acute bend and is very constricted. The word *Idukki* means 'constriction' in the local language. The cord–height ratio of the gorge section at the dam site is 2 : 1, which offers an ideal site for the arch dam.

The entire gorge section including the river bed and the abutments consists of hard charnockite. Exploratory work also proved the presence of hard and sound charnockite at depth. The only unfavourable feature was a hump in the left abutment that separated the underlying rock by a joint plane, creating a possibility of slide. A wing wall was provided at this part to relieve the arch from water load and to avoid the anticipated tensile stress. The hump was kept in position by an anchor bolt. Rock bolting in three rows was also undertaken in the dam site rocks to a depth of 6 m with spacing of 1.5 m to avoid any chance of movement of spherical rock bodies present in parts of the abutments (Seshagiri 1975).

A.3.2 Umiam Dam Project, Meghalaya

The Umiam hydroelectric project of Meghalaya includes a 72 m-high and 170 m-long concrete dam constructed across the river Umiam along with other appurtenant structures for generation of power. The rocks present at the dam site are quartzite and phyllite of Shillong series that occur in the syncline of a regional fold. The quartzites possess high strength but the phyllites occur in two varieties, the arenaceous type with high strength and the argillaceous type with low strength. The central part of the dam foundation is traversed by a 3 m-wide fault and several other minor faults and shear zones, open joints, and soft phyllite bands, which are the main defective features of the dam foundation.

The main fault zone and the minor faults were treated by scooping out the soft and crushed rock to a formula depth (generally two to three times the width) from the fault planes and backfilling with concrete. Grouting was carried out from both sides of the main fault zone by inclined grout holes. The highly jointed portion of the quartzites and the soft phyllites of the foundation were excavated until better rocks were available, which were then grouted to strengthen the foundation. Field load test was carried out to confirm that the soft phyllites attained sufficient strength to bear the thrust of the dam under reservoir condition. The dam site situated close to Shillong, the capital of Meghalaya, was affected by the earthquake of 1897 and the Assam earthquake of 1950. A seismic factor of $0.18g$ for horizontal and $0.12g$ for vertical was incorporated in the design of the dam (Gangopadhyay 1973).

A.4 DAMS CONSTRUCTED ON HETEROGENEOUS ROCKS

Some of the dams constructed on heterogeneous rocks are discussed in this section.

A.4.1 Hirakud Dam, Orissa

The 64 m-high Hirakud dam with a river bed spillway and nearly 4.8 km-long earthen section is constructed across the river Mahanadi in Orissa. Archaean biotite gneiss and granitoid with mica schist are present in parts of the river bed and in the left bank. The right bank consists of arkose, quartzite, slate, and shale of Cuddapahs. The contact of the Archaean and the Cuddapahs is faulted. The geological problems include deep weathering, the presence of a 6 m wide zone of shattered rock in the rocks of the left bank, and a fault that affected the rocks in a part of the spillway foundation.

The weak rocks of the foundation were excavated until sound rock was obtained. The fault zone was treated by scooping out the shattered rocks up to a depth of 3 m and further removing the clay gouge for a depth of twice its width and backfilling by concrete. A cut-off trench was provided at the upstream part. In addition, clay blanketing was adopted for a distance of 90 m on the upstream part covering the fault zone. In the earth dam section, a 4.5 m-deep cut-off trench and a curtain up to 9.5 m depth were provided from the bottom of the trench.

The dam suffered from post-construction distress due to alkali–aggregate reaction of the concrete used in the construction. Horizontal cracks up to 9 mm in width and extended to a depth of 2 m developed in the concreted vertical face of the right spillway and right foundation gallery. The river shingle containing quartzite and crushed rock aggregate were identified as reactive. The remedial measures taken up include grouting the body of the dam, sealing the cracks of the upstream face, and reinstalling porous drains in the body of the dam (Ramchandran and Gangopadhyay 1972).

A.4.2 Tenughat Dam, Bihar

At the Tenughat dam in Hazaribagh district of Bihar, a 51 m-high and 6097 m-long rolled earth-filled dam has been built across the river Damodar accommodating a 188 m-long masonry spillway in the right bank. The reservoir supplies water (25 m^3/s) to the Bokaro steel plant and ancillary structures. The dam rests partly on interbedded sandstone and shales of the Lower Gondwana formation and partly on the Archaean crystalline consisting of schist, gneiss, and amphibolite. The spillway structure in the masonry portion of the dam has been located confining mostly within a 30 m wide band of amphibolite.

The boundary fault between the Archaean and the Gondwana rocks is aligned normal to the dam axis and consists of crushed rocks with clay gouge affecting the dam foundation. Numerous clay lenses are present in the rocks of the left bank, which posed the problem of settlement by liquefaction. To prevent leakage through the fault, a 20 m-deep cut-off and a 1.5 m-thick clay blanket extending to a length of 150 m towards upstream side were provided. In addition, 18 m-deep grouting was undertaken underneath the cut-off.

In the foundation rocks of the masonry section of the dam, some soft and jointed bands of schistose rocks were present, which led to the problem of stability. To provide safety to the dam, the following remedial measures were taken up: an upstream curvature was provided to the spillway, a monolithic block spanning the entire soft mica schist band was erected, and a 9 m deep key trench was excavated and backfilled with reinforced concrete. In addition, blanket and curtain grouting were undertaken to a depth of 9 m covering a distance of 13 m of spillway length to consolidate the foundation and contain leakage. Later, high-pressure grouting was carried out from the foundation gallery to provide a single row curtain extending to a depth of 24 m (Chowdhury 1975).

A.5 DAMS BUILT IN HIMALAYAN TERRAIN

This section gives a brief description of some dams built in the Himalayan terrain.

A.5.1 Bhakra–Nangal Dam

The Bhakra–Nangal complex, with the main dam at Bhakra, a barrage at Nangal, two powerhouses, and other ancillary structures including a network of irrigation canals, is one of the largest multipurpose projects of the world. The Bhakra dam constructed across the river Sutlej is a 226 m-high, 518 m-long concrete dam. It is founded on the Lower Siwalik sandstone interbedded with siltstone and claystone dipping steeply towards downstream. A clay band known as heel claystone occurs at a depth of 150 m below the dam foundation. The rocks of the dam site have been affected by a fault and several large shear zones.

The foundation treatment included the excavation of the hill claystone and backfilling with concrete. A 15 m-thick concrete slab was provided to act as a strut to block the heel claystone. The bedding shear zones and the claystone–siltstone members were excavated to formula depth and backfilled by pre-cooled concrete. The foundation rocks were further strengthened by grouting. In the post-construction stage, some cracks developed in the heel claystone strut with consequent leakage. The gallery cracks and other defects were finally corrected by concrete plugging.

The 27 m-high Nangal barrage constructed downstream of the Bhakra dam rests on pervious terrace gravels overlying the boulder conglomerate bed of the Upper Siwalik formation. Problem of leakage through the foundation of pervious terrace gravel was checked by providing grout curtain extended down to the boulder conglomerate, which is hardened to a great extent by siliceous matrix. Drainage holes were also provided to relieve the uplift pressure (Hukku 1975).

A.5.2 Ranjit Sagar (Thein) Dam, Punjab

The Ranjit Sagar (Thein) multipurpose dam project of Gurudaspur district in Punjab consists of an earth-cum-rock-fill dam 160 m high and 600 m long, located across the river Ravi in the

Siwalik range of the Outer Himalayas. The project includes a powerhouse of 600 MW capacity and four 12 m-diameter tunnels of which two tunnels are being used as power tunnels and two as irrigation tunnels. Located along a natural saddle at the left bank, the dam is provided with a 133 m-wide chute spillway comprising an ogee crest along with a desilting basin and a roller bucket.

The dam site comprises alternate sequence of sandstone and claystone/siltstone of the Lower Siwalik formation of Miocene age. Due to the presence of soft rocks in the dam site and high seismicity of the project area (seismic zone V), an earth-cum-rock-fill dam has been constructed. A seismic coefficient of 0.15g (horizontal) and 0.075g (vertical) acting simultaneously at the centre of gravity of the load has been adopted in the design. An upstream inclined clay core of thickness 0.5H, one line grout curtain of depth one-third the dam height plus 15 m, and drainage galleries on both abutments have been provided to check piping and uplift. In addition, blanket grouting was taken up to the base of the dam core to strengthen the soft and open jointed rocks of the foundation (ISEG News 2001).

A.6 INSTANCES OF DAM FAILURES

This section provides some instances of dam failure.

A.6.1 Tigra Dam, Madhya Pradesh

The Tigra dam in Madhya Pradesh is a typical example of dam failure due to geological reasons. It was one of the oldest dams in India constructed in 1917 across the river Sank. The dam was a masonry gravity structure, 25 m high and 1.8 m long, and was made of hand-chiselled sandstone blocks. It was founded on sandstone alternated with thin beds of shale dipping at a low angle towards the downstream.

The dam was breached in several blocks immediately after filling of the reservoir. The breach was nearly 400 m across. The failure mechanism includes percolation of water through the joints in rocks that aided creation of the uplift pressure. The low dipping beds of foundation rocks lubricated by the water gave easy passage for sliding of the masonry block along the bedding planes. A flood discharge of about 8500 m³/s with a water head of 400 m length flowed over the ungated dam. The force of the rushing water was so high that several masonry blocks were bodily lifted and pushed to a distance of 14 m from the dam. The failure was purely due to the adverse geological features of the dam foundation coupled with overtopping caused by flood (Mehta and Pradhan 1972).

A.6.2 Kedarnala Dam, Madhya Pradesh

The 22 m-high Kedarnala earth dam in Raigarh district, Madhya Pradesh, failed within a few months of its construction due to the adverse geological condition of the dam site. The dam was founded on quartzitic sandstone with shale intercalation. On 6 July 1964, immediately after filling of the reservoir, muddy water flowed out from the boulder toe junction with the ground. Leakage then started from the embankment slope above the boulder toe level and the quantity of the discharge increased and the dam was breached coinciding with continuous rainfall to an extent of 408 cm within a period of 13 hours.

A porous sand zone in the foundation rock that remained untreated during the foundation treatment served as the avenue of water leakage through the foundation and created uplift pressure causing the breakage. The differential settlement in the earthwork due to poor compaction of the earth section coupled with heavy and continuous downpour is considered to be the contributory factor for the dam failure (Mehta and Pradhan 1972).

A.6.3 Khadakwasla Dam, Maharashtra

The Khadakwasla dam near Pune in Maharashtra was constructed in 1879 as a masonry gravity structure founded on hard rock. The supply from the stored water was used mainly for irrigation purpose. The dam had a height of 31.25 m above the river bed. Its crest length was 1.71 m and it had a free board of 2.74 m. The dam had a flood capacity of 2775 m^3/s and a reservoir capacity of 103 m^3.

A breach in an upstream dam (Panshet dam) and the release of excessively large volume of water from the storage of that dam to Khadakwasla reservoir caused overtopping of the dam as the volume of inflow was much above the designed flood. The battering of the incoming flood caused vibration of the dam, and the dam failed within four hours of the additional delivery of flood water from the Panshet dam. The Khadakwasla masonry gravity dam that had irrigated the land for more than eight decades (1879–1961) is considered to be the first of its kind in the world (Thandaveswara 2008).

SUMMARY

- Dams constructed in igneous rocks such as basalt encountered problems of foundation weakness due to excessive joints, shear zones, and even faults. Excavation of the shattered rocks and soft clayey materials of the shear zones and faults and backfilling by concrete could solve the problem. Blanket and curtain grouting could strengthen the rock.
- In foundation of dams consisting of sedimentary rocks, problems experienced include weathering of rocks and seepage causing upheaval and uplift pressure. Excavation until firm bedrock and consolidation grouting had to be taken up for strengthening the foundation. In Gondwana sedimentary rocks of Bhakra dam, Punjab, a shale bed from the dam foundation was excavated and backfilled with concrete.
- The problems of metamorphic rocks in dam foundation are mainly related to its fabric and structure such as joints, fault, and shear zones. Faults were treated by the removal of soft gouge of faults and shattered materials of shear zones. Blanket and curtain grouting were also undertaken to strengthen the foundation rocks.
- There are instances of failures of some dams of about 20 m to 30 m in height due to geological weakness in the dam foundation and faulty design. Founding of the dam on inclined sedimentary beds alternated with sandstone and soft shale caused slip of bedrock in the foundation. Sudden rush of excessive flood and no provision in design for emergency spilling caused overtopping and failure of dams.
- Analytical results of geological hazards in dam construction including failures of some dams in the past have raised awareness towards the need for geological investigation prior to dam construction. The investigation is taken up for evaluating the defects in the dam site and their corrective measures and also for determining appropriate design requirement towards safe construction of the dams in sites of various geological conditions.
- In dams constructed in high seismicity areas such as Meghalaya and Jammu and Kashmir, suitable earthquake factors need to be incorporated in the design of dams for their stability against seismic tremors. The Koyna dam project of Maharashtra, which was subjected to reservoir-induced seismicity, required measures such as strengthening of the dam by buttress, cabling, and grouting.

APPENDIX B

Geotechnical Problems of Tunnels and Their Solutions

This presentation includes geological weaknesses encountered and the measures adopted for their treatment towards safe construction of some tunnels in India and as such it will be proper to study this subject matter with Chapter 13. Readers interested in a detailed study may consult the publications given in the references for each case study.

B.1 HARD ROCK TUNNELLING

This section deals with some of the problems encountered in tunnelling through hard rocks.

B.1.1 Water Conductor Tunnel, Stage I, Umiam Project, Meghalaya

The 3 m-diameter and 2.1 km-long tunnel under stage I of the Umiam hydroelectric project of Meghalaya pierces through quartzite, phyllite, and metamorphosed conglomerate and meta-dolerite of the Shillong series. A thick band of extensively jointed meta-dolerite having intrusive relationship with the quartzites occurs at the central part of the tunnel. The entire rock types are folded and jointed to some extent. A fault and several shear zones also cut across the tunnel rocks.

Insufficient rock cover above the tunnel grade covering a 60 m stretch, presence of numerous shear zones, spheroidal weathering, and constant infiltration of surface water through joints in meta-dolerite of this part resulted in roof collapse. Subsequently, several slides affected the wall rocks during heavy rains. The rock slides could be finally arrested by confining the excavation and debris clearance works within the dry season (winter) and providing immediate reinforced cement concrete (RCC) cover to the roof and concrete lining to the wall rocks.

There was another major setback in tunnelling. At a distance of about 300 m from the outlet portal, porous sandstone charged with groundwater caused a large quantity of water to gush out with silt and sand that piled up and completely chocked the outlet end for six months. After the water flow was reduced to some extent in the dry season, the sandy debris was cleared off and consolidation grouting was carried out in the sandstone bed. Thick RCC lining was then provided in the affected part and the tunnel was made through (Gangopadhyay 1967).

B.1.2 Head Race Tunnel, Balimela Multipurpose Project, Orissa

The 7.6 m-diameter and 3.99 km-long water conductor (head race tunnel) of Balimela multipurpose project is located in the Eastern Ghats for carrying water impounded behind a 73 m-high earth dam, constructed across the river Sileru in Koraput district in Orissa for generation of 480 MW power. The tunnel passes through hard charnockite and pyroxene granulite thinly interbanded with basic igneous rocks and some pegmatite veins.

The tunnel rocks are without any major structural defects but contain joints and shear zones at some stretches, but no treatment was initially undertaken there. However, 14 stretches of jointed and sheared rocks were later required to be treated by reinforced lining and high-pressure grouting. The single stretch of these treated portions varied between 8 m and 42 m in length. The total length of treatment required in tunnel rock was 322 m, the thickness of lining in each stretch being 30 cm to 60 cm (Sinha, Pradhan, and Singh 1971).

B.2 SOFT ROCK TUNNELLING

In this section, some of the problems that were experienced when tunnelling through soft rocks are discussed.

B.2.1 Water Conductor Tunnel, Rana Pratap Sagar Multipurpose Project, Rajasthan

The Rana Pratap Sagar project of Rajasthan is constituted of a high dam constructed across the river Chambal and a 1466 m-long and 12.9-m (base width) horseshoe-shaped tunnel. The water conductor tunnel located in the left bank of the river passes through Vindhyan sandstone containing several thin beds of ferruginous siltstone and shales. At places, 1 cm to 5 cm thick silt and clay seams occur interlayered with the sandstone. Water seeps through these interlayer zones. Heavy seepage was detected when the tunnel passed through the inlet portal where the sandstone contains numerous clay seams.

Two sets of intersected joints are the main structural elements of the sandstone. A portion of the tunnel roof experienced overbreaks due to this geological defect (joints in rock), but in other places overbreaks were the result of heavy blasting. The normal overbreak was only 2 per cent to 5 per cent. The tunnel was lined to a thickness of 45 cm by RCC only to minimize frictional losses, as otherwise the tunnel could be kept unlined. The grout consumption was negligible and has no correlation with the openness of joints in tunnel rocks (Sanganeria 1975).

B.2.2 Water Conductor Tunnel, Ramganga Project, Uttar Pradesh

Located in the Himalayan terrain in Uttar Pradesh, the Ramganga project constitutes a 124 m-high rock-fill dam across the river Ramganga and a water conductor system characterized by two tunnels. One of these is a 1 km-long and 9 m-diameter circular tunnel, and the other is a power tunnel. The two tunnels pass through alternative beds of sand-rock and clay shale of Siwalik formation. The rocks, affected by two thrusts, namely Dagshais and Krols, are loosely cemented and porous in nature, possessing low strength. The clay shale bed consisting of several thin siltstone and plastic clay seams occupies 25 per cent of the tunnel length. All the rocks are traversed by several sets of joints.

The tunnel is aligned at a right angle to the bedding strike. A pilot tunnel driven along the power tunnel indicated seepage of water from several places, especially from the contact planes of sand-rock and clay shale. During tunnelling, roof falls took place affecting the rock up to a height of 5 m from the crown. Since the rocks had very low tensile strength, immediate supports were provided to protect such large overbreaks. Close-spaced (0.6 m intervals) supports were provided covering the entire length of the clay shale rock. Unsaturated ground slope, consisting of sand-rock at the two portal ends, posed the problem of slope failure where the ground was stabilized by providing 3 m-wide berms at every 15 m height of the slope (Varma and Mehta 1971).

B.3 TUNNELLING IN HIMALAYAN TERRAIN

This section discusses the difficulties faced when tunnelling through the Himalayan terrain.

B.3.1 Head Race and Tail Race Tunnels, Jaldhaka Project, West Bengal

The Jaldhaka project of Darjeeling Himalayan region consists of schist and gneiss that occur under a thick cover of unconsolidated terrace deposit. All the rocks are tectonically disturbed due to folding, thrusting, intense shearing, and jointing. The 3.39 km-long head race tunnel pierces through schists, gneisses, and sheared quartzites. Overbreaks including huge roof collapse in the form of chimneys were of frequent nature and mainly found where quartzite contains close-spaced joints, crushed rocks, and clay gouge. Fore-polling with steel piles was adopted to cross the affected zone by tunnelling.

The 420 m-long tail race tunnel passes through biotite schist, mica gneiss, quartzite, and amphibolite traversed by several quartz veins. Overbreaks including formation of chimneys as high as 4.6 m were formed at the intersections of two major joints and where rocks were affected by shear zones. Infiltration of water through the weak planes further aggravated the problem of overbreak. Intensive rock support and adoption of fore-polling aided the completion of the tunnel.

In Jaldhaka project, heavy overbreaks from roof and collapse of side walls had taken place covering a large stretch where the tunnel passes parallel to the foliation planes. Another difficulty was constant seepage of water that lubricated the weak planes and removed the infilling material from them. Adequate drainage arrangement and cautious approach by advance providing of structurally affected zones helped averting all these tunnelling hazards and the entire tunnel works could be completed safely (Chatterjee and Gangopadhyay 1971).

B.3.2 Banihal (Jawahar) Traffic Tunnel, Jammu and Kashmir

The Jammu–Srinagar highway is connected by the Banihal tunnel, which is a combination of two tunnels, each 2.54 km long and horseshoe shaped (5.5 m base width). The Banihal tunnel pierces through the snow-covered Pir Panjal range comprising agglomeratic slate, trap rocks (basalt with tuff), and limestone. The excavation work started by driving a pilot tunnel. The trap rocks, as a result of extensive joints, resulted in overbreaks in all parts. In limestone, seepage of water was the main problem. Presence of springs along the tunnel line also led to the problem of water flowage.

Rock bolting without netting, and at places with netting, had to be undertaken covering the roof and weak zones to prevent overbreaks. Most parts comprising the trap rocks were, however, kept unlined. Due to the soft and weak nature of the agglomeratic slate, the entire portion of the tunnel comprising this rock was lined. The spring water was diverted to the sides by longitudinal drains and lining was provided as remedial measures in the tunnel stretch comprising limestone. As the area is under a high seismic zone, articulated joints comprising copper sheets with *shalite* were placed on both walls to prevent rupture of concrete lining (Ramchandran and Gangopadhyay 1972).

B.4 *SOFT GROUND TUNNELLING—KOLKATA METRO RAILWAY TUNNEL*

The 16.5 km-long railway tunnel of Kolkata passes through alluvial soil. The soil investigation indicated two types of soil, namely river channel deposit and normal Kolkata deposit. The river channel deposit type of sub-soil consists of grey coloured silt and fine- to medium-grained sand down to 55 m from surface level with top 4 m to 5m of the deposit having more clay. The soil of normal Kolkata deposit contains light brown silty clay mixed with decomposed woods at 8 m to 9 m and calcareous nodules at 5 m to 6 m depths.

The analyses of soil samples showed the average properties as density 1.34 tonnes/m³, cohesion 0.38 kg/m², and angle of internal friction 4°. The groundwater table is measured between 5 m and 6 m depths. Perched water table is recorded in some localities at shallower levels. Kolkata falls under seismic zone III and a horizontal seismic coefficient range 0.04–0.06*g* was considered in the design.

The tunnel was constructed mainly by cut and cover method and partly by shield tunnelling. In the cut and cover method, two vertical retaining walls were made 10 m apart to support the ground with adjoining structures. The soil between the walls was then dug out to depths of 12 m to 14 m and 19 m at station areas. After completion of digging, RCC box was provided in the pit and the top part was backfilled for restoration of road.

Shield tunnelling involved four operations: construction of tunnel shafts, introduction and operation of shield, excavation of earth, and grouting and lining. The shield method helped controlling the shape and size of the tunnel and ensured safety of working men and machinery. The main problem faced was ground stability. The soft soil tried to squeeze into the shield face, threatening ground collapse and subsidence. Use of compressed air helped to support the face and solved the problem (Sengupta 1991).

SUMMARY

- Insufficient rock cover and highly jointed meta-basalt in parts of Umiam stage I tunnel in Meghalaya resulted in complete collapse of a long stretch of the tunnel. Providing RCC cover and lining of the wall could help in completing the construction of this part but sudden rush of sandy material choked another stretch of the tunnel near the portal where lining immediately after debris removal had to be undertaken.
- In the hard rocks such as charnockite, basalt, and gneiss in some other tunnel projects of India, the main problems encountered were related to close jointing, shearing, and faulting of rocks as well as flow of water. The remedial measures taken up to solve these problems included providing immediate support of the affected tunnel sections and reinforced lining. In addition, shotcreting and rock bolting had to be carried out.
- The main problems of tunnelling through soft rocks such as sandstone, shale, and slate of the tunnel projects such as the Rana Pratap Sagar in Rajasthan and Ramaganga in Uttar Pradesh were frequent roof collapse and seepage through fissures that stopped the progress of tunnelling. Remedial measures such as close-spaced supports and reinforced lining of affected parts could help in safely completing the tunnelling works.
- In the Himalayan terrain, tunnel constructions of the Jaldhaka project of West Bengal and Banihal of Jammu and Kashmir encountered geological hazards related to tectonically disturbed rocks causing frequent roof collapse and seepage through the wall rocks and squeezing grounds. The tunnelling work was safely completed by providing roof support, RCC lining, drainage of seepage water, and applying fore-polling method at stretches.
- In the construction of the Kolkata Metro tunnel, traversing through thick deposits of Indo-Gangetic alluvium, cut and cover method and partly shield method were followed. The main problem in tunnelling through soft ground was the side collapse of the excavated pit. Vertical retaining walls were provided to protect side collapse and RCC box was provided in the top parts. The shield method included excavation of earth and grouting followed by lining.

Glossary

Abyssal plain An extremely smooth portion of the deep sea floor. The gradient across the abyssal plains falls within the range of 1:1000 to 1:10,000. This fact is known from deep sea photographs and high-precision sounding techniques.

Acid rock The dominant chemical constituent of igneous rocks is silica, which ranges between 35 per cent and 75 per cent (by weight). The rocks are classified into: acid > 66 per cent, intermediate 55–66 per cent, basic 45–55 per cent, and ultrabasic < 45 per cent.

Amygdale A rounded mass of mineral formed in a gas cavity in a volcanic rock, a rock that solidified before all the gas bubbled out.

Anticline An arch-shaped fold into which rock strata have been compressed, the oldest rocks occurring in the core. The strata on either side dip away from each other.

Aquifer Soil or rock layer (stratum) in which the groundwater flows easily. Typically, aquifers consist of coarse-grained soils or fractured rock.

Aquitard Soil or rock layer (stratum) that restricts or prevents the movement of subsurface water. Typically, aquitards consist of fine-grained soils such as silts and clays or sound rock without any fractures.

Asthenosphere A zone within the earth's upper mantle in which the velocity of seismic waves is considerably reduced. Movement between the earth's outer lithosphere and inner mesosphere is thought to take place along this zone, which is capable of prolonged plastic deformation.

Atterberg limits The liquid limit, plastic limit, and shrinkage limit for soil. The *water content* at which the soil behaviour changes from the liquid to the plastic state is the liquid limit, from the plastic to the semisolid state is the plastic limit, and from the semisolid to the solid state is the shrinkage limit.

Basic rock Descriptive term for igneous rocks containing 45–55 per cent silica by weight, (also called *mafic*).

Bearing capacity The pressure that can be imposed by a foundation onto the soil or rock supporting the foundation.

Bed A stratum or single distinct sheet-like layer of sedimentary rock.

Bed load   Particles dragged, rolled, or bounced along during transport by wind or water.

Brittle   Deformation of rock by fracturing.

Caisson   A large structural chamber used to keep soil and water from entering into a deep excavation or construction area. Caisson may be installed by being sunk in place or by symmetrically excavating below the bottom unit to the desired depth.

Caldera   A volcanic crater more than 1 km across formed by subsidence of floor during an eruption.

Chemical weathering   The process of weathering whereby chemical reactions such as hydration, solution, oxidation, and ion exchange break down and possibly change rock and soil materials.

Cleavage   Regular planes of weakness in a crystal that are a consequence of its atomic structure.

Compaction   The process of increasing the density or unit weight of soil (frequently fill soil) by rolling, vibrating, or other mechanical means.

Conchoidal   Describing the smoothly curved, glasslike character of a fracture (broken surface) marked by concentric arcuate ridges common to many minerals, for example, quartz.

Consolidation   The process by which compression of a newly stressed clay soil occurs simultaneously with the expulsion of water present in the soil void spaces. Initially, the newly imposed stress acting on the clay is imparted onto the water in the soil voids (pore water), and not onto the soil particles. Because of the increased pressure, water is gradually forced out of the soil. As the pore water pressure is reduced, the magnitude of the stress being imposed onto the soil particles is correspondingly increased. Compression of clay layer occurs only as rapidly as pore water can drain from the soil and this is related to the permeability of the clay layer.

Continental drift   The relative movement of continental blocks across the surface of the earth as a result of sea floor spreading. The hypothesis of continental drift was proposed in the early 1900s, but only with the advent of plate tectonics later, the theory was accepted to be a viable mechanism for the movement of continents.

Craton   A stable area of continental crust, also known as a shield, which has remained undeformed since the Archaen time.

Cross bedding   Sedimentary beds with development of internal lamination within a stratum inclined at an angle to the main bedding planes, resulting from changes in the direction of water or wind currents during deposition.

Crystal   A solid substance formed by a repeating, three-dimensional pattern of atoms of various elements. Most minerals are present in crystalline forms.

Density   The mass per unit volume. In reference to soil, the term often also indicates unit volume and is synonymous with unit weight.

Dip   The angle between the horizontal plane and a geological surface, such as a bedding plane.

Drawdown   The lowering of the level of groundwater table that occurs in the vicinity of a water well (on dewatering equipment) when it is pumped.

Ductile   Deformation of rock by bending or squeezing, but without fracturing.

Dyke   A body of igneous rock, intruded vertically and discordant to the structure of the rocks through which it passes.

Earth pressure   Normally used in reference to the lateral pressure or force imposed by a soil mass against an earth-supporting structure such as a retaining wall or basement wall located within the soil mass.

Earthquake   The shaking of the ground caused by a sudden movement within the earth due to the release of energy that occurs as a result of deep rock fracturing or shifting, volcanic eruption, or a large explosion.

Epicentre   The point on the earth's surface directly above an earthquake's focus.

Expansive clay   Clay soils that experience significant volume expansion in the presence of water and shrink upon drying. Clays including the montmorillonite minerals are especially noted for their volume change characteristics.

Fault   A fracture in the earth's crust along the plane of which there has been displacement of rock on one side relative to the other.

Floodplain   Flat low-lying ground besides a river over which it floods after heavy rainfall. The course of a river tends to migrate to and fro across its flood plain over tens of thousands of years.

Focus   The breakpoint at which earthquake motion begins and where most of the energy is released.

Fold   Buckling of bedded sedimentary rocks due to deformation processes.

Foliation   A banded or laminated structure within a metamorphic rock caused by the alignment of minerals in layers by regional metamorphism.

Footing   A type of foundation typically installed as a shallow depth and constructed to provide a relatively large area of bearing onto the supporting soil.

Footwall   The surface of rock beneath a fault plane.

Gabions   Stone-filled steel wire baskets that can be assembled or stalked like building blocks to act as retaining walls or provide slope and erosion protection.

Geothermal gradient   The rate at which temperature increases with depth inside the earth.

Graben   A down-dropped crust segment, such as the great rift in Central Africa.

Groundwater table   The surface of the underground supply of water. It is also referred to as the phreatic surface.

Hanging wall   The surface of rock above a fault plane.

Heave   Upward movement of the soil and the foundations supported on soil, caused by expansion occurring in the soil as a result of factors such as freezing or swelling due to increased water content. Frost heave refers to the vertical soil movement that occurs in freezing temperature as ice layers or lenses form within the soil and cause the soil mass to expand.

Hydraulic gradient   Mathematical term indicating the difference in pressure head existing between two locations divided by the distance between these locations. It is designated by the symbol *i*.

Hydrolysis   Chemical breakdown of minerals to dry particles by the action of slightly acidic water during weathering.

Intrusive   Describing magma (molten rock) or rock formed from magma that hardened before reaching the surface, forced into cracks or layers between existing rocks and characterized by larger crystals than extrusive rock.

Joint   A smooth fracture through a body of rock caused by release of pressure or cooling. There is no movement between the opposite sides of a joint.

Karst   Topography that develops over soluble limestone and dolomite, characterized by sinkholes and fluted limestone remains.

Laccolith   A concordant intrusive igneous body with a dyke-like feeder, usually forming small lens-like features less than 5 km in diameter. It has a flat base but the upper surface is convex.

Liquefaction   Loss of strength occurring in saturated cohesionless soil exposed to shock or vibration when the soil particles momentarily loose contact. The materials then behave as a fluid.

Magma   Molten rocks; usually used to refer to molten rock at depth, whereas molten rock at the surface is called lava.

Malleable   Can be flattened out by pounding, a characteristic of native metals and of the metals freed from ores.

Matrix   The fine-grained material that surrounds larger crystals, pebbles, or fossils in rock, especially sedimentary rock.

Mechanical weathering   The process of weathering whereby physical forces, such as frost action and temperature changes, break down or reduce rock to smaller fragments without involving chemical change.

Metamorphism   The process of changing pre-existing igneous and sedimentary rocks without melting under condition of high temperature and/or pressure. Thermal or contact metamorphism is a result of proximity to an igneous intrusion. Regional metamorphism is a result of burial to great depth.

Mineral   A naturally occurring crystalline substance with a well-defined chemical composition.

Monocline   A bending of rock strata produced in sedimentary sequence that has deformed under conditions favouring the formation of normal fault.

Normal fault   A fault showing displacement in the down-dip direction of the fault. This results from a stress condition in which the principal maximum stress is vertical and the other two principal stresses are horizontal.

Orogeny   A mountain building episode, resulting from a collision between two continents or between a continent and an island arc.

Outcrop   The area on the earth's surface where a particular rock type or rock body is present.

Overlap   The deposition of beds lying above an unconformity, where they were deposited by the transgressing sea.

Oxbow lake   A crescent-shaped lake occupying an abandoned channel (oxbow) that was formerly part of a channel.

Permeability   The ability of a rock, sediment, or soil to allow pore fluids and gases to pass through it.

Pile   The relatively long, slender column-like type of foundation that obtain supporting capacity from the soil or rock some distance below the ground surface.

Plate tectonics   A theory arising from a series of ideas developed in the early 1960s proposing that the surface of the earth is composed of relatively thin plates of rigid materials. These tectonic plates extend down to the low-velocity zone of upper mantle. They are all in motion relative to one another and it is through these movements and the consequent collisions between the plates that the present distribution of almost all volcanic, seismic, and orogenic activities are controlled.

Poisson's ratio   The ratio of lateral unit strain to the longitudinal unit strain in a body that has been stressed longitudinally within its elastic limit.

Pore pressure   Water pressure developed in the voids of a soil mass. *Excess pore pressure* refers to pressure greater than the normal hydraulic pressure expected as a result of position bellow the water table.

Porosity   The extent to which a body of soil, rock, or sediment is permeated with cavities between grains, usually expressed as a percentage of volume. These pores are filled by air and water, which impede air movement and allow water to move only by capillarity.

Pyroclastic rock   A rock formed by accumulation of fragmented materials thrown out by volcanic explosion. It may be associated with ash flow of exploded lava that became pulverized as it came down a slope to form a deposit of fragmented rocks and dust.

Reverse fault   A type of fault in which the movement along the inclined fault plane is towards the up-dip. The rock on the down slip side of the fault has moved upwards.

Rip-rap   The layer of boulders or crushed rocks typically ranging from 15 cm to 60 cm in size, placed as a covering to protect the surface of an earth dam and earth slopes against erosion.

Seismic wave   A wave generated by an explosion or earthquake within the earth or on its surface. There are four main types of seismic waves, namely primary wave, secondary wave, Rayleigh wave, and love wave.

Seismogram   The oscillating line record of ground movements measured by a seismograph during an earthquake.

Seismograph   An instrument to measure the horizontal and vertical movements or vibrations that occur within the earth or at the surface because of an earthquake or other seismic waves.

Shear strength   The ability of a soil to resist shearing stresses within a soil mass as a result of loading imposed on the soil.

Sill   A near-horizontal sheet-like igneous intrusion, usually dolerite, of roughly uniform thickness but thin relative to its area.

Slickenside   Small parallel grooves or striations formed on the surface of a fault as result of movement of rocks against each other. The surface is smooth in one direction but rough in the opposite direction.

Strike   The direction along a rock stratum at right angle to the true dip.

Subduction   The process describing one tectonic plate descending at an angle below another, which happens at a destructive plate boundary.

Syncline   A basin-shaped fold in which the beds dip towards each other.

Thrust   A low-angle reverse fault that extends over a large distance pushing older rocks over the younger rock.

Tide   The regular rising and falling of water level in ocean from the gravitational attraction that exists between the earth, the sun, and the moon. Depending upon their positions, either *spring tides* or *neap tides* occur.

Tsunami   A seismic sea wave generated in the ocean by submarine earthquake, volcanic eruptions, or mass slides underwater. The height of the waves is low while travelling through deep water but it gradually increases, and the waves become extremely dangerous as they approach the shore.

Unconformity   A surface representing a period of non-deposition or erosion separating rocks of different age.

Vadose zone   The subsurface soil zone that lies directly above the position of the groundwater table. Usually, this zone is partially saturated with water that has migrated upward via capillary movement from the water table.

Vein   A more or less upright sheet deposit of minerals, cutting other rocks and formed from solution rather than from a molten magma-like dyke.

Void ratio   The total volume occupied by a soil mass includes the soil particles plus void spaces. The void ration is the ratio of the void space volume to the volume of soil solids.

Volcano   A fissure or vent on the earth's surface connected by a conduit to the earth's interior from which lava, gas, and pyroclastic materials erupt.

Wave-cut platform   An irregular gently sloping bare rock platform extending out to ocean and usually backed by cliffs. It is the wearing of the cliff that causes the enlargement of the platform. As this will not develop in areas where there is a covering of beach material, the most favourable sites are headlands.

Weathering   The process of slow decay of rock and its constituent minerals upon exposure to the earth's surface by either mechanical or chemical attack.

References

Acharya, P. K., trans. (1973). *Manasara on Architecture and Sculpture.* 7 vols, pt. 21. London. Published by Munshiram Manoharlal Publications Pvt Ltd.

Anand, S. K. (1991). *Construction Material Studies for Dhaleswari Hydoelectric Project, Aizol District, Mizoram.* Proceedings of Seminar on Trend in Geotechnical Investigation in Last Twenty Five Years, ISEG, 20 (1–4): 12–5.

Anbalagan, R., Chakraborty, D., and Kohli D. (2008). 'Landslide Hazard Zonation (LMZ) Mapping on Meso-scale for Systematic Town Planning in Mountain Terrain', *Journal of Scientific and Industrial Research*, 67: pp. 486–497.

Attewell, P. B., and Farmer, I. W. (1976). *Principles of Engineering Geology.* New York: John Wiley and Sons.

Auden, J. B. (1942). 'A Geological Investigation of Tunnel Alignment for the Jumna Hydroelectric Scheme', *Records of the Geological Survey of India*, Professional Paper (2): 78.

Auden, J. B. (1971). *Reconnaissance Investigation of Three Alternative Alignments for Diversion Tunnel in Eastern Nepal.* Proceedings of Seminar on Engineering and Geological Problems in Tunnelling, ISEG. Proceedings Vol. (pt 2): 72–90.

Awasthi, R. K., and Singh, Joginder (1992). 'Gumti Hydroelectric Project, Tripura: The Nature and Causes of the Damages in Water Conductor System', *Journal of Engineering Geology, ISEG*, 21 (3 & 4): 95–104.

Bagchi, A. (1991). *Basic Concept of Pumped Storage Scheme and Its Application for Hydro Power Development in West Bengal.* Proceedings of Seminar on Trends in Geotechnical Investigation in Last Twenty Five Years, ISEG, 20 (1–4): 24–30.

Balasundaram, M. S., and Rao, G. S. M. (1972). 'History and Development of Engineering Geology', *Records of the Geological Survey of India*, 104, (pt 2): 1–12.

Balasundaram. (1982). 'Development of Engineering Geology in India and its future prospects', *ISEG*, 11 (3 & 4): 12–50.

Banerjee, S. L., and Midha, R. K. (1971). *Dynamic Elasticity Measurements in the Drifts at Slapper, Beas Sutlej Link Project, Himachal Pradesh.* Proceedings of Seminar on Engineering and Geological Problems in Tunnelling, 22–23 October 1971, ISEG, Proceedings Vol. (pt 2): 93–100.

Banerjee, S. L., and Rao, G. V. (1971). *Horizontal Rock Cover and Its In Situ Quality at Jaldhaka Tunnel, West Bengal.* Proceedings of Seminar on Engineering and Geological Problems in Tunnelling, 22–23 October 1971, ISEG, Proceedings Vol. (pt 2): 101–9.

Barton, N, Lien, R., and Lunde, J. (1974). 'Engineering Classification of Rock Masses for the Design of Tunnel Support', *Rock Mechanics*, 6 (4): 189–236.

Barton, N, Lien, R., and Lunde, J. (1975). *Estimation of Support Requirements for Underground Excavations.* Proceedings of 16th Symposium on Design Methods in Rock Mechanics, pp. 234–41.

Barton, N., Loset, F., Lien, R., and Lunde, J. (1980). *Application of the Q-system in Design Decisions Concerning Dimensions and Appropriate Support for Underground Installations.* International Conference on Subsurface Space, Rockstore, Stockholm, vol. 2, pp. 553–61.

Barton, N., and Grimstad, E. (1997). *The Q-system following Twenty Years of Application in NMT Support Selection.* Indo-Norwegian Workshop at NIRM on Recent Trend in Rock Mechanics.

Berkey, Dr Charles P. (1950). *Application of Geology to Engineering Practice.* New York: Geological Society of America (Berkey Volume).

Bhandari, R. K. (2006). *The Indian Landslide Scenario: Strategic Issues and Action Points.* Indian Disaster Management Congress, 20–30 November 2006, New Delhi.

Bieniawski, Z. T. (1974): *Geomechanics Classification of Rock Masses and Its Application in Tunnelling.* Proceedings of '3rd International Congress on Rock Mechanics'. ISRM, Denver, USA, vol. IIA, pp. 27–32.

Bieniawski, Z. T. (1975). *Case Studies: Predictions of Rock Mass Behavior by Geomechanics Classification.* Proceedings of 2nd Australia–New Zealand Conference on Geomechanics, Brisbane, pp. 36–41.

Bieniawski Z. T. (1979): *The Geomechanics Classification in Rock Engineering Applications.* Proceedings of 4th International Congress on Rock Mechanics, ISRM, Montreaux, vol.2, pp. 41–48.

Bieniawski, Z. T. (1989). *Engineering Rock Mass Classification.* New York: Wiley.

Billings, M. P. (1997). *Structural Geology.* New Delhi: Prentice Hall of India Pvt. Ltd.

BIS code IS 14496. (1998). Indian standard for preparation of landslide hazard zonation maps in mountainous terrain. Guidelines, Part 2: Macro zonation.

BIS code (modified) (2004). Modified BIS guidelines by GSI for macro-level landslide hazard zonation mapping.

Biswas, S. N. (1970). 'Extract of research work of a project engineer, Bihar, related to computer study on faults in foundation of gravity dams', Postgraduate thesis (personal communication). Univ Of Patna, Bihar, India.

Blyth, F. G. H. and de Freitas, M. H. (1967): *A Geology for Engineers.* London: Edward Arnold Ltd.

Bolt, Bruce A. (1999). *Earthquakes.* New York: W.H. Freeman and Company.

Bose, R. N., and Arora, C. L. (1969). 'Longitudinal wave velocities in various formations of India', *Journal ISEG*, 4 (1): 78–87.

Building Material and Technology Promotion Council (BMTPC). 2005. News bulletin of BMPTC (September 2005). Release of Landslide Hazard Zonation Atlas of India.

Casagrande, A. (1948). 'Classification and Identification of Soils', *Transactions of the American Society of Civil Engineers*, 113:901–91.

Casagrande, L. (1949). 'Electro-osmosis in Soils', *Geotechnique*, 1:159–77.

Chatterjee, B. (1983). 'Problem of upheaval of flume excavation through sheared and pulverized Daling Phyllite, Jaldhaka Stage II hydroelectric project, Darjeeling district, West Bengal', *Journal ISEG*, 12 (3 & 4): 89–94.

Chatterjee, P. K., and Gangopadhyay, S. (1971). *Some Aspects of Geotechnical Investigation for Tunnelling through Metamorphic Rocks of Eastern India.* Proceedings of Seminar on Engineering and Geotechnical Problems in Tunnelling, 22–23 October 1971, Journal ISEG, pt. 1, pp. 42–50.

Chaturvedi, S. N., and Mandwal, N. K. (1978). *Geological Problems and Remedial Measures for the Powerhouse foundation of the Ramganga River Project, Garwal District, Uttar Pradesh.* Proceedings of Seminar on Foundation Problems of Powerhouses and Related Ancillary Structures, New Delhi, 1973, ISEG, pp. 95–103.

Chibber, I. B., Chepe, A. B., and Ramachandra, H. M. (1991). *Geotechnical and Petrographic Studies Conducted for the Pier Foundation, Narmada Aqueduct Site, Tilwarsghat, Jabalpur, Madhya Pradesh, An Appraisal.* Proceedings of Silver Jubilee Seminar on Trend in Geotechnical Investigations in Last Twenty Five Years, ISEG, 20 (1–4): 68–78.

Chugh, C. P. (1971). *Diamond Drilling.* New Delhi: Oxford and JBH Publication Co.

Chugh, C. P. (1985). *Manual of Drilling Technology*. New Delhi: Oxonian Press Pvt. Ltd.

Chowdhury, A. K. (1971). *Geological Set-up and Its Influence on Tunnelling Condition, Jaldhaka Hydel Project, West Bengal*. Proceedings of Seminar on Engineering and Geotechnical Problems in Tunnelling, 22–23 October 1971, Journal ISEG, pt. 2, pp. 182–93.

Chowdhury, A. K. (1975). 'Tenughat Dam project, Bihar', *GSI Miscellaneous Publication*, 29, pt. 1, pp. 208–16.

Chowdhury, A. K. (1990). 'Construction Problems along Power Channel Alignment of Loktak Hydel Project, Manipur', *Journal ISEG*, 19 (3–4): 99–106.

Chowdhury, A. K. (1996). 'A Comprehensive Geotechnical Report on Loktak Hydel Project, Manipur', *Geological Survey of India Bulletin Series B, No. 50*.

Chowdhury, A. K. (1999). *Disaster from Dam and Embankment Failure*. Proceedings of National Workshop on Disaster Management, January 1999, Jadavpur University, Kolkata.

Chowdhury, A. K., and Biswajit Das (2009). 'Geotechnical Assessment of Diversion Tunnels, Lower Subansiri Hydro-Electric Project, Assam and Arunachal Pradesh', *Journal of Engineering Geology*, 35 (1–4): 206–11.

Chowhan, R. P. S. and Sarda, Y. P. (1991). *Rock Load Estimation and Support System for a 21.8-km Long Head Race Tunnel in High Range of the Western Himalayas*. Proceedings of Silver Jubilee Seminar on Trends in Geotechnical Investigation in Last Twenty Five Years, Kolkata, ISEG, 20 (1–4): 199–204.

Cruden, D. M., and Varnes, D. J. (1996). 'Landslide Types and Processes', in A. K. Turner and R. L. Shuster (eds), *Landslides: Investigation and Mitigation Transportation Research Board Special Report*, 247, pp. 36–75. Washington DC: National Academy Press.

Dana, E. S., and Ford, W. E. (1948). *A Text Book of Mineralogy*. London: John Wiley and Sons.

Davis, Bruce (2001). *GIS: A Visual Approach*. Nevada: Onward Press.

Davis, S. N. and Dewist, R. J. M. (1966). *Hydrology*. New York: John Wiley and Sons.

Deva, Yogendra and Srivastava, Mridul (2006). 'Grid-based Analytical Approach to Macro Landslide Hazard Zonation Mapping', ISEG, Proceedings Vol. p. 63.

Deere, D. W., Hendron, A. J., Patton, G. D., and Cording, E. J. (1967). 'Design of Surface and Subsurface Construction in Rock', in C. Fairhurst (ed.), *Failure and Breakage of Rock: Proceedings of 8th U.S. Symposium on Rock Mechanics.publ AIME,* New York, 237–302.

Deere, D. U. (1973). 'The Foliation Shear Zone—An Adverse Engineering Geologic Feature of Metamorphic Rocks', *Journal of the Boston Society of Civil Engineers*, 60 (4). pp. 11.

Deere, D. U., and Deere, D. W. (1988). 'The Rock Quality Designation (RQD) Index in Practice,' in L. Kirkaldie (ed.), *Rock Classification Symposium for Engineering Purposes*. Philadelphia: American Society for Testing and Materials Special Publication, 981, 91–101.

Demin, W., and Xun, G. (1982). *An Outline of the Evaluation of Geological Stabilities for Nuclear Power Plant Site in China*. Proceedings of 4th International Congress, New Delhi, IAEG, vol. 3, theme 6, pp. 205–9.

Dhar, Y. R. (1991). *Seismotectonic Evaluation of the Area around Uri Hydroelectric Project, Kashmir Himalayas*. Proceedings of Silver Jubilee Seminar on Trend in Geotechnical Investigation in Last Twenty Five Years, Kolkata, ISEG, 20 (1–4): 59–65.

Fookes, P. G., Dearman, W. R., and Franklin, J. A. (1971). 'Some Engineering Aspects of Rock Weathering with Field Examples from Dartmoor and Elsewhere', *Quarterly Journal of Engineering Geology*, 4:139–85.

Fumio, Ishi, RaguichiHo, Iida, and Isao, Shibata (1967). 'Mechanical Study on the Treatment of Fault in the Foundation Rock of the Gravity Dams', in *The Transaction of the 9th International Congress on Large Dams*. Queensland, Australia: Irrigation and Water Supply Commission.

Gangopadhyay, S. (1967). 'Mechanism of Rock Slides in Parts of the Tunnel of Umiam Hydel Project, Stage-1, Assam', *Journal of Engineering Geology*, ISEG, Vol 2 (Issue 1): 22–26.

Gangopadhyay, S. (1970). 'Engineering geological study of the reservoir area of Kopili Hydel Project, Assam, India', Ph.D. thesis, Univ of Calcutta, Kolkata, India. Indian Journal of Power and River Valley Development, Special number pp. 32–7.

Gangopadhyay, S. (1971). *Importance of Structural Analysis in Forecasting the Tunnelling Hazards of Umiam Hydel Project, Meghalaya, India*. Proceedings of Seminar on Engineering and Geotechnical Problems in Tunnelling, 22–23 October 1971, Journal ISEG, pt. 1, pp. 91–9.

Gangopadhyay, S. (1973). 'Foundation Geology and Grouting of the Umiam Dam, Assam', *Journal of Engineering Geology*, ISEG, December 1973.

Gangopadhyay, S. (1978). *Geotechnical Evaluation for Planning of Land Use in Sikkim, Himalayas*. Proceedings of 3rd International Congress of IAEG, September 1978, Madrid, Spain, sec. 1, vol. 1, pp. 56–63.

Gangopadhyay, S. (1980). *Petrographic Study of Suspended Particles of Some Himalayan Streams in Relation to Engineering Problems*. Proceedings of International Geological Congress, Paris, 1980, sec. 13 & 20, vol. 3, p. 1189.

Gangopadhyay, S. (1981). *Significance of Water Pumping Test in Detecting Subsurface Weak Features with Reference to Projects of Eastern India*. Proceedings of International Symposium of Weak Rock, Tokyo, Japan, September 1981, pp. 595–600.

Gangopadhyay, S. (1988). *Monuments of India Built through Ages with Special Reference to the Construction Materials*. Proceedings of Symposium of IAEG on Engineering.

Gangopadhyay, S., and Majumdar, A. K. (1989). 'Development of the Water Resources in Relation to Development Management in the Damodar Valley', *ISEG*, 8 (3 & 4). pp

Gangopadhyay, S. (1990). *Quantitative Evolution of Rock and Soil Properties*. Proceedings of Silver Jubilee Seminar on Trend in Geotechnical Investigations in Last Twenty Five Years, Kolkata, 28 October 1990, ISEG, 20 (souvenir volume) (1–4).

Gangopadhyay, S., and Mishra, P. (1991). *Geotechnical Evaluation of Rock Condition and Support Requirement for Tail Race Tunnel of North Koel Project, Bihar*. Proceedings of Silver Jubilee Seminar on Trend in Geotechnical Investigations in Last Twenty Five Years, ISEG, 20 (1–4): 239–44.

Gangopadhyay, S. (2002a). *Testimony of Stone. Vol. 1. Prehistoric Indians*. Kolkata: Das Gupta Publishers.

Gangopadhyay, S. (2002b): *Testimony of Stone. Vol. 2. Monuments of India*. Kolkata: Das Gupta Publishers.

Gupta, M. L. (1999). 'Inaugural Address in Workshop on Geohazards and Related Social Issues, Lucknow, November 1998', *Journal of Engineering Geology*, 27 (4): 1–4.

Gupta, S. K. (2003). 'Macroseismic Studies in Northeastern India during First Centenary of the Great Assam Earthquake of 1897', *Journal of Engineering Geology*, 28 (1–4): 15–28.

Gupta, Harsh K. (1992). 'Reservoir Induced Earthquakes', *Current Science*, Vol. 62 (pp. 182).

Gupta, Harsh K. (2006). 'BIS Methodology for Landslide Hazard Zonation – A Case Study from Bhimtal and Naukuchhia Tal Area, Nainital District, Uttaranchal', p. 84. Publisher: Bureau of Indian Standards, New Delhi.

Gupta, Harsh K. (2002). 'A review of recent studies of triggered earthquakes by artificial water reservoir with special emphasis on earthquakes in Koyna, India', *Earth-Science Reviews*, 58:279–310 (www. prpreinternational.org).

Grimstad, E., and Barton, N. (1993). 'Updating the Q-System for NMT', in *Proceedings of International Symposium on Sprayed Concrete – Modern Use of Wet Mix Sprayed Concrete for Underground Support, Fagernes*. Oslo: Norwegian Concrete Association, 46–66.

Hatch, F. H., Wells, A. K., and Wells M. K. (1984). *Petrology of Igneous Rocks*. New Delhi: CBS Publications and Distributors.

Hill, E. S. (1953). *Outlines of Structural Geology*. London: Methuen and Co. Ltd.

Hoek, E. and Bray, J. W. (1981). *Rock Slope Engineering*. London: Institute of Mining and Metallurgy.

Hoek E. and Brown E.T. (1980a). 'Empirical Strength Criterion for Rock Mass', *Journal of Geotechnical Engineering*, 106 (GT 9): 1013–35.

Hoek, E. and Brown, E. T. (1980b): *Underground Excavation in Rock*. London: Institute of Mining and Metallurgy, Stephen Austin and Sons Ltd.

Hoek, E. and Brown, E.T. (1997). 'Practical Estimation of Rock Mass Strength', *International Journal of Rock Mechanics and Mining Science & Geomechanics Abstracts*, 34 (8): 1165–86.

Hoek, E., Carranza-Torres, C., and Corkumn, B. (2002). *Hoek–Brown failure criterion—2002 Edition.* Proceedings of NARMS–TAC Conference, Toronto, 2002, 1:262–73.

Hoek, E. (2007). http://www.rocscience.com, Access date: 10/12/2010

Hukku, B. M., Chaturbedi, S. N., and Ashraf, Z. (1978). *Tunnelling Experience in Jammu and Kashmir Himalayas*. Proceedings on Seminar on Engineering and Geotechnical Problems in Tunnelling, 22–23 October 1971, ISEG, pt. 1, pp. 116–127.

Hukku, B. M. (1975). 'Bhakra Nangal Project, Punjab', *GSI Miscellaneous Publication*, 29(pt I): 1–10.

Hungr, O., Evans, S. G., Bovis, M., and Hutchinson, J. N. (2001). 'Review of the Classification of Landslides of the Flow Types', *Environment and Engineering Geoscience*, 7:221–38.

IAEG (1981). 'Rock and Soil Description and Classification for Engineering Geological Mapping. Report by the IAEG Commission on Engineering Geology Mapping', *IAEG Bulletin*, 24:235–74.

IAEG (1992). Homepage of IAEG, http://www.iaeg.info. Access Date: 10/12/2010

IAEG (2004). Information and Dissemination – Magnitude and Intensity (updated MSK scale). http://www.ugr.es/-iag/divulgacion/div_m_e.html. Access Date: 10/12/2010

ISEG (2001). 'Ranjit Sagar Dam, Punjab', *ISEG News*, April 2001, vol. 2, no. 1.

Jayabalan, K., Durairaj, U., and Muraleedharan, C. (2009). 'Assessment of Tsunami Impact on Landforms of West Coast between Kanyakumari and Kochi, India', *Journal IEEG*, 36 (1–4): 235–346.

Judd, W. R. (1969). *Message to ISEG.* Proceedings of Symposium on Rock Mechanics, October 1969, Journal of ISEG, 4, sec. 1–2, pp. 6–7.

Knill, J. L., Cratchley, C. R., Early, K. R., Gallois, R. W., Humphreys, J. D., Newbery, J., Price, D. G., Thurrell, R. G. (1970). 'The logging of rock cores for engineering purpose', *Quarterly Journal of Engineering Geology*, 3:1–24.

Kodkade, D. K. (1971). *Problem of Underground in Lower Himalayas with Special Reference to Yamuna Hydel Project, Stage II.* Proceedings of Seminar on Engineering and Geological Problems in Tunnelling, 22–23 October 1971, ISEG, pt. 2, pp. 75–94.

Krishnan, M. S. (1949). *Introduction to Physical Geology. Geology of India and Burma.* Chennai: Madras Law Journal Press.

Krishnamurthy, K. S., Sinha, R. G., and Bhatia, S. K. (1992). 'On the Significance of Buried Channels and Deep Overburden in Design of a Few River Valley Projects, Arunachal Pradesh', *Journal ISEG*, 21 (3 & 4): 7–18.

Krishnaswamy, V. S. (1972). 'Systematic Geotechnical Studies in the Country', *Records of the Geological Survey of* India, 104(pt 2): 13–30.

Krynine, D. R., and Judd, W. R. (1957). *Principles of Engineering Geology and Geotechnics.* New York and London: McGraw-Hill Company.

Lambe, T. W. (1953). *The Structure of Inorganic Clay.* Proceedings of American Society of Civil Engineers, Danvers, MA: ASCE.

Lewit (1982). *Hydraulics.* London: ELBS.

Lingam, A. B., Narasimulu, K., and Ramakrisna, G. (1968). 'Use of Nellore Fly Ash as Partial Replacement of Cement in Mortar and Concrete', *ISEG*, 3 (1): 140–56.

Mahendra, A. R., and Mathur, S. K. (1975). 'Ukai Project, Gujarat', *GSI Miscellaneous Publication*, 29 (pt 1): 154–63.

Majumdar, N. (1991). *Hydro-electric Project, Nagaland.* Proceedings of Silver Jubilee Seminar on Trend in Geotechnical Investigation in Last Twenty Five Years, Kolkata, ISEG, 20 (1–4): 251–60.

Mandwal, N. K. (2005). 'Problems of Locating an Atomic Power Plant on Aeolian Deposits', *ISEG*, 32 (1–4): 1–12.

Marinos, P., and Hoek, E. (2001). 'Estimating the Geological Properties of Heterogeneous Rock Masses such as Flysch', *Bulletin of Engineering Geology and Environment (IAEG)*, 60:85–92.

Mathur, S. K. (1983). 'Geotechnical Evaluation of Foundation Rocks of Bridge Sites across Chambal and Banas Rivers in Rajasthan', *Journal ISEG*, 12 (3 & 4): 49–66.

McGuire, R. K. (1974). *Seismic Structural Response and Risk Analysis Incorporating Peak Response Regression on Earthquake Magnitude and Distance*, Massachusetts Institute of Technology, Department of Civil Engineering, Research Report R74–51.

Mehta, P. N., and Pradhan, S. R. (1972). 'Geological Causes for Mishap and Failure of Engineering Structures', *Records of the Geological Survey of India*, 104, pt. 2, pp. 85–96.

Mukherjee, M., and Ghosh, R. N. (2004). 'Geotechnical Appraisal of Pier Foundation of Tikira High Level Road Bridge, Angul District, Orissa', *Journal ISEG*, 30 (1–4): 111–6.

Mukherjee, D., and Bhagwan, J. (2009). 'Geological Investigation of the Landslide Hazard Prone Hill Slopes in Mizoram', *Journal ISEG*, 36 (1–4): 275–82.

Murthy, Y. K. (1977). Life of Reservoir, Technical Report No. 19, Central Bureau of Irrigation and Power.

Murti, N. G. K. (1971). *Status in the Art of Tunnelling*. Proceedings of Seminar on Engineering and Geological Problems in Tunnelling, 22–23 October 1971, *ISEG*, pt. 2, pp. 1–14.

Mustafy, A. K. (1966). 'Some Diamond Drilling Problems on Hydel Projects', *Journal of Engineering Geology*, 1 (1): 9–108.

Narula, P. L., and Shome, S. K. (1992). 'Macroseismic Studies of Recent Earthquakes in Northwestern Himalaya—A Review', *Current Science*, 62 (1 & 2): 24–33.

Narula, P. L. (1999). 'Seismotectonic Evaluation of the Himalaya and its Bearing on Hydroelectric Power Development', *Journal of Engineering Geology*, 27 (1–4): 120–27.

Nag, P. C., Manier, Y. P., and Reddy, M. S. (1978). *Foundation for Thermal Power Stations*. Proceedings of the Seminar on Foundation Problems of Powerhouses and Related Ancillary Structures, New Delhi, 1973, ISEG, pp. 177–182.

Nath, R., and Dhawale, R. A. (1971). *Seismic and Nuclear Logging of Bore Holes for Studying the In Situ Physical Characteristics of Rock at Malabar Tunnel Project, Bombay*. Proceedings on Seminar on Engineering Geological Problems of Tunnelling, New Delhi, 22–23 October 1971, ISEG, part II, pp. 47–65.

Negi, R. S. (1992). 'Evaluation of Tunnelling Condition in Srinagar Thrust Zone and Its Vicinity, Maneri Valley Hydel Project, Stage II', *ISEG*, 21 (3 & 4): 199.

Niyogy, B. N., and Seth, N. N. (1972). 'Basin-wise Systematic Geohydrological Studies by the Geological Survey of India – a Review', *Records of the Geological Survey of India*, 104, pt. 2, pp. 115–130.

Oldham, R. D. (1899). 'Report on the Great Earthquake of 12th June, 1897', *Memoirs of the Geological Survey of India*, 29:1–349.

Pal, C., Basu, Roy B., and Mishra, P. (2009). 'Distressed Railway Bridge No.144 near Lallak-Mamalkka Railway Station on Sahabganj Loop of Eastern Railway, Bhagalpur District, Bihar', *Journal of Engineering Geology*, 36 (1–4): 245–50.

Pal, N. K., and Rao, K. K. (2009). 'Coastal Erosion at Uppada along Kakinada Coast, Andhra Pradesh—A Study from Remote Sensing'. ISEG, Publisher: Indian Society of Engineering Geology, Kolkata, India.

Pande, Pravash (1999a): *Geotechnical Assessment of Stability of Slopes*. Proceedings of the Symposium on Engineering Geology and Geo-environmental Problems in Hard Rock Terrain – HREGE '99, Bhopal, 6–7 December 1999, ISEG, 27 (1-4): 199.

Pande, Pravash (1999b): *Intraplate Seismicity of Peninsular India*. Proceedings of the Symposium on Engineering Geology and Geo-environmental Problems in Hard Rock Terrain – HREGE '99, ISEG, 27 (4): 152–68.

Pande, Pravash (2005). 'The Great Indian Ocean Tsunami', *ISEG News*, 2 (1).

Parida, K. (1999). *Rock Cover for the Water Conductor Tunnel, Indravati Hydroelectric Project, Andhra Pradesh*. Proceedings of the Symposium on Engineering Geology and Geo-environmental Problems in Hard Rock Terrain – HREGE '99, Bhopal, 6–7 December 1999, ISEG, 27 (4): 134–9.

Patel, D. T., and Joshi, H. M. (2004). 'Geological Investigation at Sardar Sarobar (Narmada) Project, Gujarat, India', *Journal of Engineering Geology*, 31 (1–4): 71–84.

Pettijohn, F. J. (1957). *Sedimentary Rocks*. Kolkata: Oxford Book Company.

Phillips, W. R. and Griffen, D. T. (1981). *Optical Mineralogy: Nonopaque Minerals*. San Francisco: Freeman and Co.

Proctor, R. V., and White, T. C. (1946). *Rock Tunnelling with Steel Support*. Publisher: Commercial Shearing & Stamping, Pennsylvania, USA.

Proctor and Richard, J. (1971). 'Mapping Geological Conditions in Tunnels', *Bulletin of the Association of Engineering Geologists*, 8(1): 1–43.

Raju, M. (2004). 'Post Construction Problems of Srisailam Dam, Andhra Pradesh', *Journal of Engineering Geology*, 30 (1–4): 11–6.

Ramachandran, B., and Gangopadhyay, S. (1972). 'Engineering Geological Features of Soft Rock Areas – Sedimentaries', *Records of the Geological Survey of India*, 104, pt. 2, pp. 65–84.

Ramachandran, B., Pradhan, S. R., and Dhanota, A. S. (1981). 'A Review of Seismicity of Eastern Himalayas', *GSI Miscellaneous Publication*, 41:41–50.

Rankine, W. J. M. (1857). 'On the Stability of Loose Earth', *Philosophical Transactions of the Royal Society*, 147, pt. 1, pp. 9–27.

Rao, Chalapathi (1975). 'Nagarjuna Sagar Project, Andhra Pradesh', *GSI Miscellaneous Publication*, 29, pt. 1, pp. 62–4.

Rao, Chalapathi, and Narasimham, C. V. L. (1975). 'Srisailam Hydroelectric Project, Andhra Pradesh', *GSI Miscellaneous Publication*, 29, pt. 1, pp. 164–78.

Rao, P. J. (1968). 'Clay as Embankment Minerals', *Journal ISEG*, 3 (1): 116–22.

Rao, G. S. M. (1975). 'Rihand Dam Project, Uttar Pradesh', *GSI Miscellaneous Publication*, 29 (pt 1): 25–30.

Ray, S., Mehta, P.N., and Ashraf, Z. (1972). 'Construction Material Available in India and Their Utilization in Engineering Projects', *Records of the Geological Survey of India*, 104, pt. 2, pp. 97–114.

Romani, Saleem (1999). 'Hydrology as an Aid in Preventing Mine-flooding in Karstic Terrain', *Journal of Engineering Geology*, 27 (4): 164–68.

Sanganeria, J. S. (1975). 'Rana Pratap Sagar Project, Chambal Valley Development Scheme, Stage II, Rajasthan', *GSI Miscellaneous Publication*, 29, pt. 1, pp. 41–9.

Sengupta, A. K. (1991). *Geotechnical Aspects of Metro Construction in Calcutta*. Proceedings of Silver Jubilee Seminar on Trend in Geotechnical Investigation in Last Twenty Five Years, Kolkata, ISEG, 20 (1–4): 328–34.

Seshagiri, D. N. (1975). 'Idukki Hydroelectric Project, Kerala', *GSI Miscellaneous Publication*, 29(pt 1): 248–256.

Shome, S. K., and Kaistha, G. K. (1992). 'Findings from Construction Material Survey in Yamuna Basin between Dakpathar and Khara District Dehradun and Saharanpur, Uttar Pradesh', Project Report: pp. 19–30.

Sinha, B. N., Pradhan, S. R., and Singh, R. P. (1971). *Geoanatomy of Balimela Head Race Tunnel*. Proceedings. of Seminar on Engineering and Geotechnical Problems in Tunnelling, ISEG, 22–23 October 1971, pt. 2, pp. 134–62.

Sinha, B. N, Pradhan, S. R., and Singh, R. P. (1978). *Influence of Open Joints on the Foundation of the Balimela Powerhouse, Koraput District, Orissa*. Proceedings of the Seminar on Foundation Problems of Powerhouses and Related Ancillary Structures, New Delhi, 1973, ISEG, pp. 165–76.

Srivastava, P. B. (1975). 'Koyna Hydroelectric project—Stage-I, Maharashtra', *GSI Miscellaneous Publication*, 29 (pt 1): 31–40.

Srivastava, K. N., and Sondhi, S. N. (1978). *Geological Problems of the Foundations on the Bassi Powerhouse District, Mandi, Himachal Pradesh*. Proceedings of the Seminar on Foundation Problems of Powerhouses and Related Ancillary Structures, New Delhi, 1973, ISEG, pp. 131–38.

Srivastava, V. K. and Rao, T. B. V. M. (1991). *Digital Analysis of Remotely Sensed Data (Landsat-TM) for Mapping of Siltation/Turbidity Levels in Maithon Reservoir*. Proceedings of Silver Jubilee Seminar on Trend in Geotechnical Investigations in Last Twenty Five Years, Kolkata, 1990, ISEG, 20 (1–4): 167–90.

Subramanian, V. (1996). *Erosion of Sediment Yield: Global and Regional Perspectives (Proceedings of the Exeter Symposium, July 1996)* IAHS Publication, no. 236. IAHS Press, Wallingford, Oxfordshire, UK.

Taylor, G. S. (1959). 'Groundwater Provinces of India', *Economic Geology*, 54. (4): 688–97.

Thandaveswara, B. S. Prof. (2008). 'Failure of Dams of India', in *Hydraulics*, Chapter 41.2, Internet edition (June 23, 2008).

Terzaghi, K. (1946). 'Rock Defects and Loads on Tunnel Supports', in R. V. B. Proctor and T. L. White (eds.) *Rock Tunnelling with Steel Supports*. Youngstown, OH: Commercial Shearing and Stamping Company, vol. 1, 17–99.

Terzaghi, K. (1950a). *Mechanism in Landslides in Engineering Geology*. New York: The Geological Society of America (Berkley Volume).

Terzaghi, K. (1950b). *Rock load Terzaghi – Geologic Aspects of Soft Ground Tunnelling*. New York: John Wiley and Sons. (pp. 374 of Krynine, D. P. & Judd, W. R. (1957). Principles of Engineering Geology and Geotechnics, New York: McGraw Hill.

Terzaghi, K., and Peck, R. B. (1967). *Soil Mechanics in Engineering Practice*. New York: John Wiley and Sons.

Tilak, N. B. G. (1999). *Geotechnical Considerations for Underground Excavations*. Proceedings of Symposium on Engineering Geology and Geo-environmental Problems in Hard Rock Terrain— HREGE '99, Bhopal, 6–7 December 1999, ISEG, 27 (4): 85–109.

Turner, F. J., and Verhoogen, J. (1987). *Igneous and Metamorphic Petrology*. New Delhi: CBS Publications and Distributors.

Twenhofel, W. H. (1950). *Principles of Sedimentation*. McGraw-Hill, New York..

UNESCO (2002). *The Great Waves*. Information brochure published by UNESCO Inter Government Oceanographic Commission, Paris, France. May 2002.

USGS (2004). *Landslides Types and Processes*. http://landslides.usgs.gov. July 2004.

Varma, R. S., and Mehta, P. N. (1971). *On Some Geotechnical Aspects of Certain Tunnels in the Himalayas*. Proceedings of Seminar on Engineering and Geotechnical Problems in Tunnelling, 22–23 October 1971, ISEG, pt. 2, pp. 60–71.

Varnes, D. J. (1978). 'Slope Movement Types and Processes', in R. L. Schuster and R. J. Krizek (eds.) *Landslides, Analysis and Control. Transportation Research Board Special Report, No.176, National Academy of Science*, pp. 11–33. Publisher: U.S. Geological Survey, Virginia, USA.

Wadia, D. N. (1966). 'Presidential Address at the Inaugural Session of Indian Society of Engineering Geology', *Journal of Engineering Geology*, 1 (1): 18–23.

Wadia, D. N. (1966). *The Geology of India*. London: Macmillan Publishers.

Wahlstrom, Ernest (1948). 'Application of Geology for Tunnel Problems', *American Society of Civil Engineers*, 113:1320.

Wahlstrom, E. E. (1973). *Tunneling in Rock*. London: ELSEVIER Scientific publishing Company.

Wahlstrom, E. E. (1974). *Dams, Dam Foundation and Reservoir Sites*. New York: ELSEVIER Scientific Publication Company.

Walters, R. C. S. (1971). *Dam Geology*. Butterworths Publishers. London, UK.

Walling, D. E., and Webb, B. W. (1983). 'Pattern of Sediment Yield', in K. J. Gregory (ed.) *Background to Palaeohydrology*. New York: Pergamon, 69–100.

Watson, Janet (1983). *Geology and Man: An Introduction to Applied Earth Science*. London: George Allen and Unwin (Publishers) Ltd.

Wikipedia (2012). Web site on Groins, Aqueduct, Structure of the Earth.

Wood, A. N., and Muir (1969). *Coastal Hydraulics*. London: Macmillan and Co. Ltd.

Yeats, Robert S. (2001). *Living with Earthquake in California: A Survivor's Guide*. Oregon: Oregon State University Press.

Index

About the Author

Subinoy Gangopadhyay, formerly Senior Director of the Geological Survey of India, has overseen the progress of more than a hundred civil engineering projects, both in India and abroad during his long association spanning 34 years with this premier institute.

A Ph D from Calcutta University, Dr Gangopadhyay obtained his M Sc degree in Geology from the prestigious Presidency College, Kolkata. He also holds a diploma in Geotechnics from the Norwegian Geotechnical Institute, Oslo, Norway.

Dr Gangopadhyay is a founder member of the Indian Society of Engineering Geology and served as its editor and Vice-President during his tenure. He has written as many as 150 technical reports on the geotechnical aspects of engineering projects of India, as well as some projects abroad where he worked as a consultant. He has also published numerous technical papers in reputed national and international journals.

Related Titles

DESIGN OF STEEL STRUCTURES | 9780195676815

N. Subramanian Consultant Engineer, Maryland, USA

Design of Steel Structures is the first book based on the limit state method of design as per the latest Indian standard code IS 800:2007.

Key Features

- Coverage of topics such as materials, concepts, loading, analysis, design, and fire and corrosion resistance makes it an invaluable guide for students and designers.
- Includes a CD containing computer programs for design, additional chapters on advanced topics, conversion tables, etc.

9780198069188 | **STRUCTURAL ANALYSIS**

T.S. Thandavamoorthy Professor, Anna University

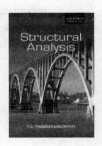

The book provides a balanced coverage of concepts, basic definitions, and analytical techniques in the field of structural analysis.

Key Features

- Explains applications of various structures.
- Presents over 500 detailed solved examples with step-by-step explanations as also chapter-wise MCQs.

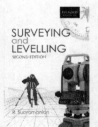

SURVEYING AND LEVELLING | 9780198085423

R. Subramanian Emeritus Professor, NITTTR, Chandigarh

The second edition of *Surveying and Levelling*, with its significantly expanded coverage, is a comprehensive textbook that covers in a single volume all the aspects of the subject that are generally covered in two successive semesters in most universities.

Key Features

- Revised and expanded sections on digital levels, digital theodolites, total stations, aerial surveying methods, image processing, remote sensing, GPS, and GIS.
- Contains a wealth of worked-out examples to convert theory into practice.

Other Related Titles

9780195671537 Santhakumar: *Concrete Technology*
9780198072393 Bhatta: *Remote Sensing and GIS*
9780198069188 Thandavamoorthy: *Structural Analysis*
9780195688177 Duggal: *Earthquake Resistant Design of Structures*
9780198061106 Subramanian: *Strength of Materials*
9780195694833 Sarkar & Saraswati: *Construction Technology*